Recent Advances in Linear Models and Related Areas

Shalabh • Christian Heumann

Recent Advances in Linear Models and Related Areas

Essays in Honour of Helge Toutenburg

Springer

Dr. Shalabh
Department of Mathematics & Statistics
Indian Institute of Technology Kanpur
Kanpur - 208016 India
shalab@iitk.ac.in

Dr. Christian Heumann
Institute of Statistics
Ludwig-Maximilians-University Munich
Akademiestr. 1
80799 Munich
chris@stat.uni-muenchen.de

ISBN 978-3-7908-2063-8 e-ISBN 978-3-7908-2064-5

Library of Congress Control Number: 2008929501

Cover design: WMXDesign GmbH, Heidelberg, Germany

Printed on acid-free paper

9 8 7 6 5 4 3 2 1

springer.com

Preface

This collection contains invited papers by distinguished statisticians to honour and acknowledge the contributions of Professor Dr. Dr. Helge Toutenburg to Statistics on the occasion of his sixty-fifth birthday. These papers present the most recent developments in the area of the linear model and its related topics.

Helge Toutenburg is an established statistician and currently a Professor in the Department of Statistics at the University of Munich (Germany) and Guest Professor at the University of Basel (Switzerland). He studied Mathematics in his early years at Berlin and specialized in Statistics. Later he completed his dissertation (Dr. rer. nat.) in 1969 on optimal prediction procedures at the University of Berlin and completed the post-doctoral thesis in 1989 at the University of Dortmund on the topic of mean squared error superiority. He taught at the Universities of Berlin, Dortmund and Regensburg before joining the University of Munich in 1991.

He has various areas of interest in which he has authored and co-authored over 130 research articles and 17 books. He has made pioneering contributions in several areas of statistics, including linear inference, linear models, regression analysis, quality engineering, Taguchi methods, analysis of variance, design of experiments, and statistics in medicine and dentistry. His most influential contributions are in the area of optimal prediction in linear models, mean squared error superiority of biased estimators, weighted mixed estimation in missing data analysis, repeated measures designs and the unification of various parameterizations of the carry-over effect in cross-over designs. His books *Prediction and Improved Estimation in Linear*

Models (Wiley) and *Prior Information in Linear Models* (Wiley) laid the foundations for further work in the field of utilization of prior information as well as in the field of prediction. Other pioneering works include *Linear Models and Generalizations: Least Squares and Alternatives* (Springer) and *Statistical Analysis of Designed Experiments* (Springer). His books in German on descriptive and inductive statistics, quality engineering, design of experiments and linear models are among the popular textbooks in several universities in Germany. He has also translated the celebrated books of Professor C.R. Rao into German. His book on statistics in dentistry is the first book in German in this area.

Helge Toutenburg maintains fruitful research collaboration with researchers in different countries like the USA, India, Korea, etc. He has hosted DAAD and Humboldt fellows. He is not only a well known researcher but also an excellent teacher. He has advised Ph.D. students from germany and abroad. His efficient working style has always been appreciated by those who had a chance to collaborate with him. He has been actively associated with the International Statistical Institute, Deutscher Hochschulverband, Deutsche Statistische Gesellschaft, Biometrical Society and Bernoulli Society for Mathematical Statistics and Probability.

Besides having a great interest in statistics, Helge Toutenburg has a great sense of humor, too. He has written several books on humor in German to the pleasure of his friends and colleagues.

This collection of invited papers brings together the recent developments in the field of linear models and its related sub-fields as well as papers from Helge Toutenburg‘s other areas of interest.

As the editors of this book, we would like to express our heartful thanks to the authors whose contributions and commitment made this book possible. We would like to thank Michael Schomaker for his immense help in the editorial process and Valentin Wimmer for his help in typing. We are also thankful to Dr. Müller of Springer for his cooperation in the publication of this book.

Munich,
June, 2008

Shalabh
Christian Heumann

Contents

List of Contributors

A.K.Md. Ehsanes Saleh
Carleton University/School of Mathematics and Statistics
Ottawa, Canada
esaleh@math.carleton.ca

Alan T.K. Wan City University of Hong Kong/Department of Management Sciences - Kowloon, Hong Kong
msawan@cityu.edu.hk

Aman Ullah University of California/Department of Economics University Avenue 900 - Riverside, CA 92521, U.S.A.
aman.ullah@ucr.edu

Angela Döring GSF-National Research Center for Environment and Health, Ingolästter Landstrasse 1, 85764 Neuherberg, Germany
doering@gsf.de

Anoop Chaturvedi University of Allahabad/Department of Statistics - Allahabad, India
anoopchaturv@gmail.com

Bettina Grün Technische Universität Wien/Institut für Statistik und Wahrscheinlichkeitstheorie Wiedner Hauptstrasse 8-10 - 1040 Vienna, Austria
Bettina.Gruen@ci.tuwien.ac.at

Burkhard Schaffrin Ohio State University/School of Earth Sciences - Columbus Ohio, U.S.A.
schaffrin.1@osu.edu

C. Radhakrishna Rao Penn State University/Department of Statistics PA 16802 - University Park, U.S.A.
crr1@psu.edu

Chandra Mani Paudel Tribhuvan University/Department of Statistics - Pokhra, Nepal
cpaudel@hotmail.com

Changli He Dalarna University - Dalarna, Sweden
chh@du.se

Christian Heumann University of Munich/Department of Statistics Akademiestrasse 1 - 80799 Munich, Germany christian.heumann@stat.uni-muenchen.de

Christoph Hanck University of Dortmund/Department of Statistics Vogelpothsweg 78 - 44221 Dortmund, Germany christoph.hanck@uni-dortmund.de

David Rummel emnos GmbH Theresienhöhe 12 - 80339 Munich, Germany d_rummel@web.de

Dietrich Trenkler University of Osnabrück/Rolandstrasse 8 - 49069 Osnabrück, Germany d.trenkler@nts6.oec.uni-osnabrueck.de

Friedrich Leisch University of Munich/Department of Statistics Ludwigstrasse 33 - 80539 Munich, Germany Friedrich.Leisch@stat.uni-muenchen.de

Guoqi Qian University of Melbourne/Department of Mathematics and Statistics VIC 3010 - Melbourne, Australia g.qian@ms.unimelb.edu.au

Gerhard Tutz University of Munich/Department of Statistics Akademiestrasse 1 - 80799 Munich, Germany tutz@stat.uni-muenchen.de

Götz Trenkler University of Dortmund/Department of Statistics Vogelpothsweg 87 - 44221 Dortmund, Germany trenkler@statistik.uni-dortmund.de

Guohua Zou University of Rochester/Department of Biostatistics and Computational Biology Rochester, NY 14642, U.S.A. Guohua_Zou@URMC.Rochester.edu

Hans Malmsten Länsförsäkringar - Stockholm, Sweden Hans.Malmsten@lansforsakringar.se

Hans Schneeweiss University of Munich/Department of Statistics Ludwigstrasse 33 - 80539 Munich, Germany schneew@stat.uni-muenchen.de

Helmut Küchenhoff University of Munich/Department of Statistics Akademiestrasse 1 - 80799 Munich, Germany helmut.kuechenhoff@stat.uni-muenchen.de

Hilmar Drygas University of Kassel/Department of Mathematics - 34109 Kassel, Germany drygas@mathematik.uni.kassel.de

Huaizhen Qin Michigan Technological University/Department of Mathematical Sciences 1400 Townsend Drive - Houghton, Michigan 49931-1295, U.S.A.
hqin@mtu.edu

Hyang Mi Kim University of Calgary/Department of Mathematics and Statistics - Calgary, Canada
hmkim@ucalgary.ca

Hyuk Joo Kim Wonkwang University/Division of Mathematics - Jeonbuk 570-749, Korea
hjkim@wonkwang.ac.kr

Iris Pigeot Bremen Institute for Prevention Research and Social Medicine Linzer Strasse 10 - 28359 Bremen, Germany
pigeot@bips.uni-bremen.de

Jae-Il Cho Dongbu Electronics/Management Innovation
Part Street, Gyeonggi 420-712, Korea
jaeil.cho@dsemi.com

Jan Ulbricht University of Munich/Department of Statistics Akademiestrasse 1 - 80799 Munich, Germany
ulbricht@stat.uni-muenchen.de

Jörg Drechsler Institute for Employment Research of the Federal Employment Agency/Department for Statistical Methods Regensburger Strasse 104 - 90478 Nürnberg, Germany
joerg.drechsler@iab.de

Ludwig Fahrmeir University of Munich/Department of Statistics Ludwigstrasse 33 - 80539 Munich, Germany
ludwig.fahrmeir@stat.uni-muenchen.de

Michael Akritas Penn State University/Department of Statistics 325, Thomas Building, University Park - PA 16802, U.S.A.
mga@stat.psu.edu

Narinder Kumar Panjab University/Department of Statistics - Chandigarh 160014, India
nkumar@pu.ac.in

Nina Wawro Bremen Institute for Prevention Research and Social Medicine Linzer Strasse 10 - 28359 Bremen, Germany
wawro@bips.uni-bremen.de

Panu Somervuo University of Helsinki/Institute of Biotechnology P.O.Box 56 - Helsinki 00014, Finland
pjsomerv@mappi.helsinki.fi

Petri Auvinen University of Helsinki/Institute of Biotechnology P.O.Box 56 - Helsinki 00014, Finland
petri.auvinen@helsinki.fi

Qing Shao Novartis Pharmaceuticals
Corporation/Biostatistics and Statistical Reporting One Health Plaza, Bldg. 435 - East Hanover, NJ 07936, U.S.A.
qing.shao@novartis.com

Ram Chandra University of Agriculture and Technology - Kanpur, India ramchandra20@rediffmail.com

Rashi Gupta University of Helsinki/Department of Mathematics and Statistics P.O. Box 68 - Helsinki 00014, Finland
rashi.gupta@helsinki.fi

Sangita Kulathinal University of Helsinki/Department of Mathematics and Statistics P.O. Box 68 - Helsinki 00014, Finland
sangita.kulathinal@inseed.org

Shalabh Indian Institute of Technology/Department of Mathematics and Statistics - Kanpur 208016, India
shalab@iitk.ac.in

Shu-Min Liao Penn State University/Department of Statistics 325, Thomas Building, University Park - PA 16802, U.S.A.
sx1340@psu.edu

Suchita Kesarwani University of Allahabad/Department of Statistics - Allahabad, India
suchitakesarwani@yahoo.co.in

Sung Hyun Park Seoul National University/
Department of Statistics - Seoul 151-747, Korea
parksh@plaza.snu.ac.kr

Susanne Rässler Otto-Friedrich-University
Bamberg/Department of Statistics and Econometrics
Feldkirchenstrasse 21 - 96045 Bamberg, Germany
susanne.raessler@sowi.uni-bamberg.de

Thomas Augustin University of Munich/Department of Statistics Ludwigstrasse 33 - 80539 Munich, Germany
thomas@stat.uni-muenchen.de

Thomas Kneib University of Munich/Department of Statistics Ludwigstrasse 33 - 80539 Munich, Germany
thomas.kneib@stat.uni-muenchen.de

Thomas Nittner UBS AG/Credit & Country Risk Controlling
Pelikanstrasse 6/8 - 8098 Zurich, Switzerland
thomas.nittner@ubs.com

Timo Teräsvirta University of Aarhus/CREATES, School of Economics and Management - Aarhus, Denmark
tterasvirta@econ.au.dk

Walter Krämer University of Dortmund/Department of Statistics Vogelpothsweg 78 - 44221 Dortmund, Germany
walterk@statistik.uni-dortmund.de

Weiqiang Qian University of California/Department of Economics University Avenue 900 - Riverside, CA 92521, U.S.A.
weiqiang.qian@ucr.edu

Yuehua Wu York University/Department of Mathematics and
Statistics 4700 Keele Street - Toronto, M3J 1P3, Canada
wuyh@yorku.ca

On the Identification of Trend and Correlation in Temporal and Spatial Regression

Ludwig Fahrmeir[1] and Thomas Kneib[2]

[1] Department of Statistics, University of Munich, Ludwigstrasse 33, 80539 Munich, Germany ludwig.fahrmeir@stat.uni-muenchen.de
[2] Department of Statistics, University of Munich, Ludwigstrasse 33, 80539 Munich, Germany thomas.kneib@stat.uni-muenchen.de

1 Introduction

In longitudinal or spatial regression problems, estimation of temporal or spatial trends is often of primary interest, while correlation itself is of secondary interest or is regarded as a nuisance component. In other situations, the stochastic process inducing the correlation may be of interest in itself. In this paper, we investigate for some simple time series and spatial regression models, how well trend and correlation can be separated if both are modeled in a flexible manner.

From a classical point of view, trends are considered as deterministic unknown functions to be estimated from the data, whereas correlation is thought to be generated from an unobservable, latent temporal or spatial process. If the focus of statistical inference is on recovering trends, then the latent error process is often only used to give some guidance in choosing reasonable correlation functions to enhance quality of trend estimation. Even more, the derived correlation structure may only be considered as working correlation such as in marginal models for longitudinal data, see e.g. Toutenburg (2003, Ch. 10) and the references therein. To make the discussion concrete, let us consider a simple nonparametric regression problem, where observations $y(t_i)$ on a process $\{y(t), t \geq 0\}$ are available at time points $t_1 < \ldots < t_n$, say. The observable process is related to an unknown trend function $f(t)$ through the additive relation

$$y(t_i) = f(t_i) + \varepsilon(t_i), \quad i = 1, \ldots, n \ , \tag{1}$$

where $\varepsilon(t), t \geq 0$ is an unobservable Gaussian error process with marginal distributions $\varepsilon(t) \sim N(0, \sigma^2)$. Defining the vectors $y =$

$(y(t_1),\dots,y(t_n))'$, $f = (f(t_1),\dots,f(t_n))'$ and $\varepsilon = (\varepsilon(t_1),\dots,\varepsilon(t_n))'$, we obtain the model in matrix notation as

$$y = f + \varepsilon, \quad \varepsilon \sim N(0, \sigma^2 R), \tag{2}$$

where the correlation matrix R has elements $r_{ij} = \rho(\varepsilon(t_i), \varepsilon(t_j))$ with some suitable correlation function ρ. In a purely parametric approach, the trend function could be approximated as a linear combination

$$f(t) = \sum_{j=1}^{p} \beta_j B_j(t) \tag{3}$$

of a few basis functions. To achieve optimality, the unknown coefficients $\beta = (\beta_1,\dots,\beta_p)'$ would then be estimated by minimizing a weighted least squares criterion based on the 'true' correlation matrix R or a consistent estimate $\hat{R}$.

Simple parametric forms like (3) are often too restrictive, at least prior to exploratory data analysis, for modelling trend functions. The most popular nonparametric alternatives are basis function approaches in combination with penalization, such as smoothing splines or penalized splines, and kernel-based local regression techniques. In case of i.i.d errors ε_t, where $R = I$, there is a close connection between both concepts, see e.g. Fahrmeir and Tutz (2001, Ch. 5), and empirical experience shows that they often lead to rather comparable estimates from a practical point of view.

It might intuitively be expected that this similarity in practical performance transfers to the case of correlated error processes as long as a good estimate of R is available. Surprisingly, this is not the case. Kohn, Schimek and Smith (2000) point out some emerging yet different consequences if correlation is neglected in estimation procedures, and they suggest some remedies. Lin and Carroll (2000) show that common kernel-based methods work best when correlation is neglected, i.e if $R = I$ is used as a working correlation matrix. Welsh, Lin and Carroll (2002) provide additional support for this result, but they also confirm that efficient spline estimates are obtained when using the true correlation structure. As a reaction to these somewhat surprising results, Wang (2003) and Linton, Mammen, Lin and Carroll (2004) constructed modified kernel-based estimates which improve upon the usual kernel estimates. Krivobokova and Kauermann (2007) investigate penalized spline estimation for time series data within a mixed model framework and provide some evidence that relatively robust nonparametric estimates are obtained when smoothing parameters are chosen as restricted maximum likelihood estimates even if the correlation structure is misspecified.

In this contribution we shed some further light on this puzzle from a Bayesian perspective. We focus on approaches with Bayesian smoothing priors for modeling trend functions, such as random walk models or extensions to Bayesian penalized (P-)splines. If the correlation-generating error process has similar stochastic structure as the smoothing prior it seems quite plausible that identifiability problems can arise. In particular, it can become difficult to separate trend from correlation. We first exemplify this using a simple time series setting in Section 2. In Section 3 we move on to the corresponding spatial situation, which arises in geostatistics. Section 4 briefly points out extensions to the general class of structured additive regression (STAR) models.

2 Trend and Correlation in Time Series Regression

Let us first revisit the classical smoothing problem already treated by Whittaker (1923), which is closely related to the nonparametric regression problem (2). Time series observations $y(t)$ on an equidistant grid of time points $t = 1, \dots, n$ are assumed to be the sum

$$y(t) = f(t) + \varepsilon(t), \quad t = 1, \dots, n \tag{4}$$

of a smooth trend function f and an irregular noise component ε with i.i.d. errors $\varepsilon \sim N(0, \sigma^2)$. Whittaker suggested to estimate f by minimizing the penalized least squares (PLS) criterion

$$\text{PLS}(f) = \sum_{t=1}^{n} (y(t) - f(t))^2 + \lambda \sum_{t=d+1}^{n} (\Delta^d f(t))^2 \tag{5}$$

where λ is a given smoothing parameter, and the sum of (squared) first ($d = 1$) or second ($d = 2$) order differences

$$\Delta^1 f(t) = f(t) - f(t-1), \qquad \Delta^2 f(t) = f(t) - 2f(t-1) + 2f(t),$$

penalizes deviations from a horizontal or a straight line, respectively. In matrix notation, the observation model becomes $y = f + \varepsilon$ as in (2), and the penalized least squares criterion (5) can be expressed as

$$\text{PLS}(f) = (y - f)'(y - f) + \lambda f' K_d f, \tag{6}$$

with penalty matrix $K_d, d = 1, 2$, given by

$$K_d = D_d' D_d \tag{7}$$

where D_1 and D_2 are first and second order difference matrices, respectively. It can be easily shown that

$$\hat{f} = (I + \lambda K_d)^{-1} y \tag{8}$$

minimizes PLS(f). The (frequentist) covariance matrix of the PLS-estimate is given by

$$\text{Cov}(\hat{f}) = \sigma^2 (I + \lambda K_d)^{-2} . \tag{9}$$

The *Bayesian version* of the smoothing problem of Whittaker can be formulated as a hierarchical model consisting of two stages. Assuming i.i.d. Gaussian errors $\varepsilon(t) \sim N(0, \sigma^2)$, the first stage is the observation model

$$y|f \sim N(f, \sigma^2 I).$$

The second stage specifies a smoothness prior for the unknown function, more exactly for the vector $f = (f(1), \ldots, f(n))'$ of function values. The stochastic analogue of first or second order difference penalties are random walk priors of first ($RW(1)$) or second ($RW(2)$) order

$$f(t) = f(t-1) + u(t)$$

or

$$f(t) = 2f(t-1) - f(t-2) + u(t),$$

for the unknown function values. The errors $u(t)$ are i.i.d. $N(0, \tau^2)$-variables, where τ^2 plays the role of an (inverse) smoothing parameter allowing for larger or enforcing smaller deviations in the development of $f(t)$. Assuming diffuse priors for initial values, i.e.,

$$p(f(1)) \propto \text{constant}$$

in case of first order random walks and

$$p(f(1)) \propto \text{constant}, \quad p(f(2)) \propto \text{constant}$$

in case of second order random walks, the joint prior for the vector $f = (f(1), \ldots, f(n))'$ is multivariate Gaussian with density

$$p(f) \propto \exp\left(-\frac{1}{2\tau^2} f' K_d f \right) \tag{10}$$

with precision matrix K_d given as in (7). Note that the random walk smoothness priors are partially improper since K_d has rank $n - d$. It can easily be shown that the posterior

$$p(f|y) \propto p(y|f)p(f) \tag{11}$$

is Gaussian with posterior mean

$$\hat{f} = E(f|y) = (I + \lambda K_d)^{-1}y, \tag{12}$$

where the smoothing parameter $\lambda = \sigma^2/\tau^2$ is defined as the noise-to-signal ratio, i.e., the ratio of error variance and variance of the random walk. The posterior covariance matrix is given by

$$\text{Cov}(f|y) = \sigma^2(I + \lambda K_d)^{-1}.$$

Thus, the Bayesian posterior mean estimate and the frequentist PLS-estimate coincide but the covariance matrices differ. To be more specific, the Bayesian posterior covariance matrix is larger (in terms of the Löwner order) than its frequentist counter part.

Since the posterior $p(f|y)$ is Gaussian, the posterior mean equals the posterior mode, which is the maximizer of the right-hand side in (11). Taking logarithms, it is straightforward to see that – up to a negative constant factor – the penalized (log-)likelihood criterion

$$l_{pen}(f) = \log p(y|f) + \log p(f)$$

is equal to the PLS criterion (6). This equivalence remains valid if we assume that errors are correlated so that the observation model is altered to

$$\varepsilon \sim N(0, \tau^2 R(\alpha)), \qquad y|f \sim N(f, \sigma^2 R(\alpha))$$

with (nonsingular) covariance matrix $R(\alpha)$, where α parameterizes the correlation structure. For example, the stochastic error process generating the correlation matrix might be a stationary autoregressive process of first or second order, i.e.,

$$\varepsilon(t) = \alpha\varepsilon(t-1) + u(t), \quad |\alpha| < 1,$$
$$\varepsilon(t) = \alpha_1\varepsilon(t-1) + \alpha_2\varepsilon(t-2) + u(t), \; |\alpha_2| < 1, |\alpha_1| < 1 + \alpha_2$$

with i.i.d. Gaussian variables $u(t) \sim N(0, \sigma^2)$. The limiting cases $\alpha \to 1$ and $\alpha_1 \to 2$, $\alpha_2 \to -1$ lead to the (nonstationary) random walk models $RW(1)$ and $RW(2)$, respectively. Defining suitable distributions for the starting values, it can be shown that

$$\varepsilon \sim N(0, \sigma^2 K_{d,\alpha}^{-1})$$

with (nonsingular) precision matrices

$$K_{1,\alpha} = \begin{pmatrix} 1 & -\alpha & & & \\ -\alpha & 1+\alpha^2 & -\alpha & & \\ & \ddots & \ddots & \ddots & \\ & & -\alpha & 1+\alpha^2 & -\alpha \\ & & & -\alpha & 1 \end{pmatrix}$$

and

$$K_{2,\alpha} = \begin{pmatrix} 1 & -\alpha_1 & -\alpha_2 & & & & & & \cdots \\ -\alpha_1 & 1+\alpha_1^2 & -\alpha_1(1-\alpha_2) & -\alpha_2 & & & & & \cdots \\ -\alpha_2 & -\alpha_1(1-\alpha_2) & 1+\alpha_1^2+\alpha_2^2 & -\alpha_1(1-\alpha_2) & -\alpha_2 & & & & \cdots \\ & \ddots & \ddots & \ddots & \ddots & & & & \\ & & & \ddots & \ddots & \ddots & \ddots & \ddots & \\ \cdots & & & -\alpha_2 & -\alpha_1(1-\alpha_2) & 1+\alpha_1^2+\alpha_2^2 & -\alpha_1(1-\alpha_2) & -\alpha_2 & \\ \cdots & & & & -\alpha_2 & -\alpha_1(1-\alpha_2) & 1+\alpha_1^2 & -\alpha_1 & \\ \cdots & & & & & -\alpha_2 & -\alpha_1 & 1 & \end{pmatrix}.$$

In the limiting cases we obtain

$$\lim_{\alpha \to 1} K_{1,\alpha} = K_1, \qquad \lim_{\alpha_1 \to 2, \alpha_2 \to -1} K_{2,\alpha} = K_2,$$

i.e., the precision matrices of the corresponding random walks. For simplicity, we only take a closer look at $AR(1)$-processes ε and $RW(1)$-priors for f. Then, the PLS criterion (6) is replaced by the penalized weighted least squares (PWLS) criterion

$$\text{PWLS}(f) = (y - f)' K_\alpha (y - f) + \lambda f' K_1 f. \tag{13}$$

The PWLS estimate is then given by

$$\hat{f}_\alpha = (K_\alpha + \lambda K_1)^{-1} K_\alpha y. \tag{14}$$

The corresponding Bayesian hierarchical model is now

$$y|f \sim N(f, \sigma^2 K_\alpha^{-1}),$$

with the same Gaussian smoothness prior for f as in (10), with precision matrix K_1. The posterior $p(f|y)$ is Gaussian, but now with posterior mean

$$E(f|y) = \hat{f}_\alpha = (K_\alpha + \lambda K_1)^{-1} K_\alpha y,$$

so that the equivalence of the frequentist and Bayesian point estimate still holds. For $\alpha = 0$, $\hat{f}_\alpha$ reduces to the unweighted PLS estimate (8). For α close to 1, we may expect identification problems, since in the limiting case $\alpha \to 1$, we get

$$P_\alpha := K_\alpha + \lambda K_1 \to (1+\lambda) K_1,$$

where K_1 is singular. These problems are reflected in the condition number

$$\kappa_\alpha = \frac{\lambda_{\max}(P_\alpha)}{\lambda_{\min}(P_\alpha)},$$

where $\lambda_{\max}(P_\alpha)$ and $\lambda_{\min}(P_\alpha)$ denote the largest and the smallest eigenvalue of P_α, respectively. Note that P_α is also the Bayesian posterior precision matrix of $\hat{f}$ thereby providing a measure for the variability of the estimate.

For large κ_α, inversion of P_α suffers from numerical instability as exemplified in Figure 1 for $n = 100$ time points. For increasing values of the autoregressive parameter α, the condition dramatically increases regardless of the value of the smoothing parameter. Small values of λ somewhat lower the effect, since the influence of K_1 on P_α is reduced, but qualitatively the effect remains the same. Note also, that the condition has been log-transformed in Figure 1 to enhance visibility. Hence, the value 10, for example, corresponds to a condition number of $\kappa_\alpha \approx 22000$.

The large condition number for values of α close to one reveals that the nonparametric function f is not well separable from the correlation and that, in particular, increasing variability of $\hat{f}$ is observed for $\alpha \to 1$. However, it seems plausible that we might still obtain a reasonable point prediction for the response vector y. To investigate this conjecture more closely, let us take a closer look at the behavior of the hat matrix P_α^{-1} projecting y on $\hat{y}$ in the limiting case $\alpha \to 1$.

Therefore we rewrite P_α^{-1} using the matrix inversion lemma Toutenburg (2003, Theorem A.18) as

$$(K_\alpha + D_1' \lambda I D_1)^{-1} = K_\alpha^{-1} - K_\alpha^{-1} D_1' \left(\frac{1}{\lambda} I + D_1 K_\alpha^{-1} D_1' \right)^{-1} D_1 K_\alpha^{-1}.$$

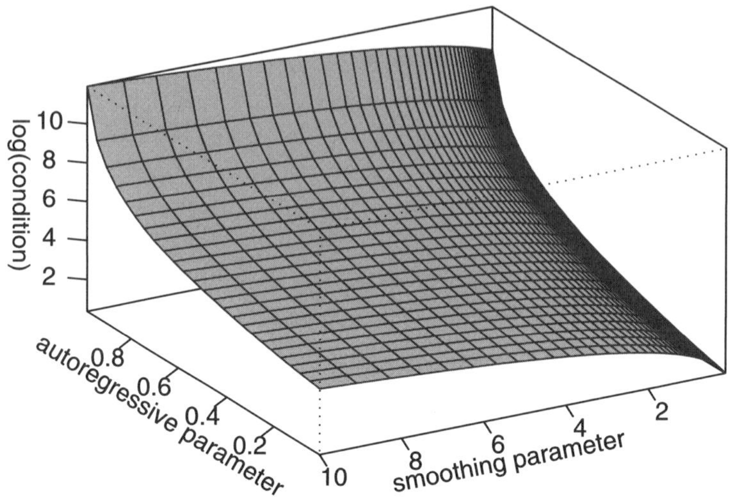

Fig. 1. Condition number κ_α for varying values of the autoregressive parameter α and the smoothing parameter λ when the nonparametric effect is modeled as first order random walk.

For $\alpha < 1$, the matrix K_α is regular and its inverse is given by

$$K_\alpha^{-1} = \frac{1}{1-\alpha^2}\begin{pmatrix} 1 & \alpha & \alpha^2 \dots & \alpha^{n-1} \\ \alpha & 1 & \alpha & \alpha^{n-2} \\ \vdots & \ddots & \ddots \quad \ddots & \vdots \\ \alpha^{n-2} & & \ddots \quad \ddots & \alpha \\ \alpha^{n-1} & \alpha^{n-2} & \dots \quad \alpha & 1 \end{pmatrix}.$$

Straightforward calculations lead to the following expression for $\hat{y}$ in the limiting case $\alpha \to 1$:

$$\hat{y} = (K_\alpha + \lambda K_1)^{-1} K_\alpha y \underset{\alpha\to 1}{\longrightarrow} \begin{pmatrix} 1-\eta & 0 & \dots\dots & 0 & \eta \\ \eta & 1-2\eta & 0 \dots & 0 & \eta \\ \vdots & & \ddots & & \vdots \\ \eta & 0 & \dots & 1-2\eta & \eta \\ \eta & 0 & \dots & 0 & 1-\eta \end{pmatrix} y,$$

i.e.,

$$\hat{y}(t) = \begin{cases} (1-\eta)y(1) + \eta y(n) & t = 1 \\ \eta y(1) + (1-2\eta)y(t) + \eta y(n) & 2 \le t \le n-1 \\ \eta y(1) + (1-\eta)y(n) & t = n, \end{cases}$$

where $\eta = 0.5\lambda/(1+\lambda)$. Therefore the prediction for $\hat{y}(t)$ is always a weighted average of $y(1)$, $y(t)$ and $y(n)$, with the influence of $y(t)$ depending on the smoothness parameter. For $\lambda \to \infty$ (and $\eta \to 0.5$ correspondingly), the influence of $y(t)$ disappears and the overall prediction is just the constant $0.5(y(1) + y(n))$. In the contrary extreme ($\lambda \to 0$ or $\eta \to 0$) the prediction simply interpolates the observed time series.

These considerations lead to the following interpretation: If we try to estimate the trend while simultaneously accounting for correlation, serious multicollinearity problems arise if $\alpha \to 1$ since both the error term and the trend function follow the same stochastic structure. The prediction $\hat{y}$ is still well-behaved as a point estimate with meaningful limiting cases as the smoothing parameter λ is varied. However, the variability of both the estimate $\hat{f}$ and therefore the prediction $\hat{y}$ dramatically increases when $\alpha \to 1$.

It seems that the multicollinearity problem arises because the random walk smoothness prior for f and the stochastic process (10) for ε become so similar with α approaching 1. We may expect less problems with other priors for the trend which imply additional smoothness properties, e.g. for (Bayesian) penalized spline regression. Then we assume that $f(t)$ is (approximated as) a linear combination

$$f(t) = \sum_{j=1}^{p} \beta_j \, B_j^l(t)$$

of B-splines of degree l, defined for an equidistant grid of knots on the time axis. The vector f of function values can then be expressed as $f = X\beta$, where the design matrix X has elements $X[t,j] = B_j^l(t)$, $t = 1, \ldots, n$, and $j = 1, \ldots, p$. To enforce smoothness, the B-spline coefficient vector β obeys the same difference penalties or – in the Bayesian version – random walk priors as before. A standard choice are cubic B-splines and a $RW(2)$-prior. Then the observation model is

$$y \sim N(X\beta, \sigma^2 K_{d,\alpha}^{-1})$$

and the smoothness prior is Gaussian and of the form (10) again. As before, the PWLS estimate and the posterior mean estimate coincide and are given as

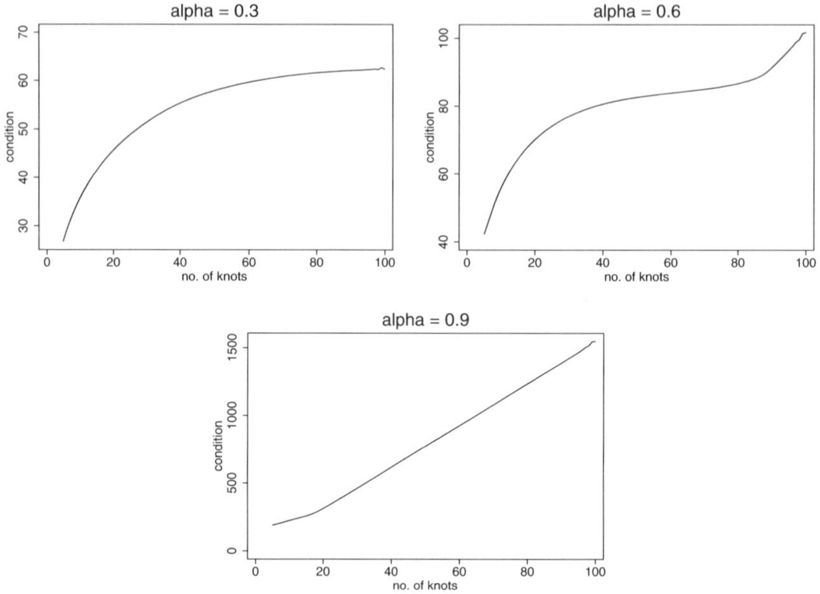

Fig. 2. Condition number κ_α for varying numbers of knots when the non-parametric effect is modeled as a cubic P-spline with second order difference penalty.

$$\hat{\beta}_\alpha = (X'K_{d,\alpha}X + \lambda K_d)^{-1} K_{d,\alpha} y.$$

For the popular choice of cubic B-splines and a $RW(2)$-prior for β, Figure 2 shows the condition number κ_α of the matrix $P_\alpha = X'K_\alpha X + K_2$ as a function of the number of knots and for different values of α. Although the shape of the condition number is quite different depending on the amount of correlation, all curves show the same qualitative behavior of an increasing condition number for larger numbers of knots. Note also the different scaling of the graphics: For high autoregressive correlation, the increase is much more dramatic than for moderate and small correlation.

We now move on a bit further and consider observation models of the form

$$y(t) = f(t) + \varepsilon(t) + \delta(t), \quad t = 1, \ldots, n \tag{15}$$

where, from a frequentist point of view, $f(t)$ is a (deterministic) trend function, $\varepsilon(t)$ is a (stationary) stochastic process inducing temporal correlation as before, and $\delta(t)$ are additional i.i.d. errors representing pure measurement noise. Models of this form are the time series version of geostatistic ("kriging") models considered in the next section and allow for the estimation (or prediction) of both the trend function and the correlated error component. Assuming Gaussian errors, we have

$$y = f + \varepsilon + \delta, \quad \delta \sim N(0, \omega^2 I)$$

in matrix notation. As before, we adopt a basis function approach and approximate the trend through $f = X\beta$, while ε follows an $AR(1)$ or $AR(2)$-process with $\text{Cov}(\varepsilon) = \sigma^2 K_{d,\alpha}^{-1}$. If the primary interest is in estimating f, inference will be based on the *marginal* distribution

$$y|f \sim N(X\beta, \sigma^2 V_\alpha^{-1})$$

where ε and δ are assumed to be independent, and

$$\sigma^2 V_\alpha^{-1} = \sigma^2 (K_\alpha^{-1} + \eta I), \quad \eta = \omega^2/\sigma^2.$$

is the covariance matrix of $\varepsilon + \delta$.

The resulting PWLS estimate for β is

$$\hat{\beta}_\alpha = (X' V_\alpha X + \lambda K_d)^{-1} V_\alpha y.$$

In the special case of B-splines of degree zero, corresponding to random walk models, we have $X = I$ and $f = \beta$, and it may again be interesting to take a closer look at

$$\hat{f}_\alpha = (V_\alpha + \lambda_1 K_1)^{-1} V_\alpha y$$

in the limiting case $\alpha \to 1$.

Problems of (weak) identifiability become quite obvious from a Bayesian perspective if we consider the *conditional* distribution of y, given the trend f and the stochastic process ε generating correlation, i.e.,

$$y|f, \varepsilon \sim N(f + \varepsilon, \sigma^2 I),$$

with $f = X\beta$. This means, we attempt to separate observation y into three components f, ε and δ differing only through their prior specifications. If the smoothness prior for f and the stochastic process prior for ε have similar stochastic structure, then it will obviously be difficult to distinguish them given a finite sample of data y. Moreover, the Bayesian interpretation also reveals, that trend and correlation are per se connected quite closely. If long range correlation is present in the data (corresponding to $\alpha \approx 1$ in the $AR(1)$ example), these correlation will almost appear as a smooth trend in the data. Vice versa, wiggly trends may be equivalently interpreted as some kind of shorter ranged correlation. Compare also the simulation result at the very end of this section.

Obviously, random walk models for f in combination with a stationary autoregressive process, where the parameters approach the boundary of the stationary region, are a simple prototype for weak identifiability. It is easily derived that the posterior $p(f, \varepsilon|y)$ is Gaussian, and the posterior mean estimates $\hat{f}$, $\hat{\varepsilon}$ satisfies

$$\begin{pmatrix} I + \lambda_1 K_d & I \\ I & I + \lambda_2 K_{d,\alpha} \end{pmatrix} \begin{pmatrix} \hat{f} \\ \hat{\varepsilon} \end{pmatrix} = \begin{pmatrix} y \\ y \end{pmatrix},$$

where $\lambda_1 = \omega^2/\tau^2$, $\lambda_2 = \omega^2/\sigma^2$. In the limiting case $\alpha \to 1$ the matrix becomes singular and an identifiability problem arises if we want to separate f from ε. In the following we explore these (weak) identifiability issues empirically through some simulation experiments, focussing on the situation where time series data are generated from models of the form (15), with

$$y(t) = \sin(t) + \varepsilon(t) + \delta(t), \quad t = 1, \ldots, 100,$$

and grid length $\Delta t = 0.25$. We set $\sigma^2 = \text{Var}(\varepsilon(t)) = \omega^2 = \text{Var}(\delta(t)) = 0.5$, which is close to the empirical variance of the sine function, so that all three components have about the same variability. All data sets were generated for $\alpha = 0.3$ (low correlation), $\alpha = 0.6$ (medium correlation), and $\alpha = 0.9$ (strong correlation).

For estimation, the true trend was approximated through penalized P-splines, varying the degree l of the spline functions, the smoothness penalty ($RW(1)$ or $RW(2)$) for B-spline coefficients, and the number of knots. For each selected combination of α-values and B-spline tuning parameters, 50 data sets were generated according to the specific model. The models were fitted either with full Bayesian inference using MCMC or empirical Bayesian inference using mixed model technology. These inference techniques are described in Fahrmeir, Kneib and Lang (2004) and Lang and Brezger (2004), and are implemented in the software BayesX (Brezger, Kneib and Lang (2007)). For each data set, goodness of fit was assessed through

$$SQ(f) = \sum_{t=1}^{100} (f(t) - \hat{f}(t))^2,$$

$$SQ(\varepsilon) = \sum_{t=1}^{100} (\varepsilon(t) - \hat{\varepsilon}(t))^2,$$

$$SQ(y) = \sum_{t=1}^{100} (y(t) - \hat{y}(t))^2,$$

and variability measured through

$$VS(f) = \sum_{t=1}^{100} \mathrm{Var}(\hat{f}(t)),$$

$$VS(\varepsilon) = \sum_{t=1}^{100} \mathrm{Var}(\hat{\varepsilon}(t)),$$

$$VS(y) = \sum_{t=1}^{100} \mathrm{Var}(\hat{y}(t)).$$

Figures 3–6 display boxplots of these characteristics, resulting from estimation with different combinations of α and B-splines for the 50 data sets, respectively.

These figures and additional ones in Eschrich (2007) provide the following empirical evidence:

- With increasing correlation, quality of estimation of the components f and ε decreases (Figure 3, c,e). In contrast, the predictions for the response $\hat{y} = \hat{f} + \hat{\varepsilon}$ for the sum remain comparably stable regardless of the amount of correlation (Figure 3, a). This confirms the results for increasing α that we discussed from a theoretical perspective earlier in this section: While separation between f and ε proves to be difficult, the overall fit remains well identified. Note that even the variance of $\hat{y}$ is relatively stable while variability of both $\hat{f}$ and $\hat{\varepsilon}$ increases.
- Figure 4 investigates the dependence of the results on the number of knots for the standard choice of cubic P-splines with $RW(2)$-priors, corresponding to a rather smooth prior. In this case, results seem to be rather insensitive to the number of knots, as opposed to what might have been guessed from the condition number displayed in Figure 2. For the goodness of fit measures $SQ(f)$ and $SQ(\varepsilon)$, we even obtain improved results for an increased number of knots and therefore a better separation of trend and correlation.

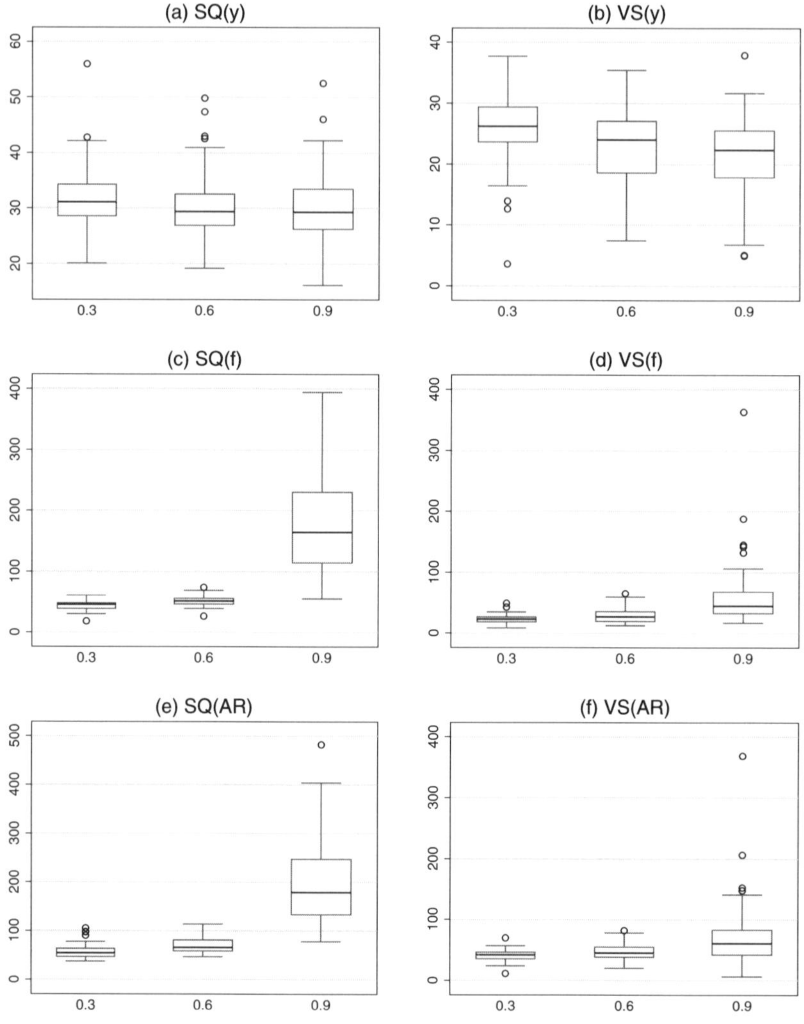

Fig. 3. Goodness of fit and variability measures for varying values of the autoregressive parameter α. The nonparametric effect is modeled as a cubic P-spline with second order random walk penalty and 40 knots.

- In contrast, if the prior for the nonparametric trend does not enforce smoothness but is closer to the $AR(1)$-process results are qualitatively different. Figures 5 and 6 show results for zero degree P-splines and a high amount of correlation for the autoregressive component. When varying the number of knots (Figure 5), both the fit of the nonparametric effect and the autoregressive component wors-

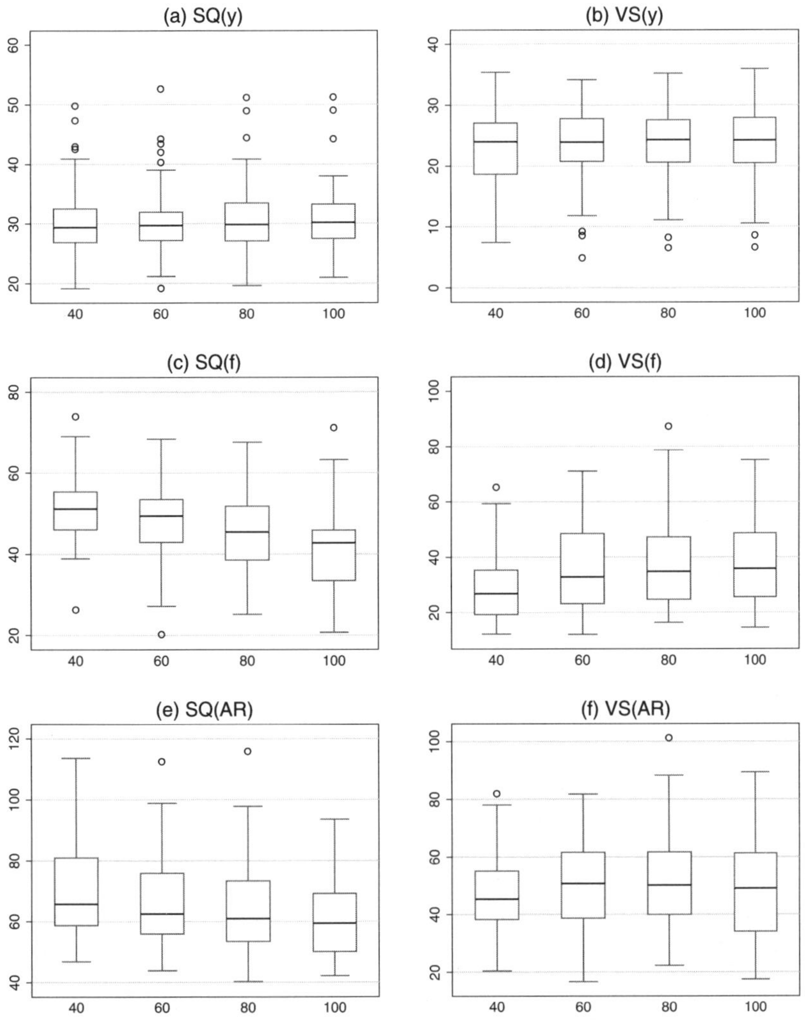

Fig. 4. Goodness of fit and variability measures for varying numbers of knots for the nonparametric effect. The autoregressive parameter is fixed at $\alpha = 0.6$ and the nonparametric effect is modeled as a cubic P-spline with second order random walk penalty.

ens, while the overall fit remains roughly the same. When comparing $RW(1)$ and $RW(2)$ priors for the nonparametric effect, identification somewhat worsens for the first order random walk, which is closer to the $AR(1)$-process than the $RW(2)$ prior (Figure 6). Overall, as expected, the worst choice in terms of identifiability is a zero

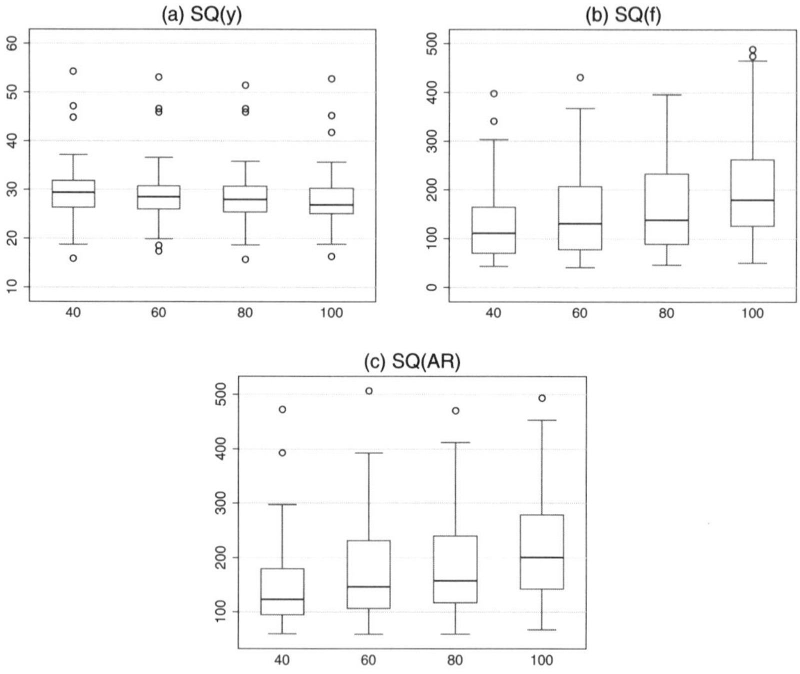

Fig. 5. Goodness of fit and variability measures for varying numbers of knots for the nonparametric effect. The autoregressive parameter is fixed at $\alpha = 0.9$ and the nonparametric effect is modeled as a piecewise constant P-spline with first order random walk penalty.

degree P-spline with random walk of first order as smoothness prior for the trend and a large number of knots.

For the results presented so far, both the autoregressive error $\varepsilon(t)$ and the independent error $\delta(t)$ have been generated anew in each simulation run. To be able to derive mean estimates averaged over the simulation runs, we repeated parts of the simulations with a fixed sequence of autoregressive errors (but still with varying independent errors, of course). Figure 7 shows one exemplary result from these simulations, where the nonparametric effect is modeled as a cubic P-spline with 20 knots and second order random walk penalty. The autocorrelation parameter is fixed at the high value, such that the generated autocorrelated error varies relatively slowly as time progresses. Therefore a large fraction of the autoregressive process is absorbed by the nonparametric effect and the original sine curve as well as the autoregressive

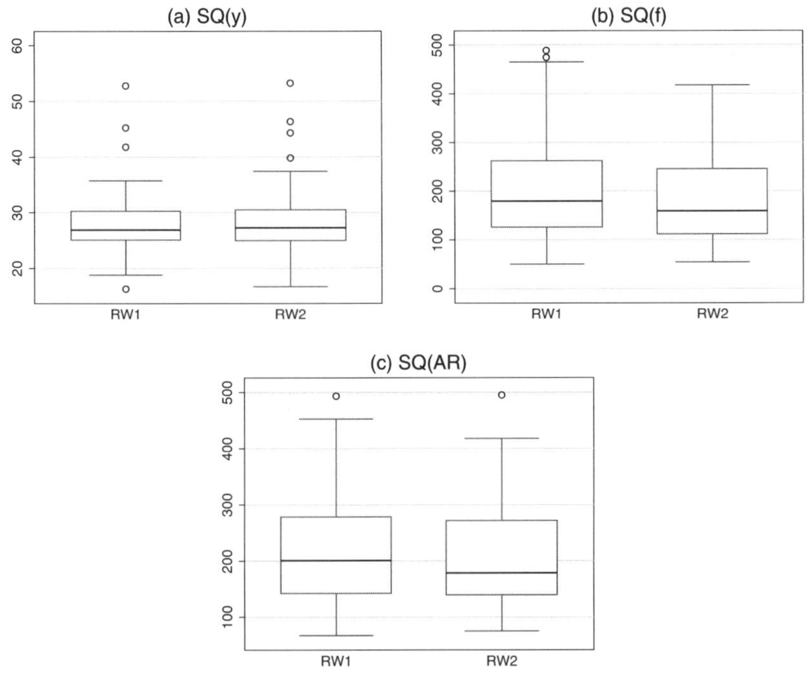

Fig. 6. Goodness of fit and variability measures for varying specifications of the prior for the nonparametric effect. The autoregressive parameter is fixed at $\alpha = 0.9$ and the nonparametric effect is modeled as a piecewise constant P-spline with 100 knots.

component are not very well identified. This again indicates, that trend estimation and modelling of correlation are not opponent concepts but overlapping areas of statistical inference.

3 Spatial Correlation

The modelling approaches for nonparametric trend estimation and temporal correlation considered in the previous section can be extended to estimation of spatial surfaces while simultaneously taking into account spatial correlation. Therefore we replace the univariate temporal model (15) with the bivariate spatial model

$$y(s) = f(s) + \varepsilon(s) + \delta(s)$$

where $s = (s_x, s_y) \in S \subset \mathbb{R}^2$ represents continuous coordinates in some suitable spatial region S, $f(s)$ models a smooth spatial trend

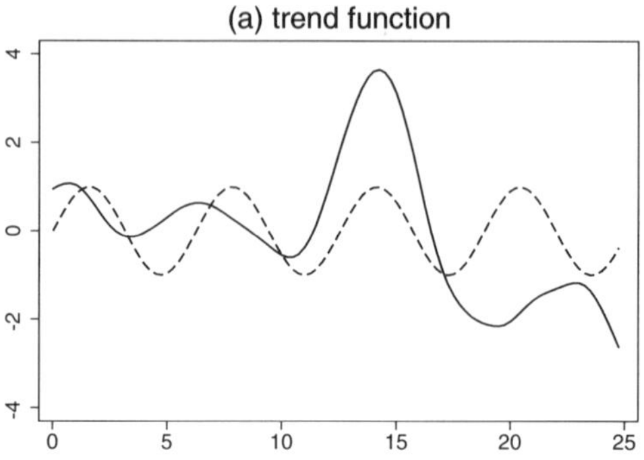

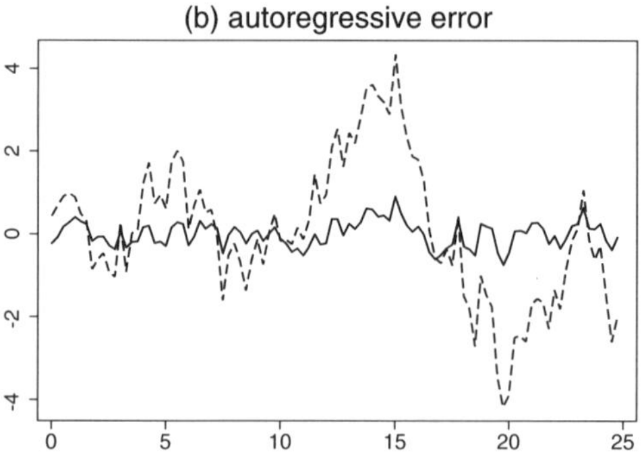

Fig. 7. Separation between the temporal trend and the autoregressive error in case of high autocorrelation ($\rho = 0.9$). The estimated effects are represented as solid lines, the true effects as dashed lines.

function, $\varepsilon \sim N(0, \sigma^2 R)$ is a spatially correlated error term, and $\delta(s)$ is an additional uncorrelated error term, usually referred to as the nugget effect in the spatial regression literature (e.g. Cressie (1993)).

To make the discussion more concrete, we will now briefly describe the necessary modifications of the temporal modelling components. For the trend function, we have considered univariate penalized splines with random walk priors or equivalent difference penalties. For bivariate trend functions we therefore have to define bivariate spline basis functions and to extend the penalty concept to two dimensions. The former can be achieved by considering tensor product B-spline basis functions

$$B^l_{j,k}(s_x, s_y) = B^l_j(s_x) B^l_k(s_y) \tag{16}$$

of degree l based on univariate B-splines in s_x- and s_y-direction, see Dierckx (1993) for a mathematically rigorous definition of tensor product splines. The vector of spatial function evaluations $f = (f(s_1), \ldots, f(s_n))'$ can then be expressed as $f = X\beta$, where X is composed of the function evaluations of the tensor product basis functions in (16) and the vector β contains the corresponding regression coefficients.

A suitable penalty matrix for the vector β can be constructed from Kronecker products of the univariate penalty matrices (see Kneib (2005) for a more detailed discussion and related approaches for the construction of penalties in bivariate smoothing). If K_x and K_y are the penalty matrices corresponding to the univariate B-spline bases, the bivariate analogue can be composed as

$$K = K_x \otimes I_y + I_x \otimes K_y$$

where I_x and I_y are identity matrices with dimensions given by the size of the univariate bases. Inflating the univariate penalty matrices in this way leads to the penalization of row-wise and column-wise differences in the field of regression coefficients corresponding to the bivariate basis functions. The resulting penalty term is of quadratic form, given by

$$\lambda\beta' K\beta.$$

As in the univariate setting, a corresponding partially improper Gaussian prior can be derived in a Bayesian formulation of the bivariate smoothing problem:

$$p(\beta) \propto \exp\left(-\frac{1}{2\tau^2}\beta' K\beta\right) \tag{17}$$

(compare Brezger and Lang (2006) for more information on the Bayesian formulation).

Note that for zero degree and a large number of basis functions (and observations on a regular lattice), tensor product P-splines reduce to an intrinsic Gaussian Markov random field prior for the spatial trend, i.e.,

$$f(s)|[f(s'), s' \neq s] \sim N\left(\frac{1}{|N(s)|}\sum_{s' \in N(s)} f(s'), \frac{\tau^2}{|N(s)|}\right), \tag{18}$$

where $N(s)$ denotes the set of spatial neighbors of $f(s)$ and $|N(s)|$ is the number of such neighbors. This is consistent with the univariate specifications where we obtained the random walk model, which is the univariate analogue to an intrinsic Markov random field, as the special case of piecewise constant P-splines. In particular, the prior distribution (17) is itself a Markov random field, specified for the field of regression coefficients.

For the error component $\varepsilon(s)$, the autoregressive prior shall be extended to the spatial case. We take the correlation function of the AR(1)-process as a starting point, which is given by

$$\rho(\varepsilon(t), \varepsilon(t')) = \alpha^{|t-t'|}.$$

The correlation function is exponentially decaying for increasing distance $|t - t'|$ and this structure can be resembled in the spatial case by assuming the exponential correlation function

$$\rho(\varepsilon(s), \varepsilon(s')) = \exp(-||s - s'||/\phi) \tag{19}$$

where $||s - s'||$ denotes Euclidean distance and ϕ is a range parameter. From the correlation function we can derive the full correlation matrix R via

$$R[i,j] = \rho(s_i, s_j) = \exp(-||s_i - s_j||/\phi).$$

Obviously this correlation function is also exponentially decaying with the speed of the decay being driven by the range parameter ϕ. Relating ϕ to the autoregressive parameter yields

$$\alpha = \exp(-1/\phi) \quad \text{or} \quad \phi = -1/\log(\alpha) \tag{20}$$

and therefore both correlation functions coincide if we only consider discrete, univariate "spatial" points t. We will use connection (20) to obtain values of the range parameter corresponding to the values considered in the temporal simulation setup later-on in this section.

Note that geostatistical models can be interpreted as bivariate smoothers, where the correlation functions play the role of radial basis functions. This provides a further hint that the joint specification of a spatial nonparametric trend surface and a correlated error process may yield identifiability problems, see Kneib (2005) for an intuitive justification and Nychka (2000) for a theoretically sound treatment.

A different extension of the temporal autoregressive model would have been possible based on so-called spatially autoregressive priors, where the following structure is assumed for the spatially correlated error component:

$$\varepsilon(s)|[\varepsilon(s'), s' \neq s] \sim N\left(\frac{\alpha}{|N(s)|} \sum_{s' \in N(s)} \varepsilon(s'), \frac{\sigma^2}{|N(s)|}\right),$$

with $0 < \alpha < 1$. This mimics an undirected representation of the temporal autoregressive prior, where

$$\varepsilon(t)|[\varepsilon(t'), t' \neq t] \sim N\left(\frac{\alpha}{2}(\varepsilon(t-1) + \varepsilon(t+1)), \frac{\sigma^2}{2}\right),$$

i.e., the (conditional) expected value at time t is a weighted average of the two neighbors in time and the variance is inverse proportional to the number of neighbors. For $\alpha \to 1$ the spatially autoregressive error component converges to the intrinsic Markov random field (18) showing the close connection to the bivariate penalized spline approach which also contained intrinsic Markov random fields as a special case. Note that, similarly as for autoregressive errors and random walks, the spatially autoregressive model leads to a proper, stationary joint distribution for the error term with full-rank precision matrix, while the precision matrix for the intrinsic Markov random field is rank-deficient. We will not pursue spatially autoregressive models in the following simulation study, partly because no suitable software has been available to estimate a spatially autoregressive model in combination with nonparametric trend estimation.

Based on the extensions for the trend function and the error term, the marginal observation model in the spatial case can be written as

$$y|\beta \sim N(X\beta, \sigma^2 V^{-1})$$

with $V^{-1} = R + \eta I$ and $\eta = \omega^2/\sigma^2$. From a Bayesian perspective, the correlated error term is just another stochastic process prior, similar to the prior of $f = X\beta$. This leads to the conditional view on the observation model:

$$y|\beta, \varepsilon \sim N(X\beta + \varepsilon, \omega^2 I).$$

If estimation of β is based on the model, we obtain a PWLS criterion

$$(y - X\beta)'V(y - X\beta) + \lambda\beta' K\beta \to \min_{\beta}$$

yielding the PWLS estimate

$$\hat{\beta} = (X'VX + \lambda K)^{-1} X'Vy$$

where $\lambda = \sigma^2/\tau^2$. In the conditional view, both β as well as the error component ε can be estimated based on the doubly penalized least squares criterion

$$(y - X\beta - \varepsilon)'(y - X\beta - \varepsilon) + \lambda_1\beta' K\beta + \lambda_2\varepsilon' R^{-1}\varepsilon$$

leading to the system of equations

$$\begin{pmatrix} X'X + \lambda_1 K & X'I \\ IX & I + \lambda_2 R^{-1} \end{pmatrix} \begin{pmatrix} \hat{\beta} \\ \hat{\varepsilon} \end{pmatrix} = \begin{pmatrix} X'y \\ Iy \end{pmatrix}$$

with smoothness parameters $\lambda_1 = \omega^2/\tau^2$ and $\lambda_2 = \omega^2/\sigma^2$.

In summary, the spatial estimators are similar in spirit and structure as for the temporal setting and it seems plausible to expect similar identification problems when trying to separate trend and correlation component. To validate this conjecture, we set up a simulation study with the simulation model

$$y(s_i) = \underbrace{\sin(s_{ix})\sin(s_{iy})}_{=f(s_i)} + \varepsilon(s_i) + \delta(s_i), \qquad i = 1, \ldots, 100,$$

where s_{ix} and s_{iy} are taken from an equidistant grid of length 10 from 0 to 6. This results in a total combination of 100 pairs (s_x, s_y). The variance of both the spatially correlated error component ε and the uncorrelated error component δ are taken to be equal to 0.2 since this value is close to the empirical variance of the spatial trend surface. The correlation function of ε was chosen to be the exponential correlation function (19). For the range parameter we considered the values $\phi = 0.8$, $\phi = 2$ and $\phi = 10$ corresponding to small, moderate and large correlation between spatially close observation points. These values have been chosen from the relation (20) to coincide approximately with the values in the simulation study on temporal correlation.

Inference in the spatial models can be performed, in analogy to the univariate setting, based on either Markov chain Monte Carlo simulation techniques or mixed model based procedures. Since numerical difficulties have been observed in the latter case, the following results have all been obtained from the MCMC analysis, see Eschrich (2007) for a description of the corresponding algorithm and the implementation.

For comparison of the estimation results, we made use of the same identification measures as in the previous section, i.e., the goodness of fit measure $SQ(\cdot)$ and the variability measure $VS(\cdot)$. Figure 8 summarizes results for cubic tensor product splines, 6 knots per univariate basis, second order difference penalty, and the three choices for the range parameter. Note that the total number of knots ($6 \times 6 = 36$) is still large enough to obtain a flexible function estimate, especially when taking the small number of observations per coordinate direction into account.

While identifiability for the spatial surface f worsens with increasing range parameter (Figure 8 c), this tendency is more or less absent for the spatially correlated error process. In general, the identification even seems to improve, although there are at least some single larger values for $SQ(\varepsilon)$. In terms of the variability measure, all three model components y, f and ε show decreasing variability with increasing correlation.

Figure 9 shows some results for varying numbers of knots when zero degree B-splines are used in combination with first order penalization. Again the results for the spatial case are qualitatively different from the results in the temporal setting (Figure 5), since there seems to be only minor dependence of the results on the number of basis functions.

In summary, the results from the temporal setting could not be completely confirmed in the spatial framework. It might be speculated that the spatial extension of the autoregressive error in our simulation study caused this changing properties. As discussed earlier in this section, there are two different ways to extend the temporal autoregressive process to the spatial situation: Spatially autoregressive models and geostatistical kriging models. Possibly both extensions have different properties with respect to the identification of spatial trend and spatial correlation, an issue that should be investigated further in future studies.

4 Extensions

In the previous sections, we have considered simple temporal or spatial models to highlight identification issues for trend and correlation. The general aspects and problems are, of course, of relevance in more complex models, e.g. for spatio-temporal regression data (y_{it}, x_{it}, s_{it}), $i = 1, \ldots, n$, $t = 1, \ldots, T$, where y_{it} is the response for individual i at time t, x_{it} is a vector of possibly time-dependent covariates, and s_{it} is the location individual i pertains to at time t. A fairly general class of space-time regression models is of the form

$$y_{it} = \eta_{it} + \varepsilon_{it}$$

with predictor

$$\eta_{it} = f_1(x_{it1}) + \ldots + f_k(x_{it}) + f_{\text{time}}(t) + f_{\text{spat}}(s) + u_{it}'\gamma$$

where $f_1, \ldots, f_k$ are nonparametric functions of continuous, possibly time-dependent covariates, f_{time} is a time trend as in Section 2, f_{spat}

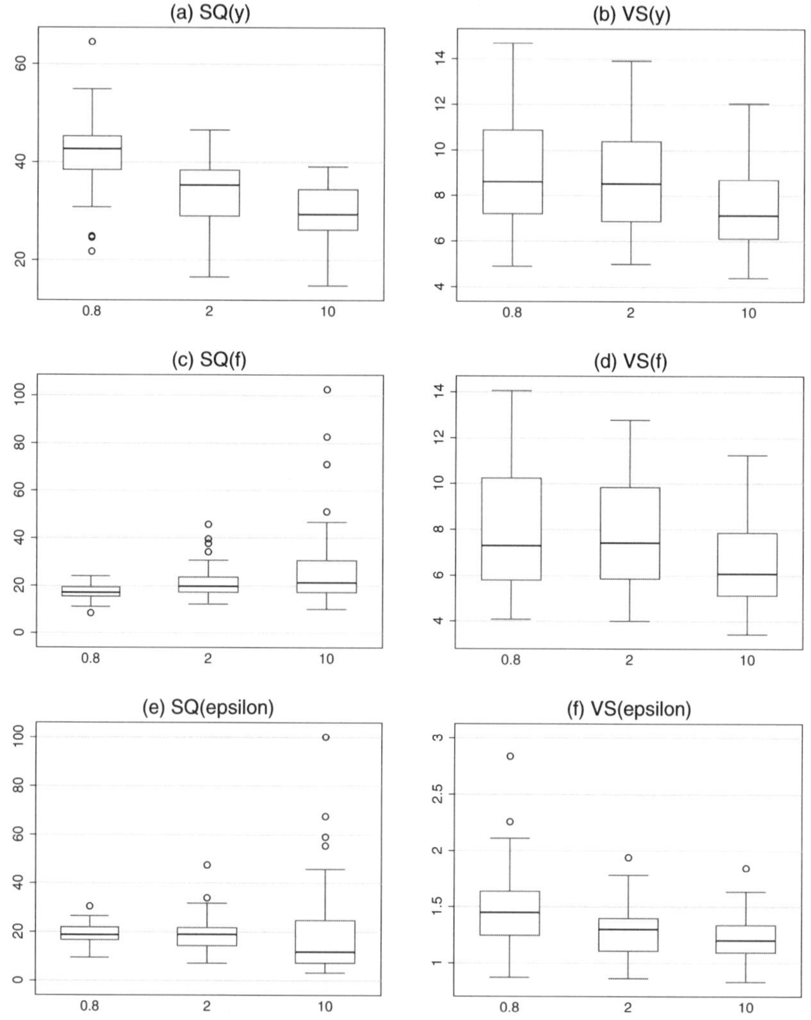

Fig. 8. Goodness of fit and variability measures for varying values of the range parameter. The spatial trend is modeled as a cubic tensor product P-spline with second order random walk penalty and 6×6 knots.

is a spatial trend as in Section 3, u_{it} are covariates with usual linear effects γ, and with with Gaussian errors, often assumed as i.i.d. variables. These models can be extended by including individual- or group-specific random effects, space-time interactions, etc. in the predictor. Furthermore, other types of responses as common in generalized linear models may be considered. This leads to the general class of structured

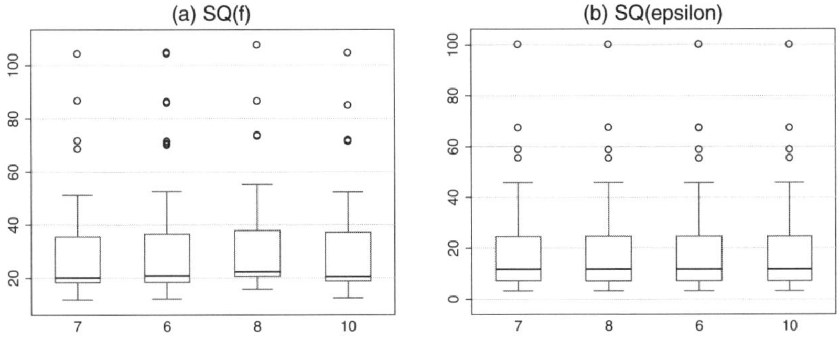

Fig. 9. Goodness of fit and variability measures for varying numbers of knots for the spatial trend. The range parameter is fixed at $\phi = 10$ and the non-parametric effect is modeled as a zero degree tensor product P-spline with first order random walk penalty.

additive regression (STAR) models presented in Fahrmeir, Kneib and Lang (2004) and Kneib (2005) from a Bayesian perspective. Collecting all observations and function values in vectors, it can be shown that the predictor can always be written in the form

$$\eta = U\gamma + X_1\beta_1 + \ldots + X_r\beta_r$$

of a large linear model, with appropriately defined design matrices X_j and random effects β_j obeying Gaussian priors. It becomes immediately clear that additional identification issues arise, for example if time-dependent covariates are highly correlated with time itself, or if one attempts to split up the spatial effect into a "deterministic" surface and a random field of spatially correlated effects. Also it can be problematic to admit (too) flexible correlation structures for the error process when correlation is already (partially) accounted for by a flexible (stochastic) model for the time trend. Finally, it is also obvious from the linear form of the predictor that a thorough knowledge of linear models, including recent developments in matrix theory, as provided by Rao, Toutenburg, Shalabh and Heumann (2008) and Toutenburg (2003) is of high relevance for methodological research in and application of more general regression models.

Acknowledgement

We thank Marco Eschrich for providing the simulation results and Christian Seiler for support in the preparation of the manuscript. This paper has been written during the summer term 2007, when the second author was visiting the Faculty of Mathematics and Economics at the University of Ulm.

References

Brezger A, Kneib T, Lang S (2007) BayesX: Software for Bayesian Inference. Available online from `http://www.stat.uni-muenchen.de/~bayesx/`.

Brezger A, Lang S (2006) Generalized additive regression based on Bayesian P-splines. Computational Statistics and Data Analysis 50:967–991

Cressie NAC (1993) Statistics for Spatial Data. Wiley, New York

Dierckx P (1993) Curve and Surface Fitting with Splines. Clarendon Press, Oxford

Eschrich M (2007) Identifikation von semiparametrischen Regressionsmodellen für korrelierte Daten. Diploma Thesis, Ludwig-Maximilians-University Munich

Fahrmeir L, Kneib T, Lang S (2004) Penalized structured additive regression for space-time data: a bayesian perspective. Statistica Sinica 14: 715-745

Fahrmeir L, Tutz G (2001) Multivariate Statistical Modelling based on Generalized Linear Models. Springer, New York

Kneib T (2005) Mixed model based inference in structured additive regression. PhD Thesis, Ludwig-Maximilians-University Munich. Available online from `http://edoc.ub.uni-muenchen.de/archive/00005011/`

Kohn R, Schimek MG, Smith M (2000) Spline and kernel regression for dependent data. In: MG Schimek (ed), Smoothing and Regression. Wiley, New York

Krivobokova T, Kauermann G (2007) A Note on Penalized Spline Smoothing with Correlated Errors. To appear in Journal of the American Statistical Association

Lang S, Brezger A (2004) Bayesian P-Splines. Journal of Computational and Graphical Statistics 13: 183-212

Lin X, Carroll RJ (2000) Nonparametric function estimation for clustered data when the predictor is measured without/with error. Journal of the American Statistical Association 95: 520-534

Linton OB, Mammen E, Lin X, Carroll RJ (2004) Correlation and marginal longitudinal kernel nonparametric regression. In D. Lin and P. Heagerty (eds), Proceedings of the Second Seattle Symposium in Biostatistics. Springer, New York

Nychka D (2002) Spatial-process estimates as smoothers. In: Schimek, MG (ed) Smoothing and Regression. Wiley, New York

Rao CR, Toutenburg H, Shalabh, Heumann C (2008) Linear Models and Generalizations - Least Squares and Alternatives (3rd edition). Springer, Berlin Heidelberg New York

Toutenburg H (2003) Lineare Modelle. Theorie und Anwendungen (2nd edition). Physica Verlag, Heidelberg

Wang N (2003) Marginal nonparametric kernel regression accounting for within-subject correlation. Biometrika 90: 43-52

Welsh AH, Lin X, Carroll RJ (2002) Marginal longitudinal semiparametric regression: Locality and effciency of spline and kernel methods. Journal of the American Statistical Association 97: 482-493

Whittaker ET (1923) On a new method of graduation. Proceedings of the Edinburgh Mathematical Society 41: 63–75

Estimating the Number of Clusters in Logistic Regression Clustering by an Information Theoretic Criterion

Guoqi Qian[1], C. Radhakrishna Rao[2], Yuehua Wu[3] and Qing Shao[4]

[1] Department of Mathcmatics and Statistics, University of Melbourne, VIC 3010, Australia g.qian@ms.unimelb.edu.au
[2] Department of Statistics, Penn State University, University Park, PA 16802, U.S.A. crr1@psu.edu
[3] Department of Mathematics and Statistics, York University, 4700 Keele Street, Toronto, Ontario, M3J 1P3, Canada wuyh@yorku.ca
[4] Biostatistics and Statistical Reporting, One Health Plaza, Bldg. 435 - 4173, Novartis Pharmaceuticals Corporation, East Hanover, NJ 07936, U.S.A. qing.shao@novartis.com

1 Introduction

It is well-known that a logistic regression model aims at finding how a response variable Y is influenced by a set of explanatory variables $\{x_1, \ldots, x_p\}$ when Y is either binary with values 0 and 1 or a proportion of values between 0 and 1. A logistic regression model consists of three components (McCullagh and Nelder (1989)):

1. A random component Y that is either binary with values 0 and 1 or a proportion with values between 0 and 1. In the latter case, $Y = Z/m$ where Z is assumed to have a binomial distribution $B(m, \pi)$ with the probability of "success" π and the number of independent "experiments" m. We have binary data if $m \equiv 1$.
2. A systematic component (*linear predictor*) $\eta = \boldsymbol{x}'\boldsymbol{\beta}$, where $\boldsymbol{x} = (x_1, \ldots, x_p)'$ and $\boldsymbol{\beta}$ is the unknown p-vector parameter of interest.
3. A function $\pi = h(\eta) = e^\eta/(1 + e^\eta)$ that relates the expectation π of Y with the linear predictor η. The inverse function $g(\pi)$ of $h(\eta)$ is named the *logistic link function*, where $g(\pi) \stackrel{\text{def}}{=} \log(\pi/(1-\pi)) = \eta$.

Logistic regression has been one of the most frequently used techniques in applications. Yet at times either the logistic curve does not describe the probability of success $\pi(\boldsymbol{x})$ adequately, or m is larger than 1 and Y

is more variable than the binomial distribution allows, which is termed *over-dispersion* in the literature. Over-dispersion relative to binomial distribution is possible if the m trials in a set are positively correlated, or an important covariate is omitted. A simple way to accommodate departures from a single logit link and over-dispersion is to introduce the logistic regression clustering model. Examples on the fitting of mixtures of logistic regression to biological and marketing data may be found in Farewell and Sprott (1988), Follmann and Lamber (1989, 1991), and Wedel and DeSarbo (1995), etc.

This paper studies the problem of estimating the number of clusters in the context of logistic regression clustering. The classification likelihood approach is employed to tackle this problem. An information theoretic criterion for selecting the number of logistic curves is proposed in the sequel and then its asymptotic property is considered.

The paper is arranged as follows: In Section 2, some notations are given and an information theoretic criterion is proposed for estimating the number of clusters. In Section 3, the small sample performance of the proposed criterion is studied by Monte Carlo simulation. In Section 4, the asymptotic property of the criterion proposed in Section 2 is investigated. Some lemmas needed in Section 4 are given in the appendix.

2 Notation and Preliminaries

Assume that we have n objects $\mathcal{O}^{(n)} = \{1, 2, \ldots, n\}$ with the associated data points $(\boldsymbol{x}_1, y_1)$, ..., $(\boldsymbol{x}_n, y_n)$, where $\boldsymbol{x}'_j = (x_{j1}, \ldots, x_{jp}) \in \mathbb{R}^p$ is a fixed explanatory p-vector and $y_j \in \mathbb{R}$ is a random dependent variable. The hidden true distributions of $y_1, \ldots, y_n$ are the binomial distributions $B(m_1, \pi_{01}), \ldots, B(m_n, \pi_{0n})$. The set of these n objects is a random sample coming from a structured population. Suppose that this population is composed of k_0 sub-populations, each of which has a distinct underlying linear predictor between the response variable and the explanatory variables. Then, there exists a hidden true partition of these n objects $\Pi^{(n)}_{k_0} = \{\mathcal{O}^{(n)}_1, \ldots, \mathcal{O}^{(n)}_{k_0}\}$, and each cluster $\mathcal{O}^{(n)}_i \triangleq \{i_1, \ldots, i_{n_i}\} \subseteq \mathcal{O}^{(n)}$ is characterized by a class-specific linear predictor

$$\eta_{j,\mathcal{O}_i} = \boldsymbol{x}'_{j,\mathcal{O}_i}\boldsymbol{\beta}_{0i}, \quad \eta_{j,\mathcal{O}_i} = \log\left(\frac{\pi_{0j,\mathcal{O}_i}}{1-\pi_{0j,\mathcal{O}_i}}\right), \quad j \in \mathcal{O}^{(n)}_i, \tag{1}$$

where $\boldsymbol{x}_{j,\mathcal{O}_i}$ and $\pi_{0j,\mathcal{O}_i}$ are just relabeled $\boldsymbol{x}_j$ and π_{0j} which indicate that the associated object is the j-th object in the i-th cluster $\mathcal{O}^{(n)}_i$

$(i = 1, \ldots, k_0)$. We will use this double-index notation throughout this paper. Let $\boldsymbol{\beta}_{0i} \in \mathbb{R}^p$, $i = 1, \ldots, k_0$, be k_0 unknown class-specific true parameter vectors, which are assumed to be pairwise distinct. For convenience, we have suppressed the n in $\mathcal{O}_i^{(n)}$ in (1).

However the true partition Π_{k_0} and the associated model (1) are not observable. Hence, based on the observed data values $(\boldsymbol{x}_j, y_j), j = 1, \ldots, n$, we need to estimate the number of clusters first, and then the model (1).

Consider any possible partition of these n objects: $\Pi_k^{(n)} = \{\mathcal{C}_1^{(n)}, \ldots, \mathcal{C}_k^{(n)}\}$, where $k \leq K$ is a positive integer. Then under the clusterwise logistic regression model, the log-likelihood function for the k parameter vectors $\boldsymbol{\beta}_s$ is

$$\begin{aligned}
&l(\boldsymbol{\beta}_1, \ldots, \boldsymbol{\beta}_k | Y_n, X_n) \\
&= \sum_{s=1}^{k} \sum_{j \in \mathcal{C}_s} \{\log \binom{m_{j,\mathcal{C}_s}}{m_{j,\mathcal{C}_s} y_{j,\mathcal{C}_s}} + m_{j,\mathcal{C}_s} y_{j,\mathcal{C}_s} \log \pi_{j,\mathcal{C}_s} \\
&\qquad\qquad + m_{j,\mathcal{C}_s}(1 - y_{j,\mathcal{C}_s}) \log(1 - \pi_{j,\mathcal{C}_s})\} \\
&= \sum_{s=1}^{k} \sum_{j \in \mathcal{C}_s} \log \binom{m_{j,\mathcal{C}_s}}{m_{j,\mathcal{C}_s} y_{j,\mathcal{C}_s}} - \sum_{s=1}^{k} \sum_{j \in \mathcal{C}_s} \xi(\pi_{j,\mathcal{C}_s}; y_{j,\mathcal{C}_s}, m_{j,\mathcal{C}_s}) \\
&= \sum_{s=1}^{k} \sum_{j \in \mathcal{C}_s} \log \binom{m_{j,\mathcal{C}_s}}{m_{j,\mathcal{C}_s} y_{j,\mathcal{C}_s}} - \sum_{s=1}^{k} \sum_{j \in \mathcal{C}_s} \xi(h(\boldsymbol{x}'_{j,\mathcal{C}_s} \boldsymbol{\beta}_i); y_{j,\mathcal{C}_s}, m_{j,\mathcal{C}_s}),
\end{aligned}$$

where $Y_n = (y_1, \ldots, y_n)'$, $X_n = (\boldsymbol{x}_1, \boldsymbol{x}_2, \ldots, \boldsymbol{x}_n)'$. Again $y_{j,\mathcal{C}_s}$, $\boldsymbol{x}_{j,\mathcal{C}_s}$, $\pi_{j,\mathcal{C}_s}$ and $m_{j,\mathcal{C}_s}$ are just relabeled $y_j, \boldsymbol{x}_j, \pi_j$ and m_j $(j = 1, \ldots, n)$ to indicate the cluster to which the associated object belongs, and

$$\xi(\pi; y, m) = -my \log \pi - m(1 - y) \log(1 - \pi).$$

Note that by convention $\xi(0; y, m) = \xi(1; y, m) = 0$. The clusterwise maximum likelihood estimator (MLE) $\widehat{\boldsymbol{\beta}}_s$ based on the partition $\Pi_k^{(n)}$ is defined to be

$$\begin{aligned}
\widehat{\boldsymbol{\beta}}_s &= \arg\max_{\boldsymbol{\beta}_s} l(\boldsymbol{\beta}_s | Y_n, X_n) \\
&\equiv \arg\min_{\boldsymbol{\beta}_s} \sum_{j \in \mathcal{C}_s} \xi(h(\boldsymbol{x}'_{j,\mathcal{C}_s} \boldsymbol{\beta}_s); y_{j,\mathcal{C}_s}, m_{j,\mathcal{C}_s}), \quad s = 1, \ldots, k.
\end{aligned}$$

We then propose an information theoretic criterion for determining the number of clusters and subsequently classifying the data as follows:

Let $q(k)$ be a strictly increasing function of k, and A_n be a sequence of constants. We define

$$D_n(\Pi_k^{(n)}) \stackrel{\text{def}}{=} \sum_{s=1}^{k} \sum_{j \in \mathcal{C}_s} \xi(h(\boldsymbol{x}'_{j,\mathcal{C}_s}\widehat{\boldsymbol{\beta}}_s); y_{j,\mathcal{C}_s}, m_{j,\mathcal{C}_s}) + q(k)A_n, \tag{2}$$

and define $\widehat{k}_n$, the estimate of k_0, to satisfy the equation

$$D_n(\widehat{k}_n) = \min_{1 \leq k \leq M} \min_{\Pi_k^{(n)}} D_n(\Pi_k^{(n)}). \tag{3}$$

It is named Criterion LG-C, which stands for *clustering by logistic regression* in this paper. It can be seen that in (2), the first term is basically the negative maximum log-likelihood; the second term is the penalty term measuring the complexity of the underlying model. In addition, Criterion LG-C in (3) shows that we determine the optimal number of clusters and the corresponding partitioning of the data simultaneously.

3 Monte Carlo Simulation

We constructed three models in the simulation study: the two-cluster case; the three-cluster case with only one covariate; and the three-cluster case with two covariates. The parameter values used to build these models are listed in Table 1. We generate the covariates as follows: for the first two cases, the covariate x is generated from $N(0, 1)$, and the two covariates x_1, x_2 in case 3 are generated from a bivariate Normal distribution with the mean of 0, variance of 1 and the covariance being 0.3.

In this simulation study, $q(k) = 3k(p + 3)$, where p is the number of regression coefficients in the model and is a constant in our study; k is the unknown number of clusters that we are seeking, and, $A_n = A_n^{(i)}, i = 1, 2, 3, 4$, where $A_n^{(i)} = (1/\lambda)((\log n)^{\lambda}) - 1$, with $\lambda_1 = 1.5, \lambda_2 = 1.8, \lambda_3 = 2$ and $\lambda_4 = 2.3$.

For reducing the exhaustive computation needed by Criterion LG-C, we adopt the approach used in Shao and Wu (2005) here. The only change is that we fit a logistic regression other than a regression model within each cluster in every iteration. Then, we first do logistic regression clustering for each k (the choice of k being the same as the previous studies), and subsequently select the best k using Criterion LG-C. We run the simulation for each model 500 times and obtain the relative frequencies of selecting every k out of these 500 repetitions. The results

Table 1. Parameter values used in the simulation study of logistic regression clustering

Case	k_0	Regression coefficients	Number of observations
1	2	$\boldsymbol{\beta}_{01} = \begin{pmatrix} 1 \\ 6 \end{pmatrix}$, $\boldsymbol{\beta}_{02} = \begin{pmatrix} 1 \\ -6 \end{pmatrix}$	$n_1 = 70$, $n_2 = 50$
2	3	$\boldsymbol{\beta}_{01} = \begin{pmatrix} 1 \\ -1 \end{pmatrix}$, $\boldsymbol{\beta}_{02} = \begin{pmatrix} -2 \\ -1 \end{pmatrix}$, $\boldsymbol{\beta}_{03} = \begin{pmatrix} -1 \\ 1 \end{pmatrix}$	$n_1 = 35$, $n_2 = 35$, $n_3 = 50$
3	3	$\boldsymbol{\beta}_{01} = \begin{pmatrix} 1 \\ -1 \\ \frac{1}{2} \end{pmatrix}$, $\boldsymbol{\beta}_{02} = \begin{pmatrix} -2 \\ -1 \\ -\frac{1}{2} \end{pmatrix}$, $\boldsymbol{\beta}_{03} = \begin{pmatrix} -1 \\ 1 \\ \frac{1}{2} \end{pmatrix}$	$n_1 = 35$, $n_2 = 35$, $n_3 = 50$

are summarized in Table 2. It can be seen that Criterion LG-C does nearly perfect a job to detect the underlying number of groups for the models considered in this simulation study.

Table 2. Relative frequencies of selecting k based on 500 simulations of logistic regression clustering

Case	B1C2 ($k_0 = 2$)				B1C3 ($k_0 = 3$)				B2C3 ($k_0 = 3$)			
Model	$A_n^{(1)}$	$A_n^{(2)}$	$A_n^{(3)}$	$A_n^{(4)}$	$A_n^{(1)}$	$A_n^{(2)}$	$A_n^{(3)}$	$A_n^{(4)}$	$A_n^{(1)}$	$A_n^{(2)}$	$A_n^{(3)}$	$A_n^{(4)}$
$k = 1$	0.000	0.000	0.000	0.000	0.000	0.000	0.000	0.000	0.000	0.000	0.000	0.000
$k = 2$	1.000	1.000	1.000	1.000	0.000	0.000	0.000	0.012	0.000	0.000	0.000	0.006
$k = 3$	0.000	0.000	0.000	0.000	1.000	1.000	1.000	0.988	1.000	1.000	1.000	0.994
$k = 4$	0.000	0.000	0.000	0.000	0.000	0.000	0.000	0.000	0.000	0.000	0.000	0.000
$k = 5$	0.000	0.000	0.000	0.000	0.000	0.000	0.000	0.000	0.000	0.000	0.000	0.000

4 Asymptotic Property of Criterion LG-C

Denote the eigenvalues of a symmetric matrix B of order p by $\lambda_1(B) \geq \ldots \geq \lambda_p(B)$. Let $\mathcal{O}_\ell = \{\ell_1, \ldots, \ell_{n_\ell}\}$ be any cluster or a subset of a cluster corresponding to the true partition $\Pi_{k_0}^{(n)}$ of $\mathcal{O}^{(n)}$, and $n_\ell = |\mathcal{O}_\ell|$. Let $X_{n_\ell} = (\boldsymbol{x}_{\ell_1,\mathcal{O}_\ell}, \ldots, \boldsymbol{x}_{\ell_{n_\ell},\mathcal{O}_\ell})'$ be the design matrix in $\mathcal{O}_\ell$. The Fisher information for the parameter $\boldsymbol{\beta}_{0\ell}$ is defined as

$$\begin{aligned}\mathcal{I}_{n_\ell}(\boldsymbol{\beta}_{0\ell}) &= -E\frac{\partial^2 l}{\partial \boldsymbol{\beta}_{0\ell}\partial \boldsymbol{\beta}_{0\ell}'} \\ &= X_{n_\ell}' M_{n_\ell} M_{\pi_\ell} X_{n_\ell},\end{aligned}$$

where

$$\begin{aligned}M_{n_\ell} &= \mathrm{diag}(m_{\ell_1,\mathcal{O}_\ell}, \ldots, m_{\ell_{n_\ell},\mathcal{O}_\ell}) \\ M_{\pi_\ell} &= \mathrm{diag}\{\pi_{0\ell_1,\mathcal{O}_\ell}(1-\pi_{0\ell_1,\mathcal{O}_\ell}), \ldots, \pi_{0\ell_{n_\ell},\mathcal{O}_\ell}(1-\pi_{0\ell_{n_\ell},\mathcal{O}_\ell})\}.\end{aligned}$$

The following assumptions are needed in the discussion on the asymptotic property of the criterion (3).

(A) For the true partition $\Pi_{k_0}^{(n)} = \{\mathcal{O}_1^{(n)}, \ldots, \mathcal{O}_{k_0}^{(n)}\}$, let $n_{0i} = |\mathcal{O}_i|$ be the number of objects in the cluster $\mathcal{O}_i^{(n)}$. Then there exists a fixed constant $a_0 > 0$ such that

$$a_0 n \leq n_{0i} \leq n, \quad \forall i = 1, \ldots, k_0. \tag{4}$$

(X1) $\lim_{n_\ell \to \infty} \lambda_\zeta\{\mathcal{I}_{n_\ell}(\boldsymbol{\beta}_{0\ell})\} = \infty$, $\zeta = 1, \ldots, p$. Also, there exists some constant $a_1 > 0$ such that $0 < \lambda_p\{\mathcal{I}_{n_\ell}(\boldsymbol{\beta}_{0\ell})\} \leq a_1 \lambda_1\{\mathcal{I}_{n_\ell}(\boldsymbol{\beta}_{0\ell})\}$.

(X2) Let $\delta_{n_\ell} = \left(\max_{j \in \mathcal{O}_\ell} m_{j,\mathcal{O}_\ell}^2 \boldsymbol{x}_{j,\mathcal{O}_\ell}' \mathcal{I}_{n_\ell}(\boldsymbol{\beta}_{0\ell})^{-1} \boldsymbol{x}_{j,\mathcal{O}_\ell}\right)^{\frac{1}{2}}$, then

$$\delta_{n_\ell}(\log\log \lambda_p\{\mathcal{I}_{n_\ell}(\boldsymbol{\beta}_{0\ell})\})^{\frac{1}{2}} = o(1).$$

(X3) $a_2 n_\ell \leq \lambda_p\{\mathcal{I}_{n_\ell}(\boldsymbol{\beta}_{0\ell})\} \leq a_3 n_\ell$ holds for some positive constants a_2 and a_3.

(X4) $a_4 n_\ell \leq \lambda\{X_{n_\ell}' M_{n_\ell} X_{n_\ell}\} \leq a_5 n_\ell$ holds for some positive constants a_4 and a_5.

(X5) Let $d_0 = \frac{1}{4}\min_{1 \leq i \neq \ell \leq k_0} |\boldsymbol{\beta}_{0i} - \boldsymbol{\beta}_{0\ell}|$. Also let

$$Q_{n_\ell} = \mathrm{diag}\{v_1, \ldots, v_{n_\ell}\},$$

where $v_i = m_{\ell_i,\mathcal{O}_\ell} e^{-d_0 ||\boldsymbol{x}_{\ell_i,\mathcal{O}_\ell}||} \pi_{0\ell_i,\mathcal{O}_\ell}(1-\pi_{0\ell_i,\mathcal{O}_\ell})$, $i = 1, \ldots, n_\ell$. Then there exists a constant $a_6 > 0$ such that $\lambda_1\{X_{n_\ell}' Q_{n_\ell} X_{n_\ell}\} \geq a_6 n_\ell$.

(Z) $n^{-1}A_n \to 0, \quad (\log\log n)^{-1}A_n \to \infty, \quad \text{as } n \to \infty.$

Remark 4.1 Assumption (A) implicitly implies that the population is comprised of k_0 sub-populations with proportions $p_1, \ldots, p_{k_0}$, where $0 < p_i \le 1$, $i = 1, \ldots, k_0$, $\sum_{i=1}^{k_0} p_i = 1$, and $a_0 = \min_{1\le i\le k_0} p_i$.

Remark 4.2 Assumptions (X1)–(X5) are essentially about the behaviour of the explanatory variables $\boldsymbol{x}$. Roughly speaking, they mean that most of the $\boldsymbol{x}$ observations should be finite and stay away from $\mathbf{0}$. In fact, as observed by Qian and Field (2002), if we assume $\boldsymbol{x}$ to be a random vector and $\boldsymbol{x} \in \mathcal{O}_i$ are i.i.d. observations within each cluster $\mathcal{O}_i$ of the true partitioning Π_{k_0}, for all $i = 1, \ldots, k_0$, then by applying the strong law of large numbers given in Chung (2001, p. 132, Theorem 5.4.1), it is easy to show that the following assumptions are sufficient for (X1) to (X5) to hold:

(S1) $P\{\boldsymbol{x}'\boldsymbol{t} \ne 0\} > 0$ for any $\boldsymbol{t} \ne 0$ in $\mathbb{R}^p$, which implies that $\mathrm{E}(\boldsymbol{x}\boldsymbol{x}')$ is positive definite.
(S2) $P\{h(\boldsymbol{x}'\boldsymbol{\beta}_{0i})(1 - h(\boldsymbol{x}'\boldsymbol{\beta}_{0i})) \ne |\boldsymbol{x}'\boldsymbol{t} \ne 0\} > 0$ for any $\boldsymbol{t} \ne 0$ in $\mathbb{R}^p$, which implies that both $\mathrm{E}(\pi_{0\mathcal{O}_i}(1-\pi_{0\mathcal{O}_i})\boldsymbol{x}\boldsymbol{x}')$ and $\mathrm{E}(e^{-d_0\|\boldsymbol{x}|}\pi_{0\mathcal{O}_i}(1-\pi_{0\mathcal{O}_i})\boldsymbol{x}\boldsymbol{x}')$ are positive definite, where $\pi_{0\mathcal{O}_i} = h(\boldsymbol{x}'\boldsymbol{\beta}_{0i})$, $\forall i = 1, \ldots, k_0$.
(S3) $\mathrm{E}\|\boldsymbol{x}\|^{2+\kappa} < \infty$ for some constant $\kappa > 0$.
(S4) $\sup_{1\le k\le n} m_k < \infty$.

Since there is no essential complexity with random $\boldsymbol{x}$, we will treat the observations $\boldsymbol{x}_1, \ldots, \boldsymbol{x}_n$ as deterministic in the sequel for ease of notation throughout the rest of this paper.
Suppose that the assumptions (A), (X1)–(X5), (Z) hold, and that $\Pi_{k_0}^{(n)} = \{\mathcal{O}_1^{(n)}, \ldots, \mathcal{O}_{k_0}^{(n)}\}$ is the underlying true classification of the n objects in $\mathcal{O}^{(n)}$. Observe that the true partition $\Pi_{k_0}^{(n)}$ is a sequence of naturally nested classifications as n increases, i.e.,

$$\mathcal{O}_i^{(n)} \subseteq \mathcal{O}_i^{(n+1)}, \quad i = 1, \ldots, k_0, \quad \text{for large } n.$$

Consider any given sequence of classifications with k clusters $\Pi_k^{(n)} = \{\mathcal{C}_1^{(n)}, \ldots, \mathcal{C}_k^{(n)}\}$ of $\mathcal{O}^{(n)}$ such that

$$\mathcal{C}_s^{(n)} \subseteq \mathcal{C}_s^{(n+1)}, \quad s = 1, \ldots, k, \quad \text{for large } n,$$

when n increases. For simplicity, when no confusion appears, n will be suppressed in $\Pi_{k_0}^{(n)}$, $\Pi_k^{(n)}$, $\mathcal{O}_i^{(n)}, 1 \le i \le k_0$, and $\mathcal{C}_s^{(n)}, 1 \le s \le k$.

Case 1: When $k_0 < k \leq K$, where $K < \infty$ is a fixed constant

First we have

$$\begin{aligned}
&D_n(\Pi_k) - D_n(\Pi_{k_0})\\
&= \sum_{s=1}^{k} \sum_{j \in \mathcal{C}_s} \xi(h(\boldsymbol{x}'_{j,\mathcal{C}_s}\widehat{\boldsymbol{\beta}}_s); y_{j,\mathcal{C}_s}, m_{j,\mathcal{C}_s})\\
&\quad - \sum_{i=1}^{k_0} \sum_{j \in \mathcal{O}_i} \xi(h(\boldsymbol{x}'_{j,\mathcal{O}_i}\widehat{\boldsymbol{\beta}}_{0i}); y_{j,\mathcal{O}_i}, m_{j,\mathcal{O}_i}) + (q(k) - q(k_0))A_n,
\end{aligned}$$

where

$$\widehat{\boldsymbol{\beta}}_s = \arg\min_{\boldsymbol{\beta}} \sum_{j \in \mathcal{C}_s} \xi(h(\boldsymbol{x}'_{j,\mathcal{C}_s}\boldsymbol{\beta}); y_{j,\mathcal{C}_s}, m_{j,\mathcal{C}_s}), \quad s = 1, \ldots, k, \tag{5}$$

$$\widehat{\boldsymbol{\beta}}_{0i} = \arg\min_{\boldsymbol{\beta}} \sum_{j \in \mathcal{O}_i} \xi(h(\boldsymbol{x}'_{j,\mathcal{O}_i}\boldsymbol{\beta}); y_{j,\mathcal{O}_i}, m_{j,\mathcal{O}_i}), \quad i = 1, \ldots, k_0. \tag{6}$$

Note that

$$\mathcal{O}^{(n)} = \bigcup_{i=1}^{k_0} \mathcal{O}_i = \bigcup_{s=1}^{k} \mathcal{C}_s = \bigcup_{s=1}^{k} \bigcup_{i=1}^{k_0} (\mathcal{C}_s \cap \mathcal{O}_i).$$

Then

$$\begin{aligned}
&D_n(\Pi_k) - D_n(\Pi_{k_0})\\
&= \sum_{s=1}^{k} \sum_{i=1}^{k_0} \sum_{j \in \mathcal{C}_s \cap \mathcal{O}_i} \Big[\xi(h(\boldsymbol{x}'_{j,\mathcal{C}_s\cap\mathcal{O}_i}\widehat{\boldsymbol{\beta}}_s); y_{j,\mathcal{C}_s\cap\mathcal{O}_i}, m_{j,\mathcal{C}_s\cap\mathcal{O}_i})\\
&\quad - \xi(h(\boldsymbol{x}'_{j,\mathcal{C}_s\cap\mathcal{O}_i}\widehat{\boldsymbol{\beta}}_{0i}); y_{j,\mathcal{C}_s\cap\mathcal{O}_i}, m_{j,\mathcal{C}_s\cap\mathcal{O}_i})\Big] + (q(k) - q(k_0))A_n\\
&= \sum_{s=1}^{k} \sum_{i=1}^{k_0} \sum_{j \in \mathcal{C}_s \cap \mathcal{O}_i} \Big[\xi(h(\boldsymbol{x}'_{j,\mathcal{C}_s\cap\mathcal{O}_i}\widehat{\boldsymbol{\beta}}_s); y_{j,\mathcal{C}_s\cap\mathcal{O}_i}, m_{j,\mathcal{C}_s\cap\mathcal{O}_i})\\
&\quad - \xi(h(\boldsymbol{x}'_{j,\mathcal{C}_s\cap\mathcal{O}_i}\widehat{\boldsymbol{\beta}}_{0si}); y_{j,\mathcal{C}_s\cap\mathcal{O}_i}, m_{j,\mathcal{C}_s\cap\mathcal{O}_i})\Big]\\
&\quad + \sum_{s=1}^{k} \sum_{i=1}^{k_0} \sum_{j \in \mathcal{C}_s \cap \mathcal{O}_i} \Big[\xi(h(\boldsymbol{x}'_{j,\mathcal{C}_s\cap\mathcal{O}_i}\widehat{\boldsymbol{\beta}}_{0si}); y_{j,\mathcal{C}_s\cap\mathcal{O}_i}, m_{j,\mathcal{C}_s\cap\mathcal{O}_i})\\
&\quad - \xi(h(\boldsymbol{x}'_{j,\mathcal{C}_s\cap\mathcal{O}_i}\widehat{\boldsymbol{\beta}}_{0i}); y_{j,\mathcal{C}_s\cap\mathcal{O}_i}, m_{j,\mathcal{C}_s\cap\mathcal{O}_i})\Big] + (q(k) - q(k_0))A_n,
\end{aligned}$$

where $\widehat{\boldsymbol{\beta}}_{0si}$ is the MLE of $\boldsymbol{\beta}$ defined by

$$\widehat{\boldsymbol{\beta}}_{0si} = \arg\min_{\boldsymbol{\beta}} \sum_{j\in\mathcal{C}_s\cap\mathcal{O}_i} \xi(h(\boldsymbol{x}'_{j,\mathcal{C}_s\cap\mathcal{O}_i}\boldsymbol{\beta}); y_{j,\mathcal{C}_s\cap\mathcal{O}_i}, m_{j,\mathcal{C}_s\cap\mathcal{O}_i}). \tag{7}$$

By (5) and (7), we have

$$\begin{aligned}\textstyle\sum_{j\in\mathcal{C}_s\cap\mathcal{O}_i} \Big[&\xi(h(\boldsymbol{x}'_{j,\mathcal{C}_s\cap\mathcal{O}_i}\widehat{\boldsymbol{\beta}}_s); y_{j,\mathcal{C}_s\cap\mathcal{O}_i}, m_{j,\mathcal{C}_s\cap\mathcal{O}_i}) \\ &-\xi(h(\boldsymbol{x}'_{j,\mathcal{C}_s\cap\mathcal{O}_i}\widehat{\boldsymbol{\beta}}_{0si}); y_{j,\mathcal{C}_s\cap\mathcal{O}_i}, m_{j,\mathcal{C}_s\cap\mathcal{O}_i})\Big] \geq 0.\end{aligned}$$

By Assumptions (X1)-(X4), (25) in Lemma 3, (6), (7) and again the fact that $\mathcal{C}_s \cap \mathcal{O}_i$ is a subset of the cluster $\mathcal{O}_i$ corresponding to the true partition Π_{k_0}, we have

$$\begin{aligned}&\sum_{s=1}^{k}\sum_{i=1}^{k_0}\sum_{j\in\mathcal{C}_s\cap\mathcal{O}_i} \Big[\xi(h(\boldsymbol{x}'_{j,\mathcal{C}_s\cap\mathcal{O}_i}\widehat{\boldsymbol{\beta}}_{0si}); y_{j,\mathcal{C}_s\cap\mathcal{O}_i}, m_{j,\mathcal{C}_s\cap\mathcal{O}_i}) \\ &-\xi(h(\boldsymbol{x}'_{j,\mathcal{C}_s\cap\mathcal{O}_i}\boldsymbol{\beta}_{0i}); y_{j,\mathcal{C}_s\cap\mathcal{O}_i}, m_{j,\mathcal{C}_s\cap\mathcal{O}_i})\Big] = O(\log\log n),\end{aligned}$$

and

$$\begin{aligned}&\sum_{i=1}^{k_0}\sum_{j\in\mathcal{O}_i} \Big[\xi(h(\boldsymbol{x}'_{j,\mathcal{O}_i}\widehat{\boldsymbol{\beta}}_{0i}); y_{j,\mathcal{O}_i}, m_{j,\mathcal{O}_i}) - \xi(h(\boldsymbol{x}'_{j,\mathcal{O}_i}\boldsymbol{\beta}_{0i}); y_{j,\mathcal{O}_i}, m_{j,\mathcal{O}_i})\Big] \\ &= O(\log\log n).\end{aligned}$$

Using the fact that

$$\begin{aligned}&\sum_{s=1}^{k}\sum_{i=1}^{k_0}\sum_{j\in\mathcal{C}_s\cap\mathcal{O}_i} \xi(h(\boldsymbol{x}'_{j,\mathcal{C}_s\cap\mathcal{O}_i}\boldsymbol{\beta}_{0i}); y_{j,\mathcal{C}_s\cap\mathcal{O}_i}, m_{j,\mathcal{C}_s\cap\mathcal{O}_i}) \\ &\equiv \sum_{i=1}^{k_0}\sum_{j\in\mathcal{O}_i} \xi(h(\boldsymbol{x}'_{j,\mathcal{O}_i}\boldsymbol{\beta}_{0i}); y_{j,\mathcal{O}_i}, m_{j,\mathcal{O}_i}),\end{aligned}$$

where

$$\begin{aligned}&\xi(h(\boldsymbol{x}'_{j,\mathcal{O}_i}\boldsymbol{\beta}_{0i}); y_{j,\mathcal{O}_i}, m_{j,\mathcal{O}_i}) \\ &= -m_{j,\mathcal{O}_i} y_{j,\mathcal{O}_i} \log \pi_{0j,\mathcal{O}_i} - m_{j,\mathcal{O}_i}(1 - y_{j,\mathcal{O}_i}) \log(1 - \pi_{0j,\mathcal{O}_i}),\end{aligned} \tag{8}$$

and $\xi(h(\boldsymbol{x}'_{j,\mathcal{C}_s\cap\mathcal{O}_i}\boldsymbol{\beta}_{0i}); y_{j,\mathcal{C}_s\cap\mathcal{O}_i}, m_{j,\mathcal{C}_s\cap\mathcal{O}_i})$ is similarly defined, we obtain

$$\sum_{s=1}^{k}\sum_{i=1}^{k_0}\sum_{j\in\mathcal{C}_s\cap\mathcal{O}_i}\Big[\xi(h(\boldsymbol{x}'_{j,\mathcal{C}_s\cap\mathcal{O}_i}\widehat{\boldsymbol{\beta}}_{0si});y_{j,\mathcal{C}_s\cap\mathcal{O}_i},m_{j,\mathcal{C}_s\cap\mathcal{O}_i})$$
$$-\xi(h(\boldsymbol{x}'_{j,\mathcal{C}_s\cap\mathcal{O}_i}\widehat{\boldsymbol{\beta}}_{0i});y_{j,\mathcal{C}_s\cap\mathcal{O}_i},m_{j,\mathcal{C}_s\cap\mathcal{O}_i})\Big]$$
$$=\sum_{s=1}^{k}\sum_{i=1}^{k_0}\sum_{j\in\mathcal{C}_s\cap\mathcal{O}_i}\Big[\xi(h(\boldsymbol{x}'_{j,\mathcal{C}_s\cap\mathcal{O}_i}\widehat{\boldsymbol{\beta}}_{0si});y_{j,\mathcal{C}_s\cap\mathcal{O}_i},m_{j,\mathcal{C}_s\cap\mathcal{O}_i})$$
$$-\xi(h(\boldsymbol{x}'_{j,\mathcal{C}_s\cap\mathcal{O}_i}\boldsymbol{\beta}_{0i});y_{j,\mathcal{C}_s\cap\mathcal{O}_i},m_{j,\mathcal{C}_s\cap\mathcal{O}_i})\Big]$$
$$-\sum_{i=1}^{k_0}\sum_{j\in\mathcal{O}_i}\Big[\xi(h(\boldsymbol{x}'_{j,\mathcal{O}_i}\widehat{\boldsymbol{\beta}}_{0i});y_{j,\mathcal{O}_i},m_{j,\mathcal{O}_i})$$
$$-\xi(h(\boldsymbol{x}'_{j,\mathcal{O}_i}\boldsymbol{\beta}_{0i});y_{j,\mathcal{O}_i},m_{j,\mathcal{O}_i})\Big]$$
$$=O(\log\log n). \tag{9}$$

Hence by (8), (9) and Assumption (Z) and the fact that $q(k)-q(k_0)>0$, we have that for large n,

$$D_n(\Pi_k)-D_n(\Pi_{k_0})\geq O(\log\log n)+(q(k)-q(k_0))A_n>0. \tag{10}$$

Case 2: When $k<k_0$

By Lemma 1, for any partition $\Pi_k^{(n)}=\{\mathcal{C}_1^{(n)},\ldots,\mathcal{C}_k^{(n)}\}$, there exist one cluster in $\Pi_k^{(n)}$ and two distinct clusters in the true partition $\Pi_{k_0}^{(n)}$, say $\mathcal{C}_1\in\Pi_k^{(n)}$ and $\mathcal{O}_1,\mathcal{O}_2\in\Pi_{k_0}^{(n)}$, such that

$$b_0n<|\mathcal{C}_1\cap\mathcal{O}_1|<n \quad\text{and}\quad b_0n<|\mathcal{C}_1\cap\mathcal{O}_2|<n, \tag{11}$$

where $b_0=a_0/k_0>0$ is a constant.

Consider

$$\sum_{j\in\mathcal{C}_1\cap\mathcal{O}_1}\xi(h(\boldsymbol{x}_{j,\mathcal{C}_1\cap\mathcal{O}_1})'\widehat{\boldsymbol{\beta}}_1;y_{j,\mathcal{C}_1\cap\mathcal{O}_1},m_{j,\mathcal{C}_1\cap\mathcal{O}_1})$$

and

$$\sum_{j\in\mathcal{C}_1\cap\mathcal{O}_2}\xi(h(\boldsymbol{x}_{j,\mathcal{C}_1\cap\mathcal{O}_2})'\widehat{\boldsymbol{\beta}}_1;y_{j,\mathcal{C}_1\cap\mathcal{O}_1},m_{j,\mathcal{C}_1\cap\mathcal{O}_2}),$$

where $\widehat{\boldsymbol{\beta}}_1$ is defined in (5) with $s=1$. Then in view of the convexity of $\xi(\cdot)$ and (5), (11) and the fact that $\boldsymbol{\beta}_{01},\boldsymbol{\beta}_{02}$ are two distinct true parameter vectors, at least one of the below two inequalities hold:

$$\sum_{j \in \mathcal{C}_1 \cap \mathcal{O}_1} \xi(h(\boldsymbol{x}_{j,\mathcal{C}_1 \cap \mathcal{O}_1})'\widehat{\boldsymbol{\beta}}_1; y_{j,\mathcal{C}_1 \cap \mathcal{O}_1}, m_{j,\mathcal{C}_1 \cap \mathcal{O}_1})$$
$$> \sum_{j \in \mathcal{C}_1 \cap \mathcal{O}_1} \xi(h(\boldsymbol{x}_{j,\mathcal{C}_1 \cap \mathcal{O}_1})'\boldsymbol{\beta}; y_{j,\mathcal{C}_1 \cap \mathcal{O}_1}, m_{j,\mathcal{C}_1 \cap \mathcal{O}_1}), \quad \forall \boldsymbol{\beta} : |\boldsymbol{\beta} - \boldsymbol{\beta}_{01}| \leq d_0, \tag{12}$$

$$\sum_{j \in \mathcal{C}_1 \cap \mathcal{O}_2} \xi(h(\boldsymbol{x}_{j,\mathcal{C}_1 \cap \mathcal{O}_2})'\widehat{\boldsymbol{\beta}}_1; y_{j,\mathcal{C}_1 \cap \mathcal{O}_2}, m_{j,\mathcal{C}_1 \cap \mathcal{O}_2})$$
$$> \sum_{j \in \mathcal{C}_1 \cap \mathcal{O}_2} \xi(h(\boldsymbol{x}_{j,\mathcal{C}_1 \cap \mathcal{O}_2})'\boldsymbol{\beta}; y_{j,\mathcal{C}_1 \cap \mathcal{O}_2}, m_{j,\mathcal{C}_1 \cap \mathcal{O}_2}), \quad \forall \boldsymbol{\beta} : |\boldsymbol{\beta} - \boldsymbol{\beta}_{02}| \leq d_0,$$

where d_0 is defined in Assumption (X5). Without loss of generality, we assume that (12) holds. Now let us focus our discussion on the set $\mathcal{C}_1 \cap \mathcal{O}_1$ first. Let $n_{11} = |\mathcal{C}_1 \cap \mathcal{O}_1|$. We want to find out the order of

$$\begin{aligned} &\textstyle\sum_{j \in \mathcal{C}_1 \cap \mathcal{O}_1} \Big[\xi(h(\boldsymbol{x}'_{j,\mathcal{C}_1 \cap \mathcal{O}_1}\widehat{\boldsymbol{\beta}}_1); y_{j,\mathcal{C}_1 \cap \mathcal{O}_1}, m_{j,\mathcal{C}_1 \cap \mathcal{O}_1}) \\ &\qquad - \xi(h(\boldsymbol{x}'_{j,\mathcal{C}_1 \cap \mathcal{O}_1}\widehat{\boldsymbol{\beta}}_{011}); y_{j,\mathcal{C}_1 \cap \mathcal{O}_1}, m_{j,\mathcal{C}_1 \cap \mathcal{O}_1})\Big] \stackrel{\text{def}}{=} T \end{aligned}$$

as n increases to infinity, where $\widehat{\boldsymbol{\beta}}_{011}$ is defined in (7). For simplicity, we will use single indices exclusively for observations in the set $\mathcal{C}_1 \cap \mathcal{O}_1$, i.e., $\boldsymbol{x}_j$, y_j, m_j and π_{0j} will respectively represent $\boldsymbol{x}_{j,\mathcal{C}_1 \cap \mathcal{O}_1}$, $y_{j,\mathcal{C}_1 \cap \mathcal{O}_1}$, $m_{j,\mathcal{C}_1 \cap \mathcal{O}_1}$ and $\pi_{0j,\mathcal{C}_1 \cap \mathcal{O}_1}$ until the equation (18).

First note that

$$\begin{aligned} T &= \sum_{j \in \mathcal{C}_1 \cap \mathcal{O}_1} \Big[\xi(h(\boldsymbol{x}'_j\widehat{\boldsymbol{\beta}}_1); y_j, m_j) - \xi(h(\boldsymbol{x}'_j\widehat{\boldsymbol{\beta}}_{011}); y_j, m_j)\Big] \\ &= \sum_{j \in \mathcal{C}_1 \cap \mathcal{O}_1} \Big[\xi(h(\boldsymbol{x}'_j\widehat{\boldsymbol{\beta}}_1); y_j, m_j) - \xi(h(\boldsymbol{x}'_j\boldsymbol{\beta}_{01}); y_j, m_j)\Big] \\ &\quad - \sum_{j \in \mathcal{C}_1 \cap \mathcal{O}_1} \Big[\xi(h(\boldsymbol{x}'_j\widehat{\boldsymbol{\beta}}_{011}); y_j, m_j) - \xi(h(\boldsymbol{x}'_j\boldsymbol{\beta}_{01}); y_j, m_j)\Big] \\ &\stackrel{\text{def}}{=} T_1 + T_2. \end{aligned}$$

By Lemma 3 and (7), we have that for large n,

$$\begin{aligned} T_2 &= \sum_{j \in \mathcal{C}_1 \cap \mathcal{O}_1} \Big[\xi(h(\boldsymbol{x}'_j\widehat{\boldsymbol{\beta}}_{011}); y_j, m_j) - \xi(h(\boldsymbol{x}'_j\boldsymbol{\beta}_{01}); y_j, m_j)\Big] \\ &= \log\log n_{11} = o(n_{11}). \end{aligned} \tag{13}$$

Now let us consider the order of T_1. For any $\boldsymbol{\beta}$, define

$$H(\boldsymbol{\beta}) = \sum_{j \in \mathcal{C}_1 \cap \mathcal{O}_1} \left\{ \xi(h(\boldsymbol{x}_j'\boldsymbol{\beta}); y_j, m_j) - \xi(h(\boldsymbol{x}_j'\boldsymbol{\beta}_{01}); y_j, m_j) \right\}.$$

From the definitions of $\xi(\pi; y, m)$ and $w(u, v)$, it follows that

$$\begin{aligned} H(\boldsymbol{\beta}) &= \sum_{j \in \mathcal{C}_1 \cap \mathcal{O}_1} \left\{ -m_j y_j \boldsymbol{x}_j'(\boldsymbol{\beta} - \boldsymbol{\beta}_{01}) - m_j \log \frac{1 - h(\boldsymbol{x}_j'\boldsymbol{\beta})}{1 - h(\boldsymbol{x}_j'\boldsymbol{\beta}_{01})} \right\} \\ &= - \sum_{j \in \mathcal{C}_1 \cap \mathcal{O}_1} m_j (y_j - \pi_{0j}) \boldsymbol{x}_j'(\boldsymbol{\beta} - \boldsymbol{\beta}_{01}) \\ &\quad + \sum_{j \in \mathcal{C}_1 \cap \mathcal{O}_1} m_j w(\boldsymbol{x}_j'\boldsymbol{\beta}, \boldsymbol{x}_j'\boldsymbol{\beta}_{01}) \stackrel{\text{def}}{=} H_1(\boldsymbol{\beta}) + H_2(\boldsymbol{\beta}). \end{aligned} \tag{14}$$

Let $A_0 = \{\boldsymbol{\beta} : \|\boldsymbol{\beta} - \boldsymbol{\beta}_{01}\| \leq d_0\}$. Then by Lemma 3 it can be shown that

$$\begin{aligned} \inf_{\boldsymbol{\beta} \in \partial A_0} H_1(\boldsymbol{\beta}) &= O(\sqrt{n_{11} \log \log n_{11}}) \inf_{\boldsymbol{\beta} \in \partial A_0} \|\boldsymbol{\beta} - \boldsymbol{\beta}_{01}\| \\ &= O(\sqrt{n_{11} \log \log n_{11}}) \qquad \text{a.s.} \end{aligned} \tag{15}$$

By (23) of Lemma 2 and Assumption (X5), we derive that

$$\begin{aligned} &\inf_{\boldsymbol{\beta} \in \partial A_0} H_2(\boldsymbol{\beta}) \\ &\geq \inf_{\boldsymbol{\beta} \in \partial A_0} \frac{1}{4} \sum_{j \in \mathcal{C}_1 \cap \mathcal{O}_1} m_j e^{-|\boldsymbol{x}_j'(\boldsymbol{\beta} - \boldsymbol{\beta}_{01})|} h(\boldsymbol{x}_j'\boldsymbol{\beta}_{01})(1 - h(\boldsymbol{x}_j'\boldsymbol{\beta}_{01})) \\ &\quad \times (\boldsymbol{x}_j'\boldsymbol{\beta} - \boldsymbol{x}_j'\boldsymbol{\beta}_{01})^2 \\ &= \frac{1}{4} \inf_{\boldsymbol{\beta} \in \partial A_0} (\boldsymbol{\beta} - \boldsymbol{\beta}_{01})' X_{\mathcal{C}_1 \cap \mathcal{O}_1}' Q_{n_{11}} X_{\mathcal{C}_1 \cap \mathcal{O}_1} (\boldsymbol{\beta} - \boldsymbol{\beta}_{01}) \\ &\geq \frac{1}{4} a_6 n_{11} \inf_{\boldsymbol{\beta} \in \partial A_0} \|\boldsymbol{\beta} - \boldsymbol{\beta}_{01}\| = \frac{1}{4} d_0 a_6 n_{11}. \end{aligned} \tag{16}$$

From (14), (15) and (16) it follows that there exists a constant $\tau > 0$ such that for large n,

$$\inf_{\boldsymbol{\beta} \in \partial A_0} H(\boldsymbol{\beta}) \geq \tau n_{11}. \tag{17}$$

By (12) and (17), we have that

$$\begin{aligned} T_1 &= \sum_{j \in \mathcal{C}_1 \cap \mathcal{O}_1} \left[\xi(h(\boldsymbol{x}_j'\widehat{\boldsymbol{\beta}}_1); y_j, m_j) - \xi(h(\boldsymbol{x}_j'\boldsymbol{\beta}_{01}); y_j, m_j) \right] \\ &\geq \inf_{\boldsymbol{\beta} \in \partial A_0} H(\boldsymbol{\beta}) \geq \tau n_{11}. \end{aligned} \tag{18}$$

Hence by combining results from (13) and (18), we have

$$\sum_{j\in\mathcal{C}_1\cap\mathcal{O}_1}\Big[\xi(h(\boldsymbol{x}'_{j,\mathcal{C}_1\cap\mathcal{O}_1}\widehat{\boldsymbol{\beta}}_1);y_{j,\mathcal{C}_1\cap\mathcal{O}_1},m_{j,\mathcal{C}_1\cap\mathcal{O}_1}) \\ -\xi(h(\boldsymbol{x}'_{j,\mathcal{C}_1\cap\mathcal{O}_1}\widehat{\boldsymbol{\beta}}_{011});y_{j,\mathcal{C}_1\cap\mathcal{O}_1},m_{j,\mathcal{C}_1\cap\mathcal{O}_1})\Big]\geq \tau n_{11}. \tag{19}$$

Note that $D_n(\Pi_k)-D_n(\Pi_{k_0})$ can be partitioned as follows:

$$\begin{aligned}
&D_n(\Pi_k)-D_n(\Pi_{k_0})\\
&=\sum_{s=1}^{k}\sum_{j\in\mathcal{C}_s}\xi(h(\boldsymbol{x}'_{j,\mathcal{C}_s}\widehat{\boldsymbol{\beta}}_s);y_{j,\mathcal{C}_s},m_{j,\mathcal{C}_s})\\
&\quad-\sum_{i=1}^{k_0}\sum_{j\in\mathcal{O}_i}\xi(h(\boldsymbol{x}'_{j,\mathcal{O}_i}\widehat{\boldsymbol{\beta}}_{0i});y_{j,\mathcal{O}_i},m_{j,\mathcal{O}_i})+(q(k)-q(k_0))A_n\\
&=\sum_{j\in\mathcal{C}_1\cap\mathcal{O}_1}\Big[\xi(h(\boldsymbol{x}'_{j,\mathcal{C}_1\cap\mathcal{O}_1}\widehat{\boldsymbol{\beta}}_1);y_{j,\mathcal{C}_1\cap\mathcal{O}_1},m_{j,\mathcal{C}_1\cap\mathcal{O}_1})\\
&\quad-\xi(h(\boldsymbol{x}'_{j,\mathcal{C}_1\cap\mathcal{O}_1}\widehat{\boldsymbol{\beta}}_{011});y_{j,\mathcal{C}_1\cap\mathcal{O}_1},m_{j,\mathcal{C}_1\cap\mathcal{O}_1})\Big]\\
&\quad+\sum_{\mathcal{J}_{is}}\sum_{j\in\mathcal{C}_s\cap\mathcal{O}_i}\Big[\xi(h(\boldsymbol{x}'_{j,\mathcal{C}_s\cap\mathcal{O}_i}\widehat{\boldsymbol{\beta}}_s);y_{j,\mathcal{C}_s\cap\mathcal{O}_i},m_{j,\mathcal{C}_s\cap\mathcal{O}_i})\\
&\quad-\xi(h(\boldsymbol{x}'_{j,\mathcal{C}_s\cap\mathcal{O}_i}\widehat{\boldsymbol{\beta}}_{0si});y_{j,\mathcal{C}_s\cap\mathcal{O}_i},m_{j,\mathcal{C}_s\cap\mathcal{O}_i})\Big]\\
&\quad+\sum_{s=1}^{k}\sum_{i=1}^{k_0}\sum_{j\in\mathcal{C}_s\cap\mathcal{O}_i}\Big[\xi(h(\boldsymbol{x}'_{j,\mathcal{C}_s\cap\mathcal{O}_i}\widehat{\boldsymbol{\beta}}_{0si});y_{j,\mathcal{C}_s\cap\mathcal{O}_i},m_{j,\mathcal{C}_s\cap\mathcal{O}_i})\\
&\quad-\xi(h(\boldsymbol{x}'_{j,\mathcal{C}_s\cap\mathcal{O}_i}\widehat{\boldsymbol{\beta}}_{0i});y_{j,\mathcal{C}_s\cap\mathcal{O}_i},m_{j,\mathcal{C}_s\cap\mathcal{O}_i})\Big]+(q(k)-q(k_0))A_n,
\end{aligned}$$

where $\mathcal{J}_{is}=\{i,s:i=1,\ldots,k;s=1,\ldots,k_0;i$ and s can not be 1 simultaneously$\}$ and hence $\mathcal{J}_{is}$ corresponds to all possible intersection sets of Π_k and Π_{k_0} excluding $\mathcal{C}_1\cap\mathcal{O}_1$; $\widehat{\boldsymbol{\beta}}_i$, $\widehat{\boldsymbol{\beta}}_{0i}$ and $\widehat{\boldsymbol{\beta}}_{0si}$ are defined in (5), (6), and (7), respectively. By (8), we obtain

$$\sum\sum_{\mathcal{J}_{is}}\sum_{j\in\mathcal{C}_s\cap\mathcal{O}_i}\Big[\xi(h(\boldsymbol{x}'_{j,\mathcal{C}_s\cap\mathcal{O}_i}\widehat{\boldsymbol{\beta}}_s);y_{j,\mathcal{C}_s\cap\mathcal{O}_i},m_{j,\mathcal{C}_s\cap\mathcal{O}_i}) \\ -\xi(h(\boldsymbol{x}'_{j,\mathcal{C}_s\cap\mathcal{O}_i}\widehat{\boldsymbol{\beta}}_{0si});y_{j,\mathcal{C}_s\cap\mathcal{O}_i},m_{j,\mathcal{C}_s\cap\mathcal{O}_i})\Big]\geq 0. \tag{20}$$

By following the same line of argument as in proving (9), we can show that

$$\sum_{s=1}^{k}\sum_{i=1}^{k_0}\sum_{j\in\mathcal{C}_s\cap\mathcal{O}_i}\Big[\xi(h(\boldsymbol{x}'_{j,\mathcal{C}_s\cap\mathcal{O}_i}\widehat{\boldsymbol{\beta}}_{0si});y_{j,\mathcal{C}_s\cap\mathcal{O}_i},m_{j,\mathcal{C}_s\cap\mathcal{O}_i})$$
$$-\xi(h(\boldsymbol{x}'_{j,\mathcal{C}_s\cap\mathcal{O}_i}\widehat{\boldsymbol{\beta}}_{0i});y_{j,\mathcal{C}_s\cap\mathcal{O}_i},m_{j,\mathcal{C}_s\cap\mathcal{O}_i})\Big] = O(\log\log n) = o(n). \quad (21)$$

Hence in terms of (11), (19), (20) and (21) and Assumption (Z), we obtain that for large n,

$$D_n(\Pi_k) - D_n(\Pi_{k_0}) \geq \tau b_0 n + o(n) + (q(k) - q(k_0))A_n > 0. \quad (22)$$

Therefore combining the results from (10) in Case 1 and (22) in Case 2, we have showed that the true classification is preferable when n increases to infinity.

Appendix

Lemma 1. *Suppose that Assumption (A) holds, for any possible partition $\Pi_k^{(n)}$ of $\mathcal{O}^{(n)}$, if $k < k_0$, where k is the number of clusters for $\Pi_k^{(n)}$ and k_0 is the true number of clusters in $\mathcal{O}^{(n)}$, there exist $\mathcal{C}_s \in \Pi_k^{(n)}$ and $\mathcal{O}_i, \mathcal{O}_l \in \Pi_{k_0}^{(n)}$ such that*

$$|\mathcal{C}_s \cap \mathcal{O}_i| > b_0 n \quad \text{and} \quad |\mathcal{C}_s \cap \mathcal{O}_l| > b_0 n,$$

where $b_0 = a_0/k_0 > 0$ is a fixed constant.

The proof can be found in Shao and Wu (2005).

Lemma 2. *Define $w(u,v) = -\log(1-h(u))/(1-h(v)) - h(v)(u-v)$, where $h(u) = e^u/(1+e^u)$. Then $w(u,v)$ is strictly convex with respect to u. Further, we have*

$$w(u,v) \geq \frac{1}{4}e^{-\zeta}h(v)(1-h(v))(u-v)^2 \quad \text{if} \quad |u-v| \leq \zeta,\ \forall \zeta > 0. \quad (23)$$

The proof can be found in Qian and Field (2002).

Lemma 3. *Suppose that Assumptions (X1)–(X4) hold. Then we have that for large n,*

$$\left.\frac{\partial l}{\partial \boldsymbol{\beta}}\right|_{\boldsymbol{\beta}=\boldsymbol{\beta}_{0\ell}} = \sum_{j\in\mathcal{O}_\ell} m_{j,\mathcal{O}_\ell}(y_{j,\mathcal{O}_\ell} - \pi_{0j,\mathcal{O}_\ell})\boldsymbol{x}_{j,\mathcal{O}_\ell}$$
$$= X'_{n_\ell}M_{n_\ell}(Y_{n_\ell} - \Pi_{0n_\ell}) = O(\sqrt{n_\ell \log\log n_\ell}), \quad (24)$$

and

$$0 \leq \sum_{j \in \mathcal{O}_\ell} \{\xi(h(\boldsymbol{x}'_{j,\mathcal{O}_\ell}\widehat{\boldsymbol{\beta}}_{n_\ell}); y_{j,\mathcal{O}_\ell}, m_{j,\mathcal{O}_\ell}) - \xi(h(\boldsymbol{x}'_{j,\mathcal{O}_\ell}\boldsymbol{\beta}_{0\ell}); y_{j,\mathcal{O}_\ell}, m_{j,\mathcal{O}_\ell})\}$$
$$= O(\log\log n_\ell), \qquad (25)$$

where $Y_{n_\ell} = (y_{\ell_1}, \ldots, y_{\ell_{n_\ell}})'$ and $\Pi_{0n_\ell} = \text{diag}\{\pi_{\ell_1}, \ldots, \pi_{\ell_{n_\ell}}\}$.
See Qian and Field (2002) for the proof. In fact, (24) and (25) are respectively the results of Lemma 2 and Theorem 2 in that paper.

Acknowledgement

The research was partially supported by the Natural Sciences and Engineering Research Council of Canada.

References

Chung KL (2001) A Course in Probability Theory (3rd edition). Academic Press

Farewell BT, Sprott D (1988) The use of a mixture model in the analysis of count data. Biometrics 44:1191–1194

Follmann DA, Lambert D (1989) Generalizing logistic regression by nonparametric mixing. Journal of the American Statistical Association 84:295–300

Follmann DA, Lambert D (1991) Identifiability for nonparametric mixtures of logistic regressions. Journal of Statistical Planning and Inference 27:375–381

McCullagh P, Nelder JA (1989) Generalized Linear Models (2nd edition). Chapman and Hall

Qian G, Field C (2002) Law of iterated logarithm and consistent model selection criterion in logistic regression. Statistics & Probability Letters 56:101–112

Shao Q, Wu Y (2005) A consistent procedure for determining the number of clusters in regression clustering. Journal of Statistical Planning and Inference 135:461–476

Wedel M, DeSarbo WS (1995) A mixture likelihood approach for generalized linear models. Journal of Classification 12:21–55

Quasi Score and Corrected Score Estimation in the Polynomial Measurement Error Model

Hans Schneeweiss

Department of Statistics, University of Munich, Akademiestrasse 1, 80799 Munich, Germany schneew@stat.uni-muenchen.de

1 Introduction

Despite the many results that have been found in recent years on the estimation of regression coefficients of a polynomial model with measurement errors in the covariable, cf., e.g., Cheng and Schneeweiss (1998), Cheng and Schneeweiss (2002), Kukush et al. (2005), Kukush and Schneeweiss (2005), Shklyar et al. (2007), some issues concerning the computation of estimators and their asymptotic covariance matrices (ACM) are still open to investigation.

The polynomial measurement error model is given by the regression equation

$$y = \zeta^\top \beta + \epsilon,$$

with $\zeta^\top = (1, \xi, \cdots, \xi^k)$, $\beta := (\beta_0, \beta_1, \cdots, \beta_k)^\top$, $\mathbb{E}\epsilon = 0$, $\mathbb{V}\epsilon = \sigma_\epsilon^2$, ϵ and ξ independent, and the measurement equation

$$x = \xi + \delta,$$

$\delta \sim N(0, \sigma_\delta^2)$ being the measurement error, which is independent of ξ and ϵ. It is assumed that σ_δ is known. In addition, we here assume that $\xi \sim N(\mu_\xi, \sigma_\xi^2)$. The problem is to estimate β from an i.i.d. sample (x_i, y_i), $i = 1, \cdots, n$.

In addition to the *naive*(N) *estimator* , we consider two consistent estimators: the (structural) *quasi score* (QS) and the (functional) *corrected score* (CS) *estimator* . The first one utilizes the distribution of ξ, the latter one does not. Both methods are based on a transformation of the powers x_i^r of the data x_i into new (artificial) data, $\mu_r(x_i)$ for QS and $t_r(x_i)$ for CS.

The first issue of this paper is to explore some, up to now unknown, properties of the variables μ_r and t_r and to reveal a peculiar duality between them. Another issue is to transform the formulas for the ACMs and their small-σ_δ approximations so that they become easier to compute, possibly with the help of a matrix oriented programming language. In particular, they should be written in terms of the observable variable x instead of the unobservable ξ. An important point in this respect is the evaluation of the terms in the ACM of QS that stem from the estimation of the nuisance parameters μ_ξ and σ_ξ^2. Contrary to what one might conclude from the original form of the ACM in Kukush et al. (2005), it turns out that these additional terms can be computed without any integration (although integration remains necessary to compute the main term of the ACM formula).

Shklyar et al. (2007) have studied a simplified version of the QS estimator, the so-called *simple score* (SS) *estimator* . Two equivalent formulas for its ACM are presented. The ACM formula has the same term originating from the estimation of the nuisance parameters as the ACM of QS.

If this term is ignored (i.e., if the nuisance parameters are taken to be known), the difference of the ACMs of the CS and SS estimators is p.s.d., cf. Shklyar et al. (2007). It is an open question whether this is still true if the nuisance parameters have to be estimated.

In Section 2, the variables μ_r and t_r are investigated. Some results on the derivatives of the μ_r are found in Section 3. Section 4 deals with the ACM of the QS estimator in the polynomial model and in particular with the terms resulting from estimating the nuisance parameters. Section 5 has a reformulation of the ACM of the CS estimator, and Section 6 deals with the SS estimator. Section 7 discusses efficiency problems. Section 8 has some concluding remarks.

2 QS and CS: The Variables μ_r and t_r

The *QS estimator* $\hat{\beta}_Q$ of the polynomial measurement error model is based on the quasi score function

$$\psi_Q(y, x, \beta) = (y - \mu^\top \beta) v^{-1} \mu,$$

where $\mu := \mathbb{E}(\zeta|x) =: (\mu_0, \mu_1, \cdots, \mu_k)^\top$ and $v := \mathbb{V}(y|x)$. The elements of the conditional mean vector μ, $\mu_r = \mathbb{E}(\xi^r|x)$, are polynomials in x of degree r. $\mu_0 = 1$ and $\mu_1 = \mu_1(x) = \mathbb{E}(\xi|x)$ is given by

$$\mu_1 = \frac{\sigma_\delta^2}{\sigma_x^2}\mu_x + \left(1 - \frac{\sigma_\delta^2}{\sigma_x^2}\right)x. \tag{1}$$

The other μ_r are polynomials of μ_1 of degree r, c.f. Thamerus (1998):

$$\mu_r = \sum_{j=0}^{r} \binom{r}{j} \mu_j^* \mu_1^{r-j} \tag{2}$$

with

$$\mu_j^* = \begin{cases} 0 & \text{if } j \text{ is odd} \\ (j-1)!!\tau^j & \text{if } j \text{ is even,} \end{cases} \tag{3}$$

$$\tau^2 := \mathbb{V}(\xi|x) = \sigma_\delta^2 \left(1 - \frac{\sigma_\delta^2}{\sigma_x^2}\right), \tag{4}$$

where $(j-1)!!$ is short for $1 \cdot 3 \cdot 5 \cdots (j-1)$ and $(-1)!! = 1$. The conditional variance v is given by

$$v = \sigma_\epsilon^2 + \beta^\top \left(M - \mu\mu^\top\right)\beta, \tag{5}$$

where $M = M(x)$ is a $(k+1)\times(k+1)$-matrix with elements $M_{rs} = \mu_{r+s}$, $r, s = 0, \cdots, k$. Note that the $\mu_r(x_i)$ can be computed from the data x_i if the nuisance parameters μ_x and σ_x^2 are given. Typically they are unknown and must be estimated from the data x_i in the usual way.

The *CS estimator* $\widehat{\beta}_C$ is based on the corrected score function

$$\psi_C(y, x, \beta) = yt - T\beta,$$

where $t = t(x)$ is such that $\mathbb{E}(t|\xi) = \zeta$. Thus $t = (t_0, t_1, \ldots, t_k)^\top$ and $\mathbb{E}(t_r|\xi) = \xi^r$. $T = T(x)$ is a $(k+1) \times (k+1)$-matrix with elements $T_{rs} = t_{r+s}$. The t_r are polynomials in x of degree r. They can be computed via the recursion formula, cf. Stefanski (1989) and Cheng and Schneeweiss (1998),

$$t_{r+1} = t_r x - r t_{r-1}\sigma_\delta^2; \quad t_0 = 1, \quad t_{-1} = 0. \tag{6}$$

Note the duality in the definitions of μ and t:

$$\mu = \mathbb{E}(\zeta|x), \quad \mathbb{E}(t|\xi) = \zeta$$

and also in the matrices M and T:

$$M = \mathbb{E}(\zeta\zeta^T|x), \quad \mathbb{E}(T|\xi) = \zeta\zeta^\top.$$

This duality reaches farther. It turns out that, although the defining formulas (2) and (6) for μ and t, respectively, are quite different, there are other ways of computing μ and t, which very much resemble (2) and (6), but with the role of μ and t interchanged.

Theorem 1. *The variables μ_r can be computed via the recursion formula*

$$\mu_{r+1} = \mu_r\mu_1 + r\mu_{r-1}\tau^2, \quad \mu_0 = 1, \quad \mu_{-1} = 0. \tag{7}$$

Proof:

According to (2)

$$\begin{aligned}
\mu_{r+1} &= \sum_{j=0}^{r+1} \binom{r+1}{j} \mu_1^{r+1-j}\mu_j^* \\
&= \sum_{j=1}^{r+1} \binom{r}{j-1} \mu_1^{r+1-j}\mu_j^* + \sum_{j=0}^{r} \binom{r}{j} \mu_1^{r+1-j}\mu_j^* \\
&= \sum_{j=0}^{r} \binom{r}{j} \mu_1^{r-j}\mu_{j+1}^* + \sum_{j=0}^{r} \binom{r}{j} \mu_1^{r+1-j}\mu_j^*.
\end{aligned}$$

In the second equation, we used the identity

$$\binom{r+1}{j} = \binom{r}{j-1} + \binom{r}{j}, \quad 1 < j < r.$$

Now again by (2), the right hand side of the recursion formula (7) is

$$\begin{aligned}
&\sum_{j=0}^{r} \binom{r}{j} \mu_1^{r+1-j}\mu_j^* + r\sum_{j=0}^{r-1} \binom{r-1}{j} \mu_1^{r-1-j}\mu_j^*\tau^2 \\
&= \sum_{j=0}^{r} \binom{r}{j} \mu_1^{r+1-j}\mu_j^* + \sum_{j=0}^{r-1} \binom{r}{j+1} \mu_1^{r-1-j}\mu_{j+2}^* \\
&= \sum_{j=0}^{r} \binom{r}{j} \mu_1^{r+1-j}\mu_j^* + \sum_{j=0}^{r} \binom{r}{j} \mu_1^{r-j}\mu_{j+1}^* = \mu_{r+1}.
\end{aligned}$$

In the second equation, the identity,

$$(j+1)\mu_j^*\tau^2 = \mu_{j+2}^*,$$

see (3), was used and in the third equation the fact that $\mu_1^* = 0$.

Remark:

The proof is similar to the proof of (6) as given in Cheng and Schneeweiss (1996).

Theorem 2. *t_r can be computed via the closed form formula*

$$t_r = \sum_{j=0}^{r} \binom{r}{j} \mu_j^+ x^{r-j}, \tag{8}$$

$$\mu_j^+ := \begin{cases} 0 & \text{if } j \text{ is odd} \\ (j-1)!!(-1)^{\frac{j}{2}} \sigma_\delta^j & \text{if } j \text{ is even.} \end{cases}$$

Proof:

If we replace μ_r, μ_1, and τ^j with t_r, x and $(-1)^{\frac{j}{2}}\sigma_\delta^j$, respectively, then (7) changes to (6) and (2) changes to (8). By Theorem 1, (7) follows from (2), and so (6) follows from (8). But as (6) defines the t_r uniquely, the t_r defined by (6) must be the same as those defined by (8). This completes the proof.

The great similarity in the construction of the variables μ_r and t_r can also be seen by looking at its values, e.g.:

$$\mu_1 = \mu_1, \quad \mu_2 = \mu_1^2 + \tau^2, \quad \mu_3 = \mu_1^3 + 3\tau^2\mu_1, \quad \mu_4 = \mu_1^4 + 6\tau^2\mu_1^2 + 3\tau^4$$

and

$$t_1 = x, \quad t_2 = x^2 - \sigma_\delta^2, \quad t_3 = x^3 - 3\sigma_\delta^2 x, \quad t_4 = x^4 - 6\sigma_\delta^2 x^2 + 3\sigma_\delta^4.$$

3 Derivatives of μ_r

By (2) and (3), μ_r is a function of μ_1 and τ^2. We can derive formulas for the derivatives of μ_r with respect to μ_1 and τ^2, which will be useful later on.

Theorem 3.

$$\frac{\partial \mu_r}{\partial \mu_1} = r\mu_{r-1}, \quad r \geq 1 \tag{9}$$

$$\frac{\partial \mu_r}{\partial \tau^2} = \binom{r}{2} \mu_{r-2}, \quad r \geq 2\,. \tag{10}$$

Proof:

Instead of (9), we will prove the stronger proposition

$$\mu_r = r \int_0^{\mu_1} \mu_{r-1} d\mu_1 + \mu_r^*.$$

Indeed, by (2) the right hand side of this equation equals

$$\begin{aligned}
& r \int_0^{\mu_1} \sum_{j=0}^{r-1} \binom{r-1}{j} \mu_j^* \mu_1^{r-1-j} d\mu_1 + \mu_r^* \\
&= r \sum_{j=0}^{r-1} \binom{r-1}{j} \mu_j^* \frac{\mu_1^{r-j}}{r-j} + \mu_r^* \\
&= \sum_{j=0}^{r-1} \binom{r}{j} \mu_j^* \mu_1^{r-j} + \mu_r^* \\
&= \sum_{j=0}^{r} \binom{r}{j} \mu_j^* \mu_1^{r-j},
\end{aligned}$$

which is equal to μ_r by (2).

To prove (10), first note that by (3), for j even and $j \geq 2$,

$$\begin{aligned}
\frac{\partial \mu_j^*}{\partial \tau^2} &= (j-1)!! \frac{j}{2} \tau^{j-2} \\
&= \binom{j}{2} (j-3)!! \tau^{j-2} = \binom{j}{2} \mu_{j-2}^*.
\end{aligned}$$

Now from (2) and the previous equation, for $r \geq 2$,

$$\begin{aligned}
\frac{\partial \mu_r}{\partial \tau^2} &= \sum_{j=2}^{r} \binom{r}{j} \binom{j}{2} \mu_{j-2}^* \mu_1^{r-j} \\
&= \frac{r(r-1)}{2} \sum_{j=2}^{r} \binom{r-2}{j-2} \mu_{j-2}^* \mu_1^{r-j} \\
&= \binom{r}{2} \sum_{j=0}^{r-2} \binom{r-2}{j} \mu_j^* \mu_1^{r-2-j} = \binom{r}{2} \mu_{r-2}.
\end{aligned}$$

This completes the proof.

By stacking the formulas (9) and (10), respectively, for $r = 0, \dots, k$, we can now give corresponding expressions for the vector μ. We introduce the $(k+1) \times (k+1)$ triangular band matrices

$$D_1 := \begin{pmatrix} 0 & & & & & \\ 1 & 0 & & & & \\ & 2 & 0 & & & \\ & & \cdots & & & \\ & & & \cdots & & \\ & & & & k & 0 \end{pmatrix}$$

$$D_2 := \begin{pmatrix} 0 & & & & & & \\ 0 & 0 & & & & & \\ \binom{2}{2} & 0 & 0 & & & & \\ & \binom{3}{2} & 0 & 0 & & & \\ & & & \cdots & & & \\ & & & & \cdots & & \\ & & & & \binom{k}{2} & 0 & 0 \end{pmatrix} \tag{11}$$

and note that

$$D_2 = \frac{1}{2} D_1^2. \tag{12}$$

Theorem 3 then, translates immediately into.

Theorem 4.

$$\frac{\partial \mu}{\partial \mu_1} = D_1 \mu \tag{13}$$

$$\frac{\partial \mu}{\partial \tau^2} = D_2 \mu. \tag{14}$$

Finally we also have

Theorem 5.

$$\mu_1 \frac{\partial \mu}{\partial \mu_1} = (D - \tau^2 D_1^2)\mu \tag{15}$$

with $D := diag(0, 1, 2, \dots, k)$.

Proof:

First note that by Theorem 1

$$\mu_1\mu = \begin{pmatrix} \mu_1 \\ \mu_2 \\ \mu_3 \\ \vdots \\ \mu_{k+1} \end{pmatrix} - \tau^2 \begin{pmatrix} 0 \\ \mu_0 \\ 2\mu_1 \\ \vdots \\ k\mu_{k-1} \end{pmatrix}.$$

The last vector equals $D_1\mu$, and the first vector on the right hand side multiplied by D_1 equals $D\mu$. Therefore

$$\mu_1 \frac{\partial \mu}{\partial \mu_1} = D_1\mu_1\mu = D\mu - \tau^2 D_1^2 \mu.$$

4 The ACM of QS

According to Kukush et al. (2005), the ACM of $\widehat{\beta}_Q$ is given by

$$\Sigma_Q = (\mathbb{E}v^{-1}\mu\mu^\top)^{-1} + (\mathbb{E}v^{-1}\mu\mu^\top)^{-1}(\sigma_x^2 F_1 F_1^\top + \frac{2}{\sigma_x^4} F_2 F_2^\top)(\mathbb{E}v^{-1}\mu\mu^\top)^{-1} \tag{16}$$

where

$$F_p = \mathbb{E}v^{-1}\mu \frac{\partial \mu^\top}{\partial \gamma_p}\beta, \quad p = 1, 2, \quad \gamma_1 = \mu_x, \quad \gamma_2 = \frac{1}{\sigma_x^2}.$$

The F-terms stem from the estimation of the nuisance parameters. The purpose of this section is to evaluate these terms so that they become computationally more accessible. It turns out that it is not necessary to compute the expected value as prescribed in the definition of F_p.

Theorem 6. *The ACM of $\widehat{\beta}_Q$ equals*

$$\Sigma_Q = (\mathbb{E}v^{-1}\mu\mu^\top)^{-1} + F, \tag{17}$$

where

$$F = \sigma_\delta^4 (G_1^\top \beta\beta^\top G_1 + 2G_2^\top \beta\beta^\top G_2),$$

$$G_1 = \frac{1}{\sigma_x} D_1,$$

$$G_2 = \frac{1}{\sigma_x^2 - \sigma_\delta^2} (\mu_X D_1 - D + \tau^2 D_2).$$

Proof:

As μ is a function of μ_1 and τ^2, we have

$$\frac{\partial \mu}{\partial \gamma_p} = \frac{\partial \mu}{\partial \mu_1}\frac{\partial \mu_1}{\partial \gamma_p} + \frac{\partial \mu}{\partial \tau^2}\frac{\partial \tau^2}{\partial \gamma_p}, \quad p = 1, 2.$$

For $p = 1$ and $p = 2$, we find because of (1) and (4)

$$\frac{\partial \mu}{\partial \gamma_1} = \frac{\partial \mu}{\partial \mu_1}\frac{\sigma_\delta^2}{\sigma_x^2},$$
$$\frac{\partial \mu}{\partial \gamma_2} = \left[\frac{\partial \mu}{\partial \mu_1}(\mu_x - x) - \frac{\partial \mu}{\partial \tau^2}\sigma_\delta^2\right]\sigma_\delta^2.$$

With

$$\mu_x - x = \frac{\sigma_x^2}{\sigma_x^2 - \sigma_\delta^2}(\mu_x - \mu_1),$$

which follows from (1), the latter becomes

$$\frac{\partial \mu}{\partial \gamma_2} = \sigma_\delta^2\left[\frac{\partial \mu}{\partial \mu_1}\frac{\sigma_x^2}{\sigma_x^2 - \sigma_\delta^2}(\mu_x - \mu_1) - \frac{\partial \mu}{\partial \tau^2}\sigma_\delta^2\right].$$

Finally, by (13) to (15),

$$\frac{\partial \mu}{\partial \gamma_1} = \frac{\sigma_\delta^2}{\sigma_x^2}D_1\mu = \frac{\sigma_\delta^2}{\sigma_x}G_1\mu,$$
$$\frac{\partial \mu}{\partial \gamma_2} = \sigma_\delta^2\left[\frac{\sigma_x^2}{\sigma_x^2 - \sigma_\delta^2}(\mu_x D_1 - D + \tau^2 D_1^2)\mu - \sigma_\delta^2 D_2\mu\right].$$

Because of (12) and (4), the latter becomes

$$\frac{\partial \mu}{\partial \gamma_2} = \sigma_\delta^2\frac{\sigma_x^2}{\sigma_x^2 - \sigma_\delta^2}(\mu_x D_1 - D + \tau^2 D_2)\mu = \sigma_\delta^2\sigma_x^2 G_2\mu.$$

We thus have

$$F_1 = \frac{\sigma_\delta^2}{\sigma_x}\mathbb{E}v^{-1}\mu\mu^\top G_1^\top\beta,$$
$$F_2 = \sigma_\delta^2\sigma_x^2\mathbb{E}v^{-1}\mu\mu^\top G_2^\top\beta.$$

By substituting F_1 and F_2 in (16) we finally obtain (17). This completes the proof.

For $k = 2$ the two matrices G_1 and G_2 are, respectively,

$$G_1 = \frac{1}{\sigma_x} \begin{pmatrix} 0\,0\,0 \\ 1\,0\,0 \\ 0\,2\,0 \end{pmatrix}, \tag{18}$$

$$G_2 = \frac{1}{\sigma_x^2 - \sigma_\delta^2} \begin{pmatrix} 0 & 0 & 0 \\ \mu_x & -1 & 0 \\ \tau^2 & 2\mu_x & -2 \end{pmatrix}. \tag{19}$$

An approximation to Σ_Q can be derived for small σ_δ^2 . The general formula in Kukush and Schneeweiss (2005) can be specialized to the polynomial case and yields

$$\begin{aligned}\Sigma_Q &= \sigma_\epsilon^2(\mathbb{E}Z)^{-1} \\ &\quad +\sigma_\delta^2(\mathbb{E}Z)^{-1}\mathbb{E}\left\{\left(\frac{\partial z^\top}{\partial x}\beta\right)^2 Z + \sigma_\epsilon^2\left(\frac{1}{2}\frac{\partial^2 Z}{\partial x^2} + \frac{\partial z}{\partial x}\frac{\partial z^\top}{\partial x}\right)\right\}(\mathbb{E}Z)^{-1} \\ &\quad +O(\sigma_\delta^4),\end{aligned}$$

where $z := (1, x, \ldots, x^k)^\top$ and $Z := zz^\top$. By noting that

$$\frac{\partial z}{\partial x} = D_1 z, \quad \frac{\partial^2 Z}{\partial x^2} = D_1^2 Z + 2D_1 Z D_1^\top + Z D_1^{\top 2},$$

this can be written as

$$\begin{aligned}\Sigma_Q = \sigma_\epsilon^2(\mathbb{E}Z)^{-1} &+ \sigma_\delta^2(\mathbb{E}Z)^{-1}\mathbb{E}\{(\beta^\top D_1 Z D_1^\top \beta)Z \\ &+\sigma_\epsilon^2(D_2 Z + Z D_2^\top + 2D_1 Z D_1^\top)\}(\mathbb{E}Z)^{-1} + O(\sigma_\delta^4).\end{aligned} \tag{20}$$

It may be noted that, contrary to (17), the expectations involved simply yield moments of x and are therefore easy to compute.

From Kukush et al. (2005) a similar formula can be derived, which however is stated in terms of ξ rather than x. Both formulas differ in value but the difference is of the order σ_δ^4.

5 The ACM of CS

In Kukush et al. (2005) a formula for the ACM of $\widehat{\beta}_C$ has been derived:

$$\Sigma_C = (\mathbb{E}\zeta\zeta^\top)^{-1}\{\sigma_\epsilon^2\mathbb{E}tt^\top + \mathbb{E}(T - t\zeta^\top)\beta\beta^\top(T - \zeta t^\top)\}(\mathbb{E}\zeta\zeta^\top)^{-1}. \tag{21}$$

This is a hybrid formula in so far as t and T are functions of x, whereas ζ is a function of ξ. With (5) and with the help of the identity

$$\begin{aligned}
&\mathbb{E}[(T - t\zeta^\top)\beta\beta^\top(T - \zeta t^\top)|x] \\
&= T\beta\beta^\top T - t\mu^\top\beta\beta^\top T - T\beta\beta^\top\mu t^\top + t\beta^\top M\beta t^\top \\
&= (T - t\mu^\top)\beta\beta^\top(T - \mu t^\top) + t\beta^\top(M - \mu\mu^\top)\beta t^\top,
\end{aligned}$$

(21) can be written as

$$\Sigma_C = (\mathbb{E}T)^{-1}\mathbb{E}\{(T - t\mu^\top)\beta\beta^\top(T - \mu t^\top) + vtt^\top\}(\mathbb{E}T)^{-1}. \quad (22)$$

Again only moments of x are needed in order to compute the ACM of $\widehat{\beta}_C$. We have several options to evaluate $\mathbb{E}T$ because, cf. Shklyar et al. (2007),

$$\mathbb{E}T = \mathbb{E}M = \mathbb{E}t\mu^\top = \mathbb{E}\zeta\zeta^\top.$$

In passing, it might be worthwile to mention the ACM of the *naive* (N) *estimator* $\hat{\beta}_N := (\sum_1^n z_i z_i^\top)^{-1}\sum_1^n z_i y_i$. A hybrid formula for its ACM is given in Kukush et al. (2005). It can be "improved" to a formula that is based on the observed variables x_i solely:

$$\Sigma_N = (\mathbb{E}Z)^{-1}\mathbb{E}vZ(\mathbb{E}Z)^{-1}.$$

6 SS and its ACM

Another structural estimator can be constructed as a simplified version of QS. It is called *simple score* (SS) *estimator* and is based on the simplified score function

$$\psi_S(y, x, \beta) = (y - \mu^\top\beta)t.$$

An equivalent score function for SS is

$$\psi_S^*(y, x, \beta) = (y - \mu^\top\beta)\mu,$$

cf. Shklyar et al. (2007), which differs from ψ_Q just by the omission of the factor v^{-1}.

The merit of the SS estimator is that it is much simpler to compute than the QS estimator. It is, however, (slightly) less efficient than the latter, but it is still more efficient than the CS estimator as long as μ_ξ and σ_ξ^2 are known and need not be estimated, see Section 7. It serves as an intermediate estimator between QS and CS and is useful if one wants to compare the relative efficiencies of the latter two.

The ACM of the SS estimator is given by two equivalent formulas depending on whether it is derived from ψ_S or ψ_S^*:

$$\begin{aligned}\Sigma_S &= (\mathbb{E}T)^{-1}\mathbb{E}vtt^\top(\mathbb{E}T)^{-1} + F \\ &= (\mathbb{E}\mu\mu^\top)^{-1}Ev\mu\mu^\top(\mathbb{E}\mu\mu^\top)^{-1} + F,\end{aligned} \tag{23}$$

where F is the same as in (17).

The first formula (23) is implicitly given in Shklyar et al. (2007), the second one follows in a similar way from ψ_S^*. Their equivalence can be directly seen by noting that $t = K\mu$ with some nonsingular matrix K and that $\mathbb{E}(\mu t^\top) = \mathbb{E}\zeta\zeta^\top = \mathbb{E}T$, cf. Shklyar et al. (2007).

7 Efficiency Comparison

One can show that $\Sigma_Q \leq \Sigma_S$, cf. Shklyar et al. (2007). Indeed, since the term F in (17) and (23) is the same, one needs only to compare the first terms in (17) and (23), respectively, and for this comparison one can use the Cauchy-Schwarz inequality.

These arguments do not hold for an efficiency comparison of CS and SS. The difference of their ACMs is

$$\Sigma_C - \Sigma_S = (\mathbb{E}T)^{-1}\mathbb{E}(T - t\mu^\top)\beta\beta^\top(T - \mu t^\top)(\mathbb{E}T)^{-1} - F. \tag{24}$$

It is not clear at the outset whether this difference is always ≥ 0. (It is, of course, ≥ 0 and, indeed, even > 0 if the last term vanishes, which occurs when the nuisance parameters need not be estimated: $\Sigma_C > \Sigma_S$ if μ_ξ and σ_ξ^2 are both known and $(\beta_1, \beta_2) \neq (0, 0)$, cf. Shklyar et al. (2007)).

There are cases where $\Sigma_C - \Sigma_S$ is singular if nuisance parameters are present. E.g., in a quadratic model, a detailed algebraic calculation shows that $\det(\Sigma_C - \Sigma_S) = 0$, implying that SS is not strictly more efficient than CS if nuisance parameters have to be estimated. But we still have $\Sigma_C \geq \Sigma_S$ in a quadratic model. In particular, all the diagonal elements of $\Sigma_C - \Sigma_S$ are positive and tend to ∞ for $\sigma_\delta^2 \to \infty$.

On the other hand, it is known that in a polynomial model of degree $k > 2$ with or without nuisance parameters, QS is always strictly more efficient than CS as long as $\beta_k \neq 0$, see Kukush et al. (2006).

8 Conclusion

The ACMs of three estimators (CS, QS, and SS) have been studied for the polynomial measurement error model. Some alternative formulas that are based solely on the observable variables have been presented. The ACMs of QS and SS (and also of other structural estimators) have a term that stems from the estimation of the nuisance parameters. This term has been evaluated.

The presence of this term in the ACMs of the QS and SS estimators diminishes the efficiency of QS and SS, which would be greater if the nuisance parameters were known. In particular, for a polynomial model, the efficiency of SS is so much reduced that, at least for the quadratic model, it is not strictly higher than the efficiency of CS anymore (as it would be if the nuisance parameters were known).

In the polynomial model, the CS and QS estimators are constructed with the help of transformed variables $t_r(x_i)$ and $\mu_r(x_i)$, respectively. New formulas for the computation of these variables have been derived.

Acknowledgement

Support by the German Science Foundation is gratefully acknowledged.

References

Cheng CL, Schneeweiss H (1996) The Polynomial Regression with Errors in the Variables Discussion Paper 42, Sonderforschungsbereich 386, University of Munich

Cheng CL, Schneeweiss H (1998) Polynomial regression with errors in the variables. Journal of Royal Statistical Society B 60:189–199

Cheng CL, Schneeweiss H (2002) On the polynomial measurement error model. In: van Huffel S and Lemmerling P (eds) Total Least Squares and Errors-in-Variables Modeling. Kluwer, Dordrecht, 131–143

Dhrymes PJ (1984) Mathematics for Econometrics (2nd edition). Springer, New York

Kukush A, Schneeweiss H (2005) Comparing different estimators in a nonlinear measurement error model I and II. Mathematical Methods of Statistics 14:53-79 and 203–223

Kukush A, Schneeweiss H, Wolf R (2005) Relative efficiency of three estimators in a polynomial regression with measurement errors. Journ. Statistical Planning and Inference 127:179–203

Kukush A, Malenko A, Schneeweiss H (2006) Optimality of the quasi-score estimator in a mean-variance model with applications to measurement error models. Discussion Paper 494, Sonderforschungsbereich 386, University of Munich

Shklyar S, Schneeweiss H, Kukush A (2007) Quasi Score is more efficient than corrected Score in a polynomial measurement error model. Metrika 65:275–295

Stefanski LA (1989) Unbiased estimation of a nonlinear function of a normal mean with application to measurement error models. Communications in Statistics, Part A - Theory and Methods 18:4335–4358

Thamerus M (1998) Different nonlinear regression models with incorrectly observed covariates. In: Galata R and Kuechenhoff H (eds) Econometrics in Theory and Practice, Festschrift for Hans Schneeweiss. Physica, Heidelberg, New York

Estimation and Finite Sample Bias and MSE of FGLS Estimator of Paired Data Model

Weiqiang Qian[1] and Aman Ullah[2]

[1] Department of Economics, University of California, 900 University Ave., Riverside, CA 92521, U.S.A. weiqiang.qian@ucr.edu

[2] Department of Economics, University of California, 900 University Ave., Riverside, CA 92521, U.S.A. aman.ullah@ucr.edu

1 Introduction

There is a growing interest in treating the cross sectional dependence in panel data models. The need to control the intracluster dependence was demonstrated in Kloek (1981) and Moulton (1990). When the cross sectional dependence is ignored, the estimated standard errors computed without considering clustering can be understated for OLS estimator, as shown in Cameron and Golotvina (2005). Recent work on treating cross-sectional dependence can be found in Pesaran (2006).

In this paper, we consider a paired data model where the dependent variable is measured according to different pairs of cross sectional units. The cross sectional dependence is introduced by each unit's influence on the paired data. Examples of such paired data can be exchange rates and trade data on countries. Cameron and Golotvina (2005) considered feasible generalized least square estimator (FGLS) for a paired data model. We consider a similar model to theirs and give a tractable FGLS estimator and investigate its finite sample bias and mean square error (MSE). Our estimator uses OLS and fixed effect (FE) residuals to estimate the covariance matrix of composite errors. Under the assumption of normal disturbances, we derive the finite sample bias and MSE of the slope estimator up to orders $O(n^{-2})$ and $O(n^{-4})$, respectively. We conducted simulation studies to investigate the influence of number of cross section units on bias and MSE of our FGLS estimator and the influence of changing variances of clustering effects and individual effects. We found that the change in variance of individual effects has a much bigger effect on MSE than that of variance of clustering effect. The finite sample MSE becomes close to asymptotic MSE when n is relatively large and exhibit downward correction from asymptotic MSE for large n and upward correction for small n.

The paper is organized as follows: Section 2 introduces the model; Section 3 develops a FGLS estimator and states the main results of its finite sample bias and MSE under normality; Section 4 provides the derivations of main results; Section 5 reports the simulation results and Section 6 concludes.

2 Model with Paired Data

Let us consider the cross-sectional paired data model as

$$y_{i,j} = x'_{i,j}\beta + \alpha_i + \alpha_j + \varepsilon_{i,j} \tag{1}$$

where $i = 1, 2, \ldots, n-1$ and $j = i+1, \ldots, n$ are the pair of cross-sectional clusters, $y_{i,j}$ is the dependent variable, $x_{i,j}$ is a vector of $1 \times k$ variables, α_i and α_j are the cluster effects for the ith and jth clusters, $\varepsilon_{i,j}$ is the individual effect, $y_{i,j} = y_{j,i}$ and $x_{i,j} = x_{j,i}$. Here the clusters could be taken as countries. For example, in the case of trade volume between countries, $y_{i,j}$ represents the trade volume between the ith and jth countries and is equal to $y_{j,i}$.

Stacking over j for cluster i gives us

$$\begin{bmatrix} y_{i,i+1} \\ \vdots \\ y_{i,n} \end{bmatrix} = \begin{bmatrix} x'_{i,i+1}\beta \\ \vdots \\ x'_{i,n}\beta \end{bmatrix} + \begin{bmatrix} \alpha_i \\ \vdots \\ \alpha_i \end{bmatrix} + \begin{bmatrix} \alpha_{i+1} \\ \vdots \\ \alpha_n \end{bmatrix} + \begin{bmatrix} \varepsilon_{i,i+1} \\ \vdots \\ \varepsilon_{i,n} \end{bmatrix}$$

or

$$y_i = X_i\beta + P_i\alpha + N_i\alpha + \varepsilon_i \tag{2}$$

where y_i and ε_i are $(n-i) \times 1$ vectors, X_i is an $(n-i) \times k$ matrix, β is a $k \times 1$ parameter vector and α is an $n \times 1$ vector of cluster effects. The matrices P_i and N_i are $(n-i) \times n$ matrices of zeroes and ones with

$$P_i = \begin{bmatrix} 0 \cdots 0\,1\,0 \cdots 0 \\ \vdots \quad\; \vdots\,\vdots\,\vdots \quad\; \vdots \\ 0 \cdots 0\,1\,0 \cdots 0 \end{bmatrix} = \left[0_{(n-i)\times(i-1)}\ l_{n-i}\ 0_{(n-i)\times(n-i)}\right] \qquad N_i = \begin{bmatrix} 0 \cdots 0 & 1 & 0 & \cdots & 0 \\ \vdots & \vdots & 0 & 1 & \cdots & \vdots \\ \vdots & \vdots & \vdots & & \ddots & 0 \\ 0 \cdots 0 & 0 & \cdots & & 0 & 1 \end{bmatrix} = \left[0_{(n-i)\times i}\ I_{n-i}\right]$$

where $0_{j\times i}$ represents a matrix of zeroes with j rows and i columns,

l_i represents a vector of ones of i rows, I_i is an identity matrix of dimension i.

Stacking over all i,

$$\begin{aligned} Y &= X\beta + P\alpha + N\alpha + \varepsilon \\ &= X\beta + L\alpha + \varepsilon \\ &= X\beta + \omega \end{aligned} \tag{3}$$

where Y is a $T \times 1$ vector, X is a $T \times k$ nonstochastic matrix that exhibits variation within each column, $\omega = L\alpha + \varepsilon$, $L = P + N$ and L, P and N are $T \times n$ matrices with $T = n(n-1)/2$. In deriving our results on the bias and MSE of FGLS estimator of slope coefficients, we assume that ε_{ij} is uncorrelated with α_i and

$$\varepsilon_{ij} \sim i.i.d.n. \left[0, \sigma_\varepsilon^2\right], \alpha_i \sim i.i.d.n. \left[0, \sigma_\alpha^2\right] \tag{4}$$

where $i.i.d.n.$ stands for independent and identically distributed with normal distribution.

Under assumption of (4),

$$Cov\left[\omega_{ij}, \omega_{kl}\right] = \begin{cases} 2\sigma_\alpha^2 + \sigma_\varepsilon^2, & \text{if } i = k, j = l \\ 0, & \text{if } i \neq k \neq j \neq l \\ \sigma_\alpha^2, & \text{otherwise.} \end{cases}$$

This gives the error variance matrix as

$$\Omega = E\left[\omega\omega'\right] = \sigma_\varepsilon^2 \left[I_T + \frac{1}{\lambda} LL'\right] \tag{5}$$

where $\lambda = \sigma_\varepsilon^2 / \sigma_\alpha^2$.

3 Main Results

Under model (3), when ignoring the clustering effects, the usual ordinary least squares (OLS) estimator is

$$\hat{\beta}_{OLS} = \left(X'X\right)^{-1} X'Y \tag{6}$$

which is an unbiased, consistent but not efficient estimator for β. Using (5), it is straightforward to show that the variance of OLS estimator is

$$V\left[\hat{\beta}_{OLS}\right] = \sigma_\varepsilon^2 \left(X'X\right)^{-1} \left[I + \frac{1}{\lambda} X'LL'X \left(X'X\right)^{-1}\right].$$

Taking the cluster effect α to be fixed, the FE estimator of β can be obtained as follows.

We define the differencing matrix $Q = I_T - \bar{Q}, \bar{Q} = L\left(L'L\right)^{-1}L'$ such that $QL = 0$ hence $Q\omega = Q\varepsilon$. Then we can transform the model (3) into following by multiplying both sides by Q

$$QY = QX\beta + Q\varepsilon. \tag{7}$$

The FE estimator is therefore obtained as

$$\hat{\beta}_{FE} = \left(X'QX\right)^{-1}X'QY$$

and one can obtain its variance as

$$V\left(\hat{\beta}_{FE}\right) = \sigma_\varepsilon^2\left(X'QX\right)^{-1}.$$

The FE estimator is efficient when α is indeed nonrandom and not so when α is random under our assumptions, in which case, we can construct a GLS estimator

$$\hat{\beta}_{GLS} = \left(X'\Omega^{-1}X\right)^{-1}X'\Omega^{-1}Y$$

where $\Omega^{-1} = \sigma_\varepsilon^{-2}\left[I_T - L\left(\lambda I_n + L'L\right)^{-1}L'\right]$ from (5), which can also be written as:

$$\Omega^{-1} = \sigma_\varepsilon^{-2}\left[I_T - \frac{1}{\lambda + (n-2)}\left(LL' - \frac{4}{\lambda + 2(n-1)}l_T l_T'\right)\right] \tag{8}$$

by using $l_T'L = (n-1)\,l_n'$, $Ll_n = 2l_T$, $L'L = (n-2)\,I_n + l_n l_n'$.

Using estimate of $\hat{\Omega}^{-1}$, we can construct a FGLS estimator of the coefficients β as

$$\hat{\beta}_{FGLS} = (X'\hat{\Omega}^{-1}X)^{-1}X'\hat{\Omega}^{-1}Y$$

where $\hat{\Omega} = \hat{\sigma}_\varepsilon^2\left[I_T + \left(\hat{\sigma}_\alpha^2/\hat{\sigma}_\varepsilon^2\right)LL'\right]$ and $\hat{\Omega}^{-1}$ is given as in (8) with σ_ε^2 and σ_α^2 replaced by their respective consistent estimators obtained as follows.

The estimate of σ_ε^2 can be obtained using the FE residuals:

$$\hat{\sigma}_\varepsilon^2 = \frac{\varepsilon'(Q - QX\left(X'QX\right)^{-1}X'Q)\varepsilon}{T - n - k} \tag{9}$$

while the estimator of σ_α^2 can be obtained from OLS residuals $\hat{\omega}_{OLS}$ from (6) and using $\hat{\sigma}_\varepsilon^2$.

Consider

$$\hat{\sigma}_{\omega}^{2} = \frac{\hat{\omega}_{OLS}'\hat{\omega}_{OLS}}{tr\left(G\right)}, \tag{10}$$

since $E\left(\hat{\sigma}_{\omega}^{2}\right) = \sigma_{\alpha}^{2} + \frac{T-k}{tr(G)}\sigma_{\varepsilon}^{2}$, we can obtain an estimator for σ_{α}^{2}:

$$\hat{\sigma}_{\alpha}^{2} = \hat{\sigma}_{\omega}^{2} - \frac{T-k}{tr\left(G\right)}\hat{\sigma}_{\varepsilon}^{2}. \tag{11}$$

Note that (11) follows from

$$\begin{aligned} E\left(\hat{\omega}_{OLS}'\hat{\omega}_{OLS}\right) &= E\left(Y'\left(I_T - X\left(X'X\right)^{-1}X'\right)Y\right) \\ &= tr\left(G\right)\sigma_{\alpha}^{2} + \left(T-k\right)\sigma_{\varepsilon}^{2} \end{aligned}$$

where $G = L'ML$ and $M = I_T - X\left(X'X\right)^{-1}X'$.

Theorem 1. *Under assumption (4), the finite sample approximations for the bias vector $E(\hat{\beta}_{FGLS} - \beta)$ up to $O(n^{-2})$ and MSE matrix $E(\hat{\beta}_{FGLS} - \beta)(\hat{\beta}_{FGLS} - \beta)'$ up to $O(n^{-4})$ are given by*

$$\begin{aligned} &Bias = 0 \\ &MSE = n^{-2}\left(X'\Omega^{-1}X/n^{2}\right)^{-1} + n^{-3}\left[4P_1\left(\frac{\lambda\left(1-h\right)\sigma_{\alpha}^{2}}{h^{2}}\right)\right] \\ &\quad + n^{-4}\left[2P_2\left(\frac{\lambda^{2}}{h^{2}}\sigma_{\alpha}^{2}\right) + 4P_3\left(\frac{\lambda}{h^{2}}I_T\sigma_{\varepsilon}^{2}\right) + 4P_4\left(\frac{\lambda\left(1-h\right)}{h^{2}}\sigma_{\alpha}^{2}\right)\right. \\ &\quad \left. - 4P_5\left(\frac{\lambda}{h^{2}}I_T\sigma_{\varepsilon}^{2}\right) + 8P_1\left(\frac{\lambda\left(2-\lambda h\right)}{h^{2}}\sigma_{\alpha}^{2}\right)\right] \end{aligned}$$

where $h = tr\left(G\right)/n^{2}$, P_1, P_2, P_3, P_4 and P_5 are given in (45), in which $A = X'\Omega^{-1}X/n^{2}$, $B = X'\Phi X/n^{2}$, $\Phi = \left[a\left(LL'/n\right) - \frac{3}{2}ab\left(l_T l_T'/n^{2}\right)\right]/\sigma_{\varepsilon}^{2}$, $a = 1/(1+\frac{\lambda-2}{n})$ and $b = 4/(2+\frac{\lambda-2}{n})$.

We note that the first term in MSE is of order $O(n^{-2})$, because the model in (3) is based on $T = n\left(n-1\right)/2$ observations. Further remarks on the results are given below.

Remark 1. When $\sigma_{\alpha}^{2} = 0, \lambda = \infty$, P_1, P_2, P_3, P_4, P_5 are $\mathbf{0}$, $\Omega^{-1} = \sigma_{\varepsilon}^{-2}I_T$, the finite sample approximate MSE of FGLS estimator is $n^{-2}\left(X'X/n^{2}\right)^{-1}\sigma_{\varepsilon}^{2}$; which is the usual variance of the OLS estimator of β; when $\sigma_{\varepsilon}^{2} = 0, \lambda = 0, \Omega^{-1} = \sigma_{\alpha}^{-2}\left(LL'\right)^{-1}$, finite sample MSE of FGLS estimator is $n^{-2}\left(X'\left(LL'\right)^{-1}X/n^{2}\right)^{-1}\sigma_{\alpha}^{2}$. The finite sample MSE are the same as asymptotic values under these two situations.

Remark 2. When λ is constant, Φ is constant, increasing in σ_ε^2 and σ_α^2 proportionally leads to decrease in Ω^{-1}, increase in P_1, P_2, P_3, P_4, P_5, and proportional increase in the MSE.

Remark 3. When σ_α^2 is held constant, an increase in σ_ε^2 leads to decrease in Ω^{-1}, increase in P_1, P_2, P_3, P_4, P_5, and increase in the MSE; similar increase in MSE will result when increasing σ_α^2 while holding σ_ε^2 constant. Also noting

$$\begin{aligned}\frac{d\Omega^{-1}}{d\sigma_\varepsilon^2} &= -\frac{1}{\sigma_\varepsilon^4} I_T + \frac{1}{e^2}[2\lambda + (n-2)]LL' \\ &\quad - \frac{4}{f^2}\left[3\lambda^2 + (6n-8)\lambda + 2(n-2)(n-1)\right] l_T l_T' \\ \frac{d\Omega^{-1}}{d\sigma_\alpha^2} &= -\frac{1}{e^2}\lambda^2 LL' + \frac{4}{f^2}\lambda^2 [2\lambda + 3n - 4] l_T l_T'\end{aligned}$$

where $e = \sigma_\varepsilon^2 [\lambda + (n-2)]$ is $O(n)$, $f = \sigma_\varepsilon^2 [\lambda + (n-2)] [\lambda + 2(n-1)]$ is $O(n^2)$. Hence $\frac{d\Omega^{-1}}{d\sigma_\varepsilon^2} < 0$ is of $O(1)$ and $\frac{d\Omega^{-1}}{d\sigma_\alpha^2} < 0$ is of $O(n^{-2})$. The ratio $\frac{dMSE}{d\sigma_\varepsilon^2} / \frac{dMSE}{d\sigma_\alpha^2}$ is of $O(n^2)$. That is, a change in variance of individual effect has a much bigger influence on the MSE of β than a comparable change in variance of cluster effect.

4 Derivation

To obtain the finite sample expansion of bias and MSE of $\hat{\beta}_{FGLS}$, we need the finite sample expansion of $\hat{\Omega}^{-1}$. To obtain this, we first look at the expansion of $\hat{\sigma}_\varepsilon^2$ and $\hat{\sigma}_\omega^2$ and use them to obtain the expansion of $\hat{\lambda} = \hat{\sigma}_\varepsilon^2 / \hat{\sigma}_\alpha^2$.

For $\hat{\sigma}_\varepsilon^2$, from (9), up to $O(n^{-2})$,

$$\begin{aligned}\hat{\sigma}_\varepsilon^2 &= \frac{\varepsilon'(Q - QX(X'QX)^{-1} X'Q)\varepsilon}{T-n-k} \\ &= 2\frac{\varepsilon'Q\varepsilon - \varepsilon'QX(X'QX)^{-1} X'Q\varepsilon}{n^2 - 3n - 2k} \\ &= \frac{2}{n^2-3n}\left[1 - \frac{2k}{n^2-3n}\right]^{-1} \left[\sigma_\varepsilon^2 \left(\frac{n^2-3n}{2}\right)\left(1 + \frac{\nu_\varepsilon}{\sqrt{T}}\right) - \sigma_\varepsilon^2 \nu_\varepsilon^*\right] \\ &= \frac{2}{n^2-3n}\left[1 + \frac{2k}{n^2-3n} + \left(\frac{2k}{n^2-3n}\right)^2 + \dots\right] \\ &\quad \times \left[\sigma_\varepsilon^2 \left(\frac{n^2-3n}{2}\right)\left(1 + \frac{\nu_\varepsilon}{\sqrt{T}}\right) - \sigma_\varepsilon^2 \nu_\varepsilon^*\right]\end{aligned}$$

$$= \sigma_\varepsilon^2 \left[1 + \frac{\nu_\varepsilon}{\sqrt{T}} + \frac{2\left(k - \nu_\varepsilon^*\right)}{n^2 - 3n} \right]$$

$$= \sigma_\varepsilon^2 \left[1 + \left(\frac{\sqrt{2}}{n} + \frac{\sqrt{2}}{2n^2} \right) \nu_\varepsilon + \frac{2\left(k - \nu_\varepsilon^*\right)}{n^2} \right] \tag{12}$$

where

$$\nu_\varepsilon = \sqrt{T} \left(\frac{\varepsilon' Q \varepsilon}{(T-n)\,\sigma_\varepsilon^2} - 1 \right),$$

$$\nu_\varepsilon^* = \varepsilon' Q X \left(X' Q X\right)^{-1} X' Q \varepsilon / \sigma_\varepsilon^2 .$$

For $\hat{\sigma}_\omega^2$, from (10),

$$\hat{\omega}'_{OLS} \hat{\omega}_{OLS} = Y' \left(I_T - X \left(X'X\right)^{-1} X' \right) Y$$

$$= \omega' \left(I_T - X \left(X'X\right)^{-1} X' \right) \omega$$

$$= \omega'\omega - \omega' X \left(X'X\right)^{-1} X' \omega \tag{13}$$

in which the first term, since $L'L = (n-2)\, I_n + l_n l_n'$,

$$\omega'\omega = (L\alpha + \varepsilon)' (L\alpha + \varepsilon)$$

$$= \alpha' L' L \alpha + \varepsilon'\varepsilon + 2\varepsilon' L \alpha$$

$$= (n-2)\, \alpha'\alpha + \alpha' l_n l_n' \alpha + \varepsilon'\varepsilon + 2\varepsilon' L\alpha$$

$$= n(n-2)\, \sigma_\alpha^2 \left(1 + \upsilon_\alpha \sqrt{n}\right) + n\left(1 + \upsilon_{\alpha\alpha}/\sqrt{n}\right)$$

$$+ T\sigma_\varepsilon^2 \left(1 + \epsilon_\varepsilon / \sqrt{T}\right) + 2\sqrt{T} \sigma_\omega^2 \upsilon_{\alpha\varepsilon}$$

where

$$\upsilon_\alpha = \sqrt{n} \left(\frac{\alpha'\alpha}{n\sigma_\alpha^2} - 1 \right), \epsilon_\varepsilon = \sqrt{T} \left(\frac{\varepsilon'\varepsilon}{T\sigma_\varepsilon^2} - 1 \right), \tag{14}$$

$$\upsilon_{\alpha\alpha} = \sqrt{n} \left(\frac{\alpha' l_n l_n' \alpha}{n\sigma_\alpha^2} - 1 \right), \upsilon_{\alpha\varepsilon} = \frac{\varepsilon' L \alpha}{\sqrt{T} \sigma_\omega^2}. \tag{15}$$

Now consider the second term in (13),

$$\omega' X \left(X'X\right)^{-1} X'\omega = (L\alpha + \varepsilon)' X \left(X'X\right)^{-1} X' (L\alpha + \varepsilon) \tag{16}$$

$$= \alpha' L' X \left(X'X\right)^{-1} X' L \alpha$$

$$+ \varepsilon' X \left(X'X\right)^{-1} X' \varepsilon + 2\alpha' L' X \left(X'X\right)^{-1} X' \varepsilon$$

$$= \sigma_\alpha^2 tr\,(Z) \left(1 + \upsilon_\alpha^* / \sqrt{n}\right)$$

$$+ \sigma_\omega^2 \left(\epsilon_\varepsilon^* + 2\sqrt{tr\,(Z)}\, \upsilon_{\alpha\varepsilon}^* \right)$$

where

$$Z = L'X\left(X'X\right)^{-1}X'L,$$
$$\upsilon_{\alpha}^{*} = \sqrt{n}\left(\frac{\alpha' Z\alpha}{tr\left(Z\right)\sigma_{\alpha}^{2}} - 1\right),$$
$$\epsilon_{\varepsilon}^{*} = \varepsilon' X\left(X'X\right)^{-1}X'\varepsilon/\sigma_{\omega}^{2},$$
$$\upsilon_{\alpha\varepsilon}^{*} = \frac{\alpha' L' X\left(X'X\right)^{-1}X'\varepsilon}{\sigma_{\omega}^{2}\sqrt{tr\left(Z\right)}}.$$

We now have

$$\begin{aligned}
tr\left(G\right)\hat{\sigma}_{\omega}^{2} &= \omega'\omega - \omega' X\left(X'X\right)^{-1}X'\omega \\
&= n\left(n-2\right)\sigma_{\alpha}^{2}\left(1+\upsilon_{\alpha}/\sqrt{n}\right) + n\sigma_{\alpha}^{2}\left(1+\upsilon_{\alpha\alpha}/\sqrt{n}\right) \\
&\quad + T\sigma_{\varepsilon}^{2}\left(1+\epsilon_{\varepsilon}/\sqrt{T}\right) + 2\sqrt{T}\sigma_{\omega}^{2}\upsilon_{\alpha\varepsilon} \\
&\quad - \left[\sigma_{\alpha}^{2} tr\left(Z\right)\left(1+\upsilon_{\alpha}^{*}/\sqrt{n}\right) + \sigma_{\omega}^{2}\left(\epsilon_{\varepsilon}^{*} + 2\sqrt{tr\left(Z\right)}\upsilon_{\alpha\varepsilon}^{*}\right)\right] \\
&= \left(T-k\right)\sigma_{\varepsilon}^{2} + tr\left(G\right)\sigma_{\alpha}^{2} + n\left(n-2\right)\sigma_{\alpha}^{2}\upsilon_{\alpha}/\sqrt{n} \\
&\quad + n\sigma_{\alpha}^{2}\upsilon_{\alpha\alpha}/\sqrt{n} + T\sigma_{\varepsilon}^{2}\epsilon_{\varepsilon}/\sqrt{T} + 2\sqrt{T}\sigma_{\omega}^{2}\upsilon_{\alpha\varepsilon} \\
&\quad - \sigma_{\alpha}^{2} tr\left(Z\right)\upsilon_{\alpha}^{*}/\sqrt{n} - \sigma_{\omega}^{2}\left(\epsilon_{\varepsilon}^{*} + 2\sqrt{tr\left(Z\right)}\upsilon_{\alpha\varepsilon}^{*}\right) + k\sigma_{\varepsilon}^{2}.
\end{aligned}$$

Dividing both sides by $tr\left(G\right)$, we have

$$\begin{aligned}
\hat{\sigma}_{\omega}^{2} = \sigma_{\omega}^{2}\Bigg[1 &+ \frac{\sqrt{n}\left(n-2\right)}{tr\left(G\right)}\frac{\sigma_{\alpha}^{2}}{\sigma_{\omega}^{2}}\upsilon_{\alpha} + \frac{\sigma_{\alpha}^{2}}{\sigma_{\omega}^{2}}\frac{\sqrt{n}\upsilon_{\alpha\alpha}}{tr\left(G\right)} + \frac{\sigma_{\varepsilon}^{2}}{\sigma_{\omega}^{2}}\frac{\sqrt{T}\epsilon_{\varepsilon}}{tr\left(G\right)} + 2\frac{\sqrt{T}}{tr\left(G\right)}\upsilon_{\alpha\varepsilon} \\
&- \frac{\sigma_{\alpha}^{2}}{\sigma_{\omega}^{2}}\frac{tr\left(Z\right)}{tr\left(G\right)}\frac{\upsilon_{\alpha}^{*}}{\sqrt{n}} - \frac{\left(\epsilon_{\varepsilon}^{*} + 2\sqrt{tr\left(Z\right)}\upsilon_{\alpha\varepsilon}^{*}\right)}{tr\left(G\right)} + \frac{k}{tr\left(G\right)}\frac{\sigma_{\varepsilon}^{2}}{\sigma_{\omega}^{2}}\Bigg].
\end{aligned}$$

For the orders of $tr\left(Z\right)$ and $tr\left(G\right)$, by Cholesky Decomposition, $X\left(X'X\right)^{-1}X' = AA'$, where A is a $T \times T$ matrix and

$$tr\left(AA'\right) = \sum_{i=1}^{T}\sum_{j=1}^{T} a_{ij}^{2} = tr\left(X\left(X'X\right)^{-1}X'\right) = k,$$

thus a_{ij} is $O\left(T^{-1}\right) = O\left(n^{-2}\right)$ and each element of $L'A$ is $O(n^{-1})$ since for any $T \times m$ matrix X, $X'L = (n-1)\left[\bar{x}_1 \ldots \bar{x}_n\right]$, where

$$\bar{x}_i = \left[\left(x_{i,i+1} + \ldots + x_{i,n}\right) + \left(x_{1,i} + \ldots + x_{i-1,i}\right)\right]/\left(n-1\right).$$

Thus

$$tr\,(Z) = tr\,(L'AA'L) = O\,(n)\,,$$
$$tr\,(G) = tr(L'L - Z) = n\,(n-2) + n - tr\,(Z) = O\,(n^2)\,.$$

Now writing

$$\frac{\sigma_\alpha^2}{\sigma_\omega^2} = \frac{1}{1+d\lambda}, \qquad \frac{\sigma_\varepsilon^2}{\sigma_\omega^2} = \frac{\lambda}{1+d\lambda}, \tag{17}$$

where

$$d = \frac{T-k}{tr\,(G)} = O\,(1)\,,$$

we obtain, up to $O(n^{-2})$,

$$\begin{aligned}
\hat{\sigma}_\omega^2 = \sigma_\omega^2 \Bigg[& 1 + \frac{\sqrt{n}\,(n-2)}{n^2 h}\frac{1}{1+d\lambda}\upsilon_\alpha + \frac{1}{1+d\lambda}\frac{\sqrt{n}\upsilon_{\alpha\alpha}}{n^2 h} + \frac{\lambda}{1+d\lambda}\frac{\sqrt{T}\epsilon_\varepsilon}{n^2 h} \\
& + 2\frac{\sqrt{T}}{n^2 h}\upsilon_{\alpha\varepsilon} - \frac{1}{1+d\lambda}\frac{g}{nh}\frac{\upsilon_\alpha^*}{\sqrt{n}} - \frac{\epsilon_\varepsilon^*}{n^2 h} - 2\frac{1}{n^{3/2}}\frac{\sqrt{g}}{h}\upsilon_{\alpha\varepsilon}^* \\
& + \frac{k}{n^2 h}\frac{\lambda}{1+d\lambda}\Bigg] \\
= \sigma_\omega^2 \Bigg[& 1 + n^{-1/2}\frac{1}{h\,(1+d\lambda)}\upsilon_\alpha + n^{-1}\left(\frac{1}{\sqrt{2}h}\frac{\lambda}{1+d\lambda}\epsilon_\varepsilon + \frac{\sqrt{2}}{h}\upsilon_{\alpha\varepsilon}\right) \\
& + n^{-3/2}\left(-2\frac{1}{h\,(1+d\lambda)}\upsilon_\alpha + \frac{1}{h\,(1+d\lambda)}\upsilon_{\alpha\alpha} - \frac{g}{h}\frac{1}{1+d\lambda}\upsilon_\alpha^* \right. \\
& \left. - 2\frac{\sqrt{g}}{h}\upsilon_{\alpha\varepsilon}^*\right) + n^{-2}\left(\frac{\lambda}{2\sqrt{2}\,(1+d\lambda)}\epsilon_\varepsilon + \frac{1}{\sqrt{2}h}\upsilon_{\alpha\varepsilon} + \frac{k\lambda}{h\,(1+d\lambda)}\right)\Bigg]
\end{aligned} \tag{18}$$

where

$$g = \frac{tr(Z)}{n} = O\,(1)\,, h = \frac{tr\,(G)}{n^2} = O\,(1)\,,$$

and we use

$$\frac{\sqrt{T}}{n^2 h} = \frac{1}{\sqrt{2}nh}\sqrt{1-\frac{1}{n}} = \frac{1}{\sqrt{2}nh}\left(1 + \frac{1}{2n} + \frac{1}{8n^2} + \frac{1}{16n^3} + \ldots\right).$$

Now using the expansions of $\hat{\sigma}_\varepsilon^2$ and $\hat{\sigma}_\omega^2$ in (12) and (18), we can obtain the expansion of $\hat{\lambda}$, estimator of $\lambda = 1/c$, where $c = \sigma_\alpha^2/\sigma_\varepsilon^2$, which is given by $\hat{\lambda} = 1/\hat{c} = \hat{\sigma}_\varepsilon^2/\hat{\sigma}_\alpha^2$, and

$$\hat{c} = \frac{\hat{\sigma}_\omega^2}{\hat{\sigma}_\varepsilon^2} - \frac{T-k}{tr\,(G)}.$$

Now define

$$C = \frac{\hat{\sigma}_\omega^2 - \sigma_\omega^2}{\sigma_\omega^2},\ D = \frac{\hat{\sigma}_\varepsilon^2 - \sigma_\varepsilon^2}{\sigma_\varepsilon^2},\ C_h = \frac{1}{h\,(1+d\lambda)},$$

then using (12) and (18), we can write, up to $O\left(n^{-1}\right)$,

$$\begin{aligned}
\hat{c} &= \frac{\hat{\sigma}_\omega^2}{\hat{\sigma}_\varepsilon^2} - \frac{T-k}{tr\,(G)} \\
&= \frac{\sigma_\omega^2}{\sigma_\varepsilon^2}\left(\frac{1+C}{1+D}\right) - \frac{T-k}{tr\,(G)} \\
&= \frac{\sigma_\omega^2}{\sigma_\varepsilon^2}\left[1 + (C-D) - CD + D^2\right] - \frac{T-k}{tr\,(G)} \\
&= c + \frac{1+d\lambda}{\lambda}\left[n^{-1/2}C_h\upsilon_\alpha + n^{-1}\left(\frac{\lambda C_h\epsilon_\varepsilon}{\sqrt{2}} + \frac{\sqrt{2}}{h}\upsilon_{\alpha\varepsilon} - \sqrt{2}\nu_\varepsilon\right)\right].
\end{aligned} \tag{19}$$

Further, we have

$$\hat{\lambda} = \frac{1}{\hat{c}} = \frac{1}{c + \Delta_c} = \lambda\left(1 - \lambda\Delta_c + \lambda^2\Delta_c^2\right)$$

where $\Delta_c = \hat{c} - c$. Using (19), up to $O\left(n^{-1}\right)$,

$$\begin{aligned}
\hat{\lambda} &= \lambda\left\{1 - (1+d\lambda)\left[n^{-1/2}C_h\upsilon_\alpha + n^{-1}\left(\frac{\lambda C_h\epsilon_\varepsilon}{\sqrt{2}} + \frac{\sqrt{2}}{h}\upsilon_{\alpha\varepsilon} - \sqrt{2}\nu_\varepsilon\right)\right]\right. \\
&\qquad \left. + (1+d\lambda)^2\frac{1}{n}C_h^2\upsilon_\alpha^2\right\} \\
&= \lambda\left\{1 - (1+d\lambda)\,n^{-1/2}C_h\upsilon_\alpha + (1+d\lambda)\right. \\
&\qquad \left. \times n^{-1}\left[(1+d\lambda)\,C_h^2\upsilon_\alpha^2 - \left(\frac{\lambda C_h\epsilon_\varepsilon}{\sqrt{2}} + \frac{\sqrt{2}}{h}\upsilon_{\alpha\varepsilon} - \sqrt{2}\nu_\varepsilon\right)\right]\right\} \\
&= \lambda + n^{-1/2}\left(-\frac{\lambda}{h}\upsilon_\alpha\right) \\
&\qquad + n^{-1}\left(\frac{\lambda}{h^2}\upsilon_\alpha^2 - \frac{\lambda^2}{\sqrt{2}h}\epsilon_\varepsilon - \frac{\sqrt{2}C_d}{h}\upsilon_{\alpha\varepsilon} + \sqrt{2}C_d\nu_\varepsilon\right)
\end{aligned} \tag{20}$$

where $C_d = \lambda\,(1+d\lambda)$.

Consider the terms of $\hat{\Omega}^{-1}$ in (8), we note that up to $O\left(n^{-2}\right)$,

$$
\begin{aligned}
&\frac{1}{1+\frac{\hat{\lambda}-2}{n}} \\
=&\frac{1}{1+\frac{\lambda-2}{n}}\{1+\lambda\left(1+d\lambda\right)n^{-3/2}C_h\upsilon_\alpha \\
&-\lambda\left(1+d\lambda\right)n^{-2}\left[\left(1+d\lambda\right)C_h^2\upsilon_\alpha^2-\left(\frac{\lambda C_h\epsilon_\varepsilon}{\sqrt{2}}+\frac{\sqrt{2}}{h}\upsilon_{\alpha\varepsilon}-\sqrt{2}\nu_\varepsilon\right)\right]\}
\end{aligned} \tag{21}
$$

and

$$
\begin{aligned}
&\frac{4}{2+\frac{\hat{\lambda}-2}{n}} \\
=&\frac{4}{2+\frac{\lambda-2}{n}}\left\{1+\frac{\lambda\left(1+d\lambda\right)}{2}n^{-3/2}C_h\upsilon_\alpha\right. \\
&\left.-\frac{\lambda\left(1+d\lambda\right)}{2}n^{-2}\left[\left(1+d\lambda\right)C_h^2\upsilon_\alpha^2-\left(\frac{\lambda C_h\epsilon_\varepsilon}{\sqrt{2}}+\frac{\sqrt{2}}{h}\upsilon_{\alpha\varepsilon}-\sqrt{2}\nu_\varepsilon\right)\right]\right\}.
\end{aligned} \tag{22}
$$

Therefore, using (21) and (22) in (8), we can write $\hat{\Omega}^{-1}$, up to $O\left(n^{-2}\right)$ as

$$\hat{\Omega}^{-1}=\Omega^{-1}+\Phi\frac{\Delta_\lambda}{n}$$

where, with $a=\left(1+\frac{\lambda-2}{n}\right)^{-1}$ and $b=4\left(2+\frac{\lambda-2}{n}\right)^{-1}$, define

$$\Phi=\frac{\left[a\left(LL'/n\right)-\frac{3}{2}ab\left(l_Tl_T'/n^2\right)\right]}{\sigma_\varepsilon^2} \tag{23}$$

$$\Delta_\lambda=\hat{\lambda}-\lambda. \tag{24}$$

Further, up to $O\left(n^{-2}\right)$,

$$\frac{X'\hat{\Omega}^{-1}X}{n^2}=A+B\frac{\Delta_\lambda}{n} \tag{25}$$

where

$$A=\frac{X'\Omega^{-1}X}{n^2},\quad B=\frac{X'\Phi X}{n^2} \tag{26}$$

and using property of $X'L$, it's easy to see that $X'LL'X/n^3$ is $O(1)$, thus A and B are $O(1)$, respectively.

This gives

$$\begin{aligned}
n\left(\hat{\beta}_{FGLS}-\beta\right) &= \left(A+B\frac{\Delta_\lambda}{n}+O_p\left(n^{-5/2}\right)\right)^{-1} \\
&\quad \times\left(X'\left(\Omega^{-1}+\Phi\frac{\Delta_\lambda}{n}+O_p\left(n^{-5/2}\right)\right)\right)\omega/n \\
&= \zeta_0+\zeta_{-1}+\zeta_{-3/2} \qquad (27)
\end{aligned}$$

where

$$\begin{aligned}
\zeta_0 &= A^{-1}\left(\frac{X'\Omega^{-1}\omega}{n}\right), \\
\zeta_{-1} &= \left(\frac{A^{-1}X'\Phi L\alpha}{n^{3/2}}\right)\frac{\Delta\lambda}{\sqrt{n}}, \\
\zeta_{-3/2} &= \left(\frac{A^{-1}X'\Phi\varepsilon}{n}-\frac{A^{-1}BA^{-1}X'\Omega^{-1}\omega}{n}\right)\frac{\Delta\lambda}{n},
\end{aligned}$$

and the subscript of ζ represents its order in terms of power of n. From (27)

$$E\left(n\left(\hat{\beta}_{FGLS}-\beta\right)\right)=A^{-1}\left(X'\Phi-BA^{-1}X'\Omega^{-1}\right)E\left(\omega\frac{\Delta_\lambda}{n}\right) \qquad (28)$$

where using assumption (4), it can be easily verified that $E\left(\omega\frac{\Delta_\lambda}{n}\right)=0$ by noting that $E\left[(L\alpha+\varepsilon)\upsilon_\alpha\right]=0$, $E\left[(L\alpha+\varepsilon)\upsilon_\alpha^2\right]=0$, $E\left[(L\alpha+\varepsilon)\epsilon_\varepsilon\right]=0$ and $E\left[(L\alpha+\varepsilon)\upsilon_{\alpha\varepsilon}\right]=0$. Hence the bias of $\hat{\beta}_{FGLS}$ multiplied by n is

$$E\left(n\left(\hat{\beta}_{FGLS}-\beta\right)\right)=0.$$

The MSE of $\hat{\beta}_{FGLS}$ multiplied by n^2 is

$$\begin{aligned}
&E\left(n^2\left(\hat{\beta}_{FGLS}-\beta\right)\left(\hat{\beta}_{FGLS}-\beta\right)'\right) \\
&= E\left(\zeta_0\zeta_0'+\zeta_0\zeta_{-1}'+\zeta_0\zeta_{-3/2}'+\zeta_{-1}\zeta_0'+\zeta_{-1}\zeta_{-1}'+\zeta_{-3/2}\zeta_0'\right) \qquad (29)
\end{aligned}$$

where

$$E\left(\zeta_0\zeta_0'\right) = A^{-1} = \left(\frac{X'\Omega^{-1}X}{n^2}\right)^{-1}, \tag{30}$$

$$E\left(\zeta_0\zeta_{-1}'\right) = \frac{A^{-1}X'\Omega^{-1}E\left(\Delta\lambda\omega\alpha'\right)L'\Phi XA^{-1}}{n^3}, \tag{31}$$

$$E\left(\zeta_{-1}\zeta_{-1}'\right) = \frac{A^{-1}X'\Phi LE\left(\alpha\alpha'\frac{(\Delta\lambda)^2}{n}\right)L'\Phi XA^{-1}}{n^3}, \tag{32}$$

$$E\left(\zeta_0\zeta_{-3/2}'\right) = \frac{A^{-1}X'\Omega^{-1}E\left(\Delta\lambda\omega\varepsilon'\right)\Phi XA^{-1}}{n^3} - \frac{A^{-1}X'\Omega^{-1}E\left(\Delta\lambda\omega\omega'\right)\Omega^{-1}XA^{-1}BA^{-1}}{n^3}. \tag{33}$$

The terms involving expectations in (31) to (33) are derived below, and they are obtained by using the results in Ullah (2004, p. 187) for the normal case.

We first look at $E\left(\Delta\lambda\omega\omega'\right)$ in (33):

$$E\left(\Delta\lambda\omega\omega'\right) = LE\left(\alpha\alpha'\Delta\lambda\right)L' + LE\left(\alpha\varepsilon'\Delta\lambda\right) + E\left(\varepsilon\varepsilon'\Delta\lambda\right) + E\left(\varepsilon\alpha'\Delta\lambda\right)L'. \tag{34}$$

Using (20), we can get firstly,

$$\begin{aligned} &E\left(\alpha\alpha'\Delta\lambda\right) \\ =&n^{-1/2}\left(-\frac{\lambda}{h}E\left(\alpha\alpha'\upsilon_\alpha\right)\right) + n^{-1}\left(\frac{\lambda}{h^2}E\left(\alpha\alpha'\upsilon_\alpha^2\right) - \frac{\lambda^2}{\sqrt{2}h}E\left(\alpha\alpha'\epsilon_\varepsilon\right)\right. \\ &\left. -\frac{\sqrt{2}C_b}{h}E\left(\alpha\alpha'\upsilon_{\alpha\varepsilon}\right) + \sqrt{2}C_bE\left(\alpha\alpha'\upsilon_\varepsilon\right)\right) \\ =&n^{-1}\left(-\frac{2\lambda\left(1-h\right)}{h^2}I_n\sigma_\alpha^2\right) + n^{-2}\frac{8\lambda}{h^2}I_n\sigma_\alpha^2. \end{aligned} \tag{35}$$

The above result is obtained by noting:

$$\begin{aligned} E\left(\alpha\alpha'\upsilon_\alpha\right) &= E\left(\alpha\alpha'\sqrt{n}\left(\frac{\alpha'\alpha}{n\sigma_\alpha^2}-1\right)\right) \\ &= \sqrt{n}\left(\frac{E\alpha\alpha'\alpha'\alpha}{n\sigma_\alpha^2} - E\left(\alpha\alpha'\right)\right) \\ &= \frac{2}{\sqrt{n}}\sigma_\alpha^2 I_n, \end{aligned}$$

$$E\left(\alpha\alpha' \upsilon_\alpha^2\right) = nE\left(\frac{E\alpha\alpha'\alpha'\alpha\alpha'\alpha}{n^2\sigma^4} - 2\frac{E\alpha\alpha'\alpha'\alpha}{n\sigma_\alpha^2} + E\alpha\alpha'\right)$$
$$= \frac{1}{n}\left(n+2\right)\left(n+4\right) I_n\sigma_\alpha^2 - 2\left(n+2\right) I_n\sigma_\alpha^2 + nI_n\sigma_\alpha^2,$$
$$= \frac{2\left(n+4\right)}{n} I_n\sigma_\alpha^2,$$
$$E\left(\alpha\alpha'\epsilon_\varepsilon\right) = 0, E\left(\alpha\alpha'\upsilon_{\alpha\varepsilon}\right) = 0, E\left(\alpha\alpha'\upsilon_\varepsilon\right) = 0.$$

Secondly,

$$E\left(\alpha\varepsilon'\Delta\lambda\right) = n^{-1/2}\left(-\frac{\lambda}{h}E\left(\alpha\varepsilon'\upsilon_\alpha\right)\right) + n^{-1}\left[\frac{\lambda}{h^2}E\left(\alpha\varepsilon'\upsilon_\alpha^2\right)\right.$$
$$\left. -\frac{\lambda^2}{\sqrt{2}h}E\left(\alpha\varepsilon'\epsilon_\varepsilon\right) - \frac{\sqrt{2}C_b}{h}E\left(\alpha\varepsilon'\upsilon_{\alpha\varepsilon}\right) + \sqrt{2}C_b E\left(\alpha\varepsilon'\upsilon_\varepsilon\right)\right]$$
$$= -n^{-1}\left(\frac{\sqrt{2}C_b}{h}\frac{\sigma_\alpha^2\sigma_\varepsilon^2}{\sigma_\omega^2}\right)L', \tag{36}$$

since

$$E\left(\alpha\varepsilon'\upsilon_\alpha\right) = 0, E\left(\alpha\varepsilon'\upsilon_\alpha^2\right) = 0, E\left(\alpha\varepsilon'\epsilon_\varepsilon\right) = 0, E\left(\alpha\varepsilon'\upsilon_\varepsilon\right) = 0$$
$$E\left(\alpha\varepsilon'\upsilon_{\alpha\varepsilon}\right) = E\left(\alpha\varepsilon'\frac{\varepsilon' L\alpha}{\sqrt{T}\sigma_\omega^2}\right) = \frac{1}{\sqrt{T}\sigma_\omega^2}\sigma_\alpha^2\sigma_\varepsilon^2 L'.$$

Thirdly,

$$E\left(\varepsilon\varepsilon'\Delta\lambda\right) = n^{-1/2}\left(-\frac{\lambda}{h}E\left(\varepsilon\varepsilon'\upsilon_\alpha\right)\right) + n^{-1}\left(\frac{\lambda}{h^2}E\left(\varepsilon\varepsilon'\upsilon_\alpha^2\right)\right)$$
$$-\frac{\lambda^2}{\sqrt{2}h}E\left(\varepsilon\varepsilon'\epsilon_\varepsilon\right) - \frac{\sqrt{2}C_b}{h}E\left(\varepsilon\varepsilon'\upsilon_{\alpha\varepsilon}\right) + \sqrt{2}C_b E\left(\varepsilon\varepsilon'\upsilon_\varepsilon\right)$$
$$= n^{-1}\left(\frac{2\lambda}{h^2}I_T\sigma_\varepsilon^2 - \frac{\sqrt{2}\lambda^2}{h\sqrt{T}}I_T\sigma_\varepsilon^2 + 2\sqrt{2}C_b\frac{\sqrt{T}}{T-n}I_T\sigma_\varepsilon^2\right), \tag{37}$$

which is obtained by using

$$E\left(\varepsilon\varepsilon'\upsilon_\alpha\right) = 0,$$
$$E\left(\varepsilon\varepsilon'\upsilon_\alpha^2\right) = \sigma_\varepsilon^2 I_T E\left(n\left(\frac{E\alpha'\alpha\alpha'\alpha}{n^2\sigma_\alpha^4} - 2\frac{\alpha'\alpha}{n\sigma_\alpha^2} + 1\right)\right)$$
$$= 2\sigma_\varepsilon^2 I_T,$$

$$E\left(\varepsilon\varepsilon'\epsilon_{\varepsilon}\right) = E\left(\left(\varepsilon\varepsilon'\right)\sqrt{T}\left(\frac{\varepsilon'\varepsilon}{T\sigma_{\varepsilon}^{2}} - 1\right)\right)$$
$$= \sqrt{T}E\left(\frac{\varepsilon\varepsilon'\varepsilon'\varepsilon}{T\sigma_{\varepsilon}^{2}} - \varepsilon\varepsilon'\right)$$
$$= \frac{2}{\sqrt{T}}I_{T}\sigma_{\varepsilon}^{2},$$

$$E\left(\varepsilon\varepsilon'\upsilon_{\alpha\varepsilon}\right) = 0,$$
$$E\left(\varepsilon\varepsilon'\upsilon_{\varepsilon}\right) = \sqrt{T}E\left(\frac{\varepsilon\varepsilon'\varepsilon'Q\varepsilon}{(T-n)\,\sigma_{\varepsilon}^{2}} - \varepsilon\varepsilon'\right)$$
$$= \frac{2\sqrt{T}}{T-n}I_{T}\sigma_{\varepsilon}^{2}.$$

Using (35), (36) and (37) in (34), and noting

$$\frac{1}{\sqrt{T}} = \frac{\sqrt{2}}{n}\left(1 + \frac{1}{2n} + \frac{1}{8n^{2}} + \frac{1}{16n^{3}} + \dots\right),$$

we obtain

$$E\left(\Delta\lambda\omega\omega'\right) = n^{-1}\left(-\frac{2\lambda\,(1-h)}{h^{2}}\sigma_{\alpha}^{2}LL'\right)$$
$$+n^{-2}\left(\frac{2\lambda - h\lambda^{2}}{h^{2}}\right)4\sigma_{\alpha}^{2}LL' + n^{-1}\left(\frac{2\lambda}{h^{2}}I_{T}\sigma_{\varepsilon}^{2}\right). \quad (38)$$

Similarly, other expectation terms in (31) to (33) are

$$E\left(\Delta\lambda\omega\alpha'\right) = LE\left(\alpha\alpha'\Delta\lambda\right) + E\left(\varepsilon\alpha'\Delta\lambda\right)$$
$$= n^{-1}\left(-\frac{2\lambda\,(1-h)}{h^{2}}\right)\sigma_{\alpha}^{2}L + n^{-2}\left(\frac{2\lambda}{h}\left(\frac{4}{h} - \lambda\right)\right)\sigma_{\alpha}^{2}L, \quad (39)$$

$$E\left(\Delta\lambda\omega\varepsilon'\right) = LE\left(\Delta\lambda\alpha\varepsilon'\right) + E\left(\Delta\lambda\varepsilon\varepsilon'\right)$$
$$= n^{-1}\frac{2\lambda}{h^{2}}I_{T}\sigma_{\varepsilon}^{2} + n^{-2}\left(4C_{b} - \frac{2\lambda^{2}}{h}\right)I_{T}\sigma_{\varepsilon}^{2}$$
$$-n^{-2}\left(\frac{2\lambda^{2}}{h}\right)\sigma_{\alpha}^{2}LL', \quad (40)$$

$$E\left(\alpha\alpha'(\Delta\lambda)^2\right) = E\left(\alpha\alpha' n^{-1}\frac{\lambda^2}{h^2}v_\alpha^2\right) = n^{-1}\frac{\lambda^2}{h^2}E\left(\alpha\alpha' v_\alpha^2\right)$$
$$= n^{-1}\frac{2\lambda^2}{h^2}I_n\sigma_\alpha^2 + n^{-2}\frac{8\lambda^2}{h^2}I_n\sigma_\alpha^2. \quad (41)$$

Using (38), (39), (40) and (41) in (29), we obtain

$$E\left(\zeta_0\zeta_{-1}'\right) = \frac{A^{-1}X'\Omega^{-1}LL'\Phi X A^{-1}}{n^3}$$
$$\times\left[n^{-1}\left(\frac{2\lambda(1-h)\sigma_\alpha^2}{h^2}\right) + n^{-2}\frac{2\lambda(4-\lambda h)}{h^2}\sigma_\alpha^2\right]$$
$$= P_1\left[n^{-1}\left(\frac{2\lambda(1-h)\sigma_\alpha^2}{h^2}\right) + n^{-2}\frac{2\lambda(4-\lambda h)}{h^2}\sigma_\alpha^2\right], \quad (42)$$

$$E\left(\zeta_0\zeta_{-3/2}'\right) = \frac{A^{-1}X'\Omega^{-1}\Phi X A^{-1}}{n^3}$$
$$\times\left(n^{-1}\frac{2\lambda}{h^2}I_T\sigma_\varepsilon^2 + n^{-2}\left(4C_b - \frac{2\lambda^2}{h}\right)I_T\sigma_\varepsilon^2\right)$$
$$+\frac{A^{-1}X'\Omega^{-1}LL'\Phi X A^{-1}}{n^3}\left(-n^{-2}\left(\frac{2\lambda^2}{h}\right)\sigma_\alpha^2\right)$$
$$+\frac{A^{-1}X'\Omega^{-1}LL'\Omega^{-1}XA^{-1}BA^{-1}}{n^3}$$
$$\times\left(n^{-1}\left(\frac{2\lambda(1-h)}{h^2}\sigma_\alpha^2\right) - n^{-2}\left(\frac{2\lambda - h\lambda^2}{h^2}\right)4\sigma_\alpha^2\right)$$
$$+\frac{A^{-1}X'\Omega^{-1}\Omega^{-1}XA^{-1}BA^{-1}}{n^3}\left(-n^{-1}\left(\frac{2\lambda}{h^2}I_T\sigma_\varepsilon^2\right)\right)$$
$$= P_3\left(n^{-2}\frac{2\lambda}{h^2}I_T\sigma_\varepsilon^2 + n^{-3}\left(4C_b - \frac{2\lambda^2}{h}\right)I_T\sigma_\varepsilon^2\right)$$
$$+P_1\left(-n^{-2}\left(\frac{2\lambda^2}{h}\right)\sigma_\alpha^2\right) + P_4\left(n^{-2}\left(\frac{2\lambda(1-h)}{h^2}\sigma_\alpha^2\right.\right.$$
$$\left.\left.-n^{-3}\left(\frac{2\lambda - h\lambda^2}{h^2}\right)4\sigma_\alpha^2\right) + P_5\left(-n^{-2}\left(\frac{2\lambda}{h^2}I_T\sigma_\varepsilon^2\right)\right), \quad (43)$$

$$E\left(\zeta_{-1}\zeta_{-1}'\right) = \frac{A^{-1}X'\Phi LL'\Phi X A^{-1}}{n^3}\left(n^{-2}\frac{2\lambda^2}{h^2}\sigma_\alpha^2\right)$$
$$= P_2\left(n^{-2}\frac{2\lambda^2}{h^2}\sigma_\alpha^2\right) \quad (44)$$

where

$$\left.\begin{array}{l} P_1 = A^{-1}X'\Omega^{-1}LL'\Phi XA^{-1}/n^3, \\ P_2 = A^{-1}X'\Phi LL'\Phi XA^{-1}/n^3, \\ P_3 = A^{-1}X'\Omega^{-1}\Phi XA^{-1}/n^2, \\ P_4 = A^{-1}X'\Omega^{-1}LL'\Omega^{-1}XA^{-1}BA^{-1}/n^2, \\ P_5 = A^{-1}X'\Omega^{-1}\Omega^{-1}XA^{-1}BA^{-1}/n^2. \end{array}\right\} \qquad (45)$$

Here note that each of A, B, P_1, P_2, P_3, P_4 and P_5 is $O(1)$.

Substituting (30), (42), (43) and (44) into (29), the MSE of $(\hat{\beta}_{FGLS}-\beta)$ up to order $O(n^{-4})$ is:

$$\begin{aligned} &E\left(\hat{\beta}_{FGLS}-\beta\right)\left(\hat{\beta}_{FGLS}-\beta\right)' \\ &= n^{-2}\left(X'\Omega^{-1}X/n^2\right)^{-1} + (P_1+P_1')\left[n^{-3}\left(\frac{2\lambda(1-h)\sigma_\alpha^2}{h^2}\right)\right. \\ &\left.+n^{-4}\frac{2\lambda(4-\lambda h)}{h^2}\sigma_\alpha^2\right] + (P_3+P_3')\left(n^{-4}\frac{2\lambda}{h^2}I_T\sigma_\varepsilon^2\right) \\ &+(P_1+P_1')\left(-n^{-4}\left(\frac{2\lambda^2}{h}\right)\sigma_\alpha^2\right) + (P_4+P_4')\left(n^{-4}\left(\frac{2\lambda(1-h)}{h^2}\sigma_\alpha^2\right)\right) \\ &+(P_5+P_5')\left(-n^{-4}\left(\frac{2\lambda}{h^2}I_T\sigma_\varepsilon^2\right)\right) + P_2\left(n^{-4}\frac{2\lambda^2}{h^2}\sigma_\alpha^2\right). \end{aligned}$$

5 Simulation

In the following, we conduct simulation of a random effects model. We estimate the slope coefficient by the estimators described in the foregoing section and study the behavior of bias and MSE for different values of λ and n.

We use the following data generating process for model (3),

$$y_{ij} = 2x_{ij} + \alpha_i + \alpha_j + \varepsilon_{ij}$$

where $\alpha_i \sim i.i.d.n.[0,1]$ and $\varepsilon_{ij} \sim i.i.d.n.[0,1]$, and $x_{ij} \sim iidn[0,\sigma_x^2]$, where σ_x^2 is chosen so that $R^2 \simeq 0.5$ from OLS regression. We first pick $n = 50$ and replicate 1000 times. The results are reported in Table 1, there, the "Correct se" is the theoretical standard deviation under our simulation setup; "Simulation se" is the sample standard deviation of the estimators from 1000 replications while "Default se" refers to the usual OLS standard deviation under homoscedastic disturbances.

We can see that FGLS estimator shows significant improvement over OLS estimator in terms of bias and standard error. Using the usual form of OLS variance here leads to under-estimate of the standard error of the OLS estimator.

Then we investigate effects of the number of cross section units on bias and MSE and compare finite sample MSE values with simulated MSE and asymptotic counterparts. We start with $n = 5$ and increase n by 5 till $n = 50$. For every n, the experiment is repeated 1000 times. The results are reported in Table 2. There, n refers to number of clusters. $\hat{\beta}_{FGLS}$ is mean value of the suggested FGLS estimates of slope coefficient obtained from 1000 replications. These replication values of β are used in calculating simulation MSE and Bias. The results are reported in columns named "Simulation MSE" and "Simulation Bias". "Finite Sample MSE" column reports, up to $O\left(n^{-4}\right)$, the theoretical MSE while "Asymptotic MSE" column reports the asymptotic values of MSE of $\hat{\beta}_{FGLS}$.

As n increases, both the bias and MSE become smaller. For each n, the bias is of smaller magnitude than the square root of MSE. The finite sample MSE is closer to the simulation MSE than asymptotic values in all cases, and it demonstrates an upward correction to asymptotic value when n is small ($n = 5$) and a downward correction when n is large ($n > 5$). The correction becomes smaller as n increases.

At both $n = 20$ and $n = 50$ cases, we conduct simulation using different values of σ_ε^2 and σ_α^2, hence different λs, and examine their effects on MSE. The results are reported in table 3 and 4. An increase in σ_α^2 and/or σ_ε^2 increases MSE. When λ is held constant, increasing σ_α^2 and σ_ε^2 at the same rate leads to a proportional increase in MSE. However, the finite sample MSE of $\hat{\beta}_{FGLS}$ is governed mostly by disturbances in individual effects: the effect of an increase in variance of cluster effects, σ_α^2, on MSE is much smaller than that of variance of individual effects, σ_ε^2. For example, in $n = 20$ case, starting from $\sigma_\alpha^2 = 0.5$ and $\sigma_\varepsilon^2 = 0.5$, when σ_ε^2 is increased to 1, the increase in MSE is 0.000994332 while the increase in MSE when σ_α^2 is increased to 1 is 0.000006110. The ratio of these two is 162.74, which is of the magnitude of n^2. This result is consistent with Remark 3 in Section 3.

6 Conclusion

This paper proposes a FGLS estimator for a cross-sectional paired data model, where the cluster effects are considered random. We investigate the estimator's finite sample bias and mean square error under Gaussian effects. We report that the finite sample bias of FGLS estimator is zero up to order $O(n^{-2})$ and its MSE is mainly governed by variances of individual effects although increasing variance in cluster effects and/or individual effect increases MSE. The finite sample MSE up to order $O(n^{-4})$ has a downward correction to asymptotic values when cluster size is large and has an upward correction in case of a small cluster size.

Appendix

Table 1. Estimation results for random effects model

	True Value	Estimator	
		OLS	FGLS
Slope	2.0000	1.9995	1.9997
Correct se		0.04996	0.0029079
Simulation se		0.05009	0.029884
Default se		0.049304	

Table 2. MSE and bias under different n

n	$\hat{\beta}_{FGLS}$	Simulation MSE	Simulation bias	Finite sample MSE	Asymptotic MSE
5	2.008	0.11252	0.008	0.11074	0.09869
10	2.0006	0.0079999	0.0006	0.0082579	0.0082738
15	2.0005	0.0042634	0.0005	0.0043822	0.0043858
20	2.0001	0.001608	0.0001	0.0017099	0.0017106
25	1.99997	0.001134	-0.00003	0.0011562	0.0011564
30	2.0002	0.00098	0.0002	0.00094531	0.00094539
35	2.0001	0.00057599	0.0001	0.00060076	0.00060083
40	1.9998	0.00047991	0.0002	0.00049426	0.00049429
45	2.0004	0.00041199	0.0004	0.00037668	0.0003767
50	1.9999	0.00028349	-0.0001	0.00027909	0.0002791

Table 3. Effect of different values of σ_α^2 and σ_ε^2 on MSE of $\hat{\beta}_{FGLS}$, $n = 20$ case

		σ_ε^2				
		0.5	1	2.5	5	10
σ_α^2	0.5	0.0010170	0.0020113	0.0048876	0.0094389	0.0180599
	1	0.0010231	0.0020339	0.0050020	0.0097752	0.0188779
	2.5	0.0010269	0.002049	0.0050848	0.0100565	0.0197211
	5	0.0010282	0.0020538	0.0051154	0.0101697	0.020113
	10	0.0010289	0.0020564	0.0051313	0.0102308	0.0203394

Table 4. Effect of different values of σ_α^2 and σ_ε^2 on MSE of $\hat{\beta}_{FGLS}$, $n = 50$ case

		σ_ε^2				
		0.5	1	2.5	5	10
σ_α^2	0.5	0.00014669	0.00029312	0.00073101	0.00145680	0.00289711
	1	0.00014676	0.00029338	0.00073249	0.00146201	0.00291360
	2.5	0.00014680	0.00029354	0.00073345	0.00146561	0.00292633
	5	0.00014681	0.00029359	0.00073378	0.00146689	0.00293121
	10	0.00014682	0.00029362	0.00073395	0.00146756	0.00293379

References

Kloek T (1981) OLS Estimation in a Model where a Microvariable is Explained by Aggregates and Contemporaneous Disturbances are Equicorrelated. Econometrica 49:205–207

Cameron A, Golotvina N (2005) Estimation of Models for Country-Pair Data controlling for Clustered Errors: with Applications. UC Davis, Manuscript

Moulton BR (1986) Random Group Effects and the Precision of Regression Estimates. Journal of Econometrics 32:385–97

Pesaran H (2006) Estimation And Inference In Large Heterogeneous Panels With A Multifactor Error Structure. Econometrica 74(4):967-1012

Ullah A (2004) Finite Sample Econometrics. Oxford University Press

Prediction of Finite Population Total in Measurement Error Models

Hyang Mi Kim[1] and A.K.Md. Ehsanes Saleh[2]

[1] Department of Mathematics and Statistics, Division of Statistics and Actuarial Science, University of Calgary, Calgary, Canada hmkim@ucalgary.ca

[2] School of Mathematics and Statistics, Carleton University, Ottawa, Canada esaleh@math.carleton.ca

1 Introduction

Measurement error regression models are different from classical regression models mainly that the covariates are measured with errors. This paper deals with the prediction of finite population total based on regression models with measurement errors. General treatment of regression problems with measurement errors is considered in the pioneering book by Fuller (1987) and Cheng and Van Ness (1999). Later Sprent (1966) proposed methods based on generalized least-squares approach for estimating the regression coefficients. Lindley (1966) and Lindley and Sayad (1968) pioneered Bayesian approach to the problem. Further, contribution in Bayesian approach have been made by Zellner (1971) and Reilly and Patino-Leal (1981). Fuller (1975) points out not much research is done for problems in finite population with measurement error models. However, Bolfarine (1991) investigated the problem of predictors for finite population with errors in variable models. Recently, Kim and Saleh (2002, 2003, 2005) pioneered the application of preliminary test and shrinkage estimation methodology in measurement error models. Recent book of Saleh (2006) presents an overview on the theory of preliminary test and Shrinkage estimators. This paper contains the application of these ideas for the prediction of finite population totals using simple linear model with measurement errors .

Let the finite population is denoted by $\mathcal{P} = \{1, \cdots, N\}$ consisting of N units, where N is known. Associated with the t th unit of $\mathcal{P}$, there be the vector, $\{(Y_t, x_t)|t = 1, \cdots, N\}'$, which is a random sample of size N from the conditional distribution distribution.

$$Y_{0t} = \beta_0 + \beta_1 x_t + e_t, \quad t = 1, 2, \cdots, N$$

where $e_t \sim N(0, \sigma_{ee})$. To gain information about the population total, $T = Y_1 + \cdots + Y_N$, a sample of size n is selected at random from $\mathcal{P}$. Let the observed sample of the Y's be $\mathbf{Y}_s = (Y_{01}, \cdots, Y_{0n})'$ and the remaining $N - n$ vectors of Y's are unobserved. Let this set be denoted $r = \mathcal{P} - s$. Our object is to estimate the total, $T = \sum_{j \in s} Y_j + \sum_{j \in r} Y_j$. We assume that the sample $Y_{01}, \cdots, Y_{0n}$ satisfy the model

$$Y_{0t} = \beta_0 + \beta_1 x_t + e_t, \quad t = 1, 2, \cdots, n \tag{1}$$

such that $e_t \sim N(0, \sigma_{ee})$ and x_t are observed with error, i.e.,

$$X_{0t} = x_t + u_t, \quad u_t \sim \mathcal{N}(0, \sigma_{uu}) \tag{2}$$

and X_{0t}'s are completely known for all $t = 1, \cdots, N$.

Our problem is reduced to the estimation of $T = \sum_{j \in s} Y_j + \sum_{j \in r} Y_j$ based on the observed sample $\{(Y_{0t}, X_{0t}) | t = 1, \cdots, n\}'$ and study properties of the estimators using (1) and (2). Section 2 deals with usual classical linear model without measurement errors and section 3 deals with the simple regression model with measurement errors.

2 Prediction of Population Total Without Measurement Error

In this section, we consider the prediction of finite population totals based on the sub-sample of the model (1) and (2),

$$Y_{0t} = \beta_0 + \beta_1 X_{0t} + e_t, \quad t = 1, 2, \cdots, n$$

where X_{0t} is known explanatory variable for $t \in \mathbf{s}$ and $e_t \sim N(0, \sigma_{ee})$. The ordinary least square (OLS) estimate of β_0 and β_1 are given by

$$\tilde{\beta}_{0(OLS)} = \bar{Y}_{0s} - \tilde{\beta}_{1(OLS)} \bar{X}_{0s} \quad \text{and} \quad \tilde{\beta}_{1(OLS)} = S^{-1}_{X_0 X_0} S_{X_0 Y_0}$$

where $\bar{X}_{0s} = \frac{1}{n} \sum_{t \in \mathbf{s}} X_{0t}$, $\bar{Y}_{0s} = \frac{1}{n} \sum_{t \in \mathbf{s}} Y_{0t}$, $S_{X_0 X_0} = \sum_{t \in \mathbf{s}} (X_{0t} - \bar{X}_{0s})^2$, and $S_{X_0 Y_0} = \sum_{t \in \mathbf{s}} (X_{0t} - \bar{X}_{0s})(Y_{0t} - \bar{Y}_{0s})$. Also, $\tilde{\sigma}_{ee} = \frac{1}{n-2} \sum_{t \in \mathbf{s}} (Y_{0t} - \tilde{\beta}_{0(OLS)} - \tilde{\beta}_{1(OLS)} X_{0t})^2$.

If β_1 is suspected to be equal to 0 (zero), then we define the preliminary test estimator (PTE) and shrinkage estimator (SE) of β_1 as given below following Saleh (2006):

$$(i)\ \hat{\beta}_{1(OLS)}^{PTE} = \tilde{\beta}_{1(OLS)} - \tilde{\beta}_{1(OLS)}\, I(F < F_\alpha)$$

$$\text{and } (ii)\ \hat{\beta}_{1(OLS)}^{SE} = \tilde{\beta}_{1(OLS)} - c\frac{\sqrt{\tilde{\sigma}_{ee}}}{\sqrt{S_{X_0X_0}}}\frac{\tilde{\beta}_{1(OLS)}}{|\tilde{\beta}_{1(OLS)}|},$$

$$\text{where} \quad F = \frac{S_{X_0X_0}\tilde{\beta}^2_{1(OLS)}}{\tilde{\sigma}_{ee}},$$

is the test statistics for testing $H_0 : \ \beta_1 = 0$ against the alternative $H_A : \ \beta_1 \neq 0$ and F_α is the upper α-level critical value of the F-statistic. The distributional results follow from the fact that

$$\tilde{\beta}_{1(OLS)} \sim N\Big(\beta_1, \frac{\sigma_{ee}}{S_{X_0X_0}}\Big)$$

$$\text{and} \quad \frac{S_{X_0X_0}\tilde{\beta}^2_{1(OLS)}}{\tilde{\sigma}_{ee}} \sim F_{1,m}(\triangle^2), \quad (m = n-2)$$

where $F_{1,m}(\triangle^2)$ is a non-central F-distribution with $(1,\ m)$ degree of freedom and non-centrality parameter $\frac{1}{2}\triangle^2$ where $\triangle^2$ is defined by

$$\triangle^2 = \frac{S_{X_0X_0}\beta_1^2}{\sigma_{ee}}.$$

The power function of this test is given by

$$1 - G_{1,m}(F_\alpha;\triangle^2) \geq \alpha = 1 - G_{1,m}(F_\alpha;0).$$

This test is unbiased and most powerful for $H_0 : \beta_1 = 0$. Note that $G_{\nu_1,\nu_2}(x;\triangle^2)$ stands for the c.d.f. of the non-central F-distribution with (ν_1,ν_2) d.f. and non-centrality parameter $\frac{1}{2}\triangle^2$ with the c.d.f.

$$G_{\nu_1,\nu_2}(x;\triangle^2)\sum_{r\geq 0}\frac{e^{-\frac{\triangle^2}{2}}(\frac{\triangle^2}{2})^r}{\Gamma(r+1)}I_y\Big(\frac{\nu_1}{2}+r:\frac{\nu_2}{2}\Big),$$

where $y = \frac{\nu_1 x}{\nu_2+\nu_1 x}$ and

$$I_y\Big(\frac{\nu_1}{2}+r:\frac{\nu_2}{2}\Big) = \frac{1}{B(\frac{\nu_1}{2}+r:\frac{\nu_2}{2})}\int_0^y x^{\frac{\nu_1}{2}+r-1}(1-x)^{\frac{\nu_2}{2}-1}dx$$

is the incomplete Beta function . These results may be found in Ahsanullah and Saleh (1972), Ahmed and Saleh (1988) and Kim and Saleh (2002) among others.

Now we consider the finite population and related estimation of T as described in the introduction. Then we consider the generalized regression predictors (GREP) of the form

$$\hat{T}_G = N\bar{Y}_{0s} + (N-n)(\bar{X}_{0r} - \bar{X}_{0s})\tilde{\beta}_{1(G)}$$

where $\tilde{\beta}_{1(G)}$ is any general estimator of β_1 and $\bar{X}_{0r} = \frac{1}{N-n}\sum_{t\in\mathbf{r}} X_{0t}$. The predictor $\hat{T}_G$ has been shown to be the best unbiased predictor (BUP) of T under normal theory in Bolfarine and Zacks (1991). Thus we may define four predictors of T when β_1 may equal zero (≈ 0) as follow:

(i) unrestricted predictor (UP)

$$\hat{T}_{OLS}^{UP} = N\bar{Y}_{0s} + (N-n)(\bar{X}_{0r} - \bar{X}_{0s})\tilde{\beta}_{1(OLS)},$$

(ii) restricted predictor (RP)

$$\hat{T}_{OLS}^{RP} = N\bar{Y}_{0s},$$

(iii) preliminary test predictor (PTP)

$$\hat{T}_{OLS}^{PTP} = N\bar{Y}_{0s} + (N-n)(\bar{X}_{0r} - \bar{X}_{0s})\hat{\beta}_{1(OLS)}^{PT},$$

and
(iv) shrinkage predictor (SP)

$$\hat{T}_{OLS}^{SP} = N\bar{Y}_{0s} + (N-n)(\bar{X}_{0r} - \bar{X}_{0s})\hat{\beta}_{1(OLS)}^{SE}.$$

In order to calculate the bias and the MSE expressions of the above estimators we use the following identities:

$$(i)\ \hat{T}_{OLS}^{UP} - T = (N-n)[(\bar{X}_{0r} - \bar{X}_{0s})(\tilde{\beta}_{1(OLS)} - \beta_1) + (\bar{e}_s - \bar{e}_r)]$$

$$(ii)\ \hat{T}_{OLS}^{RP} - T = (N-n)[(\bar{X}_{0r} - \bar{X}_{0s})\beta_1 + (\bar{e}_s - \bar{e}_r)]$$

$$(iii)\ \hat{T}_{OLS}^{PTP} - T = (N-n)[(\bar{X}_{0r} - \bar{X}_{0s})(\hat{\beta}_{1(OLS)}^{PT} - \beta_1) + (\bar{e}_s - \bar{e}_r)]$$

and

$$(iv)\ \hat{T}_{OLS}^{SP} - T = (N-n)[(\bar{X}_{0r} - \bar{X}_{0s})(\hat{\beta}_{1(OLS)}^{SE} - \beta_1) + (\bar{e}_s - \bar{e}_r)],$$

respectively.

The bias and MSE expressions may be obtained as given below. First, the bias expressions:

$$
\begin{aligned}
&(i)\ b_1(\hat{T}_{OLS}^{UP}) = 0, &(3)\\
&(ii)\ b_2(\hat{T}_{OLS}^{RP}) = (N-n)(\bar{X}_{0r} - \bar{X}_{0s})\beta_1\\
&(iii)\ b_3(\hat{T}_{OLS}^{PTP}) = -(N-n)\beta_1(\bar{X}_{0r} - \bar{X}_{0s})G_{3,m}\Big(\frac{1}{3}F_\alpha;\triangle^2\Big),\ m = n-2\\
&\quad\text{and}\\
&(iv)\ b_4(\hat{T}_{OLS}^{SP}) = -c(N-n)\sqrt{\sigma_{ee}}K_n\Big\{2\Phi(\triangle^2) - 1\Big\}\sqrt{K(X)}.
\end{aligned}
$$

Similarly, the MSE expressions are given by

$$
\begin{aligned}
&(i)\ M_1(\hat{T}_{OLS}^{UP}) = N\frac{1}{f}(1-f)\sigma_{ee}[1 + Nf(1-f)K(X)]\\
&(ii)\ M_2(\hat{T}_{OLS}^{RP}) = N\frac{1-f}{f}\sigma_{ee} + (N-n)^2\sigma_{ee}K(X)\triangle^2\\
&\qquad = N\frac{1-f}{f}\sigma_{ee}\Big[1 + Nf(1-f)K(X)\triangle^2\Big]\\
&(iii)\ M_3(\hat{T}_{OLS}^{PTP}) = M_1(\hat{T}_{OLS}^{UP}) - \sigma_{ee}N^2(1-f)^2K(X)\\
&\qquad \times\Big\{G_{3,m}\Big(\frac{1}{3}F_\alpha;\triangle^2\Big) - \triangle^2\Big[2G_{3,m}\Big(\frac{1}{3}F_\alpha;\triangle^2\Big) - G_{5,m}\Big(\frac{1}{5}F_\alpha;\triangle^2\Big)\Big]\Big\}\\
&\quad\text{and}\\
&(iv)\ M_4(\hat{T}_{OLS}^{SP}) = M_1(\hat{T}_{OLS}^{UP}) - \sigma_{ee}N^2(1-f)^2K(X)\frac{2}{\pi}K_n^2\\
&\qquad \times\Big\{2e^{-\triangle^2/2} - 1\Big\}
\end{aligned}
$$

where F_α is the upper α- level critical value from the central F-distribution with $(1, m)$ d.f with $\triangle^2 = \frac{\beta_1^2\sigma_{ee}}{S_{XX}}$, $f = \frac{n}{N}$, $K_n = \sqrt{\frac{2}{n-2}}\frac{\Gamma(\frac{n-1}{2})}{\Gamma(\frac{n-2}{2})}$ and $K(X) = \frac{(\bar{X}_{0r} - \bar{X}_{0s})^2}{S_{X_0X_0}}$.

Now we compare the four estimators based on their MSE expressions.

(A) $\hat{T}_{OLS}^{RP}$ versus $\hat{T}_{OLS}^{UP}$

In this case, the relative efficiency of $\hat{T}_{OLS}^{RP}$ compared to $\hat{T}_{OLS}^{UP}$ is given by,

$$
\begin{aligned}
E(\hat{T}_{OLS}^{RP} : \hat{T}_{OLS}^{UP}) &= \Big[N\frac{1-f}{f}\Big(1 + Nf(1-f)K(X)\Big)\Big]\\
&\quad \times\Big[\frac{N}{f} + N^2(1-f)^2K(X)\triangle^2\Big]^{-1}\\
&= \Big[1 + Nf(1-f)K(X)\Big]\Big[1 + Nf(1-f)K(X)\triangle^2\Big]^{-1}.
\end{aligned}
$$

First note that the efficiency is a decreasing function of $\triangle^2$. The maximum occurs as $\triangle^2 = 0$, i.e. $\beta_1 = 0$, with the value

$$\Big[(1-f)\Big\{1+Nf(1-f)K(X)\Big\}\Big].$$

Then, it decreases crossing the 1-line at $\triangle^2 = 1$, then converges to zero as $\triangle^2 \to \infty$. Thus, $\hat{T}^{RP}_{OLS}$ is better than $\hat{T}^{UP}_{OLS}$ in the range $0 \le \triangle^2 \le 1$ and beyond this range $\hat{T}^{UP}_{OLS}$ is better than $\hat{T}^{RP}_{OLS}$.

(B) $\hat{T}^{PTP}_{OLS}$ versus $\hat{T}^{UP}_{OLS}$

In this case, the relative efficiency is given by

$$\begin{aligned} E(\hat{T}^{PTP}_{OLS} : \hat{T}^{UP}_{OLS}) = \Big[&1 + Nf(1-f)K(X)(1+Nf(1-f)K(X))^{-1} \\ &\times\Big\{\triangle^2\Big(2G_{3,m}\Big(\frac{1}{3}F_\alpha;\triangle^2\Big) - G_{5,m}\Big(\frac{1}{5}F_\alpha;\triangle^2\Big)\Big) \\ &-G_{3,m}\Big(\frac{1}{3}F_\alpha;\triangle^2\Big)\Big\}\Big]^{-1}. \end{aligned}$$

Note that $E(\hat{T}^{PTP}_{OLS} : \hat{T}^{UP}_{OLS})$ is a function of $(\alpha, \triangle)$ and for fixed α $(0 < \alpha < 1)$ it decreases crossing the 1-line to a minimum, then increases towards the 1-line as $\triangle^2$ tends to infinity. Further, $\triangle^2$-intercept is given by

(i) $E(\hat{T}^{PTP}_{OLS} : \hat{T}^{UP}_{OLS}) \gtreqless 1$ according as

$$\triangle^2 \lesseqgtr \frac{G_{3,m}(\frac{1}{3}F_\alpha;\triangle^2)}{2G_{3,m}(\frac{1}{3}F_\alpha;\triangle^2) - G_{5,m}(\frac{1}{5}F_\alpha;\triangle^2)} = K_S(\alpha,\triangle^2)(\le 1).$$

Hence $\hat{T}^{PTP}_{OLS}$ performs better than $\hat{T}^{UP}_{OLS}$ if $\triangle^2 \le K_S(\alpha : \triangle^2)$, otherwise $\hat{T}^{UP}_{OLS}$ performs better.

(ii) Maximum efficiency for fixed α is attained for $\triangle^2 = 0$ and equals

$$\Big\{1 - \frac{Nf(1-f)K(X)}{1+Nf(1-f)K(X)} G_{3,m}\Big(\frac{1}{3}F_\alpha;0\Big)\Big\}^{-1} \tag{4}$$

and as $\triangle^2 \to \infty$, the efficiency goes to unity.

(iii) PTP can not be uniformly better that $\hat{T}^{UP}_{OLS}$ but one may determine a PTP with minimum guaranteed efficiency, say, E_0 by choosing an optimum level of significance α^* by the maxmin rule

$$\max_{\alpha}\min_{\triangle^2} E(\alpha : \triangle^2) \ge E_0.$$

The efficiency may go up to (4) if β_1 is near the origin.

(iv) As $\alpha \to 1$, $\hat{T}^{PTP}_{OLS} \to \hat{T}^{UP}_{OLS}$. Figure 1 displays one graph of the efficiency of $\hat{T}^{PTP}_{OLS}$ relative to $\hat{T}^{UP}_{OLS}$ when $n = 8$, $\alpha = 0.15$, $K(X) = 3$ and $Nf(1-f) = 4$.

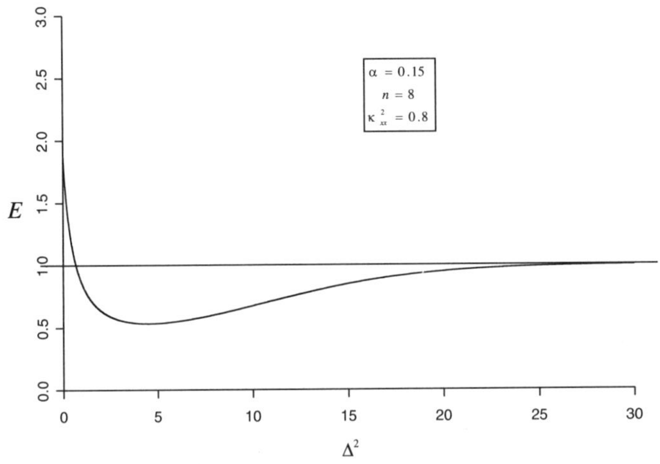

Fig. 1. Graph of $E(\hat{T}_{OLS}^{PTP} : \hat{T}_{OLS}^{UP})$

(C) $\hat{T}_{OLS}^{PTP}$ versus $\hat{T}_{OLS}^{RP}$

In this case, the efficiency expression is given by

$$E[\hat{T}_{OLS}^{PTP} : \hat{T}_{OLS}^{RP}] = E[\hat{T}_{OLS}^{RP} : \hat{T}_{OLS}^{UP}]^{-1}\Big[1 - \frac{g(\triangle^2)}{M_1(\hat{T}_{OLS}^{UP})}\Big]^{-1}$$

where

$$g(\triangle^2) = \sigma_{ee}\Big\{N^2(1-f)^2K(X)G_{3,m}\Big(\frac{1}{3}F_\alpha;\triangle^2\Big) \\ -\triangle^2\Big[2G_{3,m}\Big(\frac{1}{3}F_\alpha;\triangle^2\Big) - G_{5,m}\Big(\frac{1}{5}F_\alpha;\triangle^2\Big)\Big]\Big\}.$$

Here the efficiency is a function of $(\alpha, \triangle^2)$. For $\triangle^2 = 0$, the efficiency reduces to

$$\Big\{(1-f) + Nf(1-f)^2K(X)\Big\}^{-1}\Big\{1 - \frac{Nf(1-f)K(X)}{1+Nf(1-f)K(X)} \\ \times G_{3,m}\Big(\frac{1}{3}F_\alpha;0\Big)\Big\}^{-1}$$

and as $\triangle^2 \to \infty$ the efficiency blows up. Also, under $H_0 : \beta_1 = 0$ the MSE satisfies the following dominance picture.

$$M_2(\hat{T}_{OLS}^{RP}) < M_3(\hat{T}_{OLS}^{PTP}) < M_1(\hat{T}_{OLS}^{UP}).$$

The relationship changes as $\triangle^2$ diverts from 0 as

$$M_2(\hat{T}_{OLS}^{UP}) < M_3(\hat{T}_{OLS}^{PTP}) < M_1(\hat{T}_{OLS}^{RP})$$

beyond the intersection of $M_1(\hat{T}_{OLS}^{UP})$ and $M_3(\hat{T}_{OLS}^{PTP})$ depending on the size of the test.

(D) $\hat{T}_{OLS}^{SP}$ versus $\hat{T}_{OLS}^{UP}$
In this case, the efficiency expression is given by

$$E[\hat{T}_{OLS}^{SP} : \hat{T}_{OLS}^{UP}] = \left\{1 - \frac{2Nf(1-f)K(X)K_n^2(2e^{-\triangle^2/2} - 1)}{\pi[1 + Nf(1-f)K(X)]}\right\}^{-1}.$$

The efficiency is a decreasing function of $\triangle^2$ with a maximum value

$$\left\{1 - \frac{2Nf(1-f)K(X)}{\pi[1 + Nf(1-f)K(X)]}\right\}^{-1} \geq 1 \tag{5}$$

at $\triangle^2 = 0$ while as $\triangle^2 \to \infty$, the efficiency reduces to

$$\left\{1 + \frac{2Nf(1-f)K(X)}{\pi[1 + Nf(1-f)K(X)]}\right\}^{-1} (\leq 1) \tag{6}$$

giving a lower bound of the efficiency. Thus efficiency drops from (5) to (6) as $\triangle^2$ continues to grow from 0 to ∞. The efficiency function intersects the 1-line at $\triangle^2 = \ln 4$. Thus, $\hat{T}_{OLS}^{SP}$ is better than $\hat{T}_{OLS}^{UP}$ in the interval $0 \leq \triangle^2 \leq \ln 4$ while $\hat{T}_{OLS}^{UP}$ is better than $\hat{T}_{OLS}^{SP}$ otherwise. For more comparisons, see Saleh (2006, Chapter 3).

3 Prediction of Population Total With Measurement Error

In this section, we consider that the population t-th unit is associated with a pair $(Y_t, x_t)'$ assumed to be satisfy the model

$$\begin{aligned} Y_{0t} &= \beta_0 + \beta_1 x_t + e_t \\ X_{0t} &= x_t + u_t, \quad t = 1, 2, \cdots, N \end{aligned} \tag{7}$$

where

$$(x_t, e_t, u_t)' \sim \mathcal{N}_3\{(\mu_x, 0, 0)'; \text{diag}(\sigma_{xx}, \sigma_{ee}, \sigma_{uu})\}$$

for $t = 1, 2, \cdots, N$. The pair $(Y_{0t}, x_t)'$ is not observed directly instead $(Y_{0t}, X_{0t})', t \in s$ is observed where s is the observed sample. Further, it is assumed that the all $X's$, i.e., $X_{01}, \cdots, X_{0N}$ are available for the population. Clearly, $(Y_{0t}, X_{0t}) \sim \mathcal{N}(\beta_0 + \beta_1\mu_x, \Sigma)$, where

$$\Sigma = \begin{pmatrix} \beta_1^2\sigma_{xx} + \sigma_{uu} & \beta_1\sigma_{xx} \\ \beta_1\sigma_{xx} & \sigma_{xx} + \sigma_{uu} \end{pmatrix}.$$

Thus the conditional distribution of Y_{0t} given X_{0t} is normally distributed with mean $\nu_0 + \nu_1 X_{0t}$ and σ_{zz}, which we may write as

$$\begin{aligned} Y_{0t} &= \nu_0 + \nu_1 X_{0t} + Z_t; \quad t = 1, 2, \cdots, N \\ Z_t &\sim N(0, \sigma_{zz}), \\ \sigma_{zz} &= \sigma_{ee} + \beta_1^2\sigma_{xx}(1 - \kappa_{xx}), \\ \nu_0 &= \beta_0 + \beta_1\mu_x(1 - \kappa_{xx}), \nu_1 = \kappa_{xx}\beta_1, \\ \text{and } \kappa_{xx} &= \sigma_{xx}(\sigma_{xx} + \sigma_{uu})^{-1} \text{ (reliability ratio).} \end{aligned}$$

The reliability ratio (r.r.) κ_{xx} is typically assumed to be known. Thus, model exactly mimics the model described earlier where ν_0 is a translation of β_0 and ν_1 is a scaled version of β_1.

Consider a random sample of size n from the finite population. Let $\{(Y_{0t}, X_{0t}) | t = 1, \cdots, n\}$ be the observed pairs satisfying (7). Our objective is to predict the population total T based on the sample. In terms of the model (7) we write the expansion predictor as $\hat{T} = n\bar{Y}_0 + (N-n)\bar{Y}_{0r}$ where $\bar{Y}_{0s}$ is the average of the observed responses where $\bar{Y}_{0r}$ is the average of the unobserved responses. Thus, conditionally, we propose the four predictors defined by

$$\hat{T}_G = N\bar{Y}_{0s} + (N-n)(\bar{X}_{0r} - \bar{X}_{0s})\tilde{\nu}_{1(G)}$$

where $\tilde{\nu}_{1(G)}$ is any estimator of ν_1. However, the OLS of ν_1 is given by

$$\tilde{\nu}_{1(OLS)} = S_{X_0X_0}^{-1} S_{X_0Y_0}$$

with $S_{X_0X_0} = \sum_{t\in s}(X_{0t} - \bar{X}_{0s})^2$ and $S_{X_0Y_0} = \sum_{t\in s}(X_{0t} - \bar{X}_{0s})(Y_{0t} - \bar{Y}_{0s})$. We also consider test-statistics for testing $H_0 : \nu_1 = 0$ v.s. $H_a : \nu_1 \neq 0$ and use the conditional test

$$F = \frac{S_{X_0X_0}\tilde{\nu}_{1(OLS)}}{\tilde{\sigma}_{zz}}, \; \tilde{\sigma}_{zz} = \frac{1}{m}\sum\{(Y_{0t} - \bar{Y}_{0s}) - \tilde{\nu}_{1(OLS)}(X_{0t} - \bar{X}_{0s})\}^2$$

with $m = n - 2$. Conditionally the distribution of F under H_0 is

$$\frac{S_{X_0X_0}\tilde{\nu}^2_{1(OLS)}}{\tilde{\sigma}_{zz}} \stackrel{\mathcal{D}}{=} F_{1,m}(\triangle^{*2}) \quad m = n-2,$$

where $F_{1,m}(\triangle^{*2})$ follows a non-central F-distribution with (1, m) degree of freedom and non-centrality parameter $\frac{1}{2}\triangle^{*2}$ and $\triangle^{*2}$ defined by

$$\begin{aligned}\triangle^{*2} &= \frac{S_{X_0X_0}{\beta_1}^2\kappa^2_{xx}}{\sigma_{zz}} \frac{S_{X_0X_0}{\beta_1}^2\kappa^2_{xx}}{\sigma_{ee}+{\beta_1}^2\sigma_{xx}(1-\kappa_{xx})}\\ &= \frac{S_{X_0X_0}{\beta_1}^2\kappa^2_{xx}}{\sigma_{ee}+\beta_1^2\sigma_{uu}\kappa_{xx}} = \frac{\triangle^2\kappa^2_{xx}}{1+\beta_1^2\delta\kappa_{xx}} \le \triangle^2 = \frac{S_{X_0X_0}\beta_1^2}{\sigma_{ee}},\end{aligned}$$

where $\delta = \frac{\sigma_{uu}}{\sigma_{ee}}$. Thus, we may write the predictors $\hat{T}^{UP}_{OLS}$, $\hat{T}^{RP}_{OLS}$, $\hat{T}^{PTP}_{OLS}$ and $\hat{T}^{SP}_{OLS}$ as

$$\begin{aligned}&(i)\ \hat{T}^{UP}_{OLS} = N\bar{Y}_{0s} + (N-n)(\bar{X}_{0r}-\bar{X}_{0s})\tilde{\nu}_{1(OLS)}\\ &(ii)\ \hat{T}^{RP}_{OLS} = N\bar{Y}_{0s}\\ &(iii)\ \hat{T}^{PTP}_{OLS} = N\bar{Y}_{0s} + (N-n)(\bar{X}_{0r}-\bar{X}_{0s})\tilde{\nu}^{PT}_{1(OLS)}\\ &\quad\text{and}\\ &(iv)\ \hat{T}^{SP}_{OLS} = N\bar{Y}_{0s} + (N-n)(\bar{X}_{0r}-\bar{X}_{0s})\tilde{\nu}^{SE}_{1(OLS)},\end{aligned}$$

respectively. The bias and the MSE expressions may be obtained using the following expressions:

$$\begin{aligned}&(i)\ \hat{T}^{UP}_{OLS} - T = (N-n)[(\bar{X}_{0r}-\bar{X}_{0s})(\tilde{\nu}_{1(OLS)}-\nu_1)+(Z_s-Z_r)]\\ &(ii)\ \hat{T}^{RP}_{OLS} - T = (N-n)[(\bar{X}_{0r}-\bar{X}_{0s})\nu_1+(\bar{Z}_s-\bar{Z}_r)]\\ &(iii)\ \hat{T}^{PTP}_{OLS} - T = (N-n)[(\bar{X}_{0r}-\bar{X}_{0s})(\tilde{\nu}^{PT}_{1(OLS)}-\nu_1)+(\bar{Z}_s-\bar{Z}_r)]\\ &\quad\text{and}\\ &(iv)\ \hat{T}^{SP}_{OLS} - T = (N-n)[(\bar{X}_{0r}-\bar{X}_{0s})(\hat{\nu}^{SE}_{1(OLS)}-\nu_1)+(\bar{Z}_s-\bar{Z}_r)],\end{aligned}$$

respectively.

Next, we compute the unconditional bias and MSE expressions for the predictors. The conditional expressions are similar to the expression given by (3) and (4) with some modifications. We thus present the unconditional expression for bias and MSE's. It is easy to see that

$$b_1(\hat{T}^{UP}_{OLS}) = b_2(\hat{T}^{RP}_{OLS}) = b_3(\hat{T}^{PTP}_{OLS}) = b_4(\hat{T}^{SP}_{OLS}) = 0.$$

Consider the expressions for MSE's. First, note that $S_{X_0X_0}\sigma^{-1}_{X_0X_0}$ follows the chi-squared distribution with $(n-2)$ d.f. while $Nf(1-$

$f)\sigma^{-1}_{X_0X_0}(\bar{X}_{0r}-\bar{X}_{0s})^2$ follows chi-squared distribution with one d.f. independent of $S_{X_0X_0}\sigma^{-1}_{X_0X_0}$. Therefore,

$$\left\{1+Nf(1-f)E\left[\frac{(\bar{X}_r-\bar{X}_s)^2}{S_{X_0X_0}}\right]\right\}=\frac{n-2}{n-3}.$$

This yields the result

$$M_1(\hat{T}^{UP}_{OLS})=N\frac{1}{f}(1-f)\frac{n-2}{n-3}\sigma_{zz},$$

while

$$M_1(\hat{T}^{RP}_{OLS})=N\frac{1}{f}\left[1+\frac{1}{n-3}\triangle^{*2}_u\right]\sigma_{zz},\ \triangle^{*2}_u=\frac{\sigma_{X_0X_0}\beta_1^2\kappa^2_{xx}}{\sigma_{zz}}.$$

Hence, efficiency of $\hat{T}^{RP}_{OLS}$ compared to $\hat{T}^{UP}_{OLS}$ is

$$E^*(\hat{T}^{RP}_{OLS}:\hat{T}^{RP}_{OLS})=\left(\frac{n-2}{n-3}\right)\left[1+\frac{1}{n-3}\triangle^{*2}_u\right]^{-1}.$$

Under $H_0:\beta_1=0$, the efficiency reduces to $\left(\frac{n-2}{n-3}\right)$ (≥ 1) then increase to zero, as $\triangle^{*2}_u$ increases crossing the 1-line. The crossing point on the 1-line is $\triangle^{*2}_u=1$. Thus, $\hat{T}^{RP}_{OLS}$ is better than $\hat{T}^{UP}_{OLS}$ in the range $[0,1)$ and outside this range $\hat{T}^{UP}_{OLS}$ is better than $\hat{T}^{RP}_{OLS}$.

Now we find that

$$M_3(\hat{T}^{PTP}_{OLS})=N\frac{1}{f}(1-f)\sigma_{zz}\left[\frac{n-2}{n-3}-G^{(2)}_{3,m}\left(\frac{1}{3}F_\alpha:\triangle^{*2}_u\right)\right.$$
$$\left.+\triangle^{*2}_u\left\{2G^{(1)}_{3,m}\left(\frac{1}{3}F_\alpha:\triangle^{*2}_u\right)-G^{(1)}_{5,m}\left(\frac{1}{5}F_\alpha:\triangle^{*2}_u\right)\right\}\right]$$

where

$$G^{(j)}_{1+2i,m}\left(\frac{1}{1+2i}F_\alpha:\triangle^{*2}_u\right)$$
$$=\sum_{r\geq 0}\frac{2^{j-2}\Gamma(\frac{n-1}{2}+r+j-2)(\triangle^{*2}_u)^r}{\Gamma(r+1)\Gamma(\frac{n-1}{2})(1+\triangle^{*2}_u)^{\frac{n-1}{2}+r+j-2}}I_y\left(\frac{1+2i}{2}+r:\frac{m}{2}\right),$$

$j=1,2$, $y=\frac{F_\alpha}{n-1+F_\alpha}$ and $\triangle^{*2}_u=\frac{\beta_1^2\sigma_{XX}\kappa^2_{xx}}{\sigma_{zz}}$ and $G^{(j)}_{1,m}(x;\triangle^{*2}_u), j=1,2$ is the unconditional cdf of the test statistic under H_A.

Now we compare $\hat{T}^{PTP}_{OLS}$ relative to $\hat{T}^{UP}_{OLS}$. The efficiency expression is given by

$$E^*(\hat{T}_{OLS}^{PTP}:\hat{T}_{OLS}^{UP}) = \Big[1+\frac{n-3}{n-2}\Big\{\triangle_u^{*2}\Big(2G_{3,m}^{(1)}\Big(\frac{1}{3}F_\alpha:\triangle_u^{*2}\Big) - G_{5,m}^{(1)}\Big(\frac{1}{5}F_\alpha:\triangle_u^{*2}\Big)\Big) - G_{3,m}^{(2)}\Big(\frac{1}{3}F_\alpha:\triangle_u^{*2}\Big)\Big\}\Big]^{-1}.$$

Note that $E^*(\hat{T}_{OLS}^{PTP}:\hat{T}_{OLS}^{UP})$ is a function of $(\alpha:\triangle_u^{*2})$ and for fixed α $(0<\alpha<1)$ it decreases crossing the 1-line to a minimum, then increases towards the 1-line as $\triangle^2$ tends to infinity. Further,

(i) $E^*(\hat{T}_{OLS}^{PTP}:\hat{T}_{OLS}^{UP}) \gtreqless 1$ according as

$$\triangle_u^{*2} \lesseqgtr \frac{G_{3,m}^{(2)}(\frac{1}{3}F_\alpha;\triangle_u^{*2})}{2G_{3,m}^{(1)}(\frac{1}{3}F_\alpha;\triangle_u^{*2}) - G_{5,m}^{(1)}(\frac{1}{5}F_\alpha;\triangle_u^{*2})} = K_{OLS}^*(\alpha:\triangle_u^{*2})(\leq 1).$$

Hence $\hat{T}_{OLS}^{PTP}$ performs better than $\hat{T}_{OLS}^{UP}$ if $\triangle_u^{*2} < K_{OLS}^*(\alpha:\triangle_u^{*2})$.

(ii) Maximum attainable efficiency for fixed α is attained for $\triangle_u^{*2}=0$ and equals

$$\Big\{1-\frac{n-3}{n-2}G_{3,m}^{(2)}\Big(\frac{1}{3}F_\alpha;0\Big)\Big\}^{-1}$$

and as $\triangle_u^{*2}\to\infty$, the efficiency goes to unity.

(iii) PTP can not be uniformly better that $\hat{T}_{OLS}^{UP}$ but one may determine a PTP with minimum guaranteed efficiency, say, E_0 by choosing α^* by the maximin rule

$$\max_{\alpha}\min_{\triangle_u^{*2}} E(\alpha:\triangle_u^{*2}) \geq E_0.$$

The efficiency with this α^* goes to (8) when ν_1 is close to 0.

(iv) As $\alpha\to 1$, $\hat{T}_{OLS}^{PT}\to\hat{T}_{OLS}^{UP}$.

Figure 2 displays the graphs of the efficiency of $\hat{T}_{OLS}^{PTP}$ relative to $\hat{T}_{OLS}^{UP}$. Table 1 provides the minimum (E_{min}^*) and maximum (E_{max}^*) relative efficiency of using the PTP estimator relative to using UP estimator of the slope parameter for varying values of α and $\triangle_u^{*2}$-value at which the minimum relative efficiency occur. To choose an optimum level of significance α^* for given fixed efficiency $E^*=E_0^*$ (say), one goes through the tabular values of efficiency and α, then choose α^* corresponding the closest efficiency to E_0^*. This value α^* guarantees the minimum value of efficiency E^* which may increase if H_0 is true. For example, if the experimenter wants to have an OLS predictor which has relative efficiency no less than 0.9231, then using Table 1, one would use $\alpha^*=0.2$ because it maximizes $E^*(\alpha,0)$ and the maximum attainable relative efficiency is 1.0717.

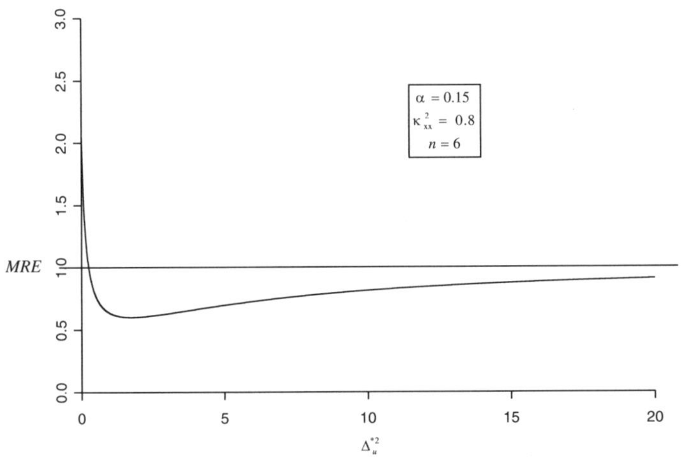

Fig. 2. Graph of $E^*(\hat{T}_{OLS}^{PTP} : \hat{T}_{OLS}^{UP})$

Table 1. Maximum and minimum efficiency of PTP relative to UP of β_1 for various values of α when $n = 8$ and $\kappa_{xx}^2 = 0.8$

α	0.05	0.1	0.15	0.2	0.25	0.3	0.35	0.4	0.45	0.5
E_{max}	1.1486	1.1155	1.0909	1.0717	1.0565	1.0442	1.0342	1.0262	1.0197	1.0145
E_{min}	0.7824	0.8563	0.8966	0.9231	0.9420	0.9561	0.9668	0.9752	0.9817	0.9867
$\triangle_{min}^2$	3.2157	2.5261	2.1900	1.9806	1.8348	1.7266	1.6429	1.5764	1.5227	1.4786

Now, we consider the MSE expressions for $\hat{T}_{OLS}^{SP}$ given by

$$M_4(\hat{T}_{OLS}^{SP}) = N\frac{1}{f}(1-f)\sigma_{zz}\Big[\frac{n-2}{n-3} - \frac{1}{n-3}\frac{2}{\pi}K_n^2\Big\{\frac{2}{(\triangle_u^{*2}+1)^{(n-1)/2-1}} - 1\Big\}\Big]$$

where $\triangle_u^{*2} = \frac{\beta_1^2 \sigma_{X_0 X_0} \kappa_{xx}^2}{\sigma_{zz}}$. Then the unconditional efficiency expression of $\hat{T}_{OLS}^{SP}$ compared to $\hat{T}_{OLS}^{UP}$ is given by

$$E^*(\hat{T}_{OLS}^{SP} : \hat{T}_{OLS}^{UP}) = \Big[1 - \frac{1}{n-2}\frac{2}{\pi}K_n^2\Big\{\frac{2}{(\triangle_u^{*2}+1)^{(n-1)/2-1}} - 1\Big\}\Big]^{-1}.$$

Note that the efficiency is a decreasing function of $\triangle_u^{*2}$. In general, $E^*(\hat{T}_{OLS}^{SP} : \hat{T}_{OLS}^{UP})$ decreases from $\left[1 - \frac{2}{\pi}\frac{1}{n-2}K_n^2\right]^{-1}$ at $\triangle_u^{*2} = 0$ and crosses the 1-line at $\triangle_u^{*2} = \sqrt[\frac{n-3}{2}]{2} - 1$ then drops to the minimum value $\left[1 + \frac{2}{\pi}\frac{1}{n-2}K_n^2\right]^{-1} (\leq 1)$ as $\triangle_u^{*2} \to \infty$. The loss of efficiency is $1 - \left[1 + \frac{2}{\pi}K_n^2\right]^{-1}$ while the gain in efficiency is $\left[1 - \frac{2}{\pi}K_n^2\right]^{-1}$. Thus, for $0 \leq \triangle_u^{*2} \leq \left(\sqrt[\frac{n-3}{2}]{2} - 1\right)$, $\hat{T}_{OLS}^{SP}$ performs better than $\hat{T}_{OLS}^{UP}$, otherwise $\hat{T}_{OLS}^{UP}$ performs better outside this interval.

Acknowledgment

The research is supported by NSERC grant of the second author.

References

Ahsanullah M, Saleh AKMdE (1972) Estimation of Intercept in A Linear Regression Model with One Dependent Variable after A Preliminary Test on The Regression Coefficient. International Statistical Review 40:139–145

Ahmed SE, Saleh AKMdE (1988) Estimation strategy using preliminary test in some normal models. Soochow Journal of Mathematics 14:135–165

Bolfarine H (1991) Finite population prediction. Canadian Journal of Statistics 19: 191–208.

Bolfarine H, Zacks S (1991) Prediction Theory for Finite Population. Springer, New York

Cheng CL, Van Ness JW (1999) Statistical Regression With Measurement Errors. Arnold Publishers, London

Fuller WA (1975) Regression Analysis from Sample Survey. Sankhya Ser. C 37: 117–132

Fuller WA (1987) Measurement Error Models. John Wiley, New York

Kim HM, Saleh AKMdE (2002) Preliminary Test Estimation of the Parameters of Simple Linear Model with Measurement Error. Metrika 57:223–251

Kim HM, Saleh AKMdE (2002) Preliminary Test Prediction of Population Total under Regression Models with Measurement Error. Pakistan Journal of Statistics 18(3): 335 –357

Kim HM, Saleh AKMdE (2003) Improved Estimation of Regression Parameters in Measurement Error Models: Finite Sample Case. Calcutta Statistical Association Bulletin 51:215–216

Kim HM, Saleh AKMdE (2005) Improved Estimation of Regression Parameters in Measurement Error Models. Journal of Multivariate Analysis 95(2): 273–300

Lindley DV (1966) Discussion on a generalized least squares approach to linear functional relationship. Journal of Royal Statistical Society 28:279–297

Lindley DV, Sayad G (1968) The Bayesian estimation of a linear functional relationship. Journal of Royal Statistical Society Serial B 30:190–202

Reilly P, Patino-Leal H (1968) The Bayesian study of the error in variable models Technometries. 28(3): 231–241

Saleh AKMdE (2006) Theory of Preliminary Test and Stein-Type Estimation with Applications. John Wiley, New York

Sprent P (1966) A generalized least squares approach to linear functional relationships. Journal of Royal Statistical Society 28:279–297

Zellner A (1971) An Introduction to Bayesian Inference in Econometrics. John Wiley, New York

The Vector Cross Product and 4×4 Skew-symmetric Matrices

Götz Trenkler[1] and Dietrich Trenkler[2]

[1] Department of Statistics, University of Dortmund, Vogelpothsweg 87, 44221 Dortmund, Germany trenkler@statistik.uni-dortmund.de
[2] University of Osnabrück, Rolandstrasse 8, 49069 Osnabrück, Germany d.trenkler@nts6.oec.uni-osnabrueck.de

1 Introduction

Let us consider 4×4 skew-symmetric matrices $\mathbf{M}$ which in general can be written as

$$\mathbf{M} = \begin{pmatrix} 0 & -a_3 & a_2 & b_1 \\ a_3 & 0 & -a_1 & b_2 \\ -a_2 & a_1 & 0 & b_3 \\ -b_1 & -b_2 & -b_3 & 0 \end{pmatrix}$$

with entries $a_1, a_2, a_3, b_1, b_2, b_3$ being real numbers.

These matrices have become important in robotics (see Selig (1996) or Murray, Li and Sastry (1994)). Furthermore, when letting $b_1 = -a_1$, $b_2 = -a_2$ and $b_3 = -a_3$, the matrix $\mathbf{M} + a_0\mathbf{I}_4$, where a_0 is a real scalar and $\mathbf{I}_n$ is the $n \times n$ identity matrix, can be used to define left multiplication of quaternions (see Altmann (2003) or Groß, Trenkler and Troschke (2001)).

Observe that $\mathbf{M}$ can be written in the partitioned form

$$\mathbf{M} = \begin{pmatrix} \mathbf{T_a} & \mathbf{b} \\ -\mathbf{b}' & 0 \end{pmatrix}, \tag{1}$$

where $\mathbf{b} = (b_1, b_2, b_3)'$ and the 3×3 skew-symmetric matrix

$$\mathbf{T_a} = \begin{pmatrix} 0 & -a_3 & a_2 \\ a_3 & 0 & -a_1 \\ -a_2 & a_1 & 0 \end{pmatrix} \tag{2}$$

corresponds to the vector $\mathbf{a} = (a_1, a_2, a_3)'$.

In the following we analyze the matrices given in (1) on the basis of the facts concerning the matrix $\mathbf{T_a}$, where use will be made of properties of partitioned matrices.

2 The Vector Cross Product

It is well-known that the matrix $\mathbf{T_a}$ in (2) can be used to define the vector cross product in $\mathbb{R}^3$ (see Rao and Mitra (1969) or Room (1952)). In fact

$$\mathbf{T_a x} = \mathbf{a} \times \mathbf{x}$$

for any vector $\mathbf{x}$ from $\mathbb{R}^3$. We report here some properties of $\mathbf{T_a}$ (see e.g. Trenkler (1998) or Trenkler (2001)). For $\mathbf{a}, \mathbf{b} \in \mathbb{R}^3$ we have:

(i) For $\alpha, \beta \in \mathbb{R}$ it holds that $\mathbf{T}_{\alpha\mathbf{a}+\beta\mathbf{b}} = \alpha\mathbf{T_a} + \beta\mathbf{T_b}$
(ii) $\mathbf{T_a b} = -\mathbf{T_b a}$
(iii) $\mathbf{T_a} = -\mathbf{T_a'}$
(iv) $\mathbf{T_a a} = \mathbf{0}$
(v) $\mathbf{T_a T_b} = \mathbf{ba'} - \mathbf{a'bI_3}$
(vi) $\mathbf{T_a T_a T_a} = -(\mathbf{a'a})\mathbf{T_a}$.

For an $m \times n$ matrix $\mathbf{A}$ the Moore-Penrose inverse (MP-inverse) of $\mathbf{A}$ is the unique matrix $\mathbf{A}^\dagger$ simultaneously satisfying the conditions

$$\mathbf{AA^\dagger A} = \mathbf{A},\ \mathbf{A^\dagger AA^\dagger} = \mathbf{A^\dagger},\ (\mathbf{A^\dagger A})' = \mathbf{A^\dagger A},\ (\mathbf{AA^\dagger})' = \mathbf{AA^\dagger},$$

(see Ben-Israel and Greville (2003)). From (v) and (vi) it easily follows that

(vii) $\mathbf{T_a^\dagger} = -(1/\mathbf{a'a})\mathbf{T_a}$, provided that $\mathbf{a} \neq \mathbf{0}$.

3 Eigenvalues and Inverses of 4×4 Skew-symmetric Matrices

As demonstrated in section 1, a 4×4 skew-symmetric matrix can be written in the form

$$\mathbf{M} = \begin{pmatrix} \mathbf{T_a} & \mathbf{b} \\ -\mathbf{b}' & 0 \end{pmatrix} \tag{3}$$

for some vectors $\mathbf{a}, \mathbf{b} \in \mathbb{R}^3$. Let us first calculate the determinant of $\mathbf{M}$. To avoid trivial cases we assume that $\mathbf{a}$ and $\mathbf{b}$ are different from the zero vector.

Theorem 1. *The determinant of the skew-symmetric matrix* $\mathbf{M}$ *given in* (3) *is*

$$\det(\mathbf{M}) = (\mathbf{a}'\mathbf{b})^2. \tag{4}$$

Proof:

According to a well-known result concerning the determinant of bordered matrices (see e.g. Meyer (2000, p. 485)) it follows that

$$\det(\mathbf{M}) = \mathbf{b}'\mathbf{T}^{\#}_{\mathbf{a}}\mathbf{b},$$

where $\mathbf{T}^{\#}_{\mathbf{a}}$ is the adjugate of $\mathbf{T_a}$, i.e., the transpose of the matrix of cofactors of $\mathbf{T_a}$. Some straightforward calculations show that $\mathbf{T}^{\#}_{\mathbf{a}} = \mathbf{a}\mathbf{a}'$. Hence $\det(\mathbf{M}) = \mathbf{b}'\mathbf{a}\mathbf{a}'\mathbf{b} = (\mathbf{a}'\mathbf{b})^2$.
This completes the proof.

Let us now proceed to the problem of determination of the eigenvalues of $\mathbf{M}$.

Theorem 2. *The characteristic polynomial of* $\mathbf{M}$ *is*

$$P_{\mathbf{M}}(\lambda) = \lambda^4 + \lambda^2(\mathbf{a}'\mathbf{a} + \mathbf{b}'\mathbf{b}) + (\mathbf{a}'\mathbf{b})^2. \tag{5}$$

Proof:

The characteristic polynomial of $\mathbf{M}$ is

$$P_{\mathbf{M}}(\lambda) = \det(\mathbf{M} - \lambda \mathbf{I}_4) = \det \begin{pmatrix} \mathbf{T_a} - \lambda \mathbf{I}_3 & \mathbf{b} \\ -\mathbf{b}' & -\lambda \end{pmatrix}.$$

If $\lambda = 0$, then $P_{\mathbf{M}}(\lambda) = \det(\mathbf{M}) = (\mathbf{a}'\mathbf{b})^2$. Suppose that $\lambda \neq 0$. Then $\mathbf{T_a} - \lambda \mathbf{I}_3$ is nonsingular and

$$(\mathbf{T_a} - \lambda \mathbf{I}_3)^{-1} = -\frac{1}{\mathbf{a}'\mathbf{a} + \lambda^2} \left(\mathbf{T_a} + \lambda \mathbf{I}_3 + \frac{1}{\lambda} \mathbf{a}\mathbf{a}' \right). \qquad (6)$$

By a well-known result concerning the determinant of bordered matrices (see Meyer (2000, p. 475)) we obtain

$$\det \begin{pmatrix} \mathbf{T_a} - \lambda \mathbf{I}_3 & \mathbf{b} \\ \mathbf{b}' & -\lambda \end{pmatrix} = \det(\mathbf{T_a} - \lambda \mathbf{I}_3) \left(-\lambda + \mathbf{b}'(\mathbf{T_a} - \lambda \mathbf{I}_3)^{-1} \mathbf{b} \right).$$

Using (6) and the identity $\det(\mathbf{T}_a - \lambda \mathbf{I}_3) = -\lambda^3 - \lambda \mathbf{a}'\mathbf{a}$, after some easy calculations we arrive at (5).
This completes the proof.

To obtain the eigenvalues of $\mathbf{M}$, put $\lambda^2 = x$. Then $P_{\mathbf{M}}(\lambda) = 0$ is equivalent to the equation

$$x^2 + x(\mathbf{a}'\mathbf{a} + \mathbf{b}'\mathbf{b}) + (\mathbf{a}'\mathbf{b})^2 = 0,$$

which has the two solutions

$$x_{1,2} = -\frac{\mathbf{a}'\mathbf{a} + \mathbf{b}'\mathbf{b}}{2} \pm \frac{\sqrt{(\mathbf{a}'\mathbf{a})^2 + (\mathbf{b}'\mathbf{b})^2 + 2(\mathbf{a}'\mathbf{a})(\mathbf{b}'\mathbf{b}) - 4(\mathbf{a}'\mathbf{b})^2}}{2}.$$

Since $x_1 + x_2 = -\mathbf{a}'\mathbf{a} - \mathbf{b}'\mathbf{b}$ and $x_1 x_2 = (\mathbf{a}'\mathbf{b})^2$, by Vieta's law, we see that $x_{1,2} \leq 0$. Thus the four eigenvalues $\lambda_{1,2} = \pm\sqrt{x_1}$ and $\lambda_{3,4} = \pm\sqrt{x_2}$ are purely imaginary which is clear since $\mathbf{M}$ is skew-symmetric.

As a side effect, using Gerschgorin's Theorem (see Meyer (2000, p. 498)) we find the inequality

$$\|\mathbf{a}+\mathbf{b}\| \leq \max\{|a_2|+|a_3|+|b_1|, |a_1|+|a_3|+|b_2|, |a_1|+|a_2|+|b_3|, |b_1|+|b_2|+|b_3|\}$$

for orthogonal vectors $\mathbf{a}$ and $\mathbf{b}$. Here $\|\cdot\|$ denotes the common Euclidean norm. Setting $\mathbf{a} = (1, -1, 0)'$ and $\mathbf{b} = (1, 1, 1)'$ one sees that this upper bound can be sharper than $\|\mathbf{a}\| + \|\mathbf{b}\|$, known from the triangle inequality.

Let us now proceed to the determination of the inverse or the MP-inverse of $\mathbf{M}$, depending on the determinant of $\mathbf{M}$. For this purpose we introduce the following quantities (see Meyer (1972)):

$$\mathbf{k} = \mathbf{T}_{\mathbf{a}}^{\dagger}\mathbf{b} = \frac{1}{\mathbf{a}'\mathbf{a}}\mathbf{T}_{\mathbf{a}}'\mathbf{b} = -\frac{1}{\mathbf{a}'\mathbf{a}}\mathbf{T}_{\mathbf{a}}\mathbf{b},$$

$$\mathbf{h} = (\mathbf{T}_{\mathbf{a}}^{\dagger})'(-\mathbf{b}) = \mathbf{T}_{\mathbf{a}}^{\dagger}\mathbf{b} = \mathbf{k},$$

$$\omega_1 = 1 + \mathbf{k}'\mathbf{k} = 1 + \left(\frac{1}{\mathbf{a}'\mathbf{a}}\right)^2 \mathbf{b}'\mathbf{T}_{\mathbf{a}}\mathbf{T}_{\mathbf{a}}'\mathbf{b} = 1 + \frac{(\mathbf{a}'\mathbf{a})(\mathbf{b}'\mathbf{b}) - (\mathbf{a}'\mathbf{b})^2}{(\mathbf{a}'\mathbf{a})^2},$$

$$\omega_2 = 1 + \mathbf{h}'\mathbf{h} = \omega_1,$$

$$\beta = -(-\mathbf{b}')\mathbf{T}_{\mathbf{a}}^{\dagger}\mathbf{b} = 0 \qquad \text{(since } \mathbf{T}_{\mathbf{a}}^{\dagger} \text{ is skew-symmetric)},$$

$$\mathbf{u} = (\mathbf{I}_3 - \mathbf{T}_{\mathbf{a}}\mathbf{T}_{\mathbf{a}}^{\dagger})\mathbf{b} = \mathbf{a}\mathbf{a}^{\dagger}\mathbf{b} = \frac{\mathbf{a}'\mathbf{b}}{\mathbf{a}'\mathbf{a}}\mathbf{a},$$

$$\mathbf{v} = (\mathbf{I}_3 - \mathbf{T}_{\mathbf{a}}^{\dagger}\mathbf{T}_{\mathbf{a}})(-\mathbf{b}) = -\mathbf{u}.$$

Case 1: $\det(\mathbf{M}) \neq 0$.
Then by 4 we have $\mathbf{a}'\mathbf{b} \neq 0$, and thus $\mathbf{u} \neq \mathbf{0}$. Since $\mathbf{b}$ does not belong to the column space of $\mathbf{T}_{\mathbf{a}}$ by $\mathbf{a}'\mathbf{b} \neq 0$, we have $\text{rk}(\mathbf{M}) = \text{rk}(\mathbf{T}_{\mathbf{a}}) + 2$ (see Meyer (1972)). On the other hand, the matrix $\mathbf{T}_{\mathbf{a}}$ is of rank 2, confirming the nonsingularity of $\mathbf{M}$. By condition (i) of the Theorem in Meyer (1972), we obtain

$$\mathbf{M}^{-1} = \begin{pmatrix} \mathbf{T}_{\mathbf{a}}^{\dagger} - \mathbf{k}\mathbf{u}^{\dagger} - \mathbf{v}'^{\dagger}\mathbf{h}' & \mathbf{v}'^{\dagger} \\ \mathbf{u}^{\dagger} & 0 \end{pmatrix}. \tag{7}$$

It is easy to check that

$$\mathbf{u}^{\dagger} = \frac{1}{\mathbf{a}'\mathbf{b}}\mathbf{a}', \qquad \mathbf{k}\mathbf{u}^{\dagger} = -\frac{1}{(\mathbf{a}'\mathbf{a})(\mathbf{a}'\mathbf{b})}\mathbf{T}_{\mathbf{a}}\mathbf{b}\mathbf{a}',$$

$$\mathbf{v}'^{\dagger} = -\frac{1}{\mathbf{a}'\mathbf{b}}\mathbf{a}, \qquad \mathbf{v}'^{\dagger}\mathbf{h}' = -\frac{1}{(\mathbf{a}'\mathbf{a})(\mathbf{a}'\mathbf{b})}\mathbf{a}\mathbf{b}'\mathbf{T}_{\mathbf{a}}.$$

Using these expressions, from (7) it follows that

$$\mathbf{M}^{-1} = \frac{1}{\mathbf{a}'\mathbf{b}} \begin{pmatrix} (\mathbf{a}'\mathbf{b})\mathbf{T}_{\mathbf{a}}^{\dagger} + \frac{1}{\mathbf{a}'\mathbf{a}}(\mathbf{T}_{\mathbf{a}}\mathbf{b}\mathbf{a}' + \mathbf{a}\mathbf{b}'\mathbf{T}_{\mathbf{a}}) & -\mathbf{a} \\ \mathbf{a}' & 0 \end{pmatrix}.$$

Now for vectors $\mathbf{f}$ and $\mathbf{g}$ we have the rule (see Trenkler (2001)):

$$\mathbf{g}\mathbf{f}' - \mathbf{f}\mathbf{g}' = \mathbf{T}_{\mathbf{T}_{\mathbf{f}}\mathbf{g}}. \tag{8}$$

Using this we get

$$\mathbf{T}_{\mathbf{a}}\mathbf{b}\mathbf{a}' + \mathbf{a}\mathbf{b}'\mathbf{T}_{\mathbf{a}} = \mathbf{T}_{\mathbf{a}}\mathbf{b}\mathbf{a}' - \mathbf{a}\mathbf{b}'\mathbf{T}_{\mathbf{a}}' = (\mathbf{a}'\mathbf{b})\mathbf{T}_{\mathbf{a}} - (\mathbf{a}'\mathbf{a})\mathbf{T}_{\mathbf{b}},$$

and $\mathbf{M}^{-1}$ becomes

$$\mathbf{M}^{-1} = -\frac{1}{\mathbf{a}'\mathbf{b}} \begin{pmatrix} \mathbf{T}_{\mathbf{b}} & \mathbf{a} \\ -\mathbf{a}' & 0 \end{pmatrix}.$$

This result is not surprising since the adjugate of $\mathbf{M}$ is

$$\mathbf{M}^{\#} = -\mathbf{a}'\mathbf{b} \begin{pmatrix} \mathbf{T}_{\mathbf{b}} & \mathbf{a} \\ -\mathbf{a}' & 0 \end{pmatrix},$$

and hence $\mathbf{M}^{-1} = \mathbf{M}^{\#}/\det(\mathbf{M})$.

Case 2: $\det(\mathbf{M}) = 0$.
Then again $\mathbf{k} = -\frac{1}{\mathbf{a}'\mathbf{a}}\mathbf{T}_{\mathbf{a}}\mathbf{b} = \mathbf{h}$, but ω_1 simplifies to $\omega_1 = \omega_2 = 1 + \frac{\mathbf{b}'\mathbf{b}}{\mathbf{a}'\mathbf{a}}$. Furthermore we obtain $\mathbf{u} = \mathbf{0}$ and $\mathbf{v} = \mathbf{0}$. Since $\mathbf{a}'\mathbf{b} = 0$, we have $\mathbf{b} \in \mathfrak{N}(\mathbf{a}') = \mathfrak{R}(\mathbf{T}_{\mathbf{a}})$, where $\mathfrak{N}(\cdot)$ and $\mathfrak{R}(\cdot)$ denote the null space and the column space of a matrix, respectively. Consequently, case (iii) of the Theorem in Meyer (1972) applies. It follows that

$$\mathbf{M}^{\dagger} = \frac{1}{\omega_1} \begin{pmatrix} \omega_1\mathbf{T}_{\mathbf{a}}^{\dagger} - \mathbf{h}\mathbf{h}'\mathbf{T}_{\mathbf{a}}^{\dagger} - \mathbf{T}_{\mathbf{a}}^{\dagger}\mathbf{h}\mathbf{h}' & \mathbf{T}_{\mathbf{a}}^{\dagger}\mathbf{h} \\ \mathbf{h}'\mathbf{T}_{\mathbf{a}}^{\dagger} & 0 \end{pmatrix}.$$

Straightforward calculations yield $\mathbf{M}^{\dagger}$ in the somewhat simpler form

$$\mathbf{M}^{\dagger} = \frac{1}{\omega_1} \begin{pmatrix} \omega_1\mathbf{T}_{\mathbf{a}}^{\dagger} + \left(\frac{1}{\mathbf{a}'\mathbf{a}}\right)^2 [\mathbf{T}_{\mathbf{a}}\mathbf{b}\mathbf{b}' + \mathbf{b}\mathbf{b}'\mathbf{T}_{\mathbf{a}}] & -\frac{1}{\mathbf{a}'\mathbf{a}}\mathbf{b} \\ \frac{1}{\mathbf{a}'\mathbf{a}}\mathbf{b}' & 0 \end{pmatrix}.$$

Using again rule (8) it turns out that $\mathbf{T_a bb'} + \mathbf{bb'T_a} = \mathbf{b'bT_a}$, which leads to

$$\mathbf{M}^{\dagger} = -\frac{1}{\mathbf{a'a} + \mathbf{b'b}} \begin{pmatrix} \mathbf{T_a} & \mathbf{b} \\ -\mathbf{b'} & 0 \end{pmatrix}.$$

This shows the remarkable fact that the MP-inverse of $\mathbf{M}$ is a scalar multiple of itself, sharing this property with $\mathbf{T_a}$.

Furthermore we get

$$\mathbf{MM}^{\dagger} = \frac{1}{\omega_1} \begin{pmatrix} \omega_1 \mathbf{T_a T_a^{\dagger}} - \mathbf{hh'} & \mathbf{h} \\ \mathbf{h'} & \mathbf{h'h} \end{pmatrix}$$

$$= -\frac{1}{\mathbf{a'a} + \mathbf{b'b}} \begin{pmatrix} \mathbf{T_a^2} - \mathbf{bb'} & \mathbf{T_a b} \\ -\mathbf{b'T_a} & -\mathbf{b'b} \end{pmatrix}.$$

Observe that since $\mathbf{M}$ is normal we can state that $\mathbf{M}^{\dagger}\mathbf{M} = \mathbf{MM}^{\dagger}$.

It is interesting to note that a simple generalized inverse of $\mathbf{M}$ is given by

$$\mathbf{M}^{-} = \begin{pmatrix} \mathbf{T_a^{\dagger}} & \mathbf{a} \\ -\mathbf{a'} & 0 \end{pmatrix}.$$

The condition $\mathbf{MM^{-}M} = \mathbf{M}$ can easily be shown.

To summarize our preceding results we get

Theorem 3.

(i) If $\mathbf{a'b} \neq 0$, $\mathbf{M}$ *is nonsingular, and its inverse is*

$$\mathbf{M}^{-1} = -\frac{1}{\mathbf{a'b}} \begin{pmatrix} \mathbf{T_b} & \mathbf{a} \\ -\mathbf{a'} & 0 \end{pmatrix}.$$

(ii) If $\mathbf{a}'\mathbf{b} = 0$*,* $\mathbf{M}$ *is singular, and its MP-inverse is*

$$\mathbf{M}^\dagger = -\frac{1}{\mathbf{a}'\mathbf{a} + \mathbf{b}'\mathbf{b}} \begin{pmatrix} \mathbf{T_a} & \mathbf{b} \\ -\mathbf{b}' & 0 \end{pmatrix},$$

and a generalized inverse of $\mathbf{M}$ *is*

$$\mathbf{M}^- = \begin{pmatrix} \mathbf{T_a}^\dagger & \mathbf{a} \\ -\mathbf{a}' & 0 \end{pmatrix}.$$

4 The Cayley Transform

Finally let us pay some attention to the Cayley transform of $\mathbf{M}$. In general, the Cayley transform of an $n \times n$ matrix $\mathbf{A}$ is given as

$$\mathcal{C}(\mathbf{A}) = (\mathbf{I}_n + \mathbf{A})^{-1}(\mathbf{I}_n - \mathbf{A}),$$

provided $\mathbf{I}_n + \mathbf{A}$ is nonsingular. Observe that by the identity $\mathbf{I}_n - \mathbf{A} = 2\mathbf{I}_n - (\mathbf{I}_n + \mathbf{A})$ we can write

$$\mathcal{C}(\mathbf{A}) = 2(\mathbf{I}_n + \mathbf{A})^{-1} - \mathbf{I}_n.$$

Since the eigenvalues of $\mathbf{M}$ are purely imaginary, sharing this property with any skew-symmetric matrix, it is clear that $\mathbf{I}_4 + \mathbf{M}$ is nonsingular. Hence

$$\mathcal{C}(\mathbf{M}) = 2(\mathbf{I}_4 + \mathbf{M})^{-1} - \mathbf{I}_4. \tag{9}$$

To determine $\mathcal{C}(\mathbf{M})$ in dependence of $\mathbf{T_a}$ and $\mathbf{b}$, we have to get an expression for $(\mathbf{I}_4 + \mathbf{M})^{-1}$. For $\mathbf{I} + \mathbf{M}$ we obtain

$$\mathbf{I}_4 + \mathbf{M} = \begin{pmatrix} \mathbf{T_a} + \mathbf{I}_3 & \mathbf{b} \\ -\mathbf{b}' & 1 \end{pmatrix},$$

whose upper left matrix $\mathbf{T} = \mathbf{T_a} + \mathbf{I}_3$ is nonsingular. By (6) it follows that

$$\mathbf{T}^{-1} = \frac{1}{1 + \mathbf{a}'\mathbf{a}}(-\mathbf{T_a} + \mathbf{I}_3 + \mathbf{a}\mathbf{a}').$$

Let $s = 1 + \mathbf{b}'\mathbf{T}^{-1}\mathbf{b} = (1 + \mathbf{a}'\mathbf{a} + \mathbf{b}'\mathbf{b} + (\mathbf{a}'\mathbf{b})^2)/(1 + \mathbf{a}'\mathbf{a})$ be the Schur complement of $\mathbf{T}$ in $\mathbf{I}_4 + \mathbf{M}$. Then by a well-known formula for the inverse of a partitioned matrix (see Meyer (2000, p. 123)) we obtain

$$(\mathbf{I}_4 + \mathbf{M})^{-1} = \frac{1}{s}\begin{pmatrix} s\mathbf{T}^{-1} - \mathbf{T}^{-1}\mathbf{b}\mathbf{b}'\mathbf{T}^{-1} & -\mathbf{T}^{-1}\mathbf{b} \\ \mathbf{b}'\mathbf{T}^{-1} & 1 \end{pmatrix}.$$

Using the representation (9), we finally arrive at

$$\mathcal{C}(\mathbf{M}) = \begin{pmatrix} \mathcal{C}(\mathbf{T_a}) - \frac{2}{s}\mathbf{T}^{-1}\mathbf{b}\mathbf{b}'\mathbf{T}^{-1} & -\frac{2}{s}\mathbf{T}^{-1}\mathbf{b} \\ \frac{2}{s}\mathbf{b}'\mathbf{T}^{-1} & \frac{2}{s} - 1 \end{pmatrix},$$

where

$$\mathcal{C}(\mathbf{T_a}) = 2(\mathbf{I}_3 - \mathbf{T_a})^{-1} - \mathbf{I}_3 = \frac{1}{1 + \mathbf{a}'\mathbf{a}}(-2\mathbf{T_a} + 2\mathbf{a}\mathbf{a}' + (1 - \mathbf{a}'\mathbf{a})\mathbf{I}_3)$$

is the Cayley transform of $\mathbf{T_a}$. Clearly, $\mathcal{C}(\mathbf{M})$ is an orthogonal matrix.

Let us finally have a look at the Rodrigues' formula . For the matrix $\mathbf{T_a}$ it states that (see Murray, Li and Sastry (1994, p. 28))

$$\exp[\mathbf{T_a}] = \mathbf{I}_3 + \frac{\sin\theta}{\theta}\mathbf{T_a} + \frac{1 - \cos\theta}{\theta^2}\mathbf{T_a}^2,$$

where $\theta = \sqrt{\mathbf{a}'\mathbf{a}}$. Similarly, when $\mathbf{a}'\mathbf{b} = 0$ we obtain

$$\exp[\mathbf{M}] = \mathbf{I}_4 + \frac{\sin\varepsilon}{\varepsilon}\mathbf{M} + \frac{1 - \cos\varepsilon}{\varepsilon^2}\mathbf{M}^2,$$

where $\varepsilon = \sqrt{\mathbf{a}'\mathbf{a} + \mathbf{b}'\mathbf{b}}$. This follows from the identity $\mathbf{M}^3 = -\varepsilon^2\mathbf{M}$ (see Theorem 3, (ii)).

5 Concluding Remark

Starting point of the preceding investigations was the paper by Pryce (1969), where, for instance, the determinant of $\mathbf{M}$ was given. Special attention was paid there to the Cayley transform of $\mathbf{M}$, where however, its entries were not given completely and the chosen approach was somewhat complicated.

References

Altmann SL (2003) Rotations, quaternions, and double groups. Clarendon Press, Oxford

Ben-Israel A, Greville TNE (2003) Generalized inverses: Theory and applications. Springer, New York

Groß J, Trenkler G, Troschke SO (2001) Quaternions: Further contributions to a matrix oriented approach. Linear Algebra and its Applications 326:205–213

Meyer CD (1972) The Moore-Penrose inverse of a bordered matrix. Linear Algebra and its Applications 5:375–382

Meyer CD (2000) Matrix analysis and applied linear algebra. SIAM, Philadelphia

Murray RM, Li Z, Sastry SS (1994) A mathematical introduction to robotic manipulation. CRC Press, Boca Raton

Pryce WJ (1969) Orthogonal matrices. The Mathematical Gazette 53:405–406

Rao CR, Mitra SK (1969) Generalized inverses of matrices and its applications. John Wiley, New York

Room TG (1952) The compositions of rotations in Euclidean three space. The American Mathematical Monthly 59:688–692

Selig JM (1996) Geometrical methods in robotics. Springer, New York

Trenkler G (1998) Vector equations and their solutions. International Journal of Mathematical Education in Science and Technology 29:455–459

Trenkler G (2001) The vector cross product from an algebraic point of view. Discussiones Mathematicae, General Algebra and Applications 21:67–82

Simultaneous Prediction of Actual and Average Values of Response Variable in Replicated Measurement Error Models

Shalabh[1], Chandra Mani Paudel[2] and Narinder Kumar[3]

[1] Department of Mathematics and Statistics, Indian Institute of Technology, Kanpur - 208016, India shalab@iitk.ac.in; shalabh1@yahoo.com
[2] Department of Statistics, Tribhuvan University, P.N. Campus, Pokhra, Nepal cpaudel@hotmail.com
[3] Department of Statistics, Panjab University, Chandigarh - 160014, India nkumar@pu.ac.in

1 Introduction

Prediction is an important aspect of decision-making process through statistical methodology. Linear regression modeling plays an important role in the prediction of an unknown value of study variable corresponding to a known value of explanatory variable. Usually, when the least square estimators are used to construct the predictors, they yield the best linear unbiased predictors provided the data recorded on variables is measured without any error. In practice, many applications fail to meet the assumption of error free observations due to various reasons. For example, due to indirect measurements, practical difficulties, qualitative variables and proxy measurements etc., the measurement error is induced in the data. The usual statistical tools in the context of linear regression analysis like ordinary least squares method then yields biased and inconsistent estimators, see Cheng and Van Ness (1999), Fuller (1987) for more details. Consequently, the predictors obtained through these estimators also then become invalid. Construction of good predictors for measurement error-ridden data and study of their performance properties under measurement error models is attempted in this article.

The problem of prediction in the presence of measurement errors in the data has been considered in the literature in various contexts, see e.g., Ganse, Amemiya, and Fuller (1983), Reilman and Gunst (1986), Schaalje and Butts (1993), Buonaccorsi (1995), Huwang and Hwang

(2002), Nummi and Möttönen (2004). Traditionally, the predictions are obtained either for the average value or the actual value of study variable but not for both simultaneously. In many applications, it may be imperative to obtain both of them simultaneously. For example, suppose a pharmaceutical firm develops a new medicine for pain relief, which claims to increase the number of hours of pain relief. The firm's interest lies in knowing the average number of hours for which the pain is relived after using the medicine. On the other hand, the patient, who is the user of medicine, would be more interested in knowing the actual increase in the number of hours of pain relief. Here the interest of the firm lies in the prediction of average number of pain relief hours whereas patient's interest lies in the actual number of pain relief hours. The query of only one of the persons can be answered with the classical approach of prediction. In order to protect the interest of both, we need to construct a framework which provides the prediction of average and actual values of study variable simultaneously. Shalabh (1995) proposed a target function which enables to predict the actual and average values of study variable simultaneously, see also, Rao, Toutenburg, Shalabh and Heumann (2008), Dube and Manocha (2002), Chaturvedi, Wan and Singh (2002), Toutenburg and Shalabh (1996), Toutenburg and Shalabh (2000), Toutenburg and Shalabh (2002) for the application of target function under different set ups.

In general, the measurement errors are assumed to follow a normal distribution. When measurement errors do not necessarily follow a normal distribution, the statistical inferences do not remain valid. Fuller (1987) and Cheng and Van Ness (1999) have discussed the prediction in measurement error model with the assumption of jointly normal distribution of observations as well as measurement errors. Also, if the true value of variable is not normally distributed, then the introduction of normal measurement error will destroy the linearity of the relation between observations (Kendall and Stuart (1967, p. 48)). Shalabh (1995) studied the issues related with the simultaneous prediction with not necessarily normally distributed errors under an unreplicated ultrastructural model with known variances of measurement errors. Thus, the assumption of normal distribution of measurement errors may become a cause of misleading inference and which may be violated easily in practice. What is the effect of departure from normality of measurement errors on the predictive properties of the predictors is an issue which is addressed in this article.

We have considered an ultrastructural model with replicated observations. The availability of replicated observations is utilized to estimate the measurement error variances and then used to obtain the consistent estimators of regression coefficient. Thus obtained estimators are then used to construct the predictors. The asymptotic properties of these predictors are derived under an ultrastructural model when measurement errors are not necessarily normally distributed.

The organization of this article is as follows. In next Section 2, we discuss the model and the target function of prediction. The predictors are constructed and discussed in Section 3. In Section 4, we derive and analyze the large sample asymptotic performance properties of the predictors in within and outside sample prediction cases. A Monte-Carlo simulation experiment is conducted to study the performance properties of the predictors in finite sample and its findings are reported in Section 5. Some concluding remarks are given in Section 6. The derivations of the results are given in Section 7.

2 The Model and the Target Function

Let the n observations on true values of explanatory variable X_i and study variable Y_i are linearly related as

$$Y_i = \alpha + \beta X_i \quad , \quad (i = 1, 2, \ldots, n) \tag{1}$$

where α is the intercept term and β is the slope parameter. Due to the presence of measurement errors in the observations, we can not observe true X_i and Y_i. Instead, we observe r replicated observations on X_i and Y_i as x_{ij} and y_{ij}, respectively as

$$x_{ij} = X_i + v_{ij} \tag{2}$$

$$y_{ij} = Y_i + u_{ij} \quad , \quad (j = 1, 2, \ldots, r) \tag{3}$$

where v_{ij} and u_{ij} are the measurement errors associated with x_{ij} and y_{ij}, respectively.

Further we assume that X_i has a distribution with mean m_i, so we can express

$$X_i = m_i + w_i \tag{4}$$

where w_i is the random error component associated with X_i. This completes the specification of an ultrastructural model which encompasses the functional and structural forms of measurement error models as its particular cases, see Dolby (1976).

The error terms u_{ij}, v_{ij} and w_i are assumed to be statistically independent of each other and with the true values of variables. The measurement errors u_{ij}'s are assumed to be independent and identically distributed with mean 0, variance σ_u^2, third central moment $\gamma_{1u}\sigma_u^3$ and fourth central moment $(\gamma_{2u}+3)\sigma_u^4$ where $\gamma_{1.}$ and $\gamma_{2.}$ are the Pearson's coefficients of skewness and kurtosis of distributions of errors appearing in the subscript. Similarly, v_{ij}'s are assumed to be independent and identically distributed with mean 0, variance σ_v^2, third central moment $\gamma_{1v}\sigma_v^3$ and fourth central moment $(\gamma_{2v}+3)\sigma_v^4$. The random error components w_i's are assumed to be independent and identically distributed with mean 0, variance σ_w^2 along with third and fourth central moments $\gamma_{1w}\sigma_w^3$ and $(\gamma_{2w}+3)\sigma_w^4$, respectively.

Traditionally, the predictions are obtained either for the actual value of the study variable (y) or for the average value of the study variable $(E(y))$ but not for both simultaneously. A target function, proposed by Shalabh (1995), which provides a framework for the simultaneous prediction of y and $E(y)$ is given as,

$$P = \lambda y + (1-\lambda)E(y) \quad ; \quad 0 \leq \lambda \leq 1 \tag{5}$$

where λ is a nonstochastic scalar which determines the weight to be assigned to the actual value prediction, see also Rao, Toutenburg, Shalabh and Heumann (2008). When $\lambda = 0$, the target function provides the average value prediction and when $\lambda = 1$, the target function provides the actual value prediction. For any other choice of λ between 0 and 1, it provides the simultaneous prediction of actual and average values with weight λ. The choice of weight is governed by considerations like the nature of the problem, social considerations and the preference of the practitioner.

3 Construction of Predictors

We now consider the construction of predictors for within and outside sample predictions.

3.1 Within Sample Prediction

We can have two types of predictions for the kth true value of study variable in within sample under the replicated ultrastructural model.

Prediction of $(k, j)^{th}$ Observation of Study Variable

Suppose we are interested in predicting the jth replicate of kth true value of study variable when the corresponding value of explanatory variable is known and lies within the sample. Let y_{kj} and $E(y_{kj})$ be the actual and average values of study variable to be predicted. Then the target function is

$$P_w = \lambda y_{kj} + (1-\lambda)E(y_{kj}) \quad ; \quad 0 \leq \lambda \leq 1 . \tag{6}$$

The natural linear predictor in this case is

$$\hat{y}_{kj} = \bar{y} + \beta(x_{kj} - \bar{x}) \tag{7}$$

where x_{kj} is the jth observation on k^{th} true value of explanatory variable, $\bar{x}$ and $\bar{y}$ are the sample means of explanatory and study variables, respectively. The slope parameter β is unknown and this restricts the use of this predictor. Further, the direct and reverse regression estimators of β arising by the use of (x_{ij}, y_{ij}) $(i = 1, 2, \ldots, n; j = 1, 2, \ldots, r)$ are $b_1 = S_{xy}/S_{xx}$ and $b_2 = S_{yy}/S_{xy}$, respectively which are inconsistent in the presence of measurement errors where $S_{xy} = \frac{1}{nr}\sum_{i=1}^{n}\sum_{j=1}^{r}(x_{ij}-\bar{x})(y_{ij}-\bar{y})$, $S_{xx} = \frac{1}{nr}\sum_{i=1}^{n}\sum_{j=1}^{r}(x_{ij}-\bar{x})^2$, $S_{yy} = \frac{1}{nr}\sum_{i=1}^{n}\sum_{j=1}^{r}(y_{ij}-\bar{y})^2$, $\bar{x} = \frac{1}{n}\sum_{i=1}^{n}\sum_{j=1}^{r}x_{ij}$ and $\bar{y} = \frac{1}{n}\sum_{i=1}^{n}\sum_{j=1}^{r}y_{ij}$. If we replace β by b_1 and b_2, then the resulting predictor will not be useful. Another alternative strategy is to estimate the measurement error variances σ_u^2 and σ_v^2 using the replicated observations and adjust b_1 and b_2 for their inconsistency. The respective resulting asymptotically unbiased and consistent estimators of β are

$$\begin{aligned} \hat{\beta}_1 &= \frac{S_{xy}}{S_{xx} - \hat{\sigma}_v^2} \quad ; \quad S_{xx} > \hat{\sigma}_v^2 \\ &= \frac{(r-1)S_{xy}}{rB_{xx} - S_{xx}} \end{aligned} \tag{8}$$

and

$$\begin{aligned} \hat{\beta}_2 &= \frac{S_{yy} - \hat{\sigma}_u^2}{S_{xy}} \quad ; \quad S_{yy} > \hat{\sigma}_u^2 \\ &= \frac{rB_{yy} - S_{yy}}{(r-1)S_{xy}}, \end{aligned} \tag{9}$$

respectively, where $\hat{\sigma}_v^2 = \frac{r}{r-1}(S_{xx} - B_{xx})$, $\hat{\sigma}_u^2 = \frac{r}{r-1}(S_{yy} - B_{yy})$, $B_{xx} = \frac{1}{n}\sum_{i=1}^{n}(\bar{x}_i - \bar{x})^2$, $B_{yy} = \frac{1}{n}\sum_{i=1}^{n}(\bar{y}_i - \bar{y})^2$, $\bar{x}_i = \frac{1}{r}\sum_{j}^{r}x_{ij}$ and $\bar{y}_i = \frac{1}{r}\sum_{j=1}^{r}y_{ij}$.

Now, replacing β by its consistent estimators $\hat{\beta}_1$ and $\hat{\beta}_2$ in (7) provides the following predictors for within sample prediction,

$$P_{1wD} = \bar{y} + \hat{\beta}_1(x_{kj} - \bar{x}) \tag{10}$$

and

$$P_{1wR} = \bar{y} + \hat{\beta}_2(x_{kj} - \bar{x}), \tag{11}$$

respectively.

Prediction of Average of r Replicates

Suppose the information on explanatory variable is available in the form of average taken over r replications and we want to predict the corresponding value of study variable which is an average based on r replications. Such situations may occur frequently in many applications. For example, let us consider a case of poultry farm with n sheds of poultry and each shed with r chickens. The chickens are fed with some particular type of feed and the experimenter is interested in predicting their gain in weight. Here it is very difficult to know the exact amount of food which a particular chicken has consumed whereas it is easier to know that how much, on an average, a chicken has consumed. Here the chickens in the kth shed are getting on an average $\bar{x}_k$ amount of food daily where $\bar{x}_k = \frac{1}{r}\sum_{j=1}^{r} x_{kj}$. It makes more sense to predict the gain in weight in the chickens of the kth shed in such cases. Then we have the following linear predictor,

$$\hat{\bar{y}}_k = \bar{y} + \beta(\bar{x}_k - \bar{x}). \tag{12}$$

When β is unknown, then (12) can not be used as the predictor. So we replace β by its consistent estimators.

Use of $\bar{y}_i$ and $\bar{x}_i$ $(i = 1, 2, \ldots, n)$ for the direct and reverse regression procedures yield the estimators $b_3 = B_{xy}/B_{xx}$ and $b_4 = B_{yy}/B_{xy}$, respectively as an estimator of β where $B_{xy} = \frac{1}{n}\sum_{i=1}^{n}(\bar{x}_i - \bar{x})(\bar{y}_i - \bar{y})$. Both b_3 and b_4 are inconsistent for β in the presence of measurement errors. So an adjustment for their inconsistency gives the following asymptotically unbiased and consistent estimators

$$\begin{aligned} \hat{\beta}_3 &= \frac{B_{xy}}{B_{xx} - \frac{\hat{\sigma}_v^2}{r}} \quad ; \quad B_{xx} > \frac{\hat{\sigma}_v^2}{r} \\ &= \frac{(r-1)B_{xy}}{rB_{xx} - S_{xx}} \end{aligned} \tag{13}$$

and

$$\hat{\beta}_4 = \frac{B_{yy} - \frac{\hat{\sigma}_u^2}{r}}{B_{xy}} \quad ; \quad B_{yy} > \frac{\hat{\sigma}_u^2}{r}$$
$$= \frac{rB_{yy} - S_{yy}}{(r-1)B_{xy}}, \tag{14}$$

respectively. Replacing β in (12) by $\hat{\beta}_3$ and $\hat{\beta}_4$, we obtain the following predictors

$$P_{2wD} = \bar{y} + \hat{\beta}_3(\bar{x}_k - \bar{x}) \tag{15}$$

and

$$P_{2wR} = \bar{y} + \hat{\beta}_4(\bar{x}_k - \bar{x}), \tag{16}$$

respectively. The corresponding target function in this case is given by

$$P_w^* = \lambda\bar{y}_k + (1-\lambda)E(\bar{y}_k) \quad ; \quad 0 \leq \lambda \leq 1. \tag{17}$$

3.2 Outside Sample Prediction

We assume that beside a sample of n observations, some additional observations on explanatory variable are available but the corresponding values on study variable are not known. Here we assume that the additional observations are independent of the existing sample observations. Further, we assume that the additional observations also satisfy the underlying model (1)-(4).

Prediction of $(f,j)^{th}$ Value Outside the Sample

Let x_{fj} be the value of explanatory variable, which lies outside the sample and is independent of given n sample units and let y_{fj} be the corresponding value of study variable, which is to be predicted. The target function in such a case can be formulated as:

$$P_o = \lambda y_{fj} + (1-\lambda)E(y_{fj}) \quad ; \quad 0 \leq \lambda \leq 1 \tag{18}$$

and the natural linear predictor for outside sample prediction in this case is

$$\hat{y}_{fj} = \bar{y} + \beta(x_{fj} - \bar{x}) \tag{19}$$

where β is unknown. We propose to replace β by its consistent estimators $\hat{\beta}_1$ and $\hat{\beta}_2$, which provides the following predictors for outside sample prediction:

$$P_{1oD} = \bar{y} + \hat{\beta}_1(x_{fj} - \bar{x}) \tag{20}$$

and

$$P_{1oR} = \bar{y} + \hat{\beta}_2(x_{fj} - \bar{x}), \tag{21}$$

respectively.

Prediction of Average of r Replicates Outside the Sample

Let us suppose that the additional information on explanatory variable is in the form of average taken over r replicates, which are independent and outside the existing sample values. The corresponding value of study variable is not known and is to be predicted. We assume that this additional observations satisfy the underlying model (1)-(4). Let $\bar{x}_f$ be the known value of on explanatory variable from outside the sample and we wish to predict the corresponding value $\bar{y}_f$ of study variable. Then the target function in this case can be written as

$$P_o^* = \lambda \bar{y}_f + (1-\lambda)E(\bar{y}_f) \quad ; \quad 0 \leq \lambda \leq 1 \tag{22}$$

and the linear predictor in this case is given by

$$\hat{\bar{y}}_f = \bar{y} + \beta(\bar{x}_f - \bar{x}). \tag{23}$$

Now we replace the unknown β by its consistent estimator $\hat{\beta}_3$ and $\hat{\beta}_4$, which yield the following predictors:

$$P_{2oD} = \bar{y} + \hat{\beta}_3(\bar{x}_f - \bar{x}) \tag{24}$$

and

$$P_{2oR} = \bar{y} + \hat{\beta}_4(\bar{x}_f - \bar{x}), \tag{25}$$

respectively.

4 Efficiency Properties of Predictors

The exact expression of predictive bias and predictive mean squared error of the proposed predictors are difficult to derive even when the errors are distributed normally. The situation becomes more complicated when the errors are distributed non-normally and the variables have repeated observations. We, therefore, use the large sample asymptotic approximation theory to derive the efficiency properties of the predictors. We assume that $S_{mm} = \frac{1}{n}\sum_{i=1}^{n}(m_i - \bar{m})^2$ tends to a finite limit as n grows large. This assumption is needed for the application of large sample asymptotic approximation theory and rules out the possibility of any trend in the explanatory variables, see Schneeweiss (1982), Schneeweiss (1991). Further, we assume that n grows large while r remains fixed.

Let

$$d = \frac{\sigma_w^2}{S_{mm} + \sigma_w^2} \quad ; \quad 0 \leq d \leq 1,$$

$$\theta = \frac{\sigma_v^2}{S_{mm} + \sigma_w^2 + \sigma_v^2} \quad ; \quad 0 \leq \theta < 1,$$

and

$$q = \frac{\sigma_u^2}{\beta^2 \sigma_v^2} \quad ; \quad q \geq 0.$$

These quantities help in determining the nature of measurement error models. In case of functional form of the measurement error model, all X_i's are fixed and so $w_i = 0$ for all $i = 1, 2, \ldots, n$. Consequently, $\sigma_w^2 = 0$ and this implies $d = 0$. Similarly, in the case of structural model, all m_i's are same and thus $S_{mm} = 0$. This implies $d = 1$. Thus d serves as a measure of departure of ultrastructural model from its two forms, viz., the functional form and the structural form. When there is no measurement errors in the explanatory variables, then $\sigma_v^2 = 0$. This gives that $\theta = 0$ and then the ultrastructural model reduces to the classical regression model. Hence, a non-zero value of θ serves as a measure of departure of ultrastructural model from the classical regression model, see also Srivastava and Shalabh (1997a), Srivastava and Shalabh (1997b).

Now, we present the asymptotic efficiency properties of the predictors arising from the application of direct regression and reverse regression in the following Theorems.

Theorem 1. *The predictive bias* (PB) *and predictive mean squared error* (PM) *of direct regression predictor in within sample prediction up to order $O(n^{-1})$ are given by*

$$\begin{aligned} \text{PB}(P_{1wD}) &= E(P_{1wD} - P_w) \\ &= \frac{\beta\theta(m_k - \bar{m})}{nr(1-\theta)} \left[2 + \frac{2\theta}{(r-1)(1-\theta)} \right] \end{aligned} \tag{26}$$

$$\text{PB}(P_{2wD}) = E(P_{2wD} - P_w^*) = \text{PB}(P_{1wD}) \tag{27}$$

$$\begin{aligned} \text{PM}(P_{1wD}) &= E(P_{1wD} - P_w)^2 \\ &= \beta^2\sigma_v^2 \left[1 + q\lambda^2 + \frac{d(1-\lambda)^2}{\theta(1-\theta)} \right] \\ &\quad + \frac{\beta^2\sigma_v^2}{nr(1-\theta)} \Bigg[q\left\{(1-\theta) - 2\lambda(1+d)\right\} - (1+\theta) - \frac{2d}{(1-\theta)} \\ &\quad - \theta \left\{ 2(1+\lambda q) - \frac{\theta}{(1-\theta)} \left\{ q + \frac{2}{(r-1)} \right. \right. \\ &\quad \left. \left. + \frac{(1-\theta)(1+q)}{\theta} \right\} \right\} \frac{(m_k - \bar{m})^2}{\sigma_v^2} - \frac{2\theta(m_k - \bar{m})\gamma_{1v}}{\sigma_v} \Bigg] \end{aligned} \tag{28}$$

$$\begin{aligned}
\mathrm{PM}(P_{2wD}) &= E(P_{2wD} - P_w^*)^2 \\
&= \frac{\beta^2\sigma_v^2}{r}\left[1 + q\lambda^2 + \frac{d(1-\lambda)^2}{\theta(1-\theta)}\right] + \frac{\beta^2\sigma_v^2}{nr}\left[(1 - 2\lambda(1+d))q\right. \\
&\quad -1 - \frac{2\theta}{r(1-\theta)}\left(1 + \lambda q + \frac{dr}{\theta(1-\theta)}\right) \\
&\quad -\frac{\theta}{(1-\theta)}\left\{2(1+\lambda q) - \frac{\theta}{(1-\theta)}\left(\frac{q}{r} + \frac{2}{r-1}\right.\right. \\
&\quad \left.\left.\left. +\frac{(1-\theta)(1+q)}{\theta}\right)\right\}\frac{(m_k - \bar{m})^2}{\sigma_v^2} - \frac{2\theta(m_k - \bar{m})\gamma_{1v}}{r(1-\theta)\sigma_v}\right].
\end{aligned} \tag{29}$$

The proof of this Theorem is outlined in Section 7.

From (26) and (27), we observe that both the predictors P_{1wD} and P_{2wD} are positively biased if $m_k - \bar{m} > 0$, and negatively biased if $m_k - \bar{m} < 0$ for $\beta > 0$ up to the order of approximation $O(n^{-1})$. Both exhibit equal magnitude of bias and no influence of skewness and kurtosis of distribution of measurement errors as well as random error component is observed on the predictive bias of the predictors. The difference in the predictive bias and the effect of skewness and kurtosis of error terms distributions may precipitate if we consider higher order approximation.

The predictive mean squared errors of both the predictors P_{1wD} and P_{2wD} are affected by the skewness of measurement errors distribution associated with explanatory variable. There is no influence of skewness of distributions of u_i's and w_i's. Also, there is no influence of kurtosis of distribution of any of the error terms in the predictive mean squared error of the predictors at least up to the order of approximation. It may be observed that if we consider the higher order approximation, then this may precipitate.

Theorem 2. *The predictive bias* (PB) *and predictive mean squared error* (PM) *of reverse regression predictors up to order* $O(n^{-1})$ *in within sample prediction are given by*

$$\begin{aligned}
\mathrm{PB}(P_{1wR}) &= E(P_{1wR} - P_w) \\
&= \frac{(m_k - \bar{m})}{nr}\frac{q\theta\beta(3\theta - 2)}{(1-\theta)^2}
\end{aligned} \tag{30}$$

$$\begin{aligned}
\mathrm{PB}(P_{2wR}) &= E(P_{2wR} - P_w^*) \\
&= \frac{(m_k - \bar{m})}{nr^2}\frac{q\theta\beta[\theta - 2r(1-\theta)]}{(1-\theta)^2}
\end{aligned} \tag{31}$$

$$\begin{aligned}
\mathrm{PM}(P_{1wR}) &= E(P_{1wR} - P_w)^2 \\
&= \beta^2\sigma_v^2\left[1 + q\lambda^2 + \frac{d(1-\lambda)^2}{\theta(1-\theta)}\right] \\
&\quad + \frac{\beta^2\sigma_v^2}{nr}\left[\left\{1 - 2\lambda\left(1 + d - \frac{\theta}{(1-\theta)}\right)\right\} q - 1\right. \\
&\quad - \frac{\theta}{(1-\theta)}\left\{2\left(1 + (1+q)(1-\lambda)\right)\frac{d}{\theta(1-\theta)} + 2q\right. \\
&\quad - \frac{(m_k - \bar{m})^2}{\sigma_v^2}\left\{2(1+\lambda q) - \frac{\theta}{(1-\theta)}\left(q + \frac{2q^2}{(r-1)}\right.\right. \\
&\quad \left.\left.\left.\left. + \frac{(1-\theta)(1+q)}{\theta}\right)\right\}\right\} - \frac{2\theta(m_k - \bar{m})\gamma_{1v}}{(1-\theta)\sigma_v}\right]
\end{aligned} \tag{32}$$

$$\begin{aligned}
\mathrm{PM}(P_{2wR}) &= E(P_{2wR} - P_w^*)^2 \\
&= \frac{\beta^2\sigma_v^2}{r}\left[1 + q\lambda^2 + \frac{rd(1-\lambda)^2}{\theta(1-\theta)}\right] \\
&\quad + \frac{\beta^2\sigma_v^2}{nr}\left[\left\{1 - 2\lambda\left(1 + d - \frac{\theta}{r(1-\theta)}\right)\right\} q - 1\right. \\
&\quad - \frac{\theta}{(1-\theta)}\left\{2\left(1 + (1+q)(1-\lambda)\right)\frac{d}{\theta(1-\theta)} + \frac{2q}{r}\right. \\
&\quad + \frac{(m_k - \bar{m})^2}{\sigma_v^2}\left\{2(1+\lambda q) - \frac{\theta}{1-\theta}\left(\frac{q}{r} + \frac{2q^2}{(r-1)}\right.\right. \\
&\quad \left.\left.\left.\left. + \frac{(1-\theta)(1+q)}{\theta}\right)\right\}\right\} - \frac{2\theta(m_k - \bar{m})\gamma_{1v}}{r(1-\theta)\sigma_v}\right].
\end{aligned} \tag{33}$$

The proof of this Theorem is outlined in Section 7.

We observe from expression (30) and (31) that the magnitude of predictive bias of P_{2wR} is smaller than the magnitude of predictive bias of P_{1wR}. The nature of predictive bias of P_{1wR} will be positive for $(m_k - \bar{m}) > 0$ and $\beta > 0$ if $\theta > 2/3$. The predictor P_{2wR} becomes positively biased for $(m_k - \bar{m}) > 0$ and $\beta > 0$, if $\theta > 2r/(2r+1)$. No influence of skewness and kurtosis of distribution of any of the measurement error terms in the bias of the predictors is observed up to the order of approximation. It may precipitate when we consider the higher order approximation.

The predictive mean squared error of both the predictors P_{1wR} and P_{2wR} is affected by the skewness of distribution of measurement errors associated with explanatory variable. No effect of skewness of distributions of u_{ij}'s and w_i's and effect of kurtosis of distribution of any of the error terms is observed at least up to the order of approximation. When we consider the higher order approximation, the effect of skewness and kurtosis of other error distributions may precipitate.

Theorem 3. *The predictive bias* (PB) *and predictive mean squared error* (PM) *of direct regression predictor in outside sample prediction up to order $O(n^{-1})$ are given by*

$$\begin{aligned}
\mathrm{PB}(P_{1oD}) &= E(P_{1oD} - P_o) \\
&= \frac{\beta\theta(m_f - \bar{m})}{nr(1-\theta)}\left[3 + \frac{2\theta}{(r-1)(1-\theta)}\right] && (34)\\
\mathrm{PB}(P_{2oD}) &= E(P_{2oD} - P_o^*) = \mathrm{PB}(P_{1oD}) && (35)\\
\mathrm{PM}(P_{1oD}) &= E(P_{1oD} - P_o)^2 \\
&= \beta^2\sigma_v^2\left[1 + q\lambda^2 + \frac{d(1-\lambda)^2}{\theta(1-\theta)}\right] \\
&\quad + \frac{\beta^2\sigma_v^2}{nr}\left[q + 1\right. \\
&\quad + \frac{2\theta}{(1-\theta)}\left\{3 + \frac{2\theta}{(1-\theta)(r-1)}\right\}\left(1 + \frac{d(1-\lambda)}{\theta(1-\theta)}\right) \\
&\quad + \frac{\theta^2}{(1-\theta)^2}\left(\frac{(m_f - \bar{m})^2}{\sigma_v^2} + 1 + \frac{d}{\theta(1-\theta)}\right) \\
&\quad \times \left.\left\{q + \frac{2}{(r-1)} + \frac{(1-\theta)(1+q)}{\theta}\right\}\right] && (36)\\
\mathrm{PM}(P_{2oD}) &= E(P_{2oD} - P_o^*)^2 \\
&= \frac{\beta^2\sigma_v^2}{r}\left[1 + q\lambda^2 + \frac{rd(1-\lambda)^2}{\theta(1-\theta)}\right] \\
&\quad + \frac{\beta^2\sigma_v^2}{nr}\left[q + 1\right. \\
&\quad + \frac{2\theta}{(1-\theta)}\left(3 + \frac{2\theta}{(1-\theta)(r-1)}\right)\left(\frac{1}{r} + \frac{(1-\lambda)d}{\theta(1-\theta)}\right) \\
&\quad + \frac{\theta^2}{(1-\theta)^2}\left(\frac{(m_f - \bar{m})^2}{\sigma_v^2} + \frac{1}{r} + \frac{d}{\theta(1-\theta)}\right) \\
&\quad \times \left.\left\{\frac{q}{r} + \frac{2}{(r-1)} + \frac{(1-\theta)(1+q)}{\theta}\right\}\right]. && (37)
\end{aligned}$$

The proof of this Theorem is outlined in the Section 7.

From (34) and (35), we observe that the predictive bias of both the predictors P_{1oD} and P_{2oD} are equal, at least up to the order of approximation $O(n^{-1})$. Both of them are positively biased if $m_f - \bar{m} > 0$, and negatively biased if $m_f - \bar{m} < 0$ for $\beta > 0$. We do not observe any effect of skewness and kurtosis of the distribution of measurement errors and random error distributions towards the bias of predictors at least up to the order of approximation.

Again, we consider the predictive mean squared error of the direct regression predictors P_{1oD} and P_{2oD}. The effect of skewness as well as kurtosis of any of the error term distributions is not observed, at least up to $O(n^{-1})$ of approximation.

Theorem 4. *The predictive bias* (PB) *and predictive mean squared error* (PM) *of reverse regression predictors up to order* $O(n^{-1})$ *in outside sample prediction are given by*

$$\begin{aligned} \mathrm{PB}(P_{1oR}) &= E(P_{1oR} - P_o) \\ &= \frac{\theta\beta(m_f - \bar{m})}{nr(1-\theta)}\left[(1-2q) + \frac{q\theta}{(1-\theta)}\right] \end{aligned} \tag{38}$$

$$\begin{aligned} \mathrm{PB}(P_{2oR}) &= E(P_{2oR} - P_o^*) \\ &= \frac{\theta\beta(m_f - \bar{m})}{nr(1-\theta)}\left[(1-2q) + \frac{q\theta}{r(1-\theta)}\right] \end{aligned} \tag{39}$$

$$\begin{aligned} \mathrm{PM}(P_{1oR}) &= E(P_{1oR} - P_o)^2 \\ &= \beta^2\sigma_v^2\left[1 + q\lambda^2 + \frac{d(1-\lambda)^2}{\theta(1-\theta)}\right] \\ &\quad + \frac{\beta^2\sigma_v^2}{nr}\left[1 + q + \frac{\theta}{(1-\theta)}\left\{\frac{d}{\theta(1-\theta)}\{(1+q)\right.\right. \\ &\quad \left. + 2(1-\lambda)(1-2q) + \frac{q\theta}{(1-\theta)}\left(3 - 2\lambda + \frac{2q}{r-1}\right)\right\} \\ &\quad + 3(1-q) + \frac{q\theta}{(1-\theta)}\left(3 + \frac{2q}{r-1}\right) \\ &\quad \left.\left. + \frac{(m_f - \bar{m})^2}{\sigma_v^2(1-\theta)}\left(1 + q + \frac{q\theta}{(1-\theta)}\left(1 + \frac{2q}{(r-1)}\right)\right)\right\}\right] \end{aligned} \tag{40}$$

$$
\begin{aligned}
\mathrm{PM}(P_{2oR}) &= E(P_{2oR} - P_o^*)^2 \\
&= \frac{\beta^2\sigma_v^2}{r}\left[1 + q\lambda^2 + \frac{rd(1-\lambda)^2}{\theta(1-\theta)}\right] \\
&\quad + \frac{\beta^2\sigma_v^2}{nr}\left[q + 1 + \frac{d}{(1-\theta)^2}\{1 + q + 2(1-\lambda)(1-2q)\right. \\
&\quad + \frac{q\theta}{(1-\theta)}\left(\frac{(3-2\lambda)}{r} + \frac{2q}{r-1}\right)\} \\
&\quad + \frac{1}{r}\left\{3(1-q) + \frac{q\theta}{(1-\theta)}\left(\frac{3}{r} + \frac{2q}{r-1}\right)\right\} \\
&\quad \left. + \frac{(m_f - \bar{m})^2}{\sigma_v^2}\left\{1 + q + \frac{q\theta}{(1-\theta)}\left(\frac{1}{r} + \frac{2q}{r-1}\right)\right\}\right]. \qquad (41)
\end{aligned}
$$

The proof of this Theorem is outlined in Section 7.

From (38) and (39), we observe that the predictive bias of P_{2oR} is smaller than that of P_{1oR} . The predictor P_{1oR} is negatively biased when $\theta < 2/3$ for $(m_f - \bar{m}) > 0$ and $\beta > 0$ whereas P_{2oR} is negatively biased when $\theta < 2r/(2r+1)$ for $(m_f - \bar{m}) > 0$ and $\beta > 0$. The magnitude of bias of both of the predictors depend on the difference between sample mean $(\bar{m})$ and mean of the additional observations $(\bar{m}_f)$ on explanatory variable. It indicates that the predictors will have large magnitude of bias if the additional observation is far away from the mean of existing sample. We do not observe any effect of skewness and kurtosis of distribution of measurement errors as well as random error components towards the bias of the predictors at least up to the order of approximation. Such effect may precipitate if we consider the higher order approximation.

We see that the PM's of P_{1oR} and P_{2oR} are affected by the skewness of distribution of v_{ij}'s. No role of skewness of distribution of other error terms is observed in the variability of predictors up to the order of approximation. Also, there is no effect of kurtosis of any of the distribution of error terms in the PM of the predictors, at least up to the order of approximation.

Further, we observe from (28) and (30) that P_{1wR} is better than P_{1wD} when

$$
\begin{aligned}
&\frac{(m_k - \bar{m})^2}{\sigma_v^2}\left[(2\theta - 1)q + \frac{2}{(r-1)}\{\theta - (1-\theta)q^2\}\right. \\
&\left. + (1-\theta)(2\theta-1)(1+q)\right] \geq 2(1+q)\left[\theta - \frac{d(1-\lambda)}{(1-\theta)}\right] \\
&\qquad -2\lambda\left[d - 1 - \theta + q(1+d)(1-\theta)\right]. \qquad (42)
\end{aligned}
$$

The reverse holds true with a reversed inequality sign. Also, it is observed from (29) and (33) that P_{2wR} is better than P_{2wD} when

$$\frac{(m_k - \bar{m})^2}{\sigma_v^2}\left[\frac{(1+q\lambda)(1-\theta)}{\theta} - \frac{\theta(1-q^2)}{(1-\theta)(r-1)}\right] \leq \frac{(1+q)}{r} + \frac{d(1+q)(1-\lambda)}{\theta(1-\theta)}. \tag{43}$$

The reverse holds true with a reversed inequality sign. Similar conditions can also be derived for the superiority of other predictors in other cases also.

5 Monte-Carlo Simulation Study

The large sample properties of the predictors give an idea when the sample size is large. To study the performance of the predictors in finite samples, we conducted a Monte-Carlo simulation experiment. The design and procedure of the simulation experiment is as follows:

- We considered two samples of sizes 15 (small sample) and 43 (large sample) to provide the value of m_i's for which $S_{mm} = 0.23$.
- Random error component w_i's are generated from normal distribution with standard deviation 0.1. Since we are mainly interested in knowing the effect of non-normality of measurement errors distributions on the predictive behavior of the predictors, so without loss of generality, we prefer to fix the distribution of w_i's to be normal with mean 0 and standard deviation 0.1..
- The measurement errors u_{ij}'s and v_{ij}'s are generated from normal, central t, and gamma distributions with mean 0 and different combinations of standard deviations $\sigma_u = 0.1, 0.2, 0.3$ and $\sigma_v = 0.1, 0.5, 0.7$.
- The other chosen values are $\alpha = 1$, $\beta = 0.3$, $m_k = 1.87$, $m_f = 5$, $r=$ 4, $n = 15$ and 43.
- The value of λ are chosen between 0 and 1 at an interval of 0.1.

The empirical predictive bias (EPB) and empirical predictive mean squared error (EPM) of predictors are calculated based on 5000 iterations under respective target functions for selected values of λ. Only some of the results of the experiments are presented in Tables 1-8 to economize the space.

Table 1. Empirical predictive bias (EPB) and empirical predictive mean squared error (EPM) of predictors in within sample prediction through direct regression predictors with $\sigma_u = 0.1$, $\sigma_v = 0.1$, $\sigma_w = 0.1$

		n=15				n=43			
		P_{1wD}		P_{2wD}		P_{1wD}		P_{2wD}	
	λ	EPB	EPM	EPB	EPM	EPB	EPM	EPB	EPM
Normal	0.1	0.0101	0.0024	0.0098	0.0017	0.0095	0.0020	0.0090	0.0013
	0.3	0.0285	0.0030	0.0283	0.0022	0.0282	0.0027	0.0276	0.0019
	0.5	0.0469	0.0044	0.0468	0.0035	0.0468	0.0041	0.0461	0.0032
	0.7	0.0653	0.0065	0.0654	0.0055	0.0655	0.0063	0.0647	0.0052
	0.9	0.0837	0.0094	0.0839	0.0081	0.0841	0.0092	0.0832	0.0079
t	0.1	0.0093	0.0024	0.0099	0.0017	0.0095	0.0020	0.0096	0.0014
	0.3	0.0278	0.0029	0.0285	0.0022	0.0280	0.0027	0.0281	0.0020
	0.5	0.0464	0.0043	0.0471	0.0035	0.0465	0.0041	0.0465	0.0032
	0.7	0.0649	0.0064	0.0657	0.0055	0.0650	0.0063	0.0650	0.0052
	0.9	0.0834	0.0093	0.0843	0.0082	0.0835	0.0092	0.0835	0.0080
Gamma	0.1	0.0095	0.0023	0.0095	0.0017	0.0096	0.0020	0.0093	0.0013
	0.3	0.0282	0.0029	0.0282	0.0022	0.0283	0.0026	0.0278	0.0019
	0.5	0.0470	0.0042	0.0468	0.0035	0.0469	0.0040	0.0463	0.0032
	0.7	0.0657	0.0063	0.0655	0.0055	0.0655	0.0061	0.0648	0.0051
	0.9	0.0845	0.0091	0.0841	0.0082	0.0842	0.0089	0.0833	0.0079

Now, we analyze the results of simulation experiment. Some outcomes are true in general for all the predictors regarding their EPB and EPM. The magnitude of EPB and EPM of the predictors decrease as sample size increases. The values of EPB and EPM of same predictor are different under different types of distributions of measurement errors. It confirms the role of skewness and kurtosis of the error terms distributions on the EPB and EPM. The results which are specific to any predictor are stated later.

First we consider the EPB and EPM of P_{1wD} and P_{2wD} from Tables 1 and 2. From the tabulated results, we observe that the magnitude of EPB of P_{2wD} is generally smaller than that of P_{1wD}. The difference between their EPB decreases as λ increases. Also, the magnitude of EPB is increasing with the increase in the value of λ. It means that the magnitude of EPB for average value prediction is smaller than the corresponding EPB of actual value prediction. When standard deviations

Table 2. Empirical predictive bias (EPB) and empirical predictive mean squared error (EPM) of predictors in within sample prediction through direct regression predictors with $\sigma_u = 0.3$, $\sigma_v = 0.7$, $\sigma_w = 0.1$

		n=15				n=43			
		P_{1wD}		P_{2wD}		P_{1wD}		P_{2wD}	
	λ	EPB	EPM	EPB	EPM	EPB	EPM	EPB	EPM
Normal	0.1	0.1359	8.6343	0.1576	2.7603	0.0241	0.3204	0.0309	0.0321
	0.3	0.1542	8.6357	0.1762	2.7656	0.0424	0.3215	0.0494	0.0333
	0.5	0.1725	8.6384	0.1948	2.7717	0.0606	0.3239	0.0679	0.0353
	0.7	0.1908	8.6425	0.2134	2.7786	0.0789	0.3276	0.0863	0.0382
	0.9	0.2091	8.6480	0.2320	2.7865	0.0971	0.3327	0.1048	0.0420
t	0.1	0.2022	86.4926	0.1608	5.1580	0.0347	0.2376	0.0318	0.0864
	0.3	0.2208	86.4988	0.1795	5.1640	0.0535	0.2395	0.0503	0.0877
	0.5	0.2395	86.5064	0.1982	5.1710	0.0723	0.2428	0.0689	0.0898
	0.7	0.2582	86.5153	0.2169	5.1787	0.0911	0.2474	0.0875	0.0928
	0.9	0.2769	86.5256	0.2356	5.1874	0.1099	0.2534	0.1060	0.0966
Gamma	0.1	0.1169	8.2360	0.1152	3.3173	0.0364	0.1589	0.0312	0.0764
	0.3	0.1354	8.2414	0.1338	3.3216	0.0549	0.1610	0.0498	0.0776
	0.5	0.1538	8.2481	0.1523	3.3268	0.0734	0.1644	0.0685	0.0798
	0.7	0.1723	8.2561	0.1709	3.3328	0.0919	0.1693	0.0871	0.0827
	0.9	0.1908	8.2655	0.1895	3.3397	0.1104	0.1756	0.1057	0.0865

of measurement errors are lower, than P_{2wD} has generally lower bias than P_{1wD} but reverse is true when standard deviations of measurement errors are higher. The effect of non-normality of measurement errors on the EPB of predictors is more distinct when standard deviations of measurement errors become large. Now, we analyze the EPM of P_{1wD} and P_{2wD}. We observe that the predictor P_{2wD} has smaller EPM than P_{1wD} in all experimental settings. When we increase σ_v while keeping σ_u and σ_w fixed at 0.1, the EPM of both the predictors increases. The rate of increase in EPM is higher in small sample cases than in large sample cases. When we fix σ_v and σ_w at 0.1 and increase σ_u only, then we do not observe any significant changes in the EPM of both the predictors. Again, when we increase σ_u and σ_v both simultaneously, then we observe an increase in the EPM of both the predictors. The highest change in the EPM of both the predictors is observed under

Table 3. Empirical predictive bias (EPB) and empirical predictive mean squared error (EPM) of predictors in within sample prediction through direct regression predictors with $\sigma_u = 0.1$, $\sigma_v = 0.1$, $\sigma_w = 0.1$

		n=15				n=43			
		P_{1oD}		P_{2oD}		P_{1oD}		P_{2oD}	
	λ	EPB	EPM	EPB	EPM	EPB	EPM	EPB	EPM
Normal	0.1	0.0041	0.0207	0.0045	0.0190	0.0032	0.0084	0.0024	0.0073
	0.3	0.0042	0.0213	0.0046	0.0189	0.0034	0.0088	0.0025	0.0072
	0.5	0.0042	0.0227	0.0047	0.0190	0.0037	0.0101	0.0026	0.0074
	0.7	0.0043	0.0251	0.0047	0.0195	0.0039	0.0123	0.0027	0.0078
	0.9	0.0043	0.0283	0.0048	0.0202	0.0041	0.0154	0.0028	0.0085
t	0.1	0.0032	0.0216	0.0036	0.0199	0.0014	0.0084	0.0011	0.0073
	0.3	0.0039	0.0225	0.0038	0.0199	0.0012	0.0094	0.0009	0.0073
	0.5	0.0045	0.0248	0.0039	0.0204	0.0010	0.0117	0.0007	0.0077
	0.7	0.0052	0.0285	0.0041	0.0212	0.0009	0.0153	0.0005	0.0084
	0.9	0.0059	0.0335	0.0043	0.0224	0.0007	0.0202	0.0002	0.0096
Gamma	0.1	0.0020	0.0188	0.0017	0.0179	0.0030	0.0083	0.0028	0.0072
	0.3	0.0021	0.0192	0.0018	0.0178	0.0027	0.0087	0.0028	0.0071
	0.5	0.0021	0.0202	0.0019	0.018	0.0025	0.0101	0.0029	0.0073
	0.7	0.0021	0.0219	0.0020	0.0184	0.0022	0.0122	0.0029	0.0076
	0.9	0.0022	0.0244	0.0022	0.0191	0.0020	0.0151	0.0030	0.0082

t-distributed measurement errors, particularly in small samples when standard deviations of measurement errors are high.

Next we analyze the EPB and EPM of the direct regression predictors in outside sample prediction cases from Tables 3 and 4. The magnitude of EPB of P_{2oD} is smaller than the magnitude of EPB of P_{1oD}. As the standard deviation of measurement errors increases, the magnitude of EPB of both the predictors increases. The magnitude of EPB for average and actual value predictions are very close under the same experimental settings. The EPB is positive in most of the cases. In general, the EPB of P_{1oD} and P_{2oD} increases as λ increases except in case of P_{1oD} in large samples with gamma distributed measurement errors when standard deviations of measurement errors are quite small. Now we analyze the EPM of the predictors P_{1oD} and P_{2oD}. As we increase σ_v while keeping σ_u and σ_w fixed at 0.1, we observe an increase in the EPM of both the predictors. This increment rate is higher in

Table 4. Empirical predictive bias (EPB) and empirical predictive mean squared error (EPM) of predictors in within sample prediction through direct regression predictors with $\sigma_u = 0.3$, $\sigma_v = 0.7$, $\sigma_w = 0.1$

		n=15				n=43			
		P_{1oD}		P_{2oD}		P_{1oD}		P_{2oD}	
	λ	EPB	EPM	EPB	EPM	EPB	EPM	EPB	EPM
Normal	0.1	0.9405	87.8417	1.0372	85.3772	0.1978	1.0677	0.1767	0.5826
	0.3	0.9415	87.8395	1.0380	85.3897	0.1977	1.0738	0.1766	0.5842
	0.5	0.9425	87.8446	1.0388	85.4040	0.1977	1.0870	0.1764	0.5876
	0.7	0.9435	87.8568	1.0397	85.4202	0.1976	1.1072	0.1763	0.5929
	0.9	0.9445	87.8763	1.0405	85.4382	0.1975	1.1345	0.1761	0.6001
t	0.1	2.1212	1446.51	1.7906	693.818	0.2004	2.6302	0.1888	1.1968
	0.3	2.1210	1446.52	1.7908	693.847	0.2009	2.6369	0.1882	1.1998
	0.5	2.1207	1446.55	1.7911	693.878	0.2014	2.6532	0.1876	1.2052
	0.7	2.1205	1446.58	1.7913	693.913	0.202	2.6791	0.1869	1.2130
	0.9	2.1203	1446.63	1.7915	693.950	0.2025	2.7146	0.1863	1.2233
Gamma	0.1	0.8787	92.9530	0.8871	52.7379	0.1751	1.5097	0.1785	1.1196
	0.3	0.8790	92.9709	0.8873	52.7480	0.1762	1.5167	0.1791	1.1223
	0.5	0.8793	92.9959	0.8876	52.7601	0.1773	1.5297	0.1797	1.1267
	0.7	0.8797	93.0280	0.8879	52.7741	0.1784	1.5488	0.1802	1.1329
	0.9	0.8800	93.0673	0.8881	52.7900	0.1794	1.5739	0.1808	1.1408

small sample than in large sample. Next, when we increase σ_u keeping σ_v and σ_w fixed at 0.1, then there is little change in the EPM of P_{1oD} and P_{2oD} in small sample. No significant change in large sample is observed. Again when we increase σ_v and σ_u together, the EPM of both the predictors increases. In general, the EPMs are significantly affected when sample size is small and standard deviations of measurement errors are high. The highest EPM of both predictors is observed under t-distributed measurement errors.

Next, we consider the EPB and EPM of P_{1wR} and P_{2wR} under various experimental settings from Tables 5 and 6. We observe that the magnitude of EPB of P_{2wR} is smaller than the magnitude of EPB of P_{1wR} except in the case of normally distributed measurement errors in small sample with high standard deviations of measurement errors. The EPB and EPM of both the predictors increases as the value of λ increases. The EPB of actual value prediction is higher than average

Table 5. Empirical predictive bias (EPB) and empirical predictive mean squared error (EPM) of predictors in within sample prediction through direct regression predictors with $\sigma_u = 0.1$, $\sigma_v = 0.1$, $\sigma_w = 0.1$

		n=15				n=43			
		P_{1wR}		P_{2wR}		P_{1wR}		P_{2wR}	
	λ	EPB	EPM	EPB	EPM	EPB	EPM	EPB	EPM
Normal	0.1	0.0052	0.0024	0.0049	0.0017	0.0075	0.0020	0.0070	0.0014
	0.3	0.0237	0.0028	0.0235	0.0021	0.0261	0.0026	0.0255	0.0019
	0.5	0.0421	0.0040	0.0420	0.0032	0.0447	0.0039	0.0441	0.0031
	0.7	0.0605	0.0060	0.0605	0.0050	0.0634	0.0060	0.0626	0.0050
	0.9	0.0789	0.0087	0.0790	0.0075	0.0820	0.0089	0.0812	0.0076
t	0.1	0.0051	0.0024	0.0054	0.0018	0.0075	0.0021	0.0075	0.0014
	0.3	0.0236	0.0029	0.0240	0.0022	0.0260	0.0026	0.0260	0.0019
	0.5	0.0421	0.0041	0.0426	0.0033	0.0445	0.0040	0.0445	0.0031
	0.7	0.0607	0.0060	0.0612	0.0051	0.0630	0.0061	0.0629	0.0051
	0.9	0.0792	0.0088	0.0798	0.0076	0.0815	0.0089	0.0814	0.0077
Gamma	0.1	0.0049	0.0119	0.0051	0.0115	0.0080	0.0020	0.0076	0.0013
	0.3	0.0237	0.0123	0.0238	0.0119	0.0267	0.0026	0.0261	0.0019
	0.5	0.0424	0.0135	0.0424	0.0131	0.0453	0.0039	0.0446	0.0031
	0.7	0.0612	0.0155	0.0611	0.0149	0.0640	0.0060	0.0631	0.0050
	0.9	0.0799	0.0181	0.0797	0.0175	0.0826	0.0088	0.0817	0.0077

value prediction. We observe that both the predictors have small EPM for small values of σ_u, σ_v and σ_w. When we increase σ_v keeping σ_u and σ_w fixed at 0.1, we observe small change in the EPM of both the predictors. When we increase σ_u keeping σ_v and σ_w fix at 0.1, we observe a small changes in the EPM of both the predictors. Again, when we increase σ_v and σ_u simultaneously, we observe significant increase in the EPM of both the predictors in small sample cases. The rate of increment of EPM is higher in small sample cases than in large sample. The EPM of both the predictors is much higher under t-distributed measurement errors and small sample than in other distributions.

Next, we analyze the results for P_{1oR} and P_{2oR} from Tables 7 and 8. These predictors have nearly equal magnitude of EPB for average and actual value prediction in most of the cases. Both predictors are negatively biased for small values of σ_u, σ_v and σ_w. When σ_u, σ_v and σ_w are high, than the sign of EPB changes except in case of t-distributed

Table 6. Empirical predictive bias (EPB) and empirical predictive mean squared error (EPM) of predictors in within sample prediction through direct regression predictors with $\sigma_u = 0.3$, $\sigma_v = 0.7$, $\sigma_w = 0.1$

		n=15				n=43			
		P_{1wR}		P_{2wR}		P_{1wR}		P_{2wR}	
	λ	EPB	EPM	EPB	EPM	EPB	EPM	EPB	EPM
Normal	0.1	-0.0162	30.4082	0.0117	27.1988	0.0522	0.8669	0.0103	0.0472
	0.3	0.0021	30.4091	0.0303	27.2008	0.0705	0.8698	0.0288	0.0478
	0.5	0.0204	30.4113	0.0489	27.2036	0.0887	0.8739	0.0472	0.0493
	0.7	0.0387	30.4149	0.0674	27.2073	0.1070	0.8794	0.0657	0.0516
	0.9	0.0570	30.4199	0.0860	27.2119	0.1252	0.8862	0.0842	0.0548
t	0.1	0.2267	143.222	0.0642	7.4700	0.0643	20.0653	0.0049	0.0249
	0.3	0.2454	143.226	0.0829	7.4734	0.0831	20.0715	0.0234	0.0254
	0.5	0.2641	143.231	0.1016	7.4777	0.1019	20.0790	0.0420	0.0267
	0.7	0.2827	143.238	0.1203	7.4829	0.1207	20.0879	0.0606	0.0290
	0.9	0.3014	143.246	0.1389	7.4889	0.1395	20.0981	0.0791	0.0321
Gamma	0.1	0.0682	1.7113	0.0333	0.6958	0.0296	0.6129	0.0075	0.0606
	0.3	0.0867	1.7155	0.0519	0.6973	0.0481	0.6148	0.0262	0.0611
	0.5	0.1052	1.7211	0.0704	0.6996	0.0666	0.6182	0.0448	0.0625
	0.7	0.1236	1.7279	0.0890	0.7027	0.0851	0.6229	0.0634	0.0648
	0.9	0.1421	1.7361	0.1076	0.7067	0.1036	0.6291	0.0821	0.0679

measurement errors in small samples. The EPM of P_{1oR} is larger than that of P_{2oR}. When we increase σ_v keeping σ_u and σ_w fixed at 0.1, we observe an increase in the EPM of both the predictors. The rate of increment in the EPM of both the predictors is higher in small sample case than in large sample case. The effect of non-normality of measurement errors on the EPM of these predictors is more significant in small samples and when standard deviations of measurement errors are high. In particular, the t-distributed measurement errors affect more than normal and gamma distributed measurement errors. Again, when we increase σ_u while fixing σ_v and σ_w at 0.1, we observe higher EPM of both the predictors. When we increase σ_u and σ_v simultaneously and keep fix $\sigma_w = 0.1$, then the EPM of both predictions increases. The increment is higher in small sample cases than in large sample cases.

Table 7. Empirical predictive bias (EPB) and empirical predictive mean squared error (EPM) of predictors in within sample prediction through direct regression predictors with $\sigma_u = 0.1$, $\sigma_v = 0.1$, $\sigma_w = 0.1$

		n=15				n=43			
		P_{1oR}		P_{2oR}		P_{1oR}		P_{2oR}	
	λ	EPB	EPM	EPB	EPM	EPB	EPM	EPB	EPM
Normal	0.1	-0.0276	0.0303	-0.0286	0.0291	-0.0076	0.0111	-0.0082	0.0101
	0.3	-0.0275	0.0309	-0.0285	0.0290	-0.0073	0.0116	-0.0081	0.0099
	0.5	-0.0275	0.0323	-0.0285	0.0291	-0.0071	0.0129	-0.0080	0.0101
	0.7	-0.0274	0.0346	-0.0284	0.0295	-0.0068	0.0150	-0.0079	0.0105
	0.9	-0.0273	0.0378	-0.0283	0.0302	-0.0066	0.0181	-0.0077	0.0112
t	0.1	-0.0293	0.0308	-0.0295	0.0291	-0.0073	0.0113	-0.0077	0.0101
	0.3	-0.0287	0.0318	-0.0293	0.0292	-0.0074	0.0123	-0.0079	0.0101
	0.5	-0.0280	0.0341	-0.0291	0.0296	-0.0076	0.0146	-0.0082	0.0105
	0.7	-0.0273	0.0377	-0.0290	0.0304	-0.0077	0.0182	-0.0084	0.0112
	0.9	-0.0267	0.0427	-0.0288	0.0316	-0.0079	0.0231	-0.0086	0.0124
Gamma	0.1	-0.0177	0.0244	-0.0190	0.0228	-0.0076	0.0112	-0.0076	0.0103
	0.3	-0.0176	0.0246	-0.0189	0.0226	-0.0079	0.0116	-0.0075	0.0102
	0.5	-0.0176	0.0262	-0.0187	0.0227	-0.0082	0.0129	-0.0075	0.0103
	0.7	-0.0176	0.0271	-0.0186	0.0231	-0.0084	0.0150	-0.0074	0.0109
	0.9	-0.0175	0.0295	-0.0185	0.0237	-0.0087	0.0179	-0.0074	0.0112

6 Concluding Remarks

We have used the availability of replicated observations to construct the consistent estimators of regression coefficients in the presence of measurement error in the data. Thus obtained estimators are then used to construct the predictors for predicting the actual and average values of study variable simultaneously. The asymptotic properties of the predictors are derived and analyzed under the specification of an ultrastructural model with not necessarily normally distributed measurement errors using large sample asymptotic approximation theory. The effect of departure from normality is clearly present in the predictive properties of the predictors in terms of departure from symmetry and peakedness of the distributions. The results from Monte-Carlo experiment confirm such an outcome in the finite samples also. The degree of departure in the values of predictive bias and predictive mean squared

Table 8. Empirical predictive bias (EPB) and empirical predictive mean squared error (EPM) of predictors in within sample prediction through direct regression predictors with $\sigma_u = 0.3$, $\sigma_v = 0.7$, $\sigma_w = 0.1$

		n=15				n=43			
		P_{1oR}		P_{2oR}		P_{1oR}		P_{2oR}	
	λ	EPB	EPM	EPB	EPM	EPB	EPM	EPB	EPM
Normal	0.1	0.0917	362.763	0.3308	84.8731	0.2204	9.806	0.0289	1.1516
	0.3	0.0927	362.78	0.3317	84.8757	0.2203	9.8138	0.0288	1.1533
	0.5	0.0937	362.805	0.3325	84.8802	0.2202	9.8287	0.0286	1.1570
	0.7	0.0947	362.837	0.3333	84.8865	0.2201	9.8507	0.0285	1.1625
	0.9	0.0957	362.876	0.3342	84.8947	0.2200	9.8797	0.0283	1.1700
t	0.1	-0.5560	782.679	0.2224	240.014	0.1509	116.018	0.1360	46.4133
	0.3	-0.5562	782.744	0.2226	240.028	0.1514	116.054	0.1354	46.4187
	0.5	-0.5564	782.820	0.2228	240.054	0.1519	116.099	0.1348	46.4264
	0.7	-0.5567	782.908	0.2231	240.065	0.1524	116.155	0.1341	46.4366
	0.9	-0.5569	783.007	0.2234	240.100	0.1532	116.255	0.1332	46.4566
Gamma	0.1	0.2619	167.214	-0.1588	183.775	0.1684	16.9305	0.0634	2.6882
	0.3	0.2622	167.234	-0.1585	183.772	0.1695	16.937	0.0640	2.6887
	0.5	0.2625	167.244	-0.1582	183.772	0.1706	16.9496	0.0646	2.6910
	0.7	0.2628	167.269	-0.1580	183.773	0.1717	16.9682	0.0652	2.6950
	0.9	0.2631	167.302	-0.1577	183.776	0.1727	16.9929	0.0658	2.7008

error depend on the values of coefficients of skewness and kurtosis of the respective distributions, sample size and measurement error variances.

7 Derivation of Results

Let

$$\begin{aligned} X &= col.(X_1, X_2, \ldots, X_n), \\ Y &= col.(Y_1, Y_2, \ldots, Y_n), \\ u &= col.(u_{11}, \ldots, u_{1r}, u_{21}, \ldots, u_{2r}, \ldots, u_{n1}, \ldots, u_{nr}), \\ v &= col.(v_{11}, \ldots, v_{1r}, v_{21}, \ldots, v_{2r}, \ldots, v_{n1}, \ldots, v_{nr}), \end{aligned}$$

$$
\begin{aligned}
w &= col.(w_1, w_2, \ldots, w_n),\\
m &= col.(m_1, m_2, \ldots, m_n),\\
x &= col.(x_{11}, \ldots, x_{1r}, x_{21}, \ldots, x_{2r}, \ldots, x_{n1}, \ldots, x_{nr}),\\
y &= col.(y_{11}, \ldots, y_{1r}, y_{21}, \ldots, y_{2r}, \ldots, y_{n1}, \ldots, y_{nr}),\\
e_z &= col.(1, 1, \ldots, 1),\\
l_n &= col.(0, \ldots, 1, \ldots, 0),\\
l_{nr} &= col.(0, \ldots, 1, \ldots, 0)
\end{aligned}
$$

where e_z is a $(z \times 1)$ column vector with each element unity, l_n is a $(n \times 1)$ vector with 1 at i^{th} place and 0 at all other places, $(i = 1, 2, \ldots, n)$, and l_{nr} is a $(nr \times 1)$ vector with 1 at $(i \times j)^{th}$ place and 0 at all other places, $(j = 1, 2, \ldots r)$, and $p_{nr} = l_n \otimes e_r$ where $\otimes$ is the Kronecker product operator of matrices.

Then we can write $w_i = l_n' w$; $v_{ij} = l_{nr}' v$; $u_{ij} = l_{nr}' u$; $\bar{v}_i = v' p_{nr}/r$ and $\bar{u}_i = u' p_{nr}/r$. Let

$$
\begin{aligned}
A &= I_{nr} - \frac{1}{nr} e_{nr} e_{nr}',\\
B &= \frac{1}{r}\left(I_n \otimes e_r' - \frac{1}{n} e_n e_{nr}'\right),\\
C &= I_n - \frac{1}{n} e_n e_n',
\end{aligned}
$$

and

$$
D = \frac{1}{r}\left(I_n \otimes e_r e_r' - \frac{1}{n} e_{nr} e_{nr}'\right).
$$

We observe that $BB' = C/r$, $B'B = D/r$, $AD = D$, and $CB = B$.

Also A, C and D are idempotent matrices with $trA = nr - 1$, $trC = n - 1$ and $trD = n - 1$.

Further, we define the following quantities and each is of order of $O_p(1)$:

$$
Q_u = \frac{e_{nr}' u}{\sqrt{nr}} = \sqrt{n}\bar{u}, \qquad Q_v = \frac{e_{nr}' v}{\sqrt{nr}} = \sqrt{n}\bar{v}, \qquad Q_w = \frac{e_n' w}{\sqrt{n}} = \sqrt{n}\bar{w},
$$

$$
\begin{aligned}
g_{xy} &= \frac{1}{\sqrt{n}\sigma_v^2}\{\frac{1}{\beta}(m+w)' Bu + (m+w)' Bv + 2m' Cw\\
&\quad + (w' Cw - n\sigma_w^2)\},\\
g_{xx} &= \frac{1}{\sqrt{n}\sigma_v^2}\left\{2m' Cw + 2(m+w)' Bv + (w' Cw - n\sigma_w^2)\right\},\\
g_{yy} &= \frac{1}{\sqrt{n}\sigma_v^2}\{\frac{2}{\beta}(m+w)' Bu + 2m' Cw + (w' Cw - n\sigma_w^2)\},
\end{aligned}
$$

$$t_{xy} = \frac{1}{\sqrt{n}r\beta\sigma_v^2}v'Au \quad , \quad t_{xy}^* = \frac{1}{\sqrt{n}r\beta\sigma_v^2}u'Dv,$$
$$t_{xx} = \frac{1}{\sqrt{n}r\sigma_v^2}(v'Av - nr\sigma_v^2) \quad , \quad t_{xx}^* = \frac{1}{\sqrt{n}r\sigma_v^2}(v'Dv - n\sigma_v^2),$$
$$t_{yy} = \frac{1}{\sqrt{n}r\beta^2\sigma_v^2}(u'Au - nr\sigma_u^2) \text{ and } t_{yy}^* = \frac{1}{\sqrt{n}r\beta^2\sigma_v^2}(u'Du - n\sigma_u^2).$$

Now we can express,

$$S_{xy} = \frac{1}{nr}\sum_{i=1}^{n}\sum_{j=1}^{r}(x_{ij} - \bar{x})(y_{ij} - \bar{y}) = \beta\sigma_v^2\left[\frac{(1-\theta)}{\theta} + \frac{1}{\sqrt{n}}\left(g_{xy} + t_{xy}\right)\right].$$

Similarly,

$$S_{xx} = \sigma_v^2\left[\frac{1}{\theta} + \frac{1}{\sqrt{n}}\left(g_{xx} + t_{xx}\right)\right],$$
$$S_{yy} = \beta^2\sigma_v^2\left[\frac{(1-\theta)}{\theta} + q + \frac{1}{\sqrt{n}}\left(g_{yy} + t_{yy}\right)\right],$$
$$B_{xy} = \beta\sigma_v^2\left[\frac{(1-\theta)}{\theta} + \frac{1}{\sqrt{n}}\left(g_{xy} + t_{xy}^*\right)\right],$$
$$B_{xx} = \sigma_v^2\left[\frac{(1-\theta)}{\theta} + \frac{1}{r} + \frac{1}{\sqrt{n}}\left(g_{xx} + t_{xx}^*\right)\right],$$

and

$$B_{yy} = \beta^2\sigma_v^2\left[\frac{(1-\theta)}{\theta} + \frac{q}{r} + \frac{1}{\sqrt{n}}\left(g_{yy} + t_{yy}^*\right)\right].$$

Proof of Theorems:

Now, the estimation error of estimator of $\hat{\beta}_1$ can be expressed as,

$$\begin{aligned}\hat{\beta}_1 - \beta &= \frac{\theta\beta}{\sqrt{n}(1-\theta)}\left[g_{xy} + t_{xy} - g_{xx} - \frac{r}{r-1}t_{xx}^* + \frac{1}{r-1}t_{xx}\right] \\ &\quad \times \left[1 + \frac{\theta}{\sqrt{n}(1-\theta)}\left\{g_{xx} + \frac{r}{r-1}t_{xx}^* - \frac{1}{r-1}t_{xx}\right\}\right]^{-1} \\ &= \frac{1}{\sqrt{n}}\eta_{-1/2} + \frac{1}{n}\eta_{-1} + O_p(n^{-3/2})\end{aligned}$$

where

$$\eta_{-1/2} = \frac{\theta\beta}{(1-\theta)} \left\{ g_{xy} + t_{xy} - g_{xx} - \frac{r}{r-1} t^*_{xx} + \frac{1}{r-1} t_{xx} \right\},$$

$$\eta_{-1} = -\frac{\beta\theta^2}{(1-\theta)^2} \left(g_{xx} + \frac{r}{r-1} t^*_{xx} - \frac{1}{r-1} t_{xx} \right)$$
$$\times \left\{ g_{xy} + t_{xy} - g_{xx} - \frac{r}{r-1} t^*_{xx} - \frac{1}{r-1} t_{xx} \right\}.$$

Similarly, the estimation errors of $\hat{\beta}_2$, $\hat{\beta}_3$ and $\hat{\beta}_4$ can be expressed as

$$\hat{\beta}_2 - \beta = \frac{1}{\sqrt{n}} \xi_{-1/2} + \frac{1}{n} \xi_{-1} + O_p(n^{-3/2}),$$

$$\hat{\beta}_3 - \beta = \frac{1}{\sqrt{n}} \phi_{-1/2} + \frac{1}{n} \phi_{-1} + O_p(n^{-3/2}),$$
$$\hat{\beta}_4 - \beta = \frac{1}{\sqrt{n}} \psi_{-1/2} + \frac{1}{n} \psi_{-1} + O_p(n^{-3/2}),$$

where

$$\xi_{-1/2} = \frac{\theta\beta}{(1-\theta)} \left[g_{xy} + t^*_{xy} - g_{xx} - \frac{r}{r-1} t^*_{xx} + \frac{1}{r-1} t_{xx} \right],$$

$$\xi_{-1} = -\frac{\theta^2\beta}{(1-\theta)^2} \left[\left(g_{xx} + \frac{r}{r-1} t^*_{xx} - \frac{1}{r-1} t_{xx} \right) \right.$$
$$\left. \times \left\{ g_{xy} + t^*_{xy} - g_{xx} - \frac{r}{r-1} t^*_{xx} + \frac{1}{r-1} t_{xx} \right\} \right],$$

$$\phi_{-1/2} = \frac{\theta\beta}{(1-\theta)} \left(g_{yy} + \frac{r}{r-1} t^*_{yy} - \frac{1}{r-1} t_{yy} - g_{xy} - t_{xy} \right),$$

$$\phi_{-1} = -\frac{\theta^2\beta}{(1-\theta)^2} (g_{xy} + t_{xy}) \left\{ g_{yy} + \frac{r}{r-1} t^*_{yy} \right.$$
$$\left. - \frac{1}{r-1} t_{yy} - g_{xy} - t_{xy} \right\},$$

$$\psi_{-1/2} = \frac{\theta\beta}{(1-\theta)} \left(g_{yy} + \frac{r}{r-1} t^*_{yy} - \frac{1}{r-1} t_{yy} - g_{xy} - t^*_{xy} \right),$$

$$\psi_{-1} = -\frac{\theta^2\beta}{(1-\theta)^2} (g_{xy} + t^*_{xy}) \left\{ g_{yy} + \frac{r}{r-1} t^*_{yy} \right.$$
$$\left. - \frac{1}{r-1} t_{yy} - g_{xy} - t^*_{xy} \right\}.$$

Then the prediction error of predictors P_{1wD}, P_{2wD}, P_{1wR} and P_{2wR} in predicting the kth true value of study variable, when corresponding value on explanatory variable is available, can be expressed as:

$$\hat{P} - P = (\hat{\beta} - \beta)(m_k - \bar{m} + w_k + v_{kj}) - \frac{1}{\sqrt{n}}(\hat{\beta} - \beta)(Q_v + Q_w) + \frac{1}{\sqrt{n}}(Q_u - \beta Q_v) + \beta(1-\lambda)w_k + \beta v_{kj} - \lambda u_{kj}$$

where $\hat{P}$ can be substituted with P_{1wD}, P_{2wD}, P_{1wR} and P_{2wR} with P being the corresponding target function. Substituting the corresponding expression of $(\hat{\beta} - \beta)$, we can obtain different prediction errors.

Similarly, the prediction error of predictors with target function $\tilde{P}_o$ in case of outside sample prediction is given by

$$\hat{P}_o - \tilde{P}_o = (\hat{\beta} - \beta)(m_f - \bar{m} + w_f + v_{fj}) - \frac{1}{\sqrt{n}}(\hat{\beta} - \beta)(Q_v + Q_w) + \frac{1}{\sqrt{n}}(Q_u - \beta Q_v) + \beta(1-\lambda)w_k + \beta v_{fj} - \lambda u_{fj}.$$

Substituting $\hat{P}_o = P_{1oD}$, P_{2oD} with corresponding $\tilde{P}_o = P_o, P_o^*$ and $\hat{\beta} = \hat{\beta}_1, \hat{\beta}_3$, we can obtain the prediction errors of P_{1oD} and P_{2oD}.

Proof of Theorem 1

Now, the predictive bias of P_{1wD} is given by

$$\begin{aligned} \text{PB}(P_{1wD}) &= E(P_{1wD} - P_w) \\ &= E\Big[(m_k - \bar{m})(\hat{\beta}_1 - \beta) + (\hat{\beta}_1 - \beta)(w_k + v_{kj}) \\ &\quad - \frac{1}{\sqrt{n}}(\hat{\beta}_1 - \beta)(Q_w + Q_v) \\ &\quad + \frac{1}{\sqrt{n}}(Q_u - \beta Q_v) + \beta(1-\lambda)w_k + \beta v_{fj} - \lambda u_{fj}\Big] \end{aligned} \tag{44}$$

and the predictive mean squared error of P_{1wD} is given by:

$$\begin{aligned} E(P_{1wD} - P_w)^2 = E\Big[&(m_k - \bar{m})(\hat{\beta}_1 - \beta) + (\hat{\beta}_1 - \beta)(w_k + v_{kj}) \\ &- \frac{1}{\sqrt{n}}(\hat{\beta}_1 - \beta)(Q_w + Q_v) \\ &+ \frac{1}{\sqrt{n}}(Q_u - \beta Q_v) + \beta(1-\lambda)w_k + \beta v_{fj} - \lambda u_{fj}\Big]^2. \end{aligned} \tag{45}$$

Similar expressions for the predictive bias and predictive mean squared error of P_{2wD} can also be derived. Utilizing the distributional properties of of u, v, and w and following Srivastava and Tiwari (1976), we can derive different Expectations of product of vectors and matrices. Using them in (44) and (45), we derive the results stated in Theorem 1.

The results for Theorem 2, 3 and 4 can also be derived in the similar way.

Acknowledgement

The first author acknowledges the support from Department of Science and Technology, Government of India through a SERC project.

References

Buonaccorsi, John P (1995) Prediction in the presence of measurement error: General discussion and an example predicting defoliation. Biometrics 51:1562–1569

Chaturvedi Anoop, Wan ATK, Singh SP (2002) Improved multivariate prediction in a general linear model with an unknown error covariance matrix. Journal Multivariate Analysis 83(1):166–182

Cheng CL, Van Ness JW (1999) Statistical Regression with Measurement Error. Arnold Publishers, London

Dolby GR (1976) The ultrastructural relation: A synthesis of functional and structural relations. Biometrika 63(1):39–50

Dube M, Manocha V (2002) Simultaneous prediction in restricted regression models. Journal Applied Statistical Science 11(4):277–288

Fuller WA (1987) Measurement Error Model. Wiley, New York

Ganse RA, Amemiya, Y, Fuller WA (1983) Prediction when both variables are subject to error, with application to earthquake magnitude. Journal of American Statistical Association 78:761–765

Huwang L, Hwang JTG (2002) Prediction and confidence intervals for nonlinear measurement error models without identifiability information. Statistics & Probability Letters 58:355–362

Kendall M, Stuart A (1967) The Advance Theory of Statistics. Vol 2 4th edition Charles Griffin & Company, London

Nummi T, Möttönen J (2004) Estimation and prediction for low degree polynomial models under measurement errors with an application to forest harvesters. Applied Statistics 53(3):495–505

Rao CR, Toutenburg H, Shalabh, Heumann, H (2008) Linear Models and Generalizations – Least Squares and Alternatives (3rd edition). Springer, Berlin Heidelberg New York

Reilman MA, Gunst RF (1986) Least squares and maximum likelohood predictors in measurement error models. Presented at 1986 Joint Statistical Meeting Chicago

Schneeweiss H (1991) Note on a linear model with errors in the variables and with trend. Statistical Papers 32:261–264

Schneeweiss H (1982) A Simple Regression Model with Trend and Error in the Exogenous Variable. In: M Deistler, E Fuest , G Schwodiauer (eds) Games, Economic Dynamics and Time Series Analysis, Wurzburg Physica Verlag, 347–358

Schaalje GB, Butts, RA (1993) Some effects of ignoring correlated measurement error variances in straight line regression and prediction. Biometrics 49:1262–1267

Shalabh (1995) Performance of Stein–rule procedure for simultaneous prediction of actual and average values of study variable in linear regression model. Bulletin of the International Statistical Institute The Netherland 1375–1390

Srivastava AK, Shalabh (1995) Predictions in linear regression models with measurement errors. Indian Journal of Applied Economics 4(2):1–14

Srivastava AK, Shalabh (1997a) Consistent estimation for the non–normal ultrastructural model. Statistics and Probability Letters 34:67–73

Srivastava AK, Shalabh (1997b) Improved estimation of slope parameter in a linear ultrastructural model when measurement errors are not necessarily normal. Journal of Econometrics 78:153–157

Srivastava VK, Tiwari R (1976) Evaluation of expectation of product of stochastic matrices. Scandinavian Journal of Statistics 3:135–138

Toutenburg H, Shalabh (1996) Predictive performance of the methods of restricted and mixed regression estimators. Biometrical Journal 38(8):951–959

Toutenburg H, Shalabh (2000) Improved prediction in linear regression model with stochastic linear constraints. Biometrical Journal 42(1):71–86

Toutenburg H, Shalabh (2002) Prediction of response values in linear regression models from replicated experiments. Statistical Papers 43:423–433

Local Sensitivity in the Inequality Restricted Linear Model

Huaizhen Qin[1], Alan T.K. Wan[2] and Guohua Zou[3]

[1] Department of Mathematical Sciences, Michigan Technological University, 1400 Townsend Drive, Houghton, Michigan 49931-1295, U.S.A. hqin@mtu.edu
[2] Department of Management Sciences, City University of Hong Kong, Kowloon, Hong Kong msawan@cityu.edu.hk
[3] Academy of Mathematics and Systems Science, Chinese Academy of Sciences, Beijing 100080, People's Republic of China, and Department of Biostatistics and Computational Biology, University of Rochester, Rochester, NY 14642, U.S.A. Guohua_Zou@URMC.Rochester.edu

1 Introduction

Diagnostic testing has traditionally been an important aspect of statistical modeling, but in recent years, sensitivity analysis has also been drawing increasing attention from econometricians and statisticians. Essentially, a diagnostic test ascertains if the model coincides with the assumed data generating process, while sensitivity analysis investigates if it matters at all that the model deviates from what is being assumed. That is, sensitivity analysis answers the question of whether a wrong model is still useful for certain purposes, and if so, it matters little that the model may be incorrect. For example, Banerjee and Magnus (1999) pointed out that the ordinary least squares (OLS) estimator of the coefficients in a linear regression model is in fact not very sensitive to disturbances' deviation from the white noise assumption. Consequently, it is quite usual to find the estimates of the parameters not changing much after fitting the model with a more general covariance structure. However, the F and t tests based on the OLS residuals are sensitive to covariance misspecification in the sense that a small stepping away from white noise disturbances is likely to cause a substantial distortion in the significance levels of the tests (Banerjee and Magnus (2000)).

In addition to the above mentioned work, Magnus and Vasnev (2007) showed that for many problems encountered in practice, diagnostic testing and sensitivity analysis are in fact independent asymptotically. In other words, diagnostic test results do not necessarily provide insights into the sensitivity of the parameter estimates. Seen in this light it becomes important to have a set of tools that measure the sensitivity of the estimates. Banerjee and Magnus (1999) and Banerjee and Magnus (2000) introduced sensitivity statistics for measuring the effects of possibly non-white noise disturbances on the OLS coefficient and variance estimators and the usual F and t tests in a linear regression. Wan, Zou and Qin (2007) provided an analytic proof showing that the coefficient and variance sensitivity statistics given in Banerjee and Magnus (1999) are approximately uncorrelated in large samples, and generalized Banerjee and Magnus' (1999) work to the restricted regression model allowing for possibly incorrect restrictions.

The current paper continues this line of research. We are concerned with a linear regression with a possibly incorrect inequality restriction (as opposed to strict equality restrictions as in Wan, Zou and Qin (2007)) on the coefficients. In econometric applications inequality restrictions frequently arise on the parameters. Finite sample properties of the inequality constraint least squares (ICLS) estimator have been investigated by Thomson (1982), Judge and Yancey (1986), Wan (1994a), Wan (1994b), among others. Judge and Yancey (1986), Wan (1994a), Wan (1994b), Wan (1995) and Wan (1996) considered the properties of the so-called inequality pre-test (IPT) estimator which chooses between the inequality restricted and OLS estimators depending on the outcome of a one-sided t test. In this paper, we investigate the sensitivity of the ICLS and IPT estimators to deviations of the disturbances from the white noise assumption. In the spirit of Banerjee and Magnus (1999), we propose sensitivity measures on the ICLS and IPT estimators to covariance misspecification and investigate the properties of these measures allowing for both correctly and incorrectly specified constraints.

The rest of this paper is organized as follows. Section 2 gives some preliminary results and defines sensitivity statistics to measure the sensitivity of the ICLS and IPT coefficient and variance estimators to covariance misspecification. Section 3 emphasizes the case of AR(1) disturbances and derives results concerning the limiting behavior of the sensitivity statistics when the AR(1) parameter is near the unit-root. Section 4 presents numerical findings on the sensitivity of the estimators under a variety of AR(1) and MA(1) settings and Section 5 concludes. Proofs of theorems are contained in Appendix A.

2 Definitions and Preliminary Results

The model under consideration is the classical full rank linear regression model,

$$y = X\beta + u, \qquad u \sim \mathrm{N}(0, \sigma^2 \Omega(\theta)) \tag{1}$$

where y and u are $n \times 1$ vectors; X $(n \times k)$ is a non-stochastic matrix of full column rank; β $(k \times 1)$ is a vector of unknown coefficients; $\Omega(\theta)$ $(n \times n)$ is a known[1] function of $\theta = (\theta_1, ..., \theta_p)'$, positive-definite, differentiable at least in a neighborhood of $\theta = 0$ and equal to I_n when $\theta = 0$.

If θ is known, the familiar generalized least squares (GLS) estimators of β and σ^2 are $\hat{\beta}(\theta) = S^{-1}(\theta) X' \Omega^{-1}(\theta) y$ and $\hat{\sigma}^2(\theta) = (y - X\hat{\beta}(\theta))' \Omega^{-1}(\theta)(y - X\hat{\beta}(\theta))/(n-k)$ respectively, where $S(\theta) = X'\Omega^{-1}(\theta)X$. Sensitivity analysis in the context of (1) has been concerned with the question of whether there is any real difference between the OLS estimates (which assume $\theta = 0$) and the GLS estimates when θ is non-zero. If θ deviates from 0, but the GLS estimates are nevertheless close to the corresponding OLS estimates, then the OLS estimator may still be a useful tool for analysis in the face of a non-spherical error covariance structure. Indeed, Banerjee and Magnus (1999) showed that $\hat{\beta}(\theta)$ is not very sensitive to covariance misspecification even though $\hat{\sigma}^2(\theta)$ can be sensitive. Furthermore, in the case of AR(1) errors, the Durbin-Watson test only indicates the sensitivity of $\hat{\sigma}^2(\theta)$, but tells us little about how sensitive $\hat{\beta}(\theta)$ is to changes in the AR(1) parameter. Banerjee and Magnus (1999) showed via a Monte-Carlo study that the Durbin-Watson test statistic is approximately uncorrelated with the sensitivity indicator of $\hat{\beta}(\theta)$. Wan, Zou and Qin (2007) provided some theoretical support for Banerjee and Magnus' (1999) numerical findings. Sensitivity analysis therefore matters, and "sensitivity statistics" (as opposed to diagnostic test statistics) are needed to decide if an estimate is sensitive to covariance misspecification.

In this paper we are concerned with the case where additional information is available in the form of a single inequality hypothesis $H_0: R\beta \geq r$, where $R(1 \times k)$ and $r(1 \times 1)$ are both known. Adding the constraint $R\beta \geq r$ to (1) and estimating the model by the method of least squares leads to the inequality constrained GLS (ICGLS) estimators

[1] The assumption of a known structure on Ω leads to no loss of generality here given the purpose of the paper is to investigate if the estimates that incorrectly assume $\theta = 0$ are really different from the GLS estimates based on a correct error covariance structure.

$$\bar{\bar{\beta}}(\theta) = \mathrm{I}_{(-\infty,r)}\left(R\hat{\beta}(\theta)\right)\bar{\beta}(\theta) + \mathrm{I}_{[r,\infty)}\left(R\hat{\beta}(\theta)\right)\hat{\beta}(\theta) \tag{2}$$

and

$$\bar{\bar{\sigma}}^2(\theta) = \mathrm{I}_{(-\infty,r)}\left(R\hat{\beta}(\theta)\right)\bar{\sigma}^2(\theta) + \mathrm{I}_{[r,\infty)}\left(R\hat{\beta}(\theta)\right)\hat{\sigma}^2(\theta) \tag{3}$$

of β and σ^2 respectively, where $\mathrm{I}_{(.)}(.)$ is an indicator function which equals 1 if the event inside the bracket occurs and 0 otherwise, and $\bar{\beta}(\theta) = \hat{\beta}(\theta) + S^{-1}(\theta)R'(RS^{-1}(\theta)R')^{-1}(r - R\hat{\beta}(\theta))$ and $\bar{\sigma}^2(\theta) = (y - X\bar{\beta}(\theta))'\Omega^{-1}(\theta)(y - X\bar{\beta}(\theta))/(n-k+1)$ are the equality constrained GLS (ECGLS) estimators which make use of the exact prior constraint $R\beta = r$. If the inequality restriction $R\beta \geq r$ is redundant the ICGLS estimator is the corresponding GLS estimator. Alternatively if the restriction is binding the ICGLS estimator reduces to the ECGLS estimator. Consider also the problem of testing $H_0: \ R\beta \geq r$. In principle one could decide if the ICGLS estimators should be used with the aid of a one-sided t-test of H_0, leading to the inequality generalized pre-test (IGPT) estimators

$$\tilde{\beta}(\theta) = I_{(-\infty,c)}\left(t(\theta)\right)\hat{\beta}(\theta) + I_{[c,\infty)}\left(t(\theta)\right)\bar{\bar{\beta}}(\theta) \tag{4}$$

and

$$\tilde{\sigma}^2(\theta) = I_{(-\infty,c)}\left(t(\theta)\right)\hat{\sigma}^2(\theta) + I_{[c,\infty)}\left(t(\theta)\right)\bar{\bar{\sigma}}^2(\theta), \tag{5}$$

where $t(\theta) = \left(R\hat{\beta}(\theta) - r\right)/\sqrt{\hat{\sigma}^2(\theta)RS^{-1}(\theta)R'}$, and $c(<0)$ is the size-α critical value for the Student's t distribution with $n-k$ degrees of freedom[2].

In the case of $\theta = 0$, the properties of the ICGLS and IGPT estimators have been thoroughly investigated. Good discussions on inequality pre-testing in the linear model are given, for example, by Judge and Yancey (1986) and Wan, Zou and Ohtani (2006). In this paper we examine the sensitivity of the ICGLS and IGPT estimators to covariance misspecification . Specifically, assuming $\theta \neq 0$, we ascertain the question of how far the ICGLS and IGPT estimates are from the corresponding estimates which assume $\theta = 0$. That is, are the results that assume $\theta = 0$ really different from the results based on the more general error covariance structure? We will also examine the relationship between the restriction specification error and the sensitivity of the estimators. The sensitivity of the ECGLS estimators was investigated in Wan, Zou and Qin (2007).

[2] The case of $c \geq 0$ is of no interest as $t(\theta) \geq c$ would then imply $R\hat{\beta}(\theta) \geq r$. Hence the unrestricted GLS estimator is always chosen when $c \geq 0$, irrespective of whether H_0 is accepted or not.

Along the lines of Banerjee and Magnus (1999), we develop sensitivity measures in the following fashion. Let $\Lambda(\theta)$ be a generic notation of the relevant estimator of interest. The Taylor series expansion for $\Lambda(\theta)$ evaluated about the point $\theta = 0$ gives the result

$$\Lambda(\theta) = \Lambda(0) + \sum_{s=1}^{p} \theta_s \left. \frac{\partial \Lambda(\theta)}{\partial \theta_s} \right|_{\theta=0} + \dots \tag{6}$$

Obviously, if $\sum_{s=1}^{p} \theta_s \left. \frac{\partial \Lambda(\theta)}{\partial \theta_s} \right|_{\theta=0} \approx 0$ then $\Lambda(\theta)$ and $\Lambda(0)$ are very close. A sufficient condition for this is that

$$\left. \frac{\partial \Lambda(\theta)}{\partial \theta_s} \right|_{\theta=0} = 0 \quad \text{for } s = 1, 2, ..., p. \tag{7}$$

Putting this general framework in the context of our analysis, we define $\bar{\bar{z}}_s = \left. \frac{\partial \left[\bar{\bar{y}}(\theta)\right]}{\partial \theta_s} \right|_{\theta=0}$, $\bar{\bar{\lambda}}_s = \left. \frac{\partial \bar{\bar{\sigma}}^2(\theta)}{\partial \theta_s} \right|_{\theta=0}$, $\tilde{z}_s = \left. \frac{\partial \left[\tilde{y}(\theta)\right]}{\partial \theta_s} \right|_{\theta=0}$ and $\tilde{\lambda}_s = \left. \frac{\partial \tilde{\sigma}^2(\theta)}{\partial \theta_s} \right|_{\theta=0}$ as the sensitivity measures of the ICGLS and IGPT predictors and variance estimators with respect to θ_s, respectively, where $\bar{\bar{y}}(\theta) = X\bar{\bar{\beta}}(\theta)$ and $\tilde{y}(\theta) = X\tilde{\beta}(\theta)$. Corresponding to each of these measures is a statistic suitable for assessing the sensitivity of the predictor or estimator of interest. Transforming these sensitivity measures in the manner of Banerjee and Magnus (1999) leads to

$$\bar{\bar{B}}_s = \frac{\bar{\bar{z}}_s'(\bar{\bar{C}}_s\bar{\bar{C}}_s')^{+}\bar{\bar{z}}_s}{\left[n - k + I_{(-\infty,r)}\left(R\hat{\beta}(0)\right)\right]\bar{\bar{\sigma}}^2(0)}, \tag{8}$$

$$\bar{\bar{D}}_s = \frac{\bar{\bar{\lambda}}_s}{\bar{\bar{\sigma}}^2(0)} = \left. \frac{\partial \log\left[\bar{\bar{\sigma}}^2(\theta)\right]}{\partial \theta_s} \right|_{\theta=0}, \tag{9}$$

$$\tilde{B}_s = \frac{\tilde{z}_s'(\tilde{C}_s\tilde{C}_s')^{+}\tilde{z}_s}{(n - k + \kappa)\,\tilde{\sigma}^2(0)}, \tag{10}$$

and

$$\tilde{D}_s = \frac{\tilde{\lambda}_s}{\tilde{\sigma}^2(0)} = \left. \frac{\partial \log\left[\tilde{\sigma}^2(\theta)\right]}{\partial \theta_s} \right|_{\theta=0} \tag{11}$$

as the corresponding sensitivity statistics of $\bar{\bar{y}}(\theta)$, $\bar{\bar{\sigma}}^2(\theta)$, $\tilde{y}(\theta)$ and $\tilde{\sigma}^2(\theta)$, respectively, where $\bar{\bar{C}}_s = I_{(-\infty,r)}\left(R\hat{\beta}(0)\right)\bar{C}_s + I_{[r,\infty)}\left(R\hat{\beta}(0)\right)C_s$, $\bar{C}_s = (I_n - \bar{M})A_s\bar{M}$, $C_s = (I_n - M)A_sM$, $A_s = \partial\Omega(\theta)/\partial\theta_s|_{\theta=0}$, $(\bar{\bar{C}}_s\bar{\bar{C}}_s')^{+}$ is the

Moore-Penrose inverse of $\bar{\bar{C}}_s\bar{\bar{C}}_s'$, $(\tilde{C}_s\tilde{C}_s')^+$ is defined analogously, $M = I_n - XS^{-1}X'$ and $\bar{M} = M + H(H'H)^{-1}H'$ are symmetric idempotent matrices of rank $n-k$ and $n-k+1$ respectively, $H = XS^{-1}R'$, $S = X'X$, $\kappa = \mathrm{I}_{(-\infty,\, r)}\left(R\hat{\beta}(0)\right)\mathrm{I}_{[c,\infty)}(t(0))$ and $\tilde{C}_s = \mathrm{I}_{(-\infty,\, c)}(t(0))\, C_s + \mathrm{I}_{[c,\,\infty)}(t(0))\, \bar{\bar{C}}_s = (1-\kappa)\, C_s + \kappa\bar{C}_s$. Note that $\bar{M}M = M\bar{M} = M$, $\bar{C}_s\bar{M} = \bar{C}_s$ and $0 \le \bar{r}_s = \mathrm{rank}(\bar{C}_s) \le \min\{k-1,\, n-k+1\}$.

It can be observed that $\tilde{B}_s$ and $\tilde{D}_s$ collapse to $\bar{\bar{B}}_s$ and $\bar{\bar{D}}_s$ as $c \to -\infty$. Now, write $\bar{y} = y - H(H'H)^{-1}r$, $\bar{z}_s = -\bar{C}_s\bar{y}$, $z_s = -C_s y_{,}$, $\bar{\lambda}_s = -\bar{y}'\bar{M}A_s\bar{M}\bar{y}/(n-k+1)$, $\lambda_s = -y'MA_sMy/(n-k)$, $\bar{W}_s = \bar{C}_s'(\bar{C}_s\bar{C}_s')^+\bar{C}_s$, $W_s = C_s'(C_sC_s')^+C_s$, $v = u/\sigma$, $\delta = (r - R\beta)/\sigma$, $\bar{v} = v - H(H'H)^{-1}\delta$, $\bar{B}_s = \bar{y}'\bar{W}_s\bar{y}/\bar{y}'\bar{M}\bar{y}$, $B_s = y'W_sy/y'My$, $\bar{D}_s = -\bar{y}'\bar{M}A_s\bar{M}\bar{y}/\bar{y}'\bar{M}\bar{y}$ and $D_s = -y'MA_sMy/y'My$. Hence we have $\kappa = \mathrm{I}_{(-\infty,\, \delta)}(H'v)\,\mathrm{I}_{[c,\infty)}(t(0))$ and $t(0) = \left(\sqrt{n-k}(H'v - \delta)\right) / \left(\sqrt{(H'H)(v'Mv)}\right)$. After some straightforward calculations, we obtain the following theorem which provides a convenient basis upon which the sensitivity statistics may be further evaluated:

Theorem 1. *The sensitivity statistics $\bar{\bar{B}}_s$, $\bar{\bar{D}}_s$, $\tilde{B}_s$ and $\tilde{D}_s$ may be written as*

$$\bar{\bar{B}}_s = \mathrm{I}_{(-\infty,\delta)}\left(H'v\right)\frac{\bar{v}'\bar{W}_s\bar{v}}{\bar{v}'\bar{M}\bar{v}} + \mathrm{I}_{[\delta,\infty)}\left(H'v\right)\frac{v'W_sv}{v'Mv}, \tag{12}$$

$$\bar{\bar{D}}_s = -\mathrm{I}_{(-\infty,\delta)}\left(H'v\right)\frac{\bar{v}'\bar{M}A_s\bar{M}\bar{v}}{\bar{v}'\bar{M}\bar{v}} - \mathrm{I}_{[\delta,\infty)}\left(H'v\right)\frac{v'MA_sMv}{v'Mv}, \tag{13}$$

$$\tilde{B}_s = \kappa\frac{\bar{v}'\bar{W}_s\bar{v}}{\bar{v}'\bar{M}\bar{v}} + (1-\kappa)\frac{v'W_sv}{v'Mv}, \tag{14}$$

and

$$\tilde{D}_s = -\kappa\frac{\bar{v}'\bar{M}A_s\bar{M}\bar{v}}{\bar{v}'\bar{M}\bar{v}} - (1-\kappa)\frac{v'MA_sMv}{v'Mv}. \tag{15}$$

Proof:

See Appendix A.

Several aspects concerning the properties of the sensitivity statistics deserve mention. First, these stochastic representations of the sensitivity statistics depend on the unknown parameters only through δ. For a given δ value, equations (12) to (15) are distributional invariant with respect to changes in the regression and constraint parameters. This allows us to conveniently investigate their properties by varying the values of δ. Second, each of the sensitivity statistics presented in (12) to (15) has a discontinuity at precisely the point where the estimator whose sensitivity the statistic is designed to measure has a discontinuity. Third, the distributions of the sensitivity statistics are in fact data dependent; nevertheless, the stochastic representations given in (12) to (15) provide a useful basis for the subsequent Monte-Carlo evaluations on the properties of these statistics and the sensitivity of the ICGLS and IGPT estimators under varying circumstances.

3 Limiting Behavior Near Unit-root

To gain further insights into the properties of these sensitivity statistics, in this section we specialize our treatment to the AR(1) process, which is often regarded as the first step away from white noise errors. Let u_t be generated by the stationary AR(1) process $u_t = \phi_1 u_{t-1} + \varepsilon_t$, where $0 \le \phi_1 < 1$ and $\varepsilon_t' s$ are white noises. So

$$\Omega(\phi_1) = (\omega_{IJ}(\phi_1))\,, \text{ where } \omega_{IJ}(\phi_1) = \begin{cases} 1/(1-\phi_1^2) & \text{if } I = J, \\ \phi_1^{|I-J|}/(1-\phi_1^2) & \text{if } I \neq J. \end{cases} \tag{16}$$

Let $T^{(1)} = (t_{IJ})$ be the symmetric Toeplitz matrix such that $t_{IJ} = 1$ if $|I-J| = 1$, and $t_{IJ} = 0$ otherwise. Note that for AR(1) disturbances, $A_s = \partial\Omega(\phi_1)/\partial\phi_1|_{\phi_1=0} = T^{(1)}$. In conformity with the notations of Banerjee and Magnus (1999) and Wan, Zou and Qin (2007), we denote the sensitivity statistics $\bar{\bar{B}}_s$, $\bar{\bar{D}}_s$, $\tilde{B}_s$ and $\tilde{D}_s$ under the AR(1) setting as $\bar{\bar{B}}1$, $\bar{\bar{D}}1$, $\tilde{B}1$ and $\tilde{D}1$ respectively. The following theorem presents exact theoretical results on the limiting behavior of $\bar{\bar{B}}1$ and $\bar{\bar{D}}1$ as the AR(1) parameter approaches the unit-root.

Theorem 2. *Suppose that* $u \sim \mathrm{N}\left(0, \sigma^2\Omega(\phi_1)\right)$ *with* $\Omega(\phi_1)$ *given in* (16), *and* $c_{\bar{\bar{B}}1}$ *and* $c_{\bar{\bar{D}}1}$ *are arbitrary constants:*

i) If $Mi \neq 0$ *and* $H'i \neq 0$, *then*

$$\lim_{\phi_1 \to 1} \Pr(\bar{\bar{B}}1 > c_{\bar{\bar{B}}1}) = \begin{cases} 0 & \text{if } c_{\bar{\bar{B}}1} > b1 \vee \bar{b}1, \\ \frac{1}{2} & \text{if } b1 < c_{\bar{\bar{B}}1} < \bar{b}1 \text{ or } b1 > c_{\bar{\bar{B}}1} > \bar{b}1, \\ 1 & \text{if } c_{\bar{\bar{B}}1} < b1 \wedge \bar{b}1, \end{cases} \tag{17}$$

and

$$\lim_{\phi_1 \to 1} \Pr(\bar{\bar{D}}1 \le c_{\bar{\bar{D}}1}) = \begin{cases} 0 & if \quad c_{\bar{\bar{D}}1} < d1 \wedge \bar{d}1, \\ \frac{1}{2} & if \quad d1 > c_{\bar{\bar{D}}1} > \bar{d}1 \;\; or \;\; d1 < c_{\bar{\bar{D}}1} < \bar{d}1, \\ 1 & if \quad c_{\bar{\bar{D}}1} > d1 \vee \bar{d}1, \end{cases} \tag{18}$$

where $\bar{b}1 = i'\bar{W}^{(1)}i/i'\bar{M}i$, $b1 = i'W^{(1)}i/i'Mi$, $\bar{d}1 = -i'\bar{M}T^{(1)}\bar{M}i/i'\bar{M}i$, $d1 = -i'MT^{(1)}Mi/i'Mi$, $\bar{W}^{(1)} = \bar{C}^{(1)\prime}(\bar{C}^{(1)}\bar{C}^{(1)\prime})^{+}\bar{C}^{(1)}$, $\bar{C}^{(1)} = (I_n - \bar{M})T^{(1)}\bar{M}$, $W^{(1)} = C^{(1)\prime}(C^{(1)}C^{(1)\prime})^{+}C^{(1)}$, $C^{(1)} = (I_n - M)T^{(1)}M$, *and* i *is an* $n \times 1$ *vector of ones.*

ii) If $Mi \neq 0$ *and* $H'i = 0$*, then*

$$\lim_{\phi_1 \to 1} \Pr(\bar{\bar{B}}1 > c_{\bar{\bar{B}}1}) = \begin{cases} 0 & if \quad c_{\bar{\bar{B}}1} > b1 \vee \bar{b}1, \\ \Pr\left(H'\bar{P}\eta < \delta\right) & if \quad b1 < c_{\bar{\bar{B}}1} < \bar{b}1, \\ \Pr\left(H'\bar{P}\eta \ge \delta\right) & if \quad b1 > c_{\bar{\bar{B}}1} > \bar{b}1, \\ 1 & if \; c_{\bar{\bar{B}}1} < b1 \wedge \bar{b}1, \end{cases} \tag{19}$$

and

$$\lim_{\phi_1 \to 1} \Pr(\bar{\bar{D}}1 \le c_{\bar{\bar{D}}1}) = \begin{cases} 0 & if \quad c_{\bar{\bar{D}}1} < d1 \wedge \bar{d}1, \\ 1 & if \quad c_{\bar{\bar{D}}1} > d1 \vee \bar{d}1, \end{cases} \tag{20}$$

where $\bar{P} = \bar{J}P$, $\bar{J}$ *is an* $n \times (n-1)$ *matrix such that* $\bar{J}' = [0|I_{n-1}]$, P *is an* $(n-1) \times (n-1)$ *lower triangular matrix with ones on and below the diagonal and zeroes elsewhere, and* $\eta \sim \mathrm{N}(0, I_{n-1})$.

iii) If $Mi = 0$ *and* $H'i \neq 0$*, then*

$$\lim_{\phi_1 \to 1} \Pr(\bar{\bar{B}}1 > c_{\bar{\bar{B}}1}) = \begin{cases} \frac{1}{2}\Pr\left(B^{(1)}(\eta) > c_{\bar{\bar{B}}1}\right) & if \quad c_{\bar{\bar{B}}1} > \bar{b}1, \\ \frac{1}{2}\left[1 + \Pr\left(B^{(1)}(\eta) > c_{\bar{\bar{B}}1}\right)\right] & if \quad c_{\bar{\bar{B}}1} < \bar{b}1, \end{cases} \tag{21}$$

and

$$\lim_{\phi_1 \to 1} \Pr(\bar{\bar{D}}1 \le c_{\bar{\bar{D}}1}) = \begin{cases} \frac{1}{2}\Pr\left(D^{(1)}(\eta) \le c_{\bar{\bar{D}}1}\right) & if \quad c_{\bar{\bar{D}}1} < \bar{d}1, \\ \frac{1}{2}\left[1 + \Pr\left(D^{(1)}(\eta) \le c_{\bar{\bar{D}}1}\right)\right] & if \quad c_{\bar{\bar{D}}1} > \bar{d}1. \end{cases} \tag{22}$$

iv) If $\bar{M}i = 0$*, then*

$$\lim_{\phi_1 \to 1} \Pr(\bar{\bar{B}}1 > c_{\bar{\bar{B}}1}) = \Pr\left(\bar{B}^{(1)}(\eta) > c_{\bar{\bar{B}}1}, \;\; H'\bar{P}\eta < \delta\right) + \Pr\left(B^{(1)}(\eta) > c_{\bar{\bar{B}}1}, \;\; H'\bar{P}\eta \ge \delta\right) \tag{23}$$

and

$$\lim_{\phi_1 \to 1} \Pr(\bar{\bar{D}}1 \le c_{\bar{\bar{D}}1}) = \Pr\left(\bar{D}^{(1)}(\eta) \le c_{\bar{\bar{D}}1},\ H'\bar{P}\eta < \delta\right) + \Pr\left(D^{(1)}(\eta) \le c_{\bar{\bar{D}}1},\ H'\bar{P}\eta \ge \delta\right), \quad (24)$$

where $\bar{B}^{(1)}(\eta) = l'\bar{W}^{(1)}l/l'\bar{M}l$, $\bar{D}^{(1)}(\eta) = -l'\bar{M}T^{(1)}\bar{M}l/l'\bar{M}l$, $l = \bar{P}\eta + \bar{\mu}$, $\bar{\mu} = -H(H'H)^{-1}\delta$, $B^{(1)}(\eta) = \eta'\bar{P}'W^{(1)}\bar{P}\eta/\eta'\bar{P}'M\bar{P}\eta$, *and* $D^{(1)}(\eta) = -\eta'\bar{P}'MT^{(1)}M\bar{P}\eta/\eta'\bar{P}'M\bar{P}\eta$.

Proof:

See Appendix A.

Theorem 2 holds for all values of δ and the limiting results derived apply to both the cases of correct and incorrect constraints. Note that $Mi = 0$ when the model contains an intercept. Parts i) and ii) of Theorem 2 therefore correspond to the case where the model has no intercept. Here, when $H'i \ne 0$ as in part i), $\Pr(\bar{\bar{B}}1 > c_{\bar{\bar{B}}1})$ and $\Pr(\bar{\bar{D}}1 \le c_{\bar{\bar{D}}1})$ approach 0, 1/2 or 1 depending on the conditions in (17) and (18). In part ii), $H'i = 0$, and the limiting probability of $\bar{\bar{D}}1 \le c_{\bar{\bar{D}}1}$ is either 0 or 1, while the limiting probability of $\bar{\bar{B}}1 > c_{\bar{\bar{B}}1}$ is 0, 1, $\Pr\left(H'\bar{P}\eta < \delta\right)$ or $\Pr\left(H'\bar{P}\eta \ge \delta\right)$ depending on the conditions in (19). Interestingly, these limiting results are quite different from those under the unrestricted and equality restricted models, where the corresponding sensitivity statistics have limiting probabilities of either 1 or 0 when the model has no intercept, as Banerjee and Magnus (1999) and Wan, Zou and Qin (2007) demonstrated. It is also worth noting that since $H \ne 0$, the conditions $Mi \ne 0$ and $H'i = 0$ in part ii) in fact imply $\bar{d}1 = d1$ and $H'\bar{P} \ne 0$ and accordingly, $H'\bar{P}\eta \sim \mathrm{N}(0, H'\bar{P}\bar{P}'H)$. Also, in the special case of $\delta = 0$, $\Pr\left(H'\bar{P}\eta < \delta\right) = \Pr\left(H'\bar{P}\eta \ge \delta\right) = 1/2$. Parts iii) and iv) of Theorem 2 correspond to the case when there is an intercept in the model. The conditions of $Mi = 0$ and $H'i \ne 0$ in part iii) imply that the inequality restriction involves the intercept, while in part iv), the restriction does not involve the intercept since $\bar{M}i = 0$ only when $Mi = 0$ and $H'i = 0$. In both parts iii) and iv), the limiting probabilities of both $\bar{\bar{B}}1 > c_{\bar{\bar{B}}1}$ and $\bar{\bar{D}}1 \le c_{\bar{\bar{D}}1}$ tend towards some constants between 0 and 1.

The next theorem presents the analogous results on the limiting behavior of $\tilde{B}1$ and $\tilde{D}1$:

Theorem 3. *Suppose that* $u \sim \mathrm{N}\left(0, \sigma^2 \Omega(\phi_1)\right)$ *with* $\Omega(\phi_1)$ *given in* (16), *and* $c_{\tilde{B}1}$ *and* $c_{\tilde{D}1}$ *are arbitrary constants:*

i) If $Mi \neq 0$ and $H'i \neq 0$, then

$$\lim_{\phi_1 \to 1} \Pr\left(\tilde{B}1 > c_{\tilde{B}1}\right) = \begin{cases} 0 & \text{if } c_{\tilde{B}1} > b1 \vee \bar{b}1, \\ 0 & \text{if } b1 < c_{\tilde{B}1} < \bar{b}1 \text{ and } |\bar{t}(0)| > -c, \\ \frac{1}{2} & \text{if } b1 < c_{\tilde{B}1} < \bar{b}1 \text{ and } |\bar{t}(0)| < -c, \\ \frac{1}{2} & \text{if } b1 > c_{\tilde{B}1} > \bar{b}1 \text{ and } |\bar{t}(0)| > -c, \\ 1 & \text{if } b1 > c_{\tilde{B}1} > \bar{b}1 \text{ and } |\bar{t}(0)| < -c, \\ 1 & \text{if } c_{\tilde{B}1} < b1 \wedge \bar{b}1, \end{cases} \tag{25}$$

and

$$\lim_{\phi_1 \to 1} \Pr\left(\tilde{D}1 \leq c_{\tilde{D}1}\right) = \begin{cases} 0 & \text{if } c_{\tilde{D}1} < d1 \wedge \bar{d}1, \\ 0 & \text{if } d1 > c_{\tilde{D}1} > \bar{d}1 \text{ and } |\bar{t}(0)| > -c, \\ \frac{1}{2} & \text{if } d1 > c_{\tilde{D}1} > \bar{d}1 \text{ and } |\bar{t}(0)| < -c, \\ \frac{1}{2} & \text{if } d1 < c_{\tilde{D}1} < \bar{d}1 \text{ and } |\bar{t}(0)| > -c, \\ 1 & \text{if } d1 < c_{\tilde{D}1} < \bar{d}1 \text{ and } |\bar{t}(0)| < -c, \\ 1 & \text{if } c_{\tilde{D}1} > d1 \vee \bar{d}1, \end{cases} \tag{26}$$

where $\bar{t}(0) = \sqrt{(n-k)/H'H}H'i \Big/ \sqrt{i'Mi}$.

ii) If $Mi \neq 0$ and $H'i = 0$, then

$$\lim_{\phi_1 \to 1} \Pr\left(\tilde{B}1 > c_{\tilde{B}1}\right) = \begin{cases} 0 & \text{if } c_{\tilde{B}1} > b1 \vee \bar{b}1, \\ \Pr\left(H'\bar{P}\eta \leq \delta\right) & \text{if } b1 < c_{\tilde{B}1} < \bar{b}1, \\ \Pr\left(H'\bar{P}\eta > \delta\right) & \text{if } b1 > c_{\tilde{B}1} > \bar{b}1, \\ 1 & \text{if } c_{\tilde{B}1} < b1 \wedge \bar{b}1, \end{cases} \tag{27}$$

and

$$\lim_{\phi_1 \to 1} \Pr\left(\tilde{D}1 \leq c_{\tilde{D}1}\right) = \begin{cases} 0 & \text{if } c_{\tilde{D}1} < d1, \\ 1 & \text{if } c_{\tilde{D}1} > d1, \end{cases} \tag{28}$$

provided that $c_{\tilde{B}1} \neq \bar{b}1$ and $\neq b1$, and $c_{\tilde{D}1} \neq d1$.

iii) If $\bar{M}i \neq 0$ and $Mi = 0$, then

$$\lim_{\phi_1 \to 1} \Pr\left(\tilde{B}1 > c_{\tilde{B}1}\right) = \Pr\left(B^{(1)}(\eta) > c_{\tilde{B}1}\right), \tag{29}$$

and

$$\lim_{\phi_1 \to 1} \Pr\left(\tilde{D}1 \leq c_{\tilde{D}1}\right) = \Pr\left(D^{(1)}(\eta) \leq c_{\tilde{D}1}\right), \tag{30}$$

provided that $c_{\tilde{B}1} \neq \bar{b}1$ *and* $c_{\tilde{D}1} \neq \bar{d}1$.

iv) If $\bar{M}i = 0$, *then*

$$\begin{aligned} \lim_{\phi_1 \to 1} \Pr\left(\tilde{B}1 > c_{\tilde{B}1}\right) &= \Pr\left(\bar{B}^{(1)}(\eta) > c_{\tilde{B}1},\ c \leq t^{(1)}(\eta) < 0\right) \\ &+ \Pr\left(B^{(1)}(\eta) > c_{\tilde{B}1},\ t^{(1)}(\eta) < c\right) \\ &+ \Pr\left(B^{(1)}(\eta) > c_{\tilde{B}1},\ t^{(1)}(\eta) \geq 0\right), \end{aligned} \tag{31}$$

and

$$\begin{aligned} \lim_{\phi_1 \to 1} \Pr\left(\tilde{D}1 \leq c_{\tilde{D}1}\right) &= \Pr\left(\bar{D}^{(1)}(\eta) \leq c_{\tilde{D}1},\ c \leq t^{(1)}(\eta) < 0\right) \\ &+ \Pr\left(D^{(1)}(\eta) \leq c_{\tilde{D}1},\ t^{(1)}(\eta) < c\right) \\ &+ \Pr\left(D^{(1)}(\eta) \leq c_{\tilde{D}1},\ t^{(1)}(\eta) \geq 0\right), \end{aligned} \tag{32}$$

where $t^{(1)}(\eta) = \sqrt{(n-k)/H'H}\left(H'\bar{P}\eta - \delta\right) \Big/ \sqrt{\eta'\bar{P}'M\bar{P}\eta}$.

Proof:

See Appendix A.

Qualitatively, the results emerged from Theorem 3 are analogous to those observed in Theorem 2. In parts i) and ii) of Theorem 3 where the regression contains no intercept, the limiting probabilities of $\tilde{B}1 > c_{\tilde{B}1}$ and $\tilde{D}1 \leq c_{\tilde{D}1}$ are not necessarily 0 or 1 depending on the conditions involved. Again, this differs from the published results for the unrestricted and equality restricted models. Interestingly, in part iii) where the model contains an intercept which is also part of the inequality restriction, the sensitivity statistics $\tilde{B}1$ and $\tilde{D}1$ take on the same limiting probabilities as their unrestricted counterparts (Banerjee and Magnus (1999)).

4 Numerical Studies

The purpose of this section is to examine the sensitivity of the ICGLS and IGPT estimators through numerical evaluations of the behavior of the proposed sensitivity statistics. We consider the cases of AR(1) and MA(1) disturbances. In the latter case $u_t = \psi_1 \varepsilon_{t-1} + \varepsilon_t$, where $0 \le \psi_1 \le 1$ and $\Omega(\psi_1) = (1+\psi_1^2)I_n + \psi_1 T^{(1)}$. Note that under MA(1) disturbances, as in the case of AR(1) disturbances. The statistics $\bar{\bar{B}}1$, $\bar{\bar{D}}1$, $\tilde{B}1$ and $\tilde{D}1$ therefore also measure the sensitivity of the estimators in the MA(1) case. Our numerical study is based on data constructed from linear combinations of the following two data sets: the first comprises the eigenvectors $t_1, t_2, \ldots, t_n$ that correspond to the eigenvalues of the $n \times n$ Toeplitz matrix $T^{(1)}$; in the second data set, the regressors are $s_1 = i_n/\sqrt{n}$, $s_p = \left(i'_{p-1},\ 1-p,\ 0_{1\times(n-p)}\right)' \big/ \sqrt{p(p-1)}$, where i_p is a $p \times 1$ vector of ones, $2 \le p \le n$. The intercept term in the regressor matrix is represented by the constant dummy s_1. We set $n = 15$, $k = 4$, and $R = [1, 0, 0, 0]$. Table 1 gives the design matrices used in our numerical experiments. Models 1 and 2 contain no intercept term. In model 3, the regression has an intercept and an inequality restriction is placed on the intercept. In model 4, the regression contains an intercept which is not part of the inequality restriction. Some characteristics of the design matrices are also given in Table 1, where $\ell = \sqrt{H'T^{(1)}M\bar{J}\bar{J}'MT^{(1)}H}$ is the length of $H'T^{(1)}M\bar{J}$. For each design matrix $X = [X_1|X_2]$, the restriction is in the form of $\beta_1 \ge r$, where β_1 is the coefficient corresponding to X_1. All our experiments set $\sigma = 1$ and $\delta = (r - \beta_1)/\sigma$ is set to $-10, -2, 0, 2$ and 10. The inequality constraint $\beta_1 \ge r$ is correct if $\delta \le 0$, and incorrect otherwise.

Table 1. Design matrices

Model	X_1	X_2	$i'Mi$	$H'i$	$H'T^{(1)}Mi$	ℓ
1	t_3	$[t_{12}, t_{13}+t_{14}, t_{14}+t_{15}]$	7.7345	−0.8035	−1.1480	1.2818
2	s_{15}	$[s_2, s_4, s_{14}]$	15	0	0	0.2323
3	s_1	$[s_{10}+s_{13}, s_{11}+s_{14}, s_{12}+s_{15}]$	0	3.8730	0	0.1966
4	s_6	$[s_1, s_2, s_3]$	0	0	0	1.3025

The investigation of the sensitivity of the ICGLS and IGPT estimators is based on the following procedure. Take the ICGLS estimators as an example. First, we determine the values of $c_{\bar{\bar{B}}1}$ and $c_{\bar{\bar{D}}1}$ such that $\Pr(\bar{\bar{B}}1 > c_{\bar{\bar{B}}1}) = \Pr(\bar{\bar{D}}1 \le c_{\bar{\bar{D}}1}) = \alpha$ under the white noise disturbances assumption. Then for values of ϕ_1 and ψ_1 between 0 and 1, we compute $\Pr(\bar{\bar{B}}1 > c_{\bar{\bar{B}}1})$ and $\Pr(\bar{\bar{D}}1 \le c_{\bar{\bar{D}}1})$ which measure the sensitivity of $\bar{\bar{y}}(\theta)$ and $\bar{\bar{\sigma}}^2(\theta)$ with respect to θ ($= \phi_1$ or ψ_1). We call the resultant probability curves the sensitivity curves. If the probabilities on the sensitivity curves do not significantly deviate from α, the ICGLS estimators that assume $\theta = 0$ are said to be robust against non white noise disturbances. In the same way, $\Pr(\tilde{B}1 > c_{\tilde{B}1})$ and $\Pr(\tilde{D}1 \leq c_{\tilde{D}1})$ measure the sensitivity of the IGPT estimators with respect to θ. The critical values are calculated using a Monte-Carlo procedure described in Appendix B.

We report the results of the numerical studies in two parts. Firstly, the sensitivity of the ICGLS estimators is discussed. Secondly, the IGPT estimators are considered, and their sensitivity with respect to θ is discussed in an analogous manner.

4.1 The ICGLS Estimators

Figures 1a – 4d provide plots of the sensitivity curves of the ICGLS estimators for several representative cases. In all cases we set $\alpha = 0.05$. We observe, first, that the limiting characteristics of the sensitivity curves are in accord with the theoretical findings presented in Theorem 2. Models 1 and 2 involve regressions with no intercept. In the case of AR(1) disturbances, Figure 1a shows, for example, that when $\delta = 2$, the limiting probability of $\bar{\bar{B}}1 > c_{\bar{\bar{B}}1}$ in model 1 (for which $H'i \neq 0$) is 1/2. This is consistent with our theoretical findings as in this case $b1 = 0.0541 > c_{\bar{\bar{B}}1} = 0.4756 > \bar{b}1 = 0.4710$, so it may be seen from part i) of Theorem 2 that $\Pr(\bar{\bar{B}}1 > c_{\bar{\bar{B}}1}) \to 1/2$ as $\phi_1 \to 1$. Figures 1b and 2b also show that for models 1 and 2, the limiting probabilities of $\bar{\bar{D}}1 \leq c_{\bar{\bar{D}}1}$ approach one as $\phi_1 \to 1$ regardless of the value of δ. Again, this arises because under both models, $c_{\bar{\bar{D}}1} > d1 \vee \bar{d}1$ for all δ considered. Models 3 and 4 both contain intercept terms. The limiting probabilities of $\bar{\bar{B}}1 > c_{\bar{\bar{B}}1}$ and $\bar{\bar{D}}1 \leq c_{\bar{\bar{D}}1}$ therefore lie between 0 and 1 (Figures 3a, 3b, 4a and 4b) as predicted from parts iii) and iv) of Theorem 3.

Secondly, under AR(1) and MA(1) errors, the sensitivity curves reveal quite different patterns. In the former case, depending on the data matrix and the values of ϕ_1 and δ, both $\bar{\bar{y}}(\theta)$ and $\bar{\bar{\sigma}}^2(\theta)$ can be sensitive to covariance misspecification. In general, the variance estimator is more sensitive than the predictor to AR(1) disturbances. As may be seen from the figures, the $\bar{\bar{B}}1$ sensitivity curves under AR(1) errors are quite close in value to the benchmark of 0.05 for small to moderate values of ϕ_1. On the other hand, the $\bar{\bar{D}}1$ curves show that the variance estimator can be highly sensitive even for small values of ϕ_1. In the case of MA(1) disturbances, the $\bar{\bar{B}}1$ curves are often flat and do not deviate significantly in value from 0.05 while the $\bar{\bar{D}}1$ curves tend to rise above the 0.05 level as ψ_1 increases even though the deviations of the curves from 0.05 are not as marked as under AR(1) errors.

Thirdly, provided that the inequality constraint is correct (i.e., $\delta \leq 0$), the magnitude of δ appears to have little bearing on the ICGLS estimators' sensitivity. However, when $\delta > 0$, a change in the magnitude of δ can alter the results to some extent. Frequently an increase in the constraint specification error can weaken the variance estimator's sensitivity to covariance misspecification. Under MA(1) errors, $\Pr(\bar{\bar{B}}1 > c_{\bar{\bar{B}}1})$ is fairly constant irrespective of the value of δ. Generally speaking, the effect of incorrect restrictions on the sensitivity of both ICGLS predictor and variance estimator is greater for AR(1)than for MA(1) disturbances.

4.2 IGPT Estimators

Figures 5a – 8d illustrate the behavior of the sensitivity curves of the corresponding IGPT regression predictor and variance estimator. The significance level of the pre-test is set at 0.05 in each case. First, the findings on the limiting behavior of the statistics portrayed by Theorem 3 are consistent with the numerical results observed. Second, the IGPT estimator appears to be relatively insensitive to MA(1) errors but can be sensitive to AR(1) errors. Under MA(1) errors the $\tilde{B}1$ curves are invariably flat and robust to variations in ψ_1. But with AR(1) errors the results are more diverse, and the behavior of the $\tilde{B}1$ sensitivity curves can be quite different depending on the underlying data matrix and the value of δ. In Figure 6a, for example, the underlying model is model 2, which contains no intercept and $Hi = 0$. There, the sensitivity curve of $\tilde{B}1$ for $\delta = 10$ increases at first as ϕ_1 increases, then falls rapidly to zero as ϕ_1 approaches the unit-root. Our figures also reveal a feature commonly observed in other similar contexts. There is a tendency for $\tilde{\sigma}^2(\theta)$ to be more sensitive than $\tilde{y}(\theta)$ to covariance misspecification. In many circumstances, the $\tilde{D}1$ curves tend to rise significantly above the benchmark 0.05 level even for moderate values of ϕ_1 and ψ_1. With AR(1) errors, the deviation from the 0.05 level is especially noticeable. In general, the pattern of results appears to depend mainly on the underlying data and the type of covariance misspecification. The magnitude of the constraint specification errors is of less importance. Depending on the data matrix, an incorrect null can exacerbate or weaken the estimator's sensitivity to covariance misspecification.

5 Conclusions

The main purpose of this paper is to explore the sensitivity of the single inequality restricted least squares and pre-test estimators in the linear regression model to disturbance covariance misspecification. This paper gains some interesting insights into the practical question of whether it matters at all that one fails to take account of the non-spherical structure of error covariances. Some exact analytical results are derived and these are evaluated for various data sets and for the cases of AR(1) and MA(1) errors. The principal conclusions to be drawn from these results may be stated quite briefly, and in some cases they reinforce the conclusions of Banerjee and Magnus (1999) and Wan, Zou and Qin (2007). First of all, while the ICGLS and IGPT variance estimators are generally very sensitive to covariance misspecification, whether or not the corresponding predictor or estimators of the coefficients are sensitive depends on the error process and the underlying data matrix. Generally, in the case of MA(1) disturbances, both $\overline{\overline{y}}(\theta)$ and $\tilde{y}(\theta)$ are insensitive regardless of the regression matrix. If the disturbances are AR(1), then both $\overline{\overline{y}}(\theta)$ and $\tilde{y}(\theta)$ can still be quite robust against covariance misspecification for small to moderate values of the autocorrelation parameter. For highly correlated AR(1) errors, however, the results are somewhat mixed and depend largely on the underlying data matrix. Indeed, the extent to which the X matrix affects the results is a notable feature of this study. The latter issue prevails when one considers the effect of restriction misspecification on the sensitivity of the variance estimator. With AR(1) errors, depending on the underlying data, specification errors in the restriction usually weaken but sometimes also exacerbate the estimators' sensitivity. With regard to the effects of pre-testing on sensitivity, pre-testing does not have any serious detrimental effect on the sensitivity of parameter estimates. Again, the exact patterns of results depend on the underlying data in each case. Sensitivity of other econometric estimators is currently being explored by the authors, in relation in particular to other common pre-test strategies used in econometrics (Magnus (1999);Wan and Zou (2003)).

Appendix A

Proof of Theorem 1:

Using the definition of the ICGLS estimator of β and recognizing that $\Pr(R\hat{\beta}(0) \neq r) = 1$ (i.e., $R\hat{\beta}(0) \neq r$ occurs almost surely), we can write

$$\bar{\bar{z}}_s = \mathrm{I}_{(-\infty,r)}\left(R\hat{\beta}(0)\right)\bar{z}_s + \mathrm{I}_{[r,\infty)}\left(R\hat{\beta}(0)\right)z_s. \tag{A.1}$$

Hence we have

$$\bar{\bar{B}}_s = \mathrm{I}_{(-\infty,r)}\left(R\hat{\beta}(0)\right)\bar{B}_s + \mathrm{I}_{[r,\infty)}\left(R\hat{\beta}(0)\right)B_s. \tag{A.2}$$

It is straightforward to show that

$$B_s = \frac{y'W_s y}{y'My} = \frac{u'W_s u}{u'Mu} = \frac{v'W_s v}{v'Mv}, \tag{A.3}$$

$$R\hat{\beta}(0) = H'y = R\beta + H'u, \tag{A.4}$$

$$\mathrm{I}_{(-\infty,r)}(R\hat{\beta}(0)) = \mathrm{I}_{(-\infty,\delta)}(H'v), \tag{A.5}$$

and

$$\mathrm{I}_{[r,\infty)}(R\hat{\beta}(0)) = \mathrm{I}_{[\delta,\infty)}(H'v). \tag{A.6}$$

Also, recall that $\bar{M}H = H$ and $\bar{y} = y - H(H'H)^{-1}r$. Hence from (1),

$$\bar{M}\bar{y} = \bar{M}[u + H(H'H)^{-1}(R\beta - r)] = \sigma\bar{M}\bar{v}. \tag{A.7}$$

Therefore,

$$\bar{B}_s = \frac{\bar{v}'\bar{W}_s\bar{v}}{\bar{v}'\bar{M}\bar{v}}. \tag{A.8}$$

We obtain (12) by substituting (A.3), (A.5), (A.6) and (A.8) in (A.2). Equation (13) can be verified similarly. Equations (14) and (15) can be obtained by further noting that $\Pr\{t(0) \neq c\} = 1$.

Proof of Theorem 2:

We only prove the results for $\bar{\bar{B}}1$. The results for $\bar{\bar{D}}1$ can be obtained by similarity. Note that

$$\Pr(\bar{\bar{B}}1 > c_{\bar{\bar{B}}1}) \;=\; \Pr\left(\bar{B}1 > c_{\bar{\bar{B}}1}, H'v < \delta\right) \;+\; \Pr\left(B1 > c_{\bar{\bar{B}}1}, H'v \geq \delta\right). \tag{A.9}$$

Using the proof to Theorem B.1 in Banerjee and Magnus (1999), write $(1-\phi_1^2)\Omega(\phi_1)$ as $(1-\phi_1^2)\Omega(\phi_1) = LL'$, where

$$L = L_0 + \rho L_1 + O(\rho^2) \tag{A.10}$$

as $\phi_1 \to 1$, $\rho = \sqrt{1-\phi_1^2}$, $L_0 = [i|0_{n\times(n-1)}]$, $L_1 = \mathrm{diag}\,(0, P)$, and P is defined in Theorem 2. Since $\rho u = \sigma L(\xi, \eta')'$ with $(\xi,\eta')' \sim N(0, I_n)$, it follows from (A.10) that

$$\rho v = L(\xi,\eta')' = i\xi + \rho\bar{P}\eta + O_p(\rho^2). \tag{A.11}$$

Therefore,

$$\rho H'v = H'L(\xi,\eta')' = H'i\xi + \rho H'\bar{P}\eta + O_p(\rho^2). \tag{A.12}$$

Since $\Pr(H'v < \delta) = \Pr(\rho H'v < \rho\delta)$ for arbitrary $\rho > 0$, it follows from (A.12) that

$$\lim_{\phi_1 \to 1} \Pr\left(H'v < \delta\right) = \begin{cases} \frac{1}{2} & \text{if } H'i \neq 0, \\ \Pr\left(H'\bar{P}\eta < \delta\right) & \text{if } H'i = 0. \end{cases} \tag{A.13}$$

For any arbitrary $\rho > 0$, we observe from (A.8) and (A.11) that as $\phi_1 \to 1$,

$$\bar{B}1 = \frac{(\rho v + \rho\bar{\mu})'\bar{W}^{(1)}(\rho v + \rho\bar{\mu})}{(\rho v + \rho\bar{\mu})'\bar{M}(\rho v + \rho\bar{\mu})} \xrightarrow{p} \begin{cases} \bar{b}1 & \text{if} \quad \bar{M}i \neq 0, \\ \bar{B}^{(1)}(\eta) & \text{if} \quad \bar{M}i = 0. \end{cases} \tag{A.14}$$

Now, from Banerjee and Magnus (2000), as $\phi_1 \to 1$,

$$B1 = \frac{(\rho v)'W^{(1)}(\rho v)}{(\rho v)'M(\rho v)} \xrightarrow{p} \begin{cases} b1 & \text{if} \quad Mi \neq 0, \\ B^{(1)}(\eta) & \text{if} \quad Mi = 0. \end{cases} \tag{A.15}$$

Parts i) and ii)

Note that $Mi \neq 0$ implies $\bar{M}i \neq 0$. It then follows from (A.14) and (A.15) that $\bar{B}1 \xrightarrow{p} \bar{b}1$ and $B1 \xrightarrow{p} b1$ as $\phi_1 \to 1$. So if $c_{\bar{\bar{B}}1} > \bar{b}1 \vee b1$, it follows from (A.9) that

$$\Pr(\bar{\bar{B}}1 > c_{\bar{\bar{B}}1}) \leq \Pr\left(\bar{B}1 > c_{\bar{\bar{B}}1}\right) + \Pr\left(B1 > c_{\bar{\bar{B}}1}\right) \to 0. \tag{A.16}$$

This proves the first result of (17) and the first result of (19). The last result of (17) and the last result of (19) can be obtained analogously. Now, let $b1 < c_{\bar{\bar{B}}1} < \bar{b}1$. Observe that

$$\begin{aligned} \Pr\left(H'v < \delta\right) &\geq \Pr\left(\bar{B}1 > c_{\bar{\bar{B}}1},\ H'v < \delta\right) \\ &= \Pr\left(H'v < \delta\right) - \Pr\left(\bar{B}1 \leq c_{\bar{\bar{B}}1},\ H'v < \delta\right) \\ &\geq \Pr\left(H'v < \delta\right) - \Pr\left(\bar{B}1 \leq c_{\bar{\bar{B}}1}\right). \end{aligned} \tag{A.17}$$

Since $c_{\bar{\bar{B}}1} < \bar{b}1$, it follows from $\bar{B}1 \xrightarrow{p} \bar{b}1$ that $\Pr\left(\bar{B}1 \leq c_{\bar{\bar{B}}1}\right) \to 0$ as $\phi_1 \to 1$. So, by (A.17) we have

$$\lim_{\phi_1 \to 1} \Pr\left(H'v < \delta,\ \bar{B}1 > c_{\bar{\bar{B}}1}\right) = \lim_{\phi_1 \to 1} \Pr\left(H'v < \delta\right). \tag{A.18}$$

Since $b1 < c_{\bar{\bar{B}}1}$ and $\Pr\left(H'v \geq \delta,\ B1 > c_{\bar{\bar{B}}1}\right) \leq \Pr\left(B1 > c_{\bar{\bar{B}}1}\right)$, it follows from $B1 \xrightarrow{p} b1$ that

$$\lim_{\phi_1 \to 1} \Pr\left(H'v \geq \delta,\ B1 > c_{\bar{\bar{B}}1}\right) = 0. \tag{A.19}$$

The first part of the second result in (17) and the second result in (19) can be obtained by substituting (A.13), (A.18) and (A.19) in (A.9). The second part of the second result in (17) and the third result in (19) can be proven by a similar mechanism.

Part iii)

It follows from (A.9) that

$$\Pr(\bar{\bar{B}}1 > c_{\bar{\bar{B}}1}) = \Pr\left(\frac{(\rho v + \rho\bar{\mu})'\bar{W}^{(1)}(\rho v + \rho\bar{\mu})}{(\rho v + \rho\bar{\mu})'\bar{M}(\rho v + \rho\bar{\mu})} > c_{\bar{\bar{B}}1},\ \ \rho H'v < \rho\delta\right)$$
$$+ \Pr\left(\frac{(\rho v)'W^{(1)}(\rho v)}{(\rho v)'M(\rho v)} > c_{\bar{\bar{B}}1},\ \ \rho H'v \geq \rho\delta\right). \tag{A.20}$$

Consider the case of $c_{\bar{\bar{B}}1} > \bar{b}1$. Note that $\bar{M}i \neq 0$ when $Mi = 0$ and $H'i \neq 0$. Hence we obtain from (A.14), (A.15) and (A.20) that

$$\begin{aligned}\lim_{\phi_1 \to 1} \Pr(\bar{\bar{B}}1 > c_{\bar{\bar{B}}1}) &= \Pr\left(B^{(1)}(\eta) > c_{\bar{\bar{B}}1},\ \ H'i\xi \geq 0\right) \qquad \text{(A.21)}\\ &= \Pr\left(\ H'i\xi \geq 0\right) \Pr\left(B^{(1)}(\eta) > c_{\bar{\bar{B}}1}\right)\\ &= \tfrac{1}{2} \Pr\left(B^{(1)}(\eta) > c_{\bar{\bar{B}}1}\right).\end{aligned}$$

Next, consider the case of $c_{\bar{\bar{B}}1} < \bar{b}1$. It follows from (A.14), (A.15) and (A.20) that

$$\begin{aligned}\lim_{\phi_1 \to 1} \Pr(\bar{\bar{B}}1 > c_{\bar{\bar{B}}1}) &= \Pr\left(\ H'i\xi < 0\right) + \Pr\left(B^{(1)}(\eta) > c_{\bar{\bar{B}}1},\ \ H'i\xi \geq 0\right)\\ & \qquad \text{(A.22)}\\ &= \tfrac{1}{2} + \Pr\left(\ H'i\xi \geq 0\right) \Pr\left(B^{(1)}(\eta) > c_{\bar{\bar{B}}1}\right)\\ &= \tfrac{1}{2}\left[1 + \Pr\left(B^{(1)}(\eta) > c_{\bar{\bar{B}}1}\right)\right].\end{aligned}$$

This completes the proof to part iii).

Part iv)

Note that $\bar{M}i = 0$ if and only if $Mi = 0$ and $H'i = 0$. As $H \neq 0$, $H'i = 0$ implies $H'\bar{J} \neq 0$. Hence equation (23) follows straightforwardly from (A.11), (A.12) and (A.20).

Proof of Theorem 3:

Again, the proofs that follow relate to the limiting behavior of $\tilde{B}1$ only. The results on the limiting behavior of $\tilde{D}1$ can be obtained by analogy. Now, from (14),

$$\begin{aligned}\Pr\left(\tilde{B}1 > c_{\tilde{B}1}\right) = & \Pr\left(\bar{B}1 > c_{\tilde{B}1},\ \mathrm{c} \leq t(0) < 0\right) \\ & + \Pr\left(B1 > c_{\tilde{B}1},\ t(0) \geq 0\right) \\ & + \Pr\left(B1 > c_{\tilde{B}1},\ t(0) < c\right). \end{aligned} \tag{A.23}$$

From Qin et al. (2007), we have

$$t(0) \xrightarrow{p} \begin{cases} \bar{t}(0)\xi/|\xi| & \text{if } Mi \neq 0 \text{ and } H'i \neq 0, \\ 0 & \text{if } Mi \neq 0 \text{ and } H'i = 0, \\ t^{(1)}(\eta) & \text{if } Mi = 0 \text{ and } H'i = 0. \end{cases} \tag{A.24}$$

One can verify parts i), ii) and iv) of Theorem 3 using (A.14), (A.15), (A.23) and (A.24). For part iii) of Theorem 3, note that by the definition of $t(0)$ and (A.15), we have

$$\begin{aligned} & \Pr\left(B1 > c_{\tilde{B}1},\ t(0) < c\right) \\ & = \Pr\left(\frac{(\rho v)'W^{(1)}(\rho v)}{(\rho v)'M(\rho v)} > c_{\tilde{B}1},\ \frac{\sqrt{n-k}}{\sqrt{H'H}}\frac{\rho H'v - \rho\delta}{\sqrt{v'Mv}} < \rho c\right) \end{aligned} \tag{A.25}$$

for any $\rho > 0$. Since $H'i \neq 0$, we have $\rho H'v = H'i\xi + O_p(\rho)$ as $\rho = \sqrt{1-\phi_1^2} \to 0$ from (A.12). In addition, by $Mi = 0$, we observe from (A.11) that $v'Mv = \eta'\bar{P}'M\bar{P}\eta + O_p(\rho)$ and $v'W^{(1)}v = \eta'\bar{P}'W^{(1)}\bar{P}\eta + O_p(\rho)$. Using these results in (A.25), we obtain

$$\begin{aligned}\Pr\left(B1 > c_{\tilde{B}1}, t(0) < c\right) \to & \Pr\left(B^{(1)}(\eta) > c_{\tilde{B}1},\ \frac{\sqrt{n-k}}{\sqrt{H'H}}\frac{H'i\xi}{\sqrt{\eta'\bar{P}'M\bar{P}\eta}} < 0\right) \\ = & \Pr\left(B^{(1)}(\eta) > c_{\tilde{B}1}\right)\Pr\left(H'i\xi < 0\right) \\ = & \tfrac{1}{2}\Pr\left(B^{(1)}(\eta) > c_{\tilde{B}1}\right). \end{aligned} \tag{A.26}$$

Similarly,

$$\Pr\left(B1 > c_{\tilde{B}1},\ t(0) \geq 0\right) \to \tfrac{1}{2}\Pr\left(B^{(1)}(\eta) > c_{\tilde{B}1}\right). \tag{A.27}$$

Now,

$$\Pr\left(\bar{B}1 > c_{\bar{\bar{B}}1},\ c \le t(0) < 0\right) \le \Pr\left(c \le t(0) < 0\right)$$
$$= \Pr\left(t(0) < 0\right) - \Pr\left(t(0) < c\right) \to 0 \qquad \text{(A.28)}$$

Part iii) of Theorem 3 follows straightforwardly from (A.23) and (A.26) – (A.28).

Appendix B: Algorithm for Quantiles Detection

We adopted the following Monte-Carlo procedure to compute the quantiles $c_{\bar{\bar{B}}1}$ and $c_{\bar{\bar{D}}1}$. We first drew 10 000 white noise series $v_i's$, each of which contains $n = 15$ realizations from N(0, 1). Let $\delta = r - \beta_1$ and $\bar{v}_i = v_i - H(H'H)^{-1}\delta$ for $1 \le i \le 10\,000$. For an arbitrary $c_a \in [0,\ 1]$, the probability $\Pr_{\theta=0}\left(\bar{\bar{B}}1 > c_a\right)$ was approximated by

$$\hat{p}(0, c_a) = \sum_{i=1}^{10\,000} \frac{I\left(\bar{\bar{B}}(i) > c_a\right)}{10\,000}, \qquad \text{(B.1)}$$

where

$$\bar{\bar{B}}(i) = I_{(-\infty,\ \delta)}\left(H'v_i\right) \frac{\bar{v}_i'\bar{W}^{(1)}\bar{v}_i}{\bar{v}_i'\bar{M}\bar{v}_i} + I_{[\delta,\ \infty)}\left(H'v_i\right) \frac{v_i'W^{(1)}v_i}{v_i'Mv_i} \qquad \text{(B.2)}$$

(See Theorem 1). Almost surely, $\hat{p}$ decreases as c_a increases, $\hat{p}(0,0) = 1$ and $\hat{p}(0,1) = 0$. To detect $c_{\bar{\bar{B}}1}$, we first set $l_0 = 0$, $g_0 = 1$ and $c_a = (l_0 + g_0)/2$, then calculated $\hat{p}\left(0, c_a\right)$. We set $l_0 = c_a$ if $\hat{p}\left(0, c_a\right) > \alpha$, else we set $g_0 = c_a$. The revised value of $c_a$was given by $c_a = (l_0 + g_0)/2$, which provided a new estimate of $\hat{p}\left(0, c_a\right)$. This procedure was repeated until $|\hat{p}(0, c_a) - 0.05| < 10^{-5}$, and the final value of c_a was taken to be the estimate of $c_{\bar{\bar{B}}1}$. In the same manner we estimated the quantile $c_{\bar{\bar{D}}1}$.

Acknowledgements

The second and third authors gratefully acknowledge financial supports from the Hong Kong Research Grants Council and the National Natural Science Foundation of China (Nos. 70625004 and 10471043).

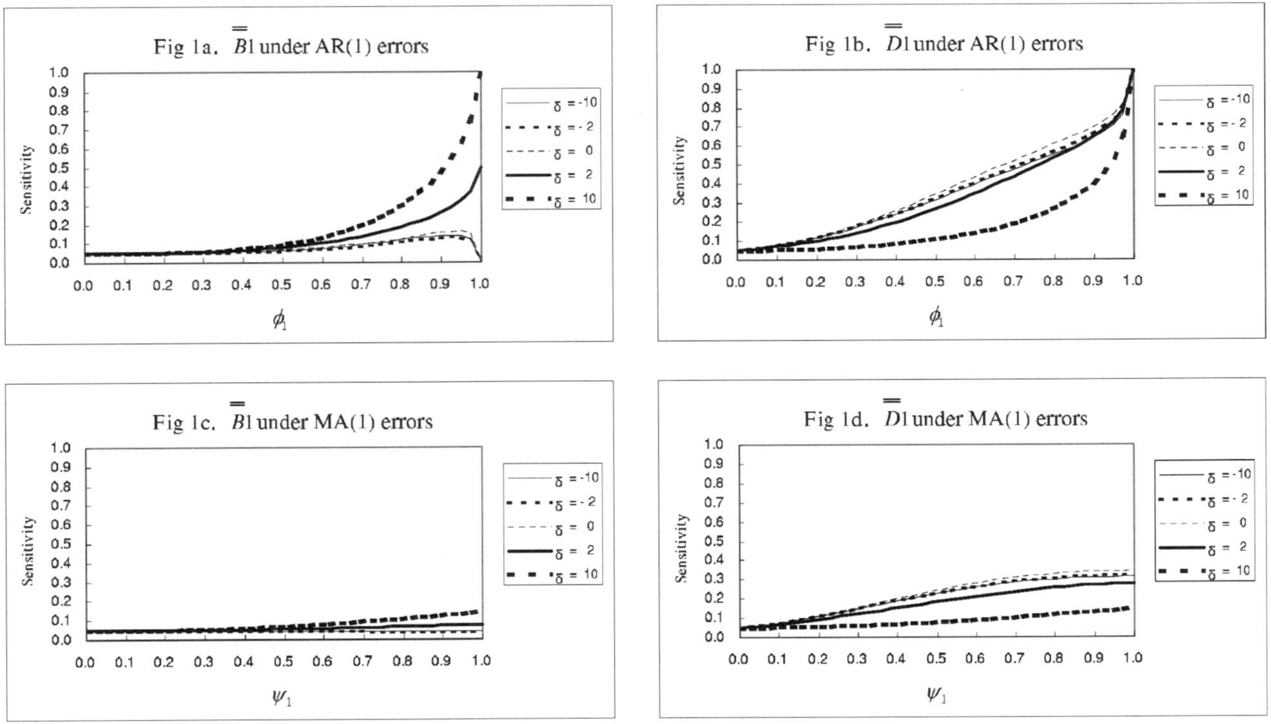

Figs 1a-1d. Sensitivity of the ICGLS estimators for model 1: $b1 = 0.5041 > \overline{b}1 = 0.4710$, $c_{\overline{\overline{B}}1} > \max\{b1, \overline{b}1\}$ for $\delta = -10, -2$ and 0, $b1 > c_{\overline{\overline{B}}1} > \overline{b}1$ for $\delta = 2$, and $c_{\overline{\overline{B}}1} < \min\{b1, \overline{b}1\}$ for $\delta = 10$; while $d1 = -1.2289 > \overline{d}1 = -1.3584$ and $c_{\overline{\overline{D}}1} > \max\{d1, \overline{d}1\}$ for all five values of δ.

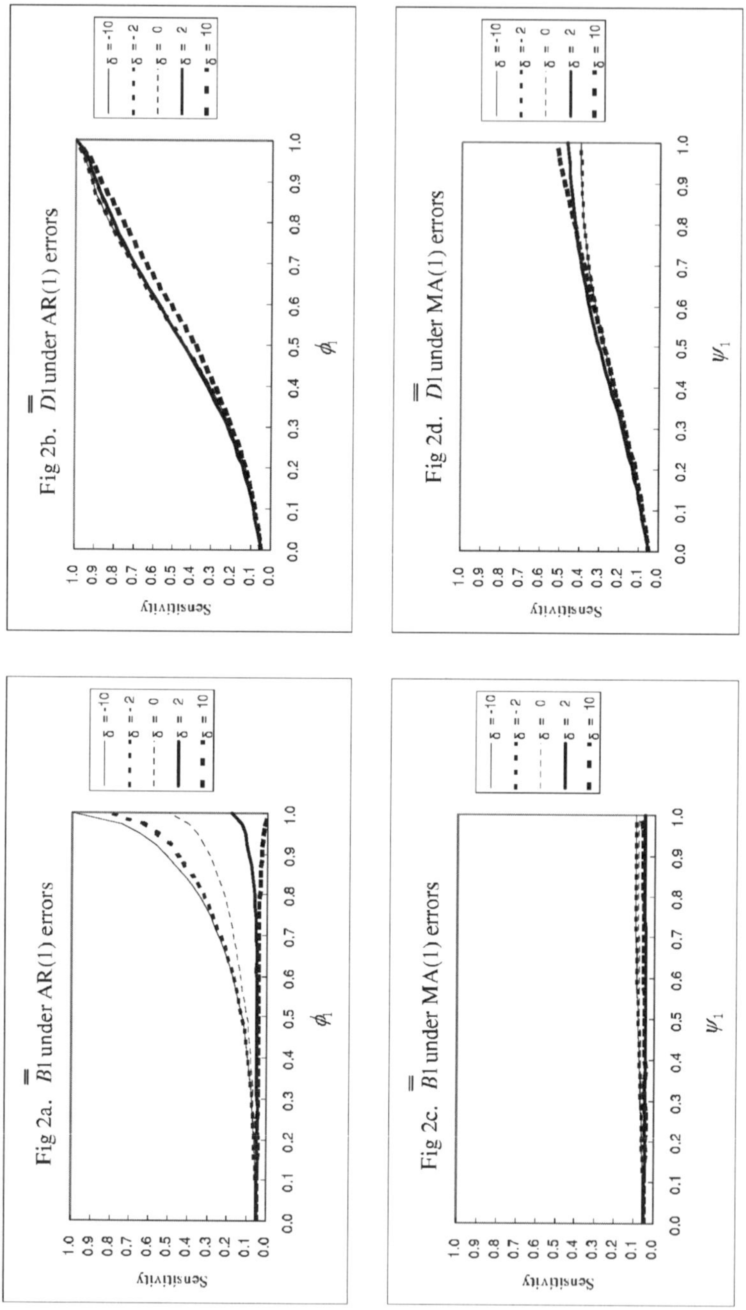

Figs 2a-2d. Sensitivity of the ICGLS estimators for model 2: $b1 = 0.9853 > c_{\bar{\bar{B}}1} > \bar{b}1 = 0.0799$, and $c_{\bar{\bar{D}}1} > d1 = \bar{d}1 = -1.8667$ for all five values of δ.

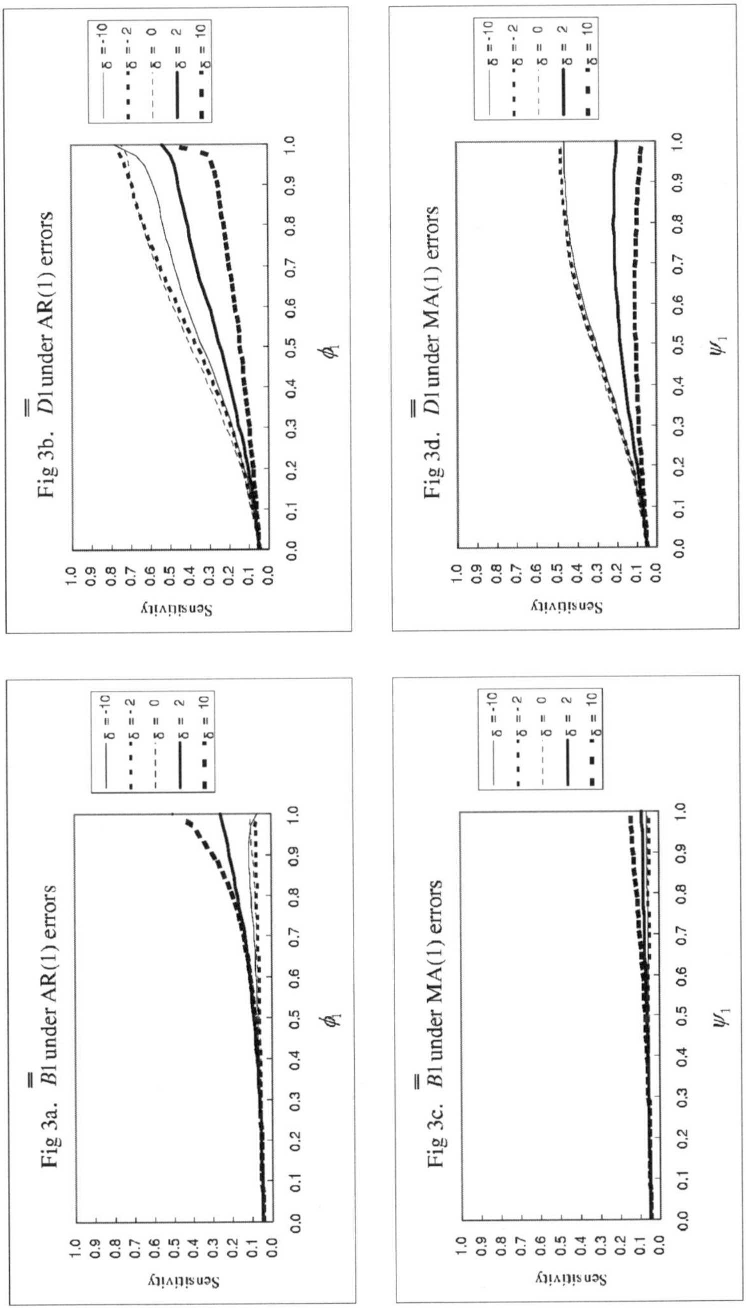

Figs 3a-3d. Sensitivity of the ICGLS estimators for model 3: $c_{\overline{\overline{B}}1} > \overline{b}1 = 0.0222$ and $c_{\overline{\overline{D}}1} > \overline{d}1 = -1.8667$ for all five values of δ.

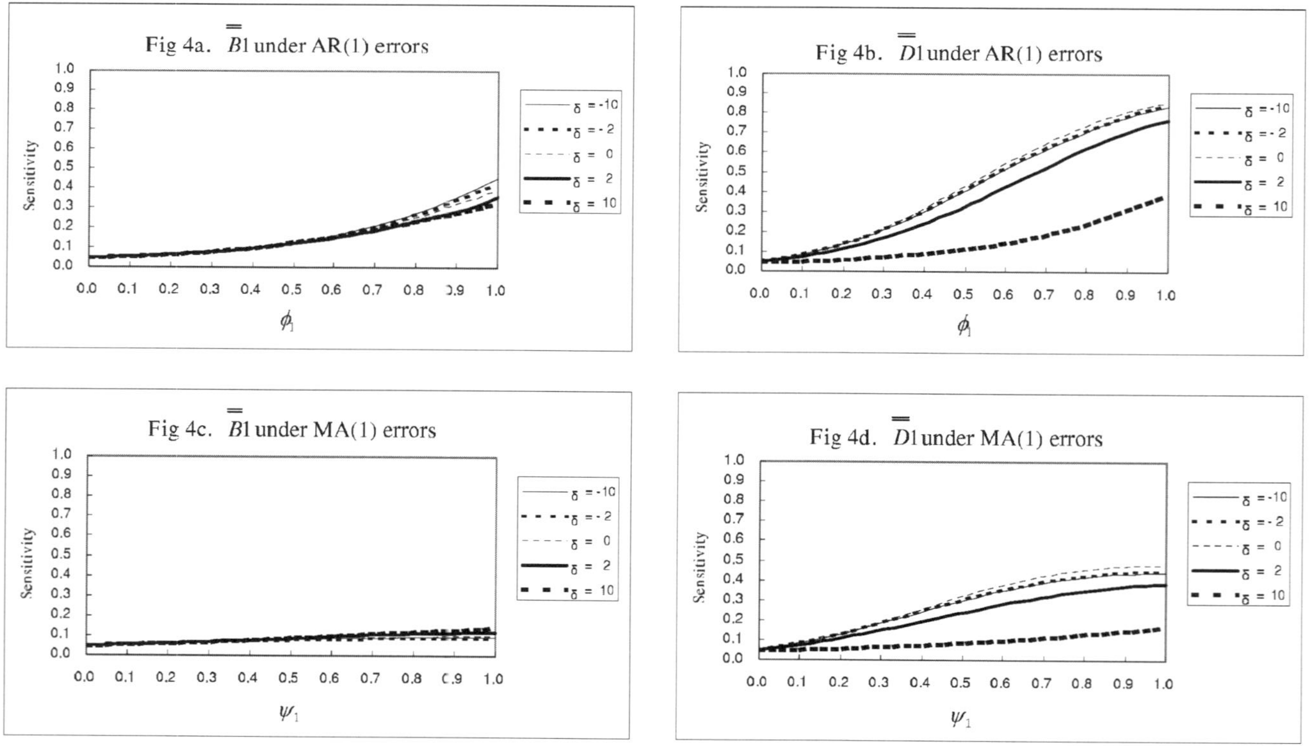

Figs 4a-4d. Sensitivity of the ICGLS estimators for model 4.

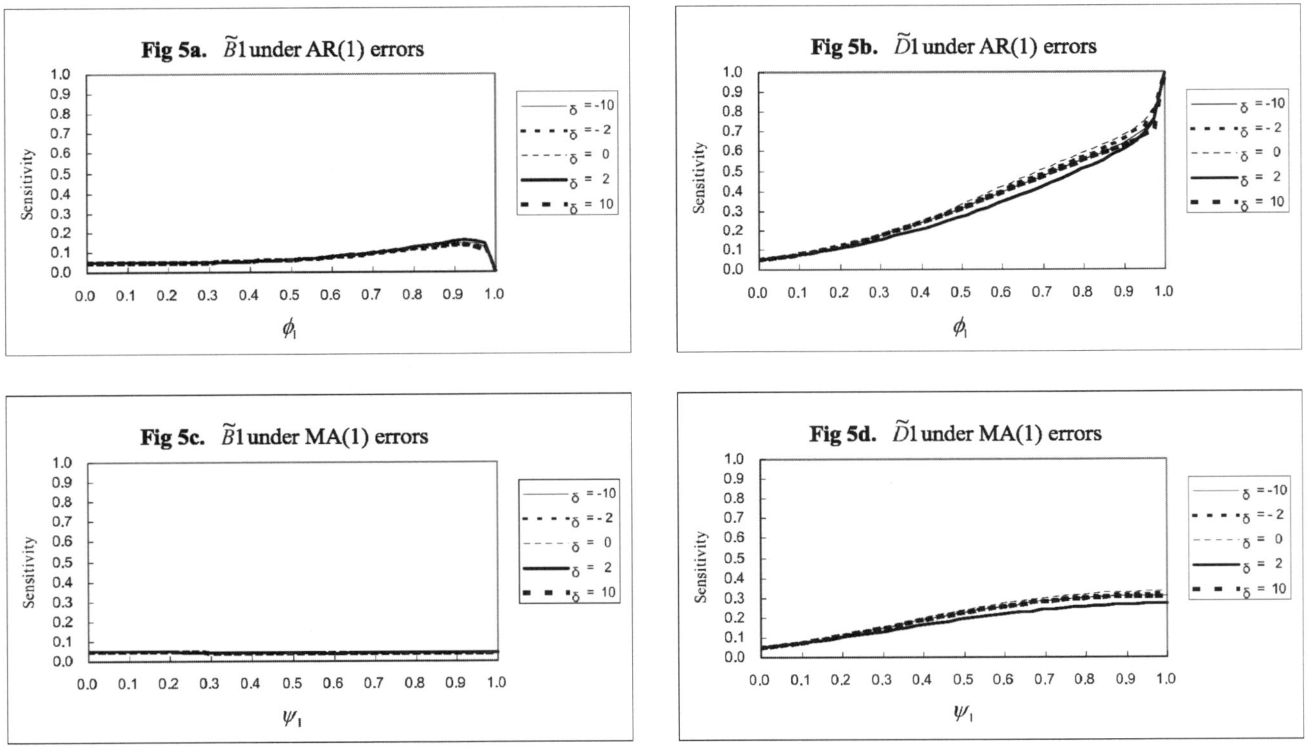

Figs 5a-5d. Sensitivity of the IGPT estimators for model 1: $c_{\widetilde{B}1} > b1 \vee \bar{b}1 = 0.5041$, and $c_{\widetilde{D}1} > d1 \vee \bar{d}1 = -1.2289$ for all five values of δ.

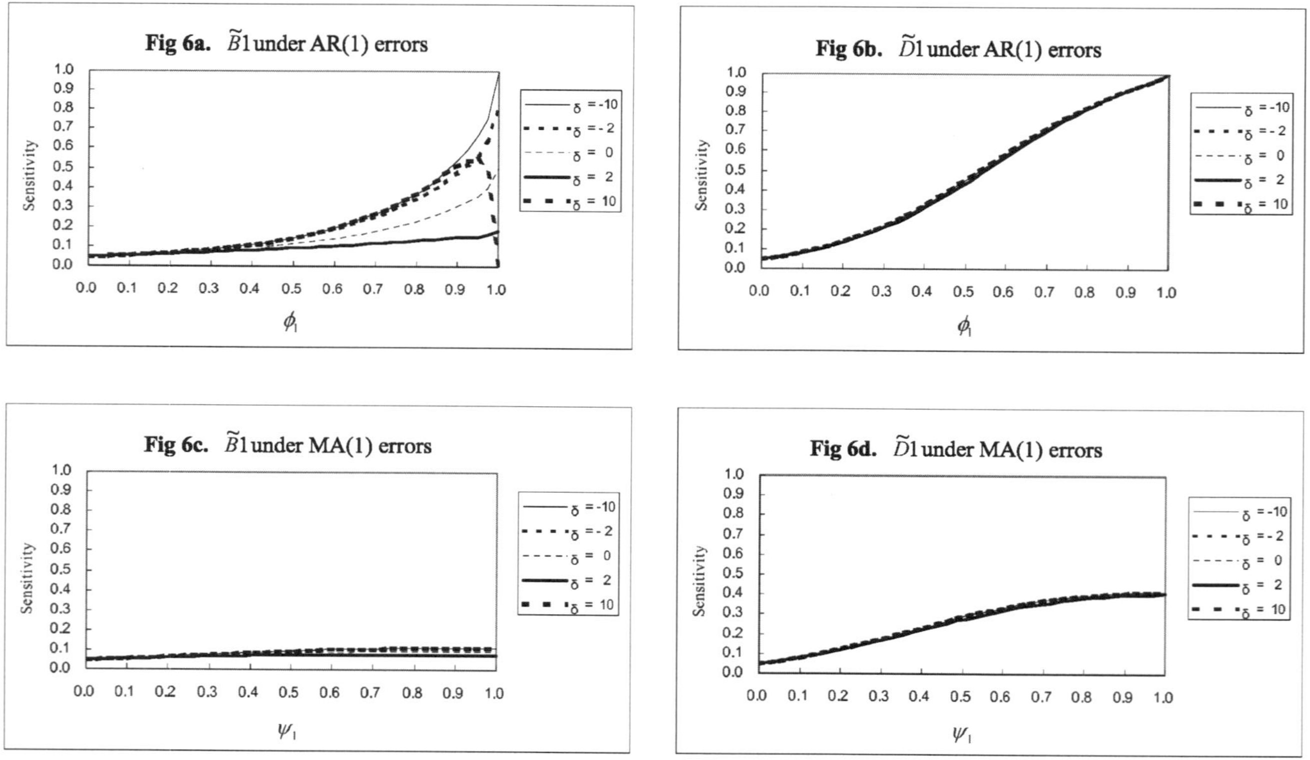

Figs 6a-6d. Sensitivity of the IGPT estimators for model 2: $b1 = 0.9853 > c_{\widetilde{B}1} > \bar{b}1 = 0.0799$, and $c_{\widetilde{D}1} > d1 = \bar{d}1 = -1.8667$ for all five values of δ.

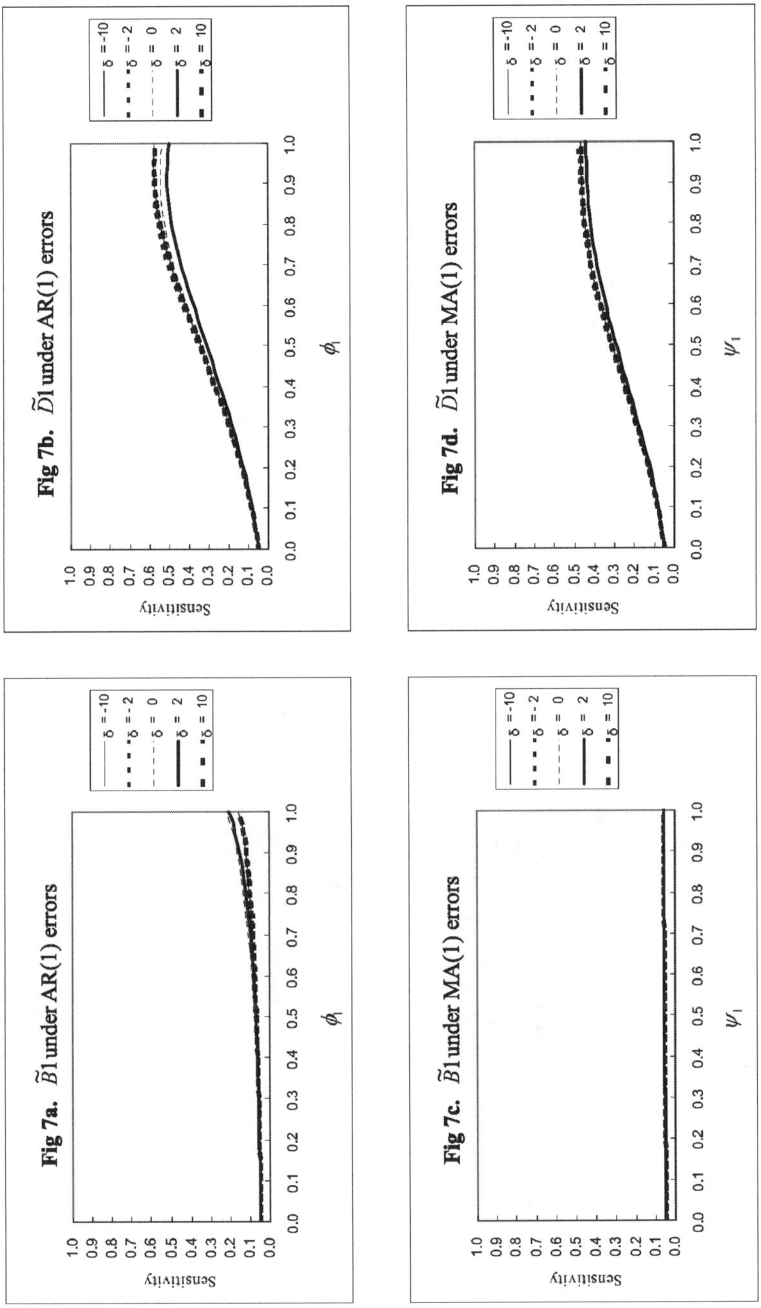

Figs 7a-7d. Sensitivity of the IGPT estimators for model 3.

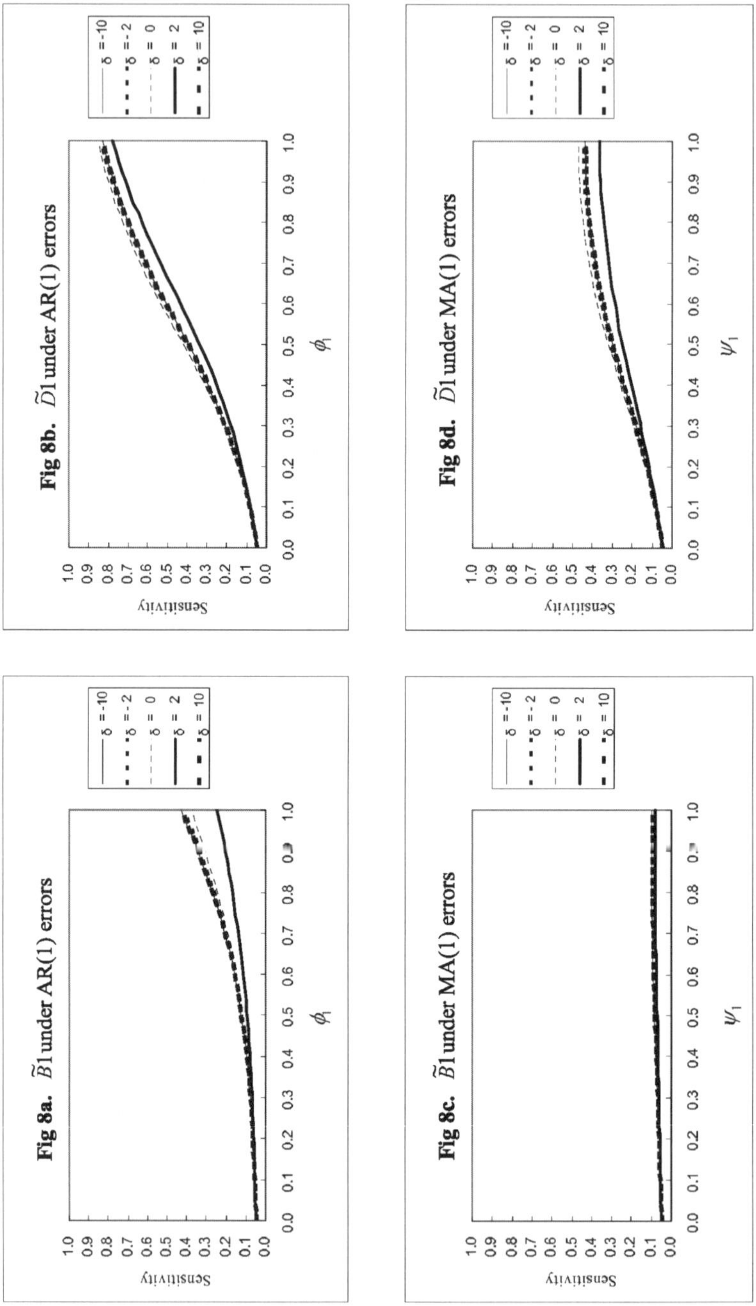

Figs 8a–8d. Sensitivity of the IGPT estimators for model 4.

References

Banerjee A, Magnus JR. (1999) The sensitivity of OLS when variance matrix is (partially) unknown. Journal of Econometrics 92:295-323

Banerjee A, Magnus JR (2000) On the sensitivity of the usual t- and F-tests to covariance misspecification. Journal of Econometrics 95:157-176

Judge GG, Yancey TA (1986) Improved methods of inference in econometrics. North-Holland, Amsterdam

Magnus JR (1999) The traditional pre-test estimators. Theory of Probability and Its Applications 44:293-308

Magnus JR, Vasnev AL (2007) Local sensitivity and diagnostic tests. Econometrics Journal 10:166-192

Qin HZ, Wan ATK, Zou GH (2007) On the sensitivity of the one-sided t-test to covariance misspecification. mimeo., City University of Hong Kong

Thomson M (1982) Some results on the statistical properties of an inequality constrained least squares estimator in a linear model with two regressors. Journal of Econometrics 19:215-231

Wan ATK (1994a) The sampling performance of inequality restricted and pre-test estimators in a mis-specified linear model. Australian Journal of Statistics 36:313-325

Wan ATK (1994b) Risk comparison of the inequality constrained least squares and other related estimators under balanced loss. Economics Letters 46:203-210

Wan ATK (1995) The optimal critical value of a pre-test for an inequality restriction in a mis-specified regression model. Australian Journal of Statistics 37:73-82

Wan ATK (1996) Estimating the error variance after a pre-test for an inequality restriction on the coefficients. Journal of Statistical Planning and Inference 52:197-213

Wan ATK, Zou GH (2003) Optimal critical values of pre-tests when estimating the regression error variance: analytical findings under a general loss structure. Journal of Econometrics 114:165-196

Wan ATK, Zou GH, Ohtani K (2006) Further results on optimal critical values of pre-test when estimating the regression error variance. Econometrics Journal 9:159-176

Wan ATK, Zou GH, Qin HZ (2007) On the sensitivity of restricted least squares estimators to covariance misspecification. Econometrics Journal 10:471-487

Boosting Correlation Based Penalization in Generalized Linear Models

Jan Ulbricht[1] and Gerhard Tutz[2]

[1] Department of Statistics, University of Munich, Akademiestrasse 1, 80799 Munich, Germany ulbricht@stat.uni-muenchen.de
[2] Department of Statistics, University of Munich, Akademiestrasse 1, 80799 Munich, Germany tutz@stat.uni-muenchen.de

1 Introduction

Linear models have a long tradition in statistics as nicely summarized in Rao, Toutenburg, Shalabh, Heumann (2008). When the number of covariates is large the estimation of unknown parameters frequently raises problems. Then the interest usually focusses on data driven subset selection of relevant regressors. The sophisticated monitoring equipment which is now routinely used in many data collection processes makes it possible to collect data with a huge amount of regressors, even with considerably more explanatory variables than observations. One example is the analysis of microarray data of gene expressions. Here the typical tasks are to select variables and to classify samples into two or more alternative categories. Binary responses of this type may be handled within the framework of generalized linear models (Nelder and Wedderburn (1972)) and are also considered in Rao, Toutenburg, Shalabh, Heumann (2008).

There are several approaches to attain subset selection in generalized linear models. Shrinkage methods with L_1 norm penalties such as the lasso estimator are one class of methods. The lasso estimator was introduced by Tibshirani (1996) for the linear model and extended to generalized linear models in Park and Hastie (2007). An alternative approach is componentwise boosting (see Bühlmann and Yu (2003)). Boosting uses an ensemble of weak learners to improve the estimator. One obtains subset selection if each learner is restricted to use a subset of covariates.

One aspect in subset selection, highlighted by Zou and Hastie (2005), is the treatment of highly correlated covariates. Instead of choosing only one representative out of a group of highly correlated variables one could encourage strongly correlated covariates to be in or out of the model together. Zou and Hastie (2005) refer to it as the grouping effect.

In this paper we propose a new regularization method and a boosted version of it, which explicitly focus on the selection of groups. To reach this target we consider a correlation based penalty which uses correlation between variables as data driven weights for penalization. See also Tutz and Ulbricht (2006) for a similar approach to linear models. This new method and some of its main properties are described in Section 2. A boosted version of it that will be presented in Section 3 allows for variable selection. In Section 4 we use simulated and real data sets to compare our new methods with existing ones.

2 Penalized Maximum Likelihood Estimation

Consider a set of n independent one-dimensional observations $y_1, \ldots, y_n$ with densities from a simple exponential family type

$$f(y|\theta, \phi) = \exp\left\{ \frac{y\theta - b(\theta)}{\phi} + c(y, \phi) \right\}, \tag{1}$$

where θ is the natural scalar parameter of the family, $\phi > 0$ is a nuisance or dispersion parameter, $b(\cdot)$ and $c(\cdot)$ are measurable functions. For each observation, also values of a set of p explanatory variables $\mathbf{x}_i = (x_{i1}, \ldots, x_{ip})^\top$ are recorded. They form a linear predictor $\eta_i = \beta_0 + \mathbf{x}_i^\top \boldsymbol{\beta}^*$, where β_0 is a constant and $\boldsymbol{\beta}^* = (\beta_1, \ldots, \beta_p)^\top$ is a p dimensional parameter vector. It is assumed that the expectation of y_i is given by $\mu_i = h(\eta_i)$, where $h(\cdot)$ is a differentiable monotone response function and μ_i is the expectation of y_i.

Assuming that the dispersion parameter ϕ is known, we are interested in finding the unknown parameter vector $\boldsymbol{\beta} = (\beta_0, \boldsymbol{\beta}^{*\top})^\top$, which maximizes the corresponding log likelihood function

$$l(\boldsymbol{\beta}) = \sum_{i=1}^{n} \left\{ \frac{y_i\theta[h(\beta_0 + \mathbf{x}_i^\top \boldsymbol{\beta}^*)] + b(\theta[h(\beta_0 + \mathbf{x}_i^\top \boldsymbol{\beta}^*)])}{\phi_i} + c(y_i, \phi_i) \right\}. \tag{2}$$

Simple derivation yields the score function

$$s(\boldsymbol{\beta}) = \frac{\partial l(\boldsymbol{\beta})}{\partial \boldsymbol{\beta}} = \sum_{i=1}^{n} \frac{y_i - b'(\theta_i)}{Var(y_i)} \frac{\partial h(\eta_i)}{\partial \eta} \mathbf{x}_i = \mathbf{X}^\top \mathbf{D} \boldsymbol{\Sigma}^{-1} (\mathbf{y} - \boldsymbol{\mu}), \tag{3}$$

where $\mathbf{X}^\top = (\mathbf{x}_1, \ldots, \mathbf{x}_n)$,

$$\mathbf{D} = \text{diag}\left\{\frac{\partial h(\eta_1)}{\partial \eta}, \ldots, \frac{\partial h(\eta_n)}{\partial \eta}\right\}, \ \boldsymbol{\Sigma} = \text{diag}\left\{Var(y_1), \ldots, Var(y_n)\right\}.$$

The Fisher matrix is given by

$$F(\boldsymbol{\beta}) = -E\left[\frac{\partial^2 l(\boldsymbol{\beta})}{\partial \boldsymbol{\beta} \partial \boldsymbol{\beta}^\top}\right] = E[s(\boldsymbol{\beta}) s(\boldsymbol{\beta})^\top] = \mathbf{X}^\top \mathbf{W} \mathbf{X}, \tag{4}$$

where $\mathbf{W} = \mathbf{D}\boldsymbol{\Sigma}^{-1}\mathbf{D}^\top$. The unknown parameter vector can be found iteratively by applying numerical methods for solving nonlinear equation systems, such as Newton-Raphson. Under weak assumptions the maximum likelihood estimator $\hat{\boldsymbol{\beta}}$ is consistent and asymptotically normal with asymptotic covariance matrix $Cov(\hat{\boldsymbol{\beta}}) = (\mathbf{X}^\top \mathbf{W} \mathbf{X})^{-1}$, see Fahrmeir and Kaufmann (1985).

In their seminal paper, Hoerl and Kennard (1970) show that the least squares estimate in the linear regression model tends to overestimate the length of the true parameter vector if the prediction vectors are not mutually orthogonal. Segerstedt (1992) shows similar effects when estimating generalized linear models. Early attempts of a generalizing ridge estimation were limited to logistic regression, see e.g. Anderson and Blair (1982), Schaefer et al. (1984) and Duffy and Santner (1989). Nyquist (1991) introduces ridge estimation of generalized linear models in the context of restricted estimation.

Since the maximum likelihood estimator of the unknown parameter vector has the tendency to overestimate length, it is advisable to fix its squared length. This restriction is formulated as constraint, so that we can use the Lagrangian approach. Formally, we solve the optimization problem

$$\hat{\boldsymbol{\beta}} = \arg\max_{\boldsymbol{\beta}} \left\{l(\boldsymbol{\beta}) - P(\boldsymbol{\beta})\right\}, \tag{5}$$

where

$$P(\boldsymbol{\beta}) = \lambda \|\boldsymbol{\beta}\|_2^2 = \lambda \sum_{j=1}^{p} \beta_j^2 \tag{6}$$

with $\|\boldsymbol{\beta}\|_2^2$ denoting the the squared L_2 norm of $\boldsymbol{\beta}$ and $\lambda > 0$ is a tuning parameter. Let $\hat{\boldsymbol{\beta}}_{ridge}(\lambda)$ denote the resulting GLM ridge estimator for given λ. Hence, $\hat{\boldsymbol{\beta}}_{ridge}(\lambda)$ is based on an L_2 penalty term.

Typically there exists a tuning parameter λ, so that the asymptotic mean squared error of the GLM ridge estimator is smaller than the asymptotic variance of the maximum likelihood estimator, for the proof see Segerstedt (1992). Nevertheless, the major drawback of $\hat{\boldsymbol{\beta}}_{ridge}(\lambda)$ is its lack in producing sparse solutions.

In the linear model setting the most important penalized regression approach that automatically includes subset selection is the lasso, as introduced by Tibshirani (1996). The L_1 based lasso penalty

$$P(\boldsymbol{\beta}) = \lambda \|\boldsymbol{\beta}\|_1 = \lambda \sum_{j=1}^{p} |\beta_j| \tag{7}$$

leads to regression fits that are sparse and interpretable, in the sense that many variables are "pruned" from the model. Shevade and Keerthi (2003) propose an L_1 penalization for logistic regression. Park and Hastie (2007) introduce a corrector-predictor algorithm for generalized linear models with lasso penalty. The main problem in using L_1 penalties within the GLM framework is the instability of coefficient estimates when some explanatory variables are strongly correlated. Furthermore, the solution might not be unique if some regressors are multicollinear. Therefore Park and Hastie (2007) modify the lasso penalty term to

$$P(\boldsymbol{\beta}) = \lambda_1 \|\boldsymbol{\beta}\|_1 + \frac{\lambda_2}{2} \|\boldsymbol{\beta}\|_2^2, \tag{8}$$

where $\lambda_1 > 0$ is an arbitrary tuning parameter and λ_2 is a fixed small positive constant. The elastic net penalty as introduced in Zou and Hastie (2005) is algebraically identical to (8), up to a rescaled tuning parameter of the L_2 penalty term. Using (8) in the way of Zou and Hastie (2005) requires simultaneous tuning parameter selection, e.g. by cross-validation, in two dimensions. This can be computationally cumbersome. One motivation Zou and Hastie (2005) give for the elastic net is its property to include groups of variables which are highly correlated. If variables are highly correlated, as for example gene expression in microarray data, the lasso selects only one out of the group whereas the elastic net catches "all the big fish", meaning that it selects the whole group.

In this paper we propose an alternative regularization procedure which aims at the selection of groups of correlated variables. In the simpler version it is based on a penalty that explicitly uses correlation between variables as weights. In the extended version boosting techniques are used for groups of variables. The correlation based penalty is introduced as

$$P_c(\boldsymbol{\beta}) = \lambda \sum_{i=1}^{p-1} \sum_{j>i} \left\{ \frac{(\beta_i - \beta_j)^2}{1-\varrho_{ij}} + \frac{(\beta_i + \beta_j)^2}{1+\varrho_{ij}} \right\}$$
$$= 2\lambda \sum_{i=1}^{p-1} \sum_{j>i} \frac{\beta_i^2 - 2\varrho_{ij}\beta_i\beta_j + \beta_j^2}{1-\varrho_{ij}^2} \tag{9}$$

where ϱ_{ij} denotes the (empirical) correlation between the ith and the jth predictor. It is designed to focus on the grouping effect, that is highly correlated effects show comparable values of estimates ($|\hat{\beta}_i| \approx |\hat{\beta}_j|$) with the sign being determined by positive or negative correlation. For strong positive correlation ($\varrho_{ij} \to 1$) the first term becomes dominant having the effect that estimates for β_i, β_j are similar ($\hat{\beta}_i \approx \hat{\beta}_j$). For strong negative correlation ($\varrho_{ij} \to -1$) the second term becomes dominant and $\hat{\beta}_i$ will be close to $-\hat{\beta}_j$. Consequently, for weakly correlated data the performance is quite close to the ridge penalty. The correlation based penalty (9) can be written as a quadratic form

$$P_c(\boldsymbol{\beta}) = \lambda \boldsymbol{\beta}^\top \mathbf{M} \boldsymbol{\beta}, \tag{10}$$

where $\mathbf{M} = (m_{ij})$ is given by

$$m_{ij} = \begin{cases} 2\sum_{s \neq i} \frac{1}{1-\varrho_{is}^2}, & i = j \\ -2\frac{\varrho_{ij}}{1-\varrho_{ij}^2}, & i \neq j. \end{cases}$$

We denote the resulting penalized maximum likelihood estimator of the unknown coefficient vector as $\hat{\boldsymbol{\beta}}_c$ and refer to it in the following as GLM PenalReg estimator.

Due to the additive structure between log likelihood function and the penalty term the computation of the correlation based penalized estimator, abbreviated by GLM PenalReg, is easily done by using the score function and Fisher matrix of the log likelihood function. For the penalized log likelihood with $P_c(\boldsymbol{\beta}) = \lambda \boldsymbol{\beta}^\top \mathbf{M} \boldsymbol{\beta}$ one obtains

$$l_p(\boldsymbol{\beta}) = l(\boldsymbol{\beta}) - \frac{\lambda}{2} \boldsymbol{\beta}^\top \mathbf{M} \boldsymbol{\beta}, \tag{11}$$

where we use a rescaling of λ for computational simplicity. Hence, the penalized score is

$$s_p(\boldsymbol{\beta}) = \frac{\partial l_p(\boldsymbol{\beta})}{\partial \boldsymbol{\beta}} = s(\boldsymbol{\beta}) - \lambda \mathbf{M} \boldsymbol{\beta} = \mathbf{X}^\top \mathbf{D} \boldsymbol{\Sigma}^{-1} (\mathbf{y} - \boldsymbol{\mu}) - \lambda \mathbf{M} \boldsymbol{\beta}, \tag{12}$$

and the penalized Fisher matrix is given by

$$F_p(\boldsymbol{\beta}) = -E\left[\frac{\partial s_p(\boldsymbol{\beta})}{\partial \boldsymbol{\beta}^\top}\right] = \mathbf{X}^\top \mathbf{W}\mathbf{X} + \lambda \mathbf{M}, \tag{13}$$

As in non-penalized maximum likelihood estimation we need to solve a nonlinear system of equations. In the same way as the GLM ridge estimator the GLM PenalReg estimator can be written as an iteratively re-weighted least squares estimator, given by

$$\hat{\boldsymbol{\beta}}_{\boldsymbol{c}}{}^{(k+1)} = (\mathbf{X}^\top \mathbf{W}\mathbf{X} + \lambda \mathbf{M})^{-1}\mathbf{X}^\top \mathbf{W}\tilde{\mathbf{y}}^{(k)}, \tag{14}$$

where $\tilde{\mathbf{y}}^{(k)} = \mathbf{X}\hat{\boldsymbol{\beta}}_{\boldsymbol{c}}{}^{(k)} + \mathbf{D}^{-1}(\mathbf{y} - \boldsymbol{\mu})$.

Based on a first order Taylor approximation one obtains the asymptotic covariance matrix

$$Cov[\hat{\boldsymbol{\beta}}_{\boldsymbol{c}}(\lambda)] = (\mathbf{X}^\top \mathbf{W}\mathbf{X} + \lambda \mathbf{M})^{-1}\mathbf{X}^\top \mathbf{W}\mathbf{X}(\mathbf{X}^\top \mathbf{W}\mathbf{X} + \lambda \mathbf{M})^{-1}. \tag{15}$$

Note that we get similar results for the generalized ridge estimator $\boldsymbol{\beta}_{ridge}(\lambda)$ when we substitute the identity matrix for the penalty matrix $\mathbf{M}$, see Segerstedt (1992) for details. A systematic report on mean squared error comparisons of competing biased estimators for the linear model is given in Trenkler and Toutenburg (1990). For performance comparisons in several simulation and practical data situations we refer to section 4.

3 Generalized Blockwise Boosting

The main drawback of the correlation based penalized estimator is its lack of sparsity. In particular when high dimensional data such as microarray data are considered one wants to select an appropriate subset of regressors. One method that is able to overcome this disadvantage is componentwise boosting as introduced by Bühlmann and Yu (2003). They propose to update in one boosting step only the component that maximally improves the fit.

Boosting methods are multiple prediction schemes that average estimated predictions from re-weighted data. With its origins in the machine learning community the first major field of applications was binary classification. The link between boosting and a gradient descent optimization technique in function space as outlined in Breiman (1998) provided the application of boosting methods in other contexts than classification. Friedman (2001) developed the L_2 Boost algorithm for a linear base learner, an optimization algorithm with squared error loss function for application in regression, which provides the foundations of componentwise boosting. For a detailed overview on boosting see e.g. Meir and Rätsch (2003). Componentwise likelihood based boosting applied to the generalized ridge estimator is described in Tutz and Binder (2007). The base learner of this boosting algorithm is the first step of the Fisher scoring algorithm.

Let $S^{(m)} \subset \{0, 1, \ldots, p\}$ denote the index set of the variables considered in the mth step, where the index 0 refers to the intercept term of the predictor. The input data to the base learner are $\{(\mathbf{x}_1, r_1), \ldots, (\mathbf{x}_n, r_n)\}$, where $r_i = y_i - \hat{\mu}_i^{(m-1)}$ $(i = 1, \ldots, n)$ denotes the residual between the origin response y_i and the estimated response from the previous boosting step.

The basic concept is to choose within the mth step of the iterative procedure the subset of variables which provides the best improvement to the fit. In componentwise maximum likelihood based boosting it is common to use the deviance as a measure of goodness-of-fit. We choose the Akaike information criterion (AIC) rather than the deviance, because it includes an automatic penalization of large subsets.

The following algorithm GenBlockBoost is a boosted version of the correlation based penalized estimate.

Algorithm GenBlockBoost

Step 1: (Initialization)

Fit the model $\mu_i = h(\beta_0)$ by iterative Fisher scoring yielding $\hat{\boldsymbol{\beta}}^{(0)} = (\hat{\beta}_0, 0, \ldots, 0)^\top$. Set $\hat{\boldsymbol{\eta}}^{(0)} = \mathbf{X}\hat{\boldsymbol{\beta}}^{(0)}, \hat{\boldsymbol{\mu}}^{(0)} = h(\hat{\boldsymbol{\eta}}^{(0)})$.

Step 2: (Iteration)

For $m = 1, 2, \ldots$

(a) *Find an appropriate order of regressors according to their improvements of fit*

For $j \in \{0, \ldots, p\}$ compute the estimates based on one step Fisher scoring

$$\hat{b}_{\{j\}} = (\mathbf{x}_{\{j\}}^\top \mathbf{W}(\hat{\boldsymbol{\eta}}^{(m-1)})\mathbf{x}_{\{j\}} + \lambda)^{-1}\mathbf{x}_{\{j\}}^\top \mathbf{W}(\hat{\boldsymbol{\eta}}^{(m-1)})\mathbf{D}(\hat{\boldsymbol{\eta}}^{(m-1)})^{-1} \times(\mathbf{y} - \hat{\boldsymbol{\mu}}^{(m-1)}),$$

yielding $\hat{b}_{j_0}, \ldots, \hat{b}_{j_p}$ such that $Dev(\hat{b}_{j_0}) \leq \ldots \leq Dev(\hat{b}_{j_p})$, where

$$Dev(\hat{b}_{j_k}) = 2\sum_{i=1}^{n}\left\{l_i(y_i) - l_i\left[h(\hat{\eta}_i^{(m-1)} + x_{ij_k}\hat{b}_{j_k})\right]\right\}, \quad k = 0, 1, \ldots, p.$$

(b) *Find a suitable number of regressors to update*
For $r = 0, \ldots, p$
With $S_r = \{j_0, \ldots, j_r\}$ we compute the estimates based on one step Fisher scoring

$$\hat{\mathbf{b}}_{S_r} = (\mathbf{X}_{S_r}^\top \mathbf{W}(\hat{\boldsymbol{\eta}}^{(m-1)})\mathbf{X}_{S_r} + \lambda_{|S_r|}\mathbf{M}_{S_r})^{-1}\mathbf{X}_{S_r}^\top \mathbf{W}(\hat{\boldsymbol{\eta}}^{(m-1)}) \\ \times(\mathbf{D}(\hat{\boldsymbol{\eta}}^{(m-1)})^{-1}\mathbf{y} - \hat{\boldsymbol{\mu}}^{(m-1)}),$$

yielding estimates $\hat{\mathbf{b}}_{S_r}$ and AIC criterion $AIC(\hat{\mathbf{b}}_{S_r})$.

(c) *Selection*
Select the subset of variables which has the best fit, yielding

$$S^{(m)} = \arg\min_{S_r} AIC(\hat{\mathbf{b}}_{S_r}).$$

(d) *Refit*
The parameter vector is updated by

$$\hat{\beta}_j^{(m)} = \begin{cases} \hat{\beta}_j^{(m-1)} + \hat{b}_j, & \text{if } j \in S^{(m)} \\ \hat{\beta}_j^{(m-1)}, & \text{otherwise,} \end{cases}$$

yielding $\hat{\boldsymbol{\beta}}^{(m)} = (\hat{\beta}_1^{(m)}, \ldots, \hat{\beta}_p^{(m)})^\top, \hat{\boldsymbol{\eta}}^{(m)} = \mathbf{X}\hat{\boldsymbol{\beta}}^{(m)}, \hat{\boldsymbol{\mu}}^{(m)} = h(\hat{\boldsymbol{\eta}}^{(m)})$.

The number of possible combinations of regressors is 2^p. Due to computational limitation we cannot apply a full search for the best subset. Therefore in a first step of each boosting iteration we order the regressors according to their individual potential improvement to the fit. This improvement is measured by the (potential) deviance

$$Dev(\hat{b}_j) = 2\sum_{i=1}^{n} \left\{ l_i(y_i) - l_i \left[h(\hat{\eta}_i^{(m-1)} + x_{ij}\hat{b}_j) \right] \right\}, \quad j = 0, \ldots, p,$$

where $x_{i0} = 1$ for all $i = 1, \ldots, n$.

For making the base learner a weak learner, so that only a small change in parameter estimates occurs within one boosting iteration, the tuning parameter λ is chosen very large. This also leads to more stable estimates. The price to pay for this choice is an increase in computation time when the value of the tuning parameter becomes larger.

For subsets S that contain only one variable the correlation based penalty (10) cannot be used directly. In those cases we define the penalty by the ridge type penalty $P_{c,\{j\}} = \lambda\beta_j^2$.

Within the algorithm the correlation based estimator is used for subsets of varying size. The tuning parameter λ that is used has to be adapted to the number of refitted regressors. If one considers the case of uncorrelated variables the penalty for all variables reduces to $P_c(\boldsymbol{\beta}) = 2\lambda(p-1)\sum_{i=1}^{p}\beta_i^2$ which equals the ridge penalty with tuning parameter $2\lambda(p-1)$. Thus $\lambda_{|S_r|}$ in step 2b of the GenBlockBoost algorithm is chosen by $\lambda_{|S_r|} = \lambda(|S_r|-1)$, where $|S_r|$ denotes the number of refitted regressors.

In order to avoid overfitting, a stopping criterion is needed for estimating the optimal number of boosting iterations. We use the AIC criterion

$$AIC(\hat{\boldsymbol{\beta}}^{(m)}) = Dev_m + 2tr(\mathbf{H}_m), \tag{16}$$

with

$$Dev_m = 2\sum_{i=1}^{n}\left[l_i(y_i) - l_i(\hat{\mu}_i^{(m)})\right].$$

An approximation of the hat matrix is given by

$$\mathbf{H}_m = \sum_{j=0}^{m}\mathbf{M}_j\prod_{i=0}^{j-1}(I-\mathbf{M}_i),$$

so that $\hat{\boldsymbol{\mu}}^{(m)} \approx \mathbf{H}_m\mathbf{y}$, where

$$\mathbf{M}_l = \boldsymbol{\Sigma}_m^{1/2}\mathbf{W}_m^{1/2}\mathbf{X}_{S^{(m)}}(\mathbf{X}_{S^{(m)}}^\top\mathbf{W}_m\mathbf{X}_{S^{(m)}}+\lambda\mathbf{M}_{S^{(m)}})^{-1}\mathbf{X}_{S^{(m)}}^\top\mathbf{W}_{(m)}^{1/2}\boldsymbol{\Sigma}_m^{-1/2}$$

and $\mathbf{M}_0 = \frac{1}{n}1_n1_n^\top$. See Tutz and Leitenstorfer (2007) for the derivation of this approximation. An estimate of the sufficient number of boosting iterations is

$$m^* = \arg\min_m AIC(\hat{\boldsymbol{\beta}}^{(m)}).$$

In the next section we investigate the performance of the correlation based penalized estimator for GLMs and the GenBlockBoost algorithm in several simulation and data settings.

4 Simulations and Real Data Example

In the simulations, we consider predictors which are given in 10 blocks, each block contains q variables, resulting in $p = 10q$ variables. All variables have unit variances. The correlations between x_i and x_j are $\varrho^{|i-j|}$ if x_i and x_j belong to the same block, otherwise they are given by a truncated $N(0, 0.1^2)$ distribution. For the true predictor η we choose the set V of all covariates that belong to three randomly chosen blocks so that

$$\eta = \mathbf{x}^\top \boldsymbol{\beta},$$

where $\mathbf{x} = (x_1, \ldots, x_p)^\top$ and $\boldsymbol{\beta} = c \cdot (\beta_1, \ldots, \beta_p)^\top$ is determined by

$$\beta_j \sim N(1,1) \text{ for } j \in V, \quad \beta_j = 0 \text{ otherwise.}$$

That means each variable included in one of the chosen blocks is considered as relevant. Note that $\beta_0 = 0$ in all simulations, but all methods are allowed to include a nonzero intercept in their vector of estimated coefficients. The final response y corresponding to the expected value of the response $\mu = E(y|x) = h(\eta)$, where $h(\eta) = \exp(\eta)/(1 + \exp(\eta))$ is drawn from a binomial distribution $B(\mu, 1)$. The constant c is chosen so that the signal-to-noise ratio

$$\text{signal-to-noise ratio} = \frac{\sum_{i=1}^{n} (\mu_i - \bar{\mu})^2}{\sum_{i=1}^{n} Var(y_i)},$$

with $\bar{\mu} = \frac{1}{n} \sum_{i=1}^{n} \mu_i$, is (approximately) equal to one. We use the Newton algorithm to find c in this case. The estimation of unknown parameters is based on 100 training data observations. The evaluation uses 1000 test data observations. We use an additional independent validation data set consisting of 100 observation to determine the tuning parameters.

We compare the GLM PenalReg estimator and the GenBlockBoost algorithm with the maximum likelihood estimator (ML), L_2 penalized maximum likelihood estimation (ridge), L_1 penalized maximum likelihood estimation (lasso) and a boosted version of the L_2 penalized maximum likelihood estimator (GenRidgeBoost). For further details on the GenRidgeBoost algorithm see Tutz and Binder (2007). The computation of the L_1 penalized maximum likelihood estimator is done with the R package glmpath by Mee Young Park and Trevor Hastie.

The performance of data fitting is measured by the deviance and the deviation between estimated and true parameter vector. The latter is defined as

$$MSE_\beta = |\hat{\boldsymbol{\beta}} - \boldsymbol{\beta}|^2. \tag{17}$$

Besides the prediction performance as an important criterion for comparison of methods the variables included into the final model are of special interest to practitioners. The final model should be as parsimonious as possible but all relevant variables should be included. We use the criteria *hits* and *false positives* to evaluate the identification of relevant variables. Hits refers to the number of correctly identified influential variables, false positives is the number of non-influential variables dubbed influential.

The simulation results are given in Table 1, 2, 3 and Figures 1 and 2. GenBlockBoost has the best prediction performance almost all the time. Considering the fit of true parameters PenalReg performs very good, but GenBlockBoost shows good results among the variable selecting procedures for small and medium sized blocks. GlmPath performs better for huge blocks. In the hits and false positives analysis GenBlockBoost clearly outperforms GlmPath and also chooses more relevant covariates than GenRidgeBoost. GenRidgeBoost generally tends to more parsimonious models, hence its median number of false positives is smaller in comparison to GenBlockBoost.

		ML	Ridge	PenalReg	GenRidgeBoost	GenBlockBoost	GlmPath (Lasso)
$q = 3$	$\varrho = 0.95$	17983.21	875.22	**866.16**	965.81	901.05	904.78
	$\varrho = 0.8$	16791.35	928.75	923.54	916.89	**907.23**	940.89
	$\varrho = 0.5$	15497.62	965.58	966.91	890.67	**881.59**	936.87
$q = 5$	$\varrho = 0.95$	20035.41	894.60	892.68	891.95	**851.78**	908.84
	$\varrho = 0.8$	21152.08	939.29	934.00	906.33	**897.05**	949.74
	$\varrho = 0.5$	19842.48	1005.99	1011.70	993.47	**958.38**	1007.23
$q = 10$	$\varrho = 0.95$	-	871.39	**854.36**	868.69	859.35	907.84
	$\varrho = 0.8$	-	970.10	947.54	937.15	**915.58**	982.49
	$\varrho = 0.5$	-	1099.91	1085.18	1119.54	1110.80	**1083.11**

Table 1. Median deviances for simulated data based on 20 replications.

		ML	Ridge	PenalReg	GenRidgeBoost	GenBlockBoost	GlmPath (Lasso)
$q = 3$	$\varrho = 0.95$	423640.00	2.19	**1.80**	2.69	2.56	3.55
	$\varrho = 0.8$	106086.80	1.98	1.68	1.89	**1.62**	1.92
	$\varrho = 0.5$	47861.17	2.00	2.04	**1.30**	1.43	1.63
$q = 5$	$\varrho = 0.95$	345348.10	1.60	**1.51**	3.62	2.07	3.71
	$\varrho = 0.8$	77118.91	2.35	1.97	2.27	**1.95**	2.80
	$\varrho = 0.5$	33738.83	2.15	2.19	2.12	**1.78**	2.18
$q = 10$	$\varrho = 0.95$	-	1.43	**1.08**	2.87	2.40	2.22
	$\varrho = 0.8$	-	2.02	1.55	2.72	**2.49**	2.46
	$\varrho = 0.5$	-	2.51	**2.38**	2.67	2.79	2.58

Table 2. Median MSE_β for simulated data based on 20 replications.

For an application to real data we use the leukemia cancer gene expression data set as described in Golub et al. (1999). In cancer treat-

		ML	Ridge	PenalReg	GenRidgeBoost	GenBlockBoost	GlmPath (Lasso)
$q = 3$	$\varrho = 0.95$	9/22	9/22	9/22	4/1	6/3	5/7
	$\varrho = 0.8$	9/22	9/22	9/22	5/1	5/2	6/8
	$\varrho = 0.5$	9/22	9/22	9/22	6/2	6/3	7/10
$q = 5$	$\varrho = 0.95$	15/36	15/36	15/36	6/3	12/4	6/9
	$\varrho = 0.8$	15/36	15/36	15/36	7/3	11/6	8/9
	$\varrho = 0.5$	15/36	15/36	15/36	8/3	9/5	9/9
$q = 10$	$\varrho = 0.95$	-	30/71	30/71	9/2	17/8	8/5
	$\varrho = 0.8$	-	30/71	30/71	10/2	16/5	12/10
	$\varrho = 0.5$	-	30/71	30/71	12/2	17/9	14/12

Table 3. Median hits/false positives for simulated data based on 20 replications.

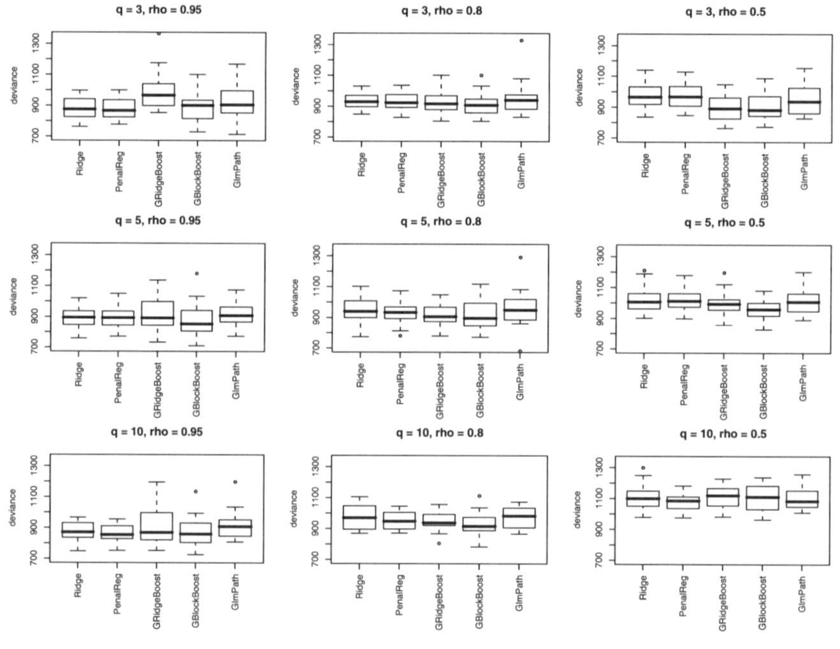

Fig. 1. Deviances for various estimators for the simulations.

ment it is important to target specific therapies to pathogenetically distinct tumor types, to gain a maximum of efficacy and a minimum of toxicity. Hence, distinguishing different tumor types is critical for successful treatment. The challenge of the leukemia data set is to classify acute leukemia into those arising from lymphoid precursors (acute lymphoblastic leukemia, ALL) and those arising from myeloid precursors (acute myeloid leukemia, AML), based on the simultaneous expression

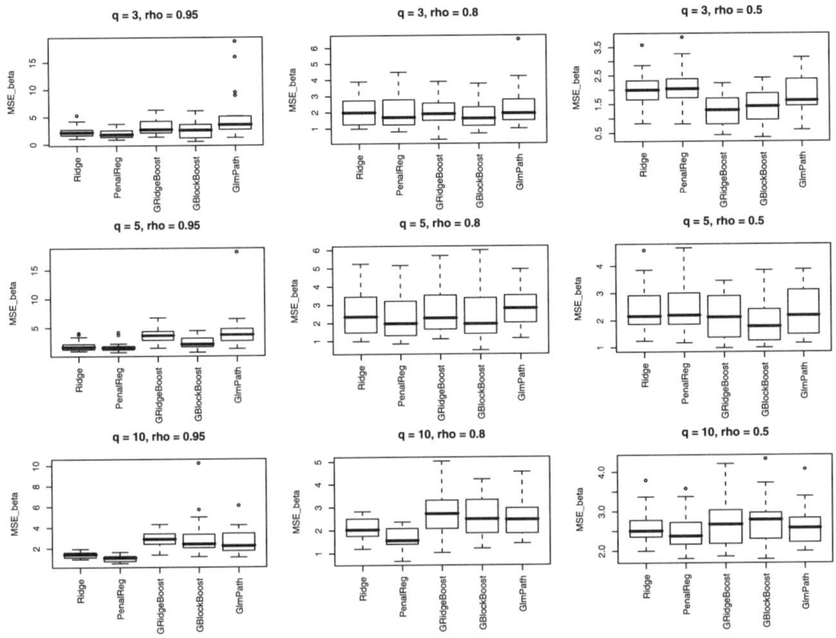

Fig. 2. MSE_β for various estimators for the simulations.

monitoring of 7129 genes using DNA microarrays. The data set consists of 72 samples, out of which 47 observations are ALL and 25 are AML. We use 20 random splits into a training and an independent test sample of sizes 38 and 34, respectively.

Besides the test deviance

$$Dev_{test} = 2 \sum_{i=1}^{n_{test}} \left[l_i(y_{i,test}) - l_i(\hat{\mu}_{i,test}) \right], \tag{18}$$

which is based on the test sample, we consider the number of genes identified as relevant variables. Since the main focus is on classification we focus on the numbers of correctly classified respective misclassified observations in the test data set as performance measures for discrimination.

Due to the 20 random splits we consider the median performance results which are given in Table 4. All three considered algorithms show quite similar performances. At the median number of correctly classified types of leukemia, GenRidgeBoost is slightly better for the ALL class, GenBlockBoost is slightly better for the AML class. When considering the overall misclassification GlmPath has the best discrimination power. Due to the test deviance, the test data fits best to the model estimated by GenBlockBoost. Here, the GenRidgeBoost estimator is only poor. When we consider the number of selected genes GenRidgeBoost is slightly more sparse than the competitors.

Performance measure	GenBlockBoost	GlmPath	GenRidgeBoost
ALL correctly classified	9	10	11
AML correctly classified	21	20	20
misclassification	5	3	5
Dev_{test}	17.75	19.11	84.98
No. of genes used	11	10	9

Table 4. Median performance results for the leukemia cancer gene expression data for 20 random splits into 38 learning data and 34 test data.

5 Concluding Remarks

We presented two approaches for parameter estimation in generalized linear models with many covariates. The GLM PenalReg estimator gives special attention to the grouping effect, the GenBlockBoost algorithm moreover put additional attention on subset selection. The simulations demonstrate the competitive data fitting performance and the small deviation between estimated and true parameter vectors. The GenBlockBoost algorithm is slightly less sparse than the GenRidge-Boost algorithm but this is a consequence of the more tightly focused grouping effect. Nevertheless the correct identification of relevant variables is quite good. As a result, the GenBlockBoost estimator can be seen as a strong competitor in the field of subset selection in generalized linear models.

Both methods may be extended to the case of multivariate generalized linear models, such as with multinomial response. Furthermore, some further theoretical aspects on MSE comparisons with the GLM ridge estimator might by interesting. Here, Trenkler and Toutenburg (1990) provides an initial point for the challenging application to generalized linear models.

References

Anderson JA, Blair V (1982) Penalized maximum likelihood estimation in logistic regression and discrimination. Biometrika 69:123–136

Breiman L (1998) Arcing classifiers. Annals of Statistics 26:801–849

Bühlmann P, Yu B (2003) Boosting with the L2 loss: Regression and classification. Journal of the American Statistical Association 98:324–339

Duffy DE, Santner TJ (1989) On the small sample properties of restricted maximum likelihood estimators for logistic regression models. Communication in Statistics, Theory & Methods 18:959–989

Fahrmeir L, Kaufmann H (1985) Consistency and asymptotic normality of the maximum likelihood estimator in generalized linear models. The Annals of Statistics 13:342–368

Friedman JH (2001) Greedy function approximation: a gradient boosting machine. Annals of Statistics 29:1189–1232

Golub TR, Slonim DK, Tamayo P, Huard C, Gaasenbeek M, Mesirov JP, Coller H, Loh ML, Downing JR, Caligiuri MA, Bloomfield CD, Lander ES (1999) Molecular classification of cancer: class discovery and class prediction by gene expression monitoring. Science 286:531–537

Hoerl AE, Kennard RW (1970) Ridge regression: Bias estimation for nonorthogonal problems. Technometrics 12:55–67

Meir R, Rätsch G (2003) An introduction to boosting and leveraging. In: Mendelson S, Smola A (eds) Advanced Lectures on Machine Learning, Springer, New York, pp 119–184

Nelder JA, Wedderburn RWM (1972) Generalized linear models. Journal of the Royal Statistical Society A 135:370–384

Nyquist H (1991) Restricted estimation of generalized linear models. Applied Statistics 40:133–141

Park MY, Hastie T (2007) An l1 regularization-path algorithm for generalized linear models. JRSS

Schaefer RL, Roi LD, Wolfe RA (1984) A ridge logistic estimate. Communication in Statistics, Theory & Methods 13:99–113

Segerstedt B (1992) On ordinary ridge regression in generalized linear models. Communication in Statistics, Theory & Methods 21:2227–2246

Shevade SK, Keerthi SS (2003) A simple and efficient algorithm for gene selection using sparse logistic regression. Bioinformatics 19:2246–2253

Tibshirani R (1996) Regression shrinkage and selection via the lasso. Journal of the Royal Statistical Society B 58:267–288

Rao CR, Toutenburg H, Shalabh, Heumann C (2008) Linear Models – Least Squares and Generalizations (3rd edition). Springer, Berlin Heidelberg New York

Trenkler G, Toutenburg H (1990) Mean squared error matrix comparisons between biased estimators – an overview of recent results. Statistical Papers 31:165–179

Tutz G, Binder H (2007) Boosting ridge regression. Computational Statistics & Data Analysis (Appearing)

Tutz G, Leitenstorfer F (2007) Generalized smooth monotonic regression in additive modeling. Journal of Computational and Graphical Statistics 16:165–188

Tutz G, Ulbricht J (2006) Penalized regression with correlation based penalty. Discussion Paper 486, SFB 386, Universität München

Zou H, Hastie T (2005) Regularization and variable selection via the elastic net. Journal of the Royal Statistical Society B 67:301–320

Simultaneous Prediction Based on Shrinkage Estimator

Anoop Chaturvedi[1], Suchita Kesarwani[2] and Ram Chandra[3]

[1] Department of Statistics, University of Allahabad, Allahabad, India
anoopchaturv@gmail.com
[2] Department of Statistics, University of Allahabad, Allahabad, India
suchitakesarwani@yahoo.co.in
[3] C.S. Azad University of Agriculture and Technology, Kanpur, India
ramchandra20@rediffmail.com

1 Introduction

In the literature of multiple regression models, the customary analysis is the estimation and hypothesis testing about the parameters of the model. However, in various applications, it is utmost important for a practitioner to predict the future values of the response variable. The most common way to tackle with such a problem is the use of Best Linear Unbiased Predictors (BLUP) discussed by Theil (1971), Hendersion (1972) and Judge, Griffiths, Hill, Lütkepohl and Lee (1985). For further details of the BLUP one can see Toyooka (1982) and Kariya and Toyooka (1985). The Stein-rule predictors and the shrinkage rules based on Stein-rule technique to forecast have also got considerable attention of the researchers in recent past. Copas (1983) considered the prediction in regression using a Stein-rule predictor. Copas and Jones (1987) applied the regression shrinkage technique for the prediction in an autoregressive model. Zellner and Hong (1989) used Bayesian shrinkage rules to forecast international growth rate. Hill and Fomby (1992) analyzed the performance of various improved estimators under an out-of-sample prediction mean square error criterion. Gotway and Cressie (1993) considered a class of linear and non-linear predictors in the context of a general linear model with known disturbances covariance matrix and observed that, under the quadratic loss function, the proposed class of predictors has uniformly smaller risk than the BLUP. Khan and Bhatti (1998) obtained the prediction distribution for a set of future responses from a multiple linear regression model following an equi-correlation structure. Tuchscheres, Herrendorfer and Tuchscheres (1998) proposed Estimated Best Linear Unbiased Predictor (EBLUP) with the help of a designated simulation experiment using MSE and GSD technique for evaluation.

In predicting the dependent variable of a regression model, a traditional practice is to obtain the prediction for either the actual values of the response variable or its average. In some circumstances, it may be desired to consider the simultaneous prediction of both the actual and the average values of a variable for the forecast period. Shalabh (1995) discussed several practical examples where one may encounter with such situations. Keeping this in view, Shalabh (1995) proposed a composite target function for the simultaneous prediction of the response variable in the context of a linear model with independent and identically distributed disturbances. He developed the predictors based on ordinary least squares and Stein-rule estimation methods and examined the bias and mean squared error of the predictors with respect to this target function, see also Srivastava and Shalabh (1996). Toutenburg and Shalabh (1996), and Rao, Toutenburg, Shalabh and Heumann (2008) utilized the same methodology for restricted regression model. Shalabh (1998) extends his results for the linear regression model with equi-correlated responses and analyzes the efficiency properties of the proposed predictors. Chaturvedi, Wan and Singh (2002) discussed the large sample asymptotic properties of a class of Stein-rule predictors based on composite target function in a general linear model. Chaturvedi and Singh (2000) estimated the MSE matrix of the proposed SR predictor when disturbances covariance matrix is known.

The present paper deals with the problem of prediction based on shrinkage estimator in a general linear model with nonspherical disturbances. A general family of predictors for the composite target function, considered by Shalabh (1995), has been proposed and its asymptotic distribution has been derived employing large sample asymptotic theory. The risk based on quadratic loss structure of the proposed family of predictors has been obtained. The performance of proposed family of predictors is compared with the feasible Best Linear Unbiased Predictor (FBLUP) under the MSE matrix criterion and Quadratic loss function criterion. Further, we obtain the expression for an estimator for the MSE matrix of the proposed predictor. The results of a numerical simulation have been presented and discussed.

2 The Model and the Estimators

Let us consider the general linear model

$$y = X\beta + u \tag{1}$$

where y is a $n \times 1$ vector of observations on the dependent variable, X is a $n \times p$ non-stochastic matrix of n observations on each of the p explanatory variables, β is a $p \times 1$ vector of unknown regression coefficients and u is a $n \times 1$ vector of disturbances.

Let y_f denotes a $T \times 1$ vector of unobservable values of the dependent variable for T forecast periods generated by the process

$$y_f = X_f\beta + u_f \tag{2}$$

where X_f is a $T \times p$ matrix of pre-specified observations on the explanatory variables for T forecast periods and u_f is a $T \times 1$ vector of disturbances for the forecast periods. Further, we assume that

$$\begin{bmatrix} u \\ u_f \end{bmatrix} \sim N(0, \sigma^2 \Sigma),$$

with

$$\Sigma = \begin{bmatrix} \Phi & V \\ V' & \Psi \end{bmatrix}.$$

Thus, $\sigma^2\Phi$ is a $n \times n$ covariance matrix of u, $\sigma^2\Psi$ is a $T \times T$ covariance matrix of u_f and $\sigma^2 V$ is a $n \times T$ matrix of covariance between u and u_f. We assume that the elements of Σ are functions of a $q \times 1$ parameter vector θ, so that we can write $\Sigma = \Sigma(\theta)$.

Let us consider the following target function discussed by Shalabh (1995) and Toutenburg (1982) which allows the prediction of both y_f and $E(y_f)$:

$$\begin{aligned} \tau &= \lambda y_f + (1-\lambda)E(y_f) \\ &= X_f\beta + \lambda u_f \end{aligned} \tag{3}$$

where λ, $(0 \le \lambda \le 1)$ is a non-stochastic scalar assigning weights to actual and expected values of y_f.

For convenience we write $\Omega = \Phi^{-1}$. If θ is known then BLUP for τ is given by

$$\tilde{\tau} = \lambda \tilde{y}_f + (1-\lambda)X_f\beta^* \tag{4}$$

where

$$\tilde{y}_f = X_f\beta^* + V'\Omega(y - X\beta^*) \tag{5}$$

is the BLUP of y_f (see Toutenburg (1982, p.138)) and

$$\beta^* = (X'\Omega X)^{-1}X'\Omega y$$

is the GLS estimator of β. Substituting (6) in (5), we obtain

$$\tilde{\tau} = X_f\beta^* + \lambda V'\Omega(y - X\beta^*). \tag{6}$$

Let us consider the following general family of shrinkage estimators for β:

$$\beta_S^* = \left[1 - \frac{k}{n-p+2}\frac{r(\eta^*)}{\eta^*}\right]\beta^*,$$

where

$$\eta^* = \frac{\beta^{*'}X'\Omega X\beta^*}{(y - X\beta^*)'\Omega(y - X\beta^*)},$$

k (≥ 0) is a characterizing scalar and $r(\eta^*)$ is a function of η^*.

If β^* is replaced by β_S^* in (4), we obtain the following general family of shrinkage regression predictor:

$$\tau_S^* = X_f\beta_S^* + \lambda V'\Omega(y - X\beta_S^*). \tag{7}$$

On the other hand, if θ is unknown and estimated by $\hat{\theta}$, then replacement of θ by $\hat{\theta}$ in (6) leads to the FBLUP for τ given by:

$$\hat{\tau} = X_f\hat{\beta} + \lambda\hat{V}'\hat{\Omega}(y - X\hat{\beta}) \tag{8}$$

where $\hat{\Omega}$ and $\hat{V}$ are obtained by replacing θ by $\hat{\theta}$ in Ω and V respectively, and $\hat{\beta} = (X'\hat{\Omega}X)^{-1}X'\hat{\Omega}y$ is the FGLS estimator of β. Notice that the first term on the right hand side of (8) is an estimator of the non-stochastic part $X_f\beta$ of τ whereas $\hat{V}'\hat{\Omega}(y - X\hat{\beta})$ is an estimator of the disturbance term u_f. Further, replacement of θ by $\hat{\theta}$ in (7) gives the following predictor based on the shrinkage estimator:

$$\hat{\tau}_S = X_f\hat{\beta}_S + \lambda\hat{V}'\hat{\Omega}(y - X\hat{\beta}_S) \tag{9}$$

where

$$\hat{\beta}_S = \left[1 - \frac{k}{n-p+2}\frac{r(\hat{\eta})}{\hat{\eta}}\right]\hat{\beta};$$

$$\hat{\eta} = \frac{\hat{\beta}'X'\hat{\Omega}X\hat{\beta}}{(y - X\hat{\beta})'\hat{\Omega}(y - X\hat{\beta})}.$$

Obviously, for $k = 0$, the predictor $\hat{\tau}_s$ reduces to $\hat{\tau}$ and for $r(\hat{\eta}) = 1$ it reduces to $\hat{\tau}_{sr}$ given by

$$\hat{\tau}_{sr} = X_f\hat{\beta}_{Sr} + \lambda\hat{V}'\hat{\Omega}(y - X\hat{\beta}_{Sr}) \tag{10}$$

where

$$\hat{\beta}_{Sr} = \left[1 - \frac{k}{n-p+2}\frac{(y - X\hat{\beta})'\hat{\Omega}(y - X\hat{\beta})}{\hat{\beta}'X'\hat{\Omega}X\hat{\beta}}\right]\hat{\beta}.$$

3 Asymptotic Distribution

In this section we obtain the asymptotic distribution of the predictor $\hat{\tau}_s$ when the sample size is large. We assume that

i) For any $n \times n$ finite matrix C with elements of order $O(1)$, the quantity $n^{-1}X'CX$ is of order $O(1)$ as $n \to \infty$.
ii) For any arbitrary matrix with elements of order $O(1)$, the quantity $n^{-1/2}X'Cu$ is of order $O_p(1)$ and
iii) The estimator $\hat{\theta}$ of θ is an even function of Mu, where

$$M = I_n - X(X'X)^{-1}X'.$$

Let us write

$$A = \frac{1}{n}X'\Omega X, \quad \hat{A} = \frac{1}{n}X'\hat{\Omega}X, \quad \tilde{\eta} = \frac{\beta' A\beta}{\upsilon\sigma^2}, \quad \alpha^* = \frac{1}{(\sigma\sqrt{n})}X'\hat{\Omega}u,$$

$$\upsilon = \frac{1}{n}\varepsilon_2'(P'\Omega^{-1}P)^{-1}P'\Omega^{-1}(\hat{\Omega} - \hat{\Omega}X(X'\hat{\Omega}X)^{-1}X'\hat{\Omega})$$
$$\times\Omega^{-1}P(P'\Omega^{-1}P)^{-1}\varepsilon_2,$$
$$\gamma = \frac{\sqrt{n}}{\sigma}(\hat{\tau}_s - \tau).$$

Since M is an idempotent matrix of rank $n - p$, there exists a $n \times (n - p)$ matrix P such that $P'P = I_{n-p}$ and $PP' = M$. Consider the transformation $\varepsilon_1 = X'\Omega u/(\sigma\sqrt{n})$ and $\varepsilon_2 = P'u/\sigma$. Then from the normality of u and observing that $P'X = 0$, it follows that ε_1 and ε_2 are independently distributed with $\varepsilon_1 \sim N(0, A)$, $\varepsilon_2 \sim N(0, P'\Omega^{-1}P)$. Further, assumption (iii) implies that $\hat{\theta}$ is an even function of ε_2.

Let us define

$$\mu(\varepsilon_2) = \frac{-k\sigma\upsilon}{\sqrt{n}\beta'\hat{A}\beta} r(\tilde{\eta})(X_f - \lambda\hat{V}'\hat{\Omega}X)\beta + \left[(X_f - \lambda\hat{V}'\hat{\Omega}X)A^{-1}\frac{X'\hat{\Omega}}{\sqrt{n}} + \lambda\sqrt{n}(\hat{V}'\hat{\Omega} - V'\Omega)\right] \times\Omega^{-1}P(P'\Omega^{-1}P)\varepsilon_2$$

and

$$\begin{aligned}\Xi(\varepsilon_2) = \lambda^2 n(\Psi - V'\Omega V) &+(X_f - \lambda V'\Omega X)A^{-1}(X_f - \lambda V'\Omega X)' \\ &-\frac{2k\sigma^2}{n\beta' A\beta} r(\tilde{\eta})(X_f - \lambda V'\Omega X)\left(A^{-1} - \frac{2\beta\beta'}{\beta' A\beta}\right)(X_f - \lambda V'\Omega X)' \\ &+2\lambda(\hat{V}'\hat{\Omega} - V'\Omega)(X_f - \lambda V'\Omega X)' \\ &+\lambda^2(\hat{V}'\hat{\Omega} - V'\Omega)XA^{-1}X'(\hat{V}'\hat{\Omega} - V'\Omega)' \\ &+\frac{4k}{n\beta' A\beta} r(\tilde{\eta})(X_f - \lambda\hat{V}'\hat{\Omega}X)\beta\beta'(X_f - \lambda\hat{V}'\hat{\Omega}X)'.\end{aligned}$$

Theorem 1. *The conditional asymptotic distribution of γ given ε_2, up to order $O_p(n^{-1})$, is normal with mean vector $\mu(\varepsilon_2)$ and variance co-variance matrix $\Xi(\varepsilon_2)$.*

Proof:

See Appendix.

Denoting the probability density function of a normal distribution with mean vector $\mu(\varepsilon_2)$ and variance-covariance matrix $\Xi(\varepsilon_2)$ by $\phi(\mu(\varepsilon_2), \Xi(\varepsilon_2))$, the asymptotic unconditional distribution of γ, up to order $O_p(n^{-1})$, is given by

$$f(\gamma) = E_{\varepsilon_2}\left[\phi(\mu(\varepsilon_2), \Xi(\varepsilon_2))\right]. \tag{11}$$

The bias vector of the predictor $\hat{\tau}_s$, up to order $O_p(n^{-1})$, is given by

$$\begin{aligned}E(\hat{\tau}_s - \tau) = &-\frac{k\sigma^2}{n}E_{\varepsilon_2}\left[\frac{1}{\beta'\hat{A}\beta}\upsilon r(\tilde{\eta})(X_f - \lambda V'\Omega X)\beta\right] \\ &+\frac{\sigma}{\sqrt{n}}E_{\varepsilon_2}\left[\left\{\frac{1}{\sqrt{n}}(X_f - \lambda\hat{V}'\hat{\Omega}X)A^{-1}X'\hat{\Omega}\right.\right. \\ &\left.\left.+\lambda\sqrt{n}(\hat{V}'\hat{\Omega} - V\Omega)\right\}\Omega^{-1}P(P'\Omega P)^{-1}\varepsilon_2\right].\end{aligned} \tag{12}$$

Since $\hat{\theta}$ is an even function of ε_2, the second term in equation (12) vanishes and the expression for the bias vector of $\hat{\tau}_s$ reduces to

$$\begin{aligned} E(\hat{\tau}_s - \tau) &= -\frac{k\sigma^2}{n} E_{\varepsilon_2}\left[\frac{1}{\beta'\hat{A}\beta} \upsilon r\left(\tilde{\eta}\right)\left(X_f - \lambda V'\Omega X\right)\beta\right] \\ &= -\frac{k\sigma^2}{n\beta' A\beta} r\left(\eta\right)\left(X_f - \lambda V'\Omega X\right)\beta + O(n^{-3/2}) \qquad (13) \end{aligned}$$

where $\eta = \beta' A\beta/\sigma^2 h$, $h = n - p$.

From (13), we observe that the bias of $\hat{\tau}_s$ increases in magnitude as $\left(\beta' A\beta/\sigma^2\right)$ decreases. The bias, in magnitude, is also a decreasing function of n. Obviously the feasible BLUP $\hat{\tau}$ is unbiased up to order $O(n^{-1})$.

4 Comparison of Predictors

The difference between the MSE matrices of the predictors $\hat{\tau}$ and $\hat{\tau}_s$, up to order $O(n^{-2})$, is given by

$$\begin{aligned} & E\left[(\hat{\tau} - \tau)(\hat{\tau} - \tau)'\right] - E\left[(\hat{\tau}_s - \tau)(\hat{\tau}_s - \tau)'\right] \\ &= \frac{2k\sigma^4}{n^2\beta' A\beta} r\left(\eta\right)\left(X_f - \lambda V'\Omega X\right)\left(A^{-1} - \frac{4 + kr\left(\eta\right)}{\beta' A\beta}\beta\beta'\right) \\ &\quad \times (X_f - \lambda V'\Omega X)' \\ &\quad + \frac{4k\sigma^2}{n^2\beta' A\beta} r'\left(\eta\right)\left(X_f - \lambda V'\Omega X\right)\beta\beta'(X_f - \lambda V'\Omega X)'. \end{aligned}$$

In order to establish the dominance of $\hat{\tau}_s$ over $\hat{\tau}$ under the MSE matrix criterion, we assume that $r'\left(\eta\right) > 0$. Then $\hat{\tau}_s$ dominates $\hat{\tau}$ iff the matrix

$$\left[A^{-1} - \frac{1}{2\beta' A\beta}\left(4 + kr\left(\eta\right)\right)\beta\beta'\right]$$

is positive semi-definite. Following Rao, Toutenburg, Shalabh and Heumann (2008, p. 506), we observe that the above matrix is positive semi-definite iff

$$\frac{1}{2}\left[4 + kr\left(\frac{\beta' A\beta}{\sigma^2 \upsilon}\right)\right] \leq 1.$$

However, it is impossible as both k and $r\left(\eta\right)$ are non-negative. Hence, under the MSE matrix criterion, the predictor $\hat{\tau}_s$ cannot uniformly dominate $\hat{\tau}$. Similarly $\hat{\tau}$ dominates $\hat{\tau}_s$ iff

$$\left[\frac{1}{2\beta' A\beta}\left(4+kr\left(\eta\right)\right)\beta\beta' - A^{-1}\right]$$

is positive semi-definite, which is also not possible, (see Rao, Toutenburg, Shalabh and Heumann (2008, p. 507)). Hence we conclude that neither of the predictors $\hat{\tau}$ nor $\hat{\tau}_s$ uniformly dominate the other under the MSE matrix criterion at least to order $O(n^{-2})$.

For comparing the performance of two predictors let us consider the risk under the following quadratic loss function

$$L(\tilde{\tau}, \tau) = (\tilde{\tau} - \tau)' Q_n (\tilde{\tau} - \tau),$$

where $\tilde{\tau}$ is a predictor of τ and Q_n is a positive definite symmetric matrix of order $O(1)$.

Let us write

$$\omega = \frac{tr\{A^{-1}(X_f - \lambda V'\Omega X)' Q_n (X_f - \lambda V'\Omega X)\}}{\mu_l\{A^{-1}(X_f - \lambda V'\Omega X)' Q_n (X_f - \lambda V'\Omega X)\}}.$$

Up to the order of our approximation, the difference between risks of two predictors is given by

$$\begin{aligned} R\left[\hat{\tau}\right] - R\left[\hat{\tau}_s\right] &= \frac{k\sigma^4}{n^2\beta' A\beta} r\left(\eta\right)\left[2tr\{A^{-1}(X_f - \lambda V'\Omega X)' Q_n (X_f - \lambda V'\Omega X)'\}\right. \\ &\quad \left. - \frac{1}{\beta' A\beta}\left(4+kr\left(\eta\right)\right)\beta'(X_f - \lambda V'\Omega X)' Q_n (X_f - \lambda V'\Omega X)\beta\right] \\ &\quad + \frac{4k}{n\beta' A\beta} r'\left(\eta\right)\beta'(X_f - \lambda V'\Omega X)' Q_n (X_f - \lambda V'\Omega X)\beta. \end{aligned} \tag{14}$$

A lower bound for the difference between risks is given by

$$\begin{aligned} R\left[\hat{\tau}\right] - R\left[\hat{\tau}_s\right] \geq \\ \frac{k\sigma^4}{n^2\beta' A\beta} r\left(\eta\right)\mu_l\left[A^{-1}(X_f - \lambda V'\Omega X)' Q_n (X_f - \lambda V'\Omega X)\right] \\ \times\left[2\omega - \left(4+kr\left(\eta\right)\right)\right]. \end{aligned}$$

Hence, up to the order of our approximation, a sufficient condition for $\hat{\tau}_s$ to dominate $\hat{\tau}$ is given by

$$0 \leq kr\left(\eta\right) \leq 2(\omega - 2), \quad \omega > 2. \tag{15}$$

5 Estimation of the MSE Matrix

Let us assume that θ is known and σ^2 is unknown. The following theorem gives an unbiased estimator for the MSE matrix of the predictor $\tilde{\tau}_s$:

Theorem 2. *An unbiased estimator of the MSE matrix of the predictor* $\tilde{\tau}_s$ *is given by*

$$\begin{aligned}\hat{M}(\tilde{\tau}_s) = & \frac{\upsilon}{h}\Delta - \frac{k}{h+2}\frac{\upsilon^2}{\beta^{*'}X'\Omega X\beta^*}W\left[\int_0^1 t^{h/2} r\left(\eta^*\right)dt\right]\left(X'\Omega X\right)^{-1}W' \\ & -\frac{2k}{h+2}W\left[\int_0^1 t^{h/2-2}\left\{\frac{\upsilon t}{\beta^{*'}X'\Omega X\beta^*}r'\left(\eta^*\right)\right.\right. \\ & \left.\left.-\left(\frac{\upsilon t}{\beta^{*'}X'\Omega X\beta^*}\right)^2 r\left(\eta^*\right)\right\}dt\right]\beta^*\beta^{*'}W' \\ & +\left(\frac{k}{h+2}\right)^2\left[\frac{r^2(\eta^*)}{\eta^{*2}}\left\{W\beta^*\beta^{*'}W'\right\}\right]W\beta^*\beta^{*'}W'. \end{aligned} \tag{16}$$

Proof:

See Appendix.

In general, the form of the estimator in equation (16) is not readily computable and requires numerical integration if it is to be calculated. In such a situation, an alternative is to derive an estimator of the scaled MSE matrix of $\tilde{\tau}_s$, defined as $M(\sigma^{-1}\tilde{\tau}_s)$.

Theorem 3. *An unbiased estimator of the scaled MSE matrix* $M\left(\sigma^{-1}\tilde{\tau}_s\right)$ *is given by*

$$\begin{aligned}\tilde{M}(\sigma^{-1}\tilde{\tau}_s) = & \frac{\upsilon}{h}\Delta - \frac{2k\upsilon}{h(h+2)}r^*\left(\eta^*\right)W\left(X'\Omega X\right)^{-1}W' \\ & +\left[\frac{4k}{h(h+2)}\left\{\frac{\upsilon}{\beta^{*'}X'\Omega X\beta^{*'}}r'\left(\eta^*\right)\right.\right. \\ & \left.-\left(\frac{\upsilon}{\beta^{*'}X'\Omega X\beta^*}\right)^2 r\left(\eta^*\right)\right\} \end{aligned}$$

$$
\begin{aligned}
&+\left(\frac{k}{h+2}\right)^2 \frac{(h-2)}{h} r^2(\eta^*) \\
&-\frac{4r^*(\eta^*)}{h}\eta^* \left\{\frac{v}{\beta^{*'}X'\Omega X\beta^*} r'(\eta^*)\right. \\
&\left.\left.-\left(\frac{v}{\beta^{*'}X'\Omega X\beta^*}\right)^2 r(\eta^*)\right\}\right] W\beta^*\beta^{*'}W'. \qquad (17)
\end{aligned}
$$

Proof:

See Appendix.

The MSE matrix of the predictor $\tilde{\tau}_s$ can be estimated by

$$\tilde{M}(\tilde{\tau}_s) = \frac{v}{h}\tilde{M}\left(\sigma^{-1}\tilde{\tau}_s\right).$$

When θ is unknown, we replace it by $\hat{\theta}$ in (16) or (17) to obtain the estimator for the MSE matrix of the predictor $\hat{\tau}_s$.

6 Monte-Carlo Study

In order to observe the performance of different predictors numerically, we carried out a Monte-Carlo simulation. The computational results are based on the model

$$y = X\beta + u$$

with the disturbance term u following an AR(1) process

$$u_t = \rho u_{t-1} + \varepsilon_t; \quad t = 1, 2, \ldots, n.$$

Here ρ is the autoregression parameter so that $|\rho| < 1$ and ε_t's identically and independently distributed random variables from $N\left(0, \sigma_\varepsilon^2\right)$ implying that $\mathrm{var}(u_t) = \sigma_\varepsilon^2/(1-\rho^2)$. The matrix $H = (X \;\; X_f)$ is chosen so that $H'H = I$. Further,

$$
\Sigma = \begin{bmatrix} \Omega^{-1} & V \\ V' & \Psi \end{bmatrix}.
$$

$$
= \begin{bmatrix}
1 & \rho & \rho^2 & \ldots & \rho^{n+T-1} \\
\rho & 1 & \rho & \ldots & \rho^{n+T-2} \\
\rho^2 & \rho & 1 & \ldots & \rho^{n+T-3} \\
\vdots & \vdots & \vdots & \ddots & \vdots \\
\rho^{n+T-1} & \rho^{n+T-2} & \rho^{n+T-3} & \ldots & 1
\end{bmatrix}.
$$

We have evaluated the relative risk for $n = 20, 50, 100$; $T = 4$; $p = 4$, 10; $\beta'\beta$= 1, 5, 10, 15, 20; ρ= -0.8, -0.6, -0.4, -0.2, 0.0, 0.2, 0.4, 0.6, 0.8; λ= 0, 0.25, 0.50, 0.75, 1.00. However in the tables 1-5, the relative risk of four common forms of Stein-type estimators, the Stein-rule estimators (SR), the double k-class of estimators (KK), the Minimum Mean Square Error estimator (MMSE) and the Admissible Feasible MMSE (AFMMSE, denoted by AMSE in the table) in comparison to the risk of predictor based on FGLS estimator are presented only for $n = 100$; $\beta'\beta = 1, 10, 20$ and all the values of T, p, ρ and λ mentioned above. The relative risks for other choices of n and $\beta'\beta$ show almost the same kind of pattern. Each of the simulation result is based on 5000 repetitions. The relative risks are evaluated by setting $Q_n = I$ and $k = p-2$. From the numerical results, we draw the following conclusions:

- For majority of cases the relative risks of predictors based on different shrinkage estimators with respect to predictor based on FGLS estimator are less than 1. Further, except for some cases with extreme values of ρ along with $\lambda \geq 0.75$, we have the following relationship between the risks of different predictors:

$$R(\hat{\tau}) > R(\hat{\tau}_{AM}) > R(\hat{\tau}_{SR}) > R(\hat{\tau}_{KK}) > R(\hat{\tau}_{MM}).$$

- As the value of $\beta'\beta$ increases, the relative risk of different estimators has a tendency to incline (or decline, whatever the case) towards 1. Hence, in general, as $\beta'\beta$ increases, the performance of different estimators goes similar in terms of risk.
- The relative risk of different predictor increases with increasing λ.
- In general, as $|\rho|$ increases, the relative risk decreases up to $|\rho| = 0.6$ and at $|\rho| = 0.8$, the relative risk increases.
- Usually, the relative risk attains maximum for ρ=0. However, there are some exceptional cases too where relative risk is maximum for $|\rho| = 0.8$ or, in few cases, for $|\rho| = 0.6$.

Appendix

Proof of Theorem 1:

We can write γ as

$$\begin{aligned}\gamma &= \frac{\sqrt{n}}{\sigma}(\hat{\tau}_s - \tau)\\ &= \frac{\sqrt{n}}{\sigma}\left[X_f(\hat{\beta}_S - \beta) + \lambda\hat{V}'\hat{\Omega}(y - X\hat{\beta}_S) - \lambda u_f\right]\end{aligned}$$

Table 1. Relative risk of estimators for $\lambda = 0.00$

		$p=4$				$p=10$			
ρ	$\beta'\beta$	SR	KK	MMSE	AMSE	SR	KK	MMSE	AMSE
-0.8	1	0.9934	0.9916	1.0005	0.9947	0.9553	0.9831	0.9557	0.9886
	10	0.9993	0.9992	1.0002	0.9995	0.9937	0.9942	0.9938	0.9986
	20	1.0000	0.9999	1.0008	0.9999	0.9975	0.9974	0.9977	0.9994
-0.6	1	0.9987	0.9932	1.0135	0.9971	0.9461	0.9830	0.9448	0.9872
	10	0.9988	0.9986	0.9993	0.9992	0.9959	0.9954	0.9967	0.9988
	20	1.0005	1.0004	1.0019	1.0001	0.9966	0.9967	0.9967	0.9992
-0.4	1	0.9931	0.9905	1.0040	0.9939	0.9434	0.9831	0.9417	0.9861
	10	0.9995	0.9992	1.0010	0.9995	0.9931	0.9937	0.9933	0.9984
	20	0.9998	0.9997	1.0006	0.9997	0.9955	0.9957	0.9954	0.9991
-0.2	1	0.9954	0.9911	1.0091	0.9949	0.9280	0.9818	0.9240	0.9836
	10	0.9991	0.9988	1.0004	0.9993	0.9929	0.9933	0.9934	0.9983
	20	0.9994	0.9993	0.9998	0.9996	0.9964	0.9963	0.9967	0.9991
0	1	0.9868	0.9886	0.9914	0.9907	0.9253	0.9816	0.9208	0.9830
	10	0.9993	0.9989	1.0007	0.9994	0.9919	0.9930	0.9921	0.9982
	20	0.9993	0.9992	0.9997	0.9995	0.9950	0.9953	0.9949	0.9989
0.2	1	0.9864	0.9891	0.9883	0.9908	0.9181	0.9814	0.9125	0.9824
	10	0.9992	0.9989	1.0002	0.9994	0.9913	0.9928	0.9912	0.9981
	20	0.9991	0.9990	0.9990	0.9994	0.9959	0.9960	0.9960	0.9991
0.4	1	0.9826	0.9893	0.9763	0.9896	0.9162	0.9815	0.9099	0.9829
	10	0.9987	0.9986	0.9986	0.9992	0.9907	0.9926	0.9902	0.9981
	20	0.9995	0.9995	0.9996	0.9997	0.9952	0.9957	0.9949	0.9990
0.6	1	0.9789	0.9893	0.9640	0.9885	0.9051	0.9815	0.8953	0.9834
	10	0.9976	0.9978	0.9958	0.9987	0.9901	0.9928	0.9889	0.9983
	20	0.9993	0.9993	0.9988	0.9996	0.9947	0.9955	0.9940	0.9991
0.8	1	0.9765	0.9889	0.9549	0.9879	0.8889	0.9811	0.8725	0.9839
	10	0.9975	0.9978	0.9951	0.9987	0.9870	0.9917	0.9841	0.9982
	20	0.9988	0.9988	0.9976	0.9994	0.9936	0.9950	0.9921	0.9991

$$= \frac{\sqrt{n}}{\sigma}\Big[(X_f - \lambda\hat{V}'\hat{\Omega}X)(\hat{\beta} - \beta) - \frac{k}{n-p+2}\frac{r(\hat{\eta})}{\hat{\eta}}(X_f - \lambda\hat{V}'\hat{\Omega}X)\hat{\beta} + \lambda\hat{V}'\hat{\Omega}u - \lambda u_f\Big].$$

Further, we have $\hat{\Omega} - \Omega = O_p(n^{-1/2})$, $\hat{V} - V = O_p(n^{-1/2})$ and $\alpha^* - \varepsilon_1 = O_p(n^{-1/2})$.

Now, following Kariya and Toyooka (1992), we observe that

$$X(X'\Omega X)^{-1}X'\Omega + \Omega^{-1}P(P'\Omega^{-1}P)P' = I_n \tag{A.1}$$

Table 2. Relative risks of estimates for λ=0.25

		$p=4$				$p=10$			
ρ	$\beta'\beta$	SR	KK	MMSE	AMSE	SR	KK	MMSE	AMSE
-0.8	1	0.9578	0.9802	0.9181	0.9783	0.8933	0.9809	0.8759	0.9851
	10	0.9910	0.9920	0.9818	0.9954	0.9641	0.9770	0.9546	0.9954
	20	0.9952	0.9955	0.9903	0.9976	0.9790	0.9837	0.9733	0.9973
-0.6	1	0.9463	0.9767	0.8969	0.9724	0.8539	0.9760	0.8318	0.9793
	10	0.9891	0.9905	0.9779	0.9944	0.9555	0.9729	0.9439	0.9942
	20	0.9943	0.9947	0.9884	0.9971	0.9747	0.9810	0.9679	0.9967
-0.4	1	0.9360	0.9743	0.8780	0.9670	0.8267	0.9738	0.8024	0.9753
	10	0.9875	0.9892	0.9745	0.9936	0.9490	0.9706	0.9358	0.9934
	20	0.9934	0.9939	0.9866	0.9966	0.9711	0.9789	0.9633	0.9963
-0.2	1	0.9301	0.9732	0.8674	0.9640	0.8100	0.9731	0.7851	0.9726
	10	0.9864	0.9885	0.9724	0.9930	0.9438	0.9687	0.9293	0.9927
	20	0.9928	0.9934	0.9853	0.9963	0.9682	0.9774	0.9596	0.9959
0	1	0.9294	0.9735	0.8666	0.9636	0.8044	0.9731	0.7796	0.9717
	10	0.9857	0.9880	0.9711	0.9927	0.9413	0.9681	0.9262	0.9924
	20	0.9925	0.9931	0.9847	0.9961	0.9664	0.9765	0.9574	0.9957
0.2	1	0.9311	0.9739	0.8694	0.9645	0.8035	0.9731	0.7787	0.9715
	10	0.9863	0.9884	0.9723	0.9930	0.9412	0.9681	0.9261	0.9924
	20	0.9927	0.9933	0.9851	0.9963	0.9664	0.9765	0.9574	0.9957
0.4	1	0.9401	0.9763	0.8856	0.9692	0.8171	0.9742	0.7932	0.9736
	10	0.9876	0.9894	0.9749	0.9937	0.9443	0.9692	0.9300	0.9928
	20	0.9934	0.9939	0.9865	0.9966	0.9682	0.9775	0.9597	0.9959
0.6	1	0.9530	0.9802	0.9094	0.9758	0.8379	0.9761	0.8152	0.9770
	10	0.9895	0.9909	0.9787	0.9946	0.9512	0.9719	0.9385	0.9937
	20	0.9943	0.9947	0.9884	0.9971	0.9718	0.9795	0.9642	0.9964
0.8	1	0.9709	0.9868	0.9433	0.9851	0.8857	0.9818	0.8678	0.9843
	10	0.9925	0.9934	0.9848	0.9962	0.9636	0.9780	0.9541	0.9953
	20	0.9957	0.9960	0.9912	0.9978	0.9775	0.9831	0.9714	0.9971

so that

$$\begin{aligned} u &= \left[X(X'\Omega X)^{-1}X'\Omega + \Omega^{-1}P(P'\Omega^{-1}P)P'\right]u \\ &= \sigma\left[\frac{1}{\sqrt{n}}XA^{-1}\varepsilon_1 + \Omega^{-1}P(P'\Omega^{-1}P)\varepsilon_2\right]. \end{aligned} \tag{A.2}$$

Hence, up to order $O_p(n^{-1})$

Table 3. Relative risks of estimates for λ=0.50

		$p=4$				$p=10$			
ρ	$\beta'\beta$	SR	KK	MMSE	AMSE	SR	KK	MMSE	AMSE
-0.8	1	0.9763	0.9891	0.9535	0.9879	0.9398	0.9898	0.9294	0.9919
	10	0.9954	0.9959	0.9906	0.9977	0.9811	0.9879	0.9761	0.9976
	20	0.9976	0.9977	0.9951	0.9988	0.9893	0.9917	0.9864	0.9986
-0.6	1	0.9676	0.9864	0.9369	0.9835	0.9060	0.9856	0.8909	0.9873
	10	0.9944	0.9951	0.9886	0.9971	0.9756	0.9852	0.9692	0.9969
	20	0.9971	0.9973	0.9940	0.9985	0.9868	0.9901	0.9833	0.9983
-0.4	1	0.9616	0.9850	0.9255	0.9804	0.8830	0.9836	0.8657	0.9842
	10	0.9935	0.9945	0.9869	0.9967	0.9717	0.9837	0.9642	0.9964
	20	0.9966	0.9969	0.9931	0.9983	0.9847	0.9888	0.9805	0.9980
-0.2	1	0.9575	0.9843	0.9180	0.9783	0.8694	0.9829	0.8514	0.9822
	10	0.9929	0.9940	0.9855	0.9964	0.9687	0.9827	0.9606	0.9960
	20	0.9963	0.9966	0.9925	0.9981	0.9830	0.9880	0.9784	0.9978
0	1	0.9558	0.9840	0.9148	0.9774	0.8603	0.9822	0.8416	0.9809
	10	0.9926	0.9938	0.9850	0.9962	0.9669	0.9821	0.9583	0.9957
	20	0.9962	0.9965	0.9922	0.9980	0.9820	0.9875	0.9772	0.9977
0.2	1	0.9572	0.9843	0.9175	0.9781	0.8599	0.9824	0.8414	0.9809
	10	0.9928	0.9939	0.9854	0.9963	0.9667	0.9821	0.9581	0.9957
	20	0.9963	0.9966	0.9924	0.9981	0.9820	0.9874	0.9771	0.9977
0.4	1	0.9625	0.9856	0.9275	0.9809	0.8719	0.9834	0.8543	0.9825
	10	0.9935	0.9945	0.9869	0.9967	0.9689	0.9829	0.9609	0.9960
	20	0.9966	0.9969	0.9931	0.9983	0.9830	0.9880	0.9784	0.9978
0.6	1	0.9706	0.9879	0.9428	0.9850	0.8910	0.9849	0.8749	0.9853
	10	0.9945	0.9952	0.9887	0.9972	0.9724	0.9841	0.9651	0.9964
	20	0.9971	0.9973	0.9941	0.9985	0.9848	0.9889	0.9807	0.9980
0.8	1	0.9813	0.9916	0.9632	0.9905	0.9296	0.9892	0.9182	0.9906
	10	0.9959	0.9964	0.9917	0.9979	0.9780	0.9867	0.9722	0.9972
	20	0.9978	0.9979	0.9954	0.9989	0.9874	0.9905	0.9839	0.9984

$$
\begin{aligned}
&\frac{1}{n}(y - X\hat{\beta})'\hat{\Omega}(y - X\hat{\beta})\\
&= \frac{1}{n}\left[u'\left(\hat{\Omega} - \hat{\Omega}X(X'\hat{\Omega}X)^{-1}X'\hat{\Omega}\right)u\right]\\
&= \frac{\sigma^2}{n}\left[\varepsilon_2'(P'\Omega^{-1}P)^{-1}P'\Omega^{-1}\left(\hat{\Omega} - \hat{\Omega}X(X'\hat{\Omega}X)^{-1}X'\hat{\Omega}\right)\right.\\
&\quad \left.\times\Omega^{-1}P(P'\Omega^{-1}P)^{-1}\varepsilon_2\right]\\
&= \sigma^2 \upsilon.
\end{aligned}
$$

Table 4. Relative risks of estimates for λ=0.75

		$p=4$				$p=10$			
ρ	$\beta'\beta$	SR	KK	MMSE	AMSE	SR	KK	MMSE	AMSE
-0.8	1	0.9840	0.9926	0.9684	0.9918	0.9575	0.9930	0.9501	0.9944
	10	0.9969	0.9972	0.9937	0.9984	0.9874	0.9919	0.9840	0.9984
	20	0.9984	0.9985	0.9967	0.9992	0.9929	0.9945	0.9910	0.9991
-0.6	1	0.9779	0.9908	0.9566	0.9887	0.9327	0.9899	0.9217	0.9911
	10	0.9962	0.9967	0.9923	0.9981	0.9837	0.9901	0.9793	0.9979
	20	0.9980	0.9982	0.9960	0.9990	0.9911	0.9933	0.9887	0.9989
-0.4	1	0.9737	0.9898	0.9487	0.9866	0.9162	0.9886	0.9035	0.9889
	10	0.9956	0.9963	0.9911	0.9978	0.9809	0.9890	0.9759	0.9975
	20	0.9978	0.9979	0.9954	0.9989	0.9897	0.9925	0.9869	0.9987
-0.2	1	0.9702	0.9891	0.9422	0.9848	0.9046	0.9879	0.8912	0.9873
	10	0.9952	0.9959	0.9902	0.9975	0.9787	0.9883	0.9731	0.9973
	20	0.9975	0.9977	0.9949	0.9987	0.9886	0.9919	0.9855	0.9985
0	1	0.9691	0.9889	0.9401	0.9843	0.8986	0.9876	0.8849	0.9865
	10	0.9951	0.9958	0.9900	0.9975	0.9775	0.9879	0.9716	0.9971
	20	0.9974	0.9976	0.9947	0.9987	0.9879	0.9915	0.9846	0.9984
0.2	1	0.9702	0.9892	0.9422	0.9848	0.9004	0.9879	0.8869	0.9867
	10	0.9952	0.9959	0.9902	0.9975	0.9773	0.9878	0.9714	0.9971
	20	0.9975	0.9977	0.9949	0.9987	0.9879	0.9916	0.9846	0.9984
0.4	1	0.9743	0.9902	0.9499	0.9869	0.9086	0.9886	0.8958	0.9878
	10	0.9956	0.9963	0.9911	0.9978	0.9787	0.9883	0.9731	0.9973
	20	0.9977	0.9979	0.9954	0.9988	0.9885	0.9919	0.9855	0.9985
0.6	1	0.9798	0.9917	0.9604	0.9897	0.9251	0.9899	0.9139	0.9901
	10	0.9963	0.9968	0.9925	0.9981	0.9812	0.9893	0.9763	0.9976
	20	0.9981	0.9982	0.9961	0.9990	0.9898	0.9926	0.9870	0.9987
0.8	1	0.9882	0.9947	0.9766	0.9940	0.9515	0.9927	0.9436	0.9936
	10	0.9974	0.9977	0.9946	0.9986	0.9850	0.9910	0.9810	0.9981
	20	0.9986	0.9987	0.9971	0.9993	0.9915	0.9936	0.9892	0.9989

Further

$$\frac{1}{n^{-1}(Y-X\hat{\beta})'\hat{\Omega}(Y-X\hat{\beta})} \equiv \frac{1}{\sigma^2 v} + O_p(n^{-1}),$$

$$\frac{1}{n-p+2} = \frac{1}{n} + O(n^{-2}),$$

Table 5. Relative risks of estimates for λ=1.00

		$p=4$				$p=10$			
ρ	$\beta'\beta$	SR	KK	MMSE	AMSE	SR	KK	MMSE	AMSE
-0.8	1	0.9879	0.9945	0.9762	0.9939	0.9673	0.9947	0.9615	0.9957
	10	0.9977	0.9979	0.9953	0.9988	0.9906	0.9940	0.9881	0.9988
	20	0.9988	0.9989	0.9975	0.9994	0.9947	0.9959	0.9933	0.9993
-0.6	1	0.9834	0.9931	0.9674	0.9916	0.9484	0.9924	0.9399	0.9932
	10	0.9972	0.9975	0.9942	0.9986	0.9877	0.9925	0.9844	0.9984
	20	0.9985	0.9986	0.9970	0.9992	0.9933	0.9950	0.9915	0.9991
-0.4	1	0.9799	0.9923	0.9607	0.9898	0.9367	0.9915	0.9270	0.9917
	10	0.9967	0.9972	0.9933	0.9983	0.9856	0.9917	0.9818	0.9982
	20	0.9983	0.9984	0.9965	0.9991	0.9922	0.9944	0.9901	0.9990
-0.2	1	0.9775	0.9918	0.9562	0.9885	0.9265	0.9908	0.9159	0.9903
	10	0.9964	0.9969	0.9926	0.9981	0.9839	0.9911	0.9797	0.9979
	20	0.9981	0.9983	0.9962	0.9990	0.9914	0.9939	0.9890	0.9989
0	1	0.9766	0.9917	0.9545	0.9881	0.9222	0.9907	0.9116	0.9897
	10	0.9963	0.9969	0.9924	0.9981	0.9829	0.9908	0.9784	0.9978
	20	0.9981	0.9982	0.9961	0.9990	0.9909	0.9936	0.9884	0.9988
0.2	1	0.9775	0.9919	0.9561	0.9885	0.9238	0.9909	0.9134	0.9900
	10	0.9964	0.9970	0.9927	0.9982	0.9829	0.9908	0.9784	0.9978
	20	0.9981	0.9983	0.9962	0.9990	0.9909	0.9936	0.9884	0.9988
0.4	1	0.9803	0.9926	0.9616	0.9900	0.9315	0.9915	0.9217	0.9910
	10	0.9967	0.9972	0.9933	0.9983	0.9839	0.9912	0.9797	0.9979
	20	0.9983	0.9984	0.9965	0.9991	0.9914	0.9939	0.9891	0.9989
0.6	1	0.9851	0.9939	0.9708	0.9924	0.9428	0.9924	0.9341	0.9925
	10	0.9973	0.9976	0.9944	0.9986	0.9858	0.9919	0.9820	0.9982
	20	0.9986	0.9987	0.9971	0.9993	0.9923	0.9944	0.9902	0.9990
0.8	1	0.9918	0.9963	0.9838	0.9958	0.9634	0.9946	0.9574	0.9952
	10	0.9981	0.9983	0.9961	0.9990	0.9886	0.9931	0.9856	0.9985
	20	0.9990	0.9990	0.9979	0.9995	0.9936	0.9952	0.9918	0.9992

$$\hat{\beta}'\hat{A}\hat{\beta}=\left(\beta+\frac{\sigma}{\sqrt{n}}\hat{A}^{-1}\alpha^*\right)'\hat{A}\left(\beta+\frac{\sigma}{\sqrt{n}}\hat{A}^{-1}\alpha^*\right)=\beta'\hat{A}\beta\left(1+\frac{2\sigma}{\sqrt{n}}\frac{\beta'\alpha^*}{\beta'\hat{A}\beta}\right)$$

so that

$$\frac{\hat{\beta}}{\hat{\beta}'\hat{A}\hat{\beta}}=\left[\left(\beta+\frac{\sigma}{\sqrt{n}}\hat{A}^{-1}\alpha^*\right)'\hat{A}\left(\beta+\frac{\sigma}{\sqrt{n}}\hat{A}^{-1}\alpha^*\right)\right]^{-1}$$
$$\times\left(\beta+\frac{\sigma}{\sqrt{n}}\hat{A}^{-1}\alpha^*\right)$$

$$= \frac{\beta}{\beta' A\beta} + \frac{\sigma}{\sqrt{n}} \frac{1}{\beta' A\beta}\left(A^{-1} - \frac{2\beta\beta'}{\beta' A\beta}\right)\varepsilon_1 + O_p\left(n^{-1}\right).$$

Again, up to order $O_p(n^{-1/2})$

$$\hat{\eta} = \frac{1}{\sigma^2 \upsilon}\beta' A\beta + \frac{2}{\sigma\sqrt{n}}\beta'\alpha^*,$$

$$r(\hat{\eta}) = r\left(\frac{\beta' A\beta}{\sigma^2\upsilon}\right) + \frac{2}{\sigma\sqrt{n}} r'\left(\frac{\beta' A\beta}{\sigma^2 \upsilon}\right)\beta'\alpha^*.$$

Hence, up to order $O_p(n^{-1})$, γ can be written as

$$\begin{aligned}\gamma = (X_f - \lambda\hat{V}'\hat{\Omega}X)\hat{A}^{-1}\alpha^* &- \frac{k\sigma\upsilon}{\sqrt{n}}\frac{1}{\beta' A\beta} r\left(\tilde{\eta}\right)(X_f - \lambda\hat{V}'\hat{\Omega}X)\beta \\ &- \frac{k\sigma^2}{n}\frac{1}{\beta' A\beta} r\left(\tilde{\eta}\right)(X_f - \lambda\hat{V}'\hat{\Omega}X)\left(A^{-1} - \frac{2\beta\beta'}{\beta' A\beta}\right)\varepsilon_1 \\ &+ \frac{2k}{n}\frac{1}{\beta' A\beta} r'\left(\tilde{\eta}\right)(X_f - \lambda\hat{V}'\hat{\Omega}X)\beta\beta'\alpha + \frac{\lambda\sqrt{n}}{\sigma}(\hat{V}'\hat{\Omega}u - u_f). \quad \text{(A.3)}\end{aligned}$$

Let us consider the transformation

$$\begin{bmatrix} u \\ u_f \end{bmatrix} = \begin{bmatrix} I_n & 0 \\ V'\Omega & I_T \end{bmatrix}\begin{bmatrix} u \\ \sigma\varepsilon_0 \end{bmatrix}. \quad \text{(A.4)}$$

We observe from the normality assumption for (u', u_f') that ε_0 is distributed independently of u (and hence independently of ε_1 and ε_2) and follows a normal distribution with mean vector 0 and covariance matrix $\Psi - V'\Omega V$. Making use of this transformation, to the order of our approximation, γ can be written as

$$\begin{aligned}\gamma =& (X_f - \lambda\hat{V}'\hat{\Omega}X)\hat{A}^{-1}\alpha^* - \frac{k\sigma\upsilon}{\sqrt{n}}\frac{1}{\beta' A\beta} r\left(\tilde{\eta}\right)(X_f - \lambda\hat{V}'\hat{\Omega}X)\beta \\ &- \frac{a\sigma^2}{n}\frac{1}{\beta' A\beta} r\left(\tilde{\eta}\right)(X_f - \lambda\hat{V}'\hat{\Omega}X)\left(A^{-1} - \frac{2\beta\beta'}{\beta' A\beta}\right)\varepsilon_1 \\ &+ \frac{2k}{n}\frac{1}{\beta' A\beta} r'\left(\tilde{\eta}\right)(X_f - \lambda\hat{V}'\hat{\Omega}X)\beta\beta'\alpha^* \\ &+ \frac{\lambda\sqrt{n}}{\sigma}(\hat{V}'\hat{\Omega} - V'\Omega)u - \lambda\sqrt{n}\varepsilon_0\end{aligned}$$

$$
\begin{aligned}
&= -\lambda\sqrt{n}\varepsilon_0 - \frac{k\sigma\upsilon}{\sqrt{n}}\frac{1}{\beta'\hat{A}\beta} r\left(\tilde{\eta}\right)(X_f - \lambda\hat{V}'\hat{\Omega}X)\beta \\
&\quad\times\left[(X_f - \lambda\hat{V}'\hat{\Omega}X)\hat{A}^{-1}\frac{X'\hat{\Omega}}{\sqrt{n}} + \frac{\lambda\sqrt{n}}{\sigma}(\hat{V}'\hat{\Omega} - V'\Omega)\right] \\
&\quad\times \Omega^{-1}P(P'\Omega^{-1}P)^{-1}\varepsilon_2 \\
&\quad+\left[(X_f - \lambda V'\Omega X)A^{-1} - \frac{k\sigma^2}{n\beta' A\beta} r\left(\tilde{\eta}\right)(X_f - \lambda V'\Omega X)\right. \\
&\quad\left.\times\left(A^{-1} - \frac{2\beta\beta'}{\beta' A\beta}\right)\right]\varepsilon_1 \\
&\quad+\lambda(\hat{V}'\hat{\Omega} - V'\Omega)XA^{-1}\varepsilon_1 + \frac{2k}{n\beta' A\beta} r'\left(\tilde{\eta}\right)(X_f - \lambda\hat{V}'\hat{\Omega}X)\beta\beta'\varepsilon_1. \quad \text{(A.5)}
\end{aligned}
$$

Since $\varepsilon_0, \varepsilon_1, \varepsilon_2$ are independently distributed and υ,$\hat{A}$,$\hat{\Omega}$ and $\hat{V}$ are functions of ε_2, we observe that up to order $O(n^{-1})$, the conditional expectation of γ given ε_2 is normal with mean vector $\mu(\varepsilon_2)$ and variance-covariance matrix $\Xi\left(\varepsilon_2\right)$.

Proof of Theorem 2:

Let us define

$$W = X_f - \lambda V'\Omega X.$$

Then, we can write $\tilde{\tau}_s$ as

$$\tilde{\tau}_s = \tilde{\tau} - \frac{k}{h+2}\frac{r(\eta^*)}{\eta^*}W\beta_S^*.$$

The MSE matrix of $\tilde{\tau}_S$ is given by

$$
\begin{aligned}
M\left(\tilde{\tau}_s\right) &= E\left[\left(\tilde{\tau}_s - \tau\right)\left(\tilde{\tau}_s - \tau\right)'\right] \\
&= E\left[\left(\tilde{\tau} - \tau\right)\left(\tilde{\tau} - \tau\right)'\right] \\
&\quad - \frac{k}{h+2}E\left[\frac{r(\eta^*)}{\eta^*}\left\{W\beta_S^*\left(\tilde{\tau} - \tau\right)' + \left(\tilde{\tau} - \tau\right)\beta_S^{*'}W'\right\}\right] \\
&\quad + \left(\frac{k}{h+2}\right)^2 E\left[\frac{r^2(\eta^*)}{\eta^{*2}}\left\{W\beta_S^*\beta_S^{*'}W'\right\}\right]. \qquad \text{(A.6)}
\end{aligned}
$$

Now

$$E\left[\left(\tilde{\tau} - \tau\right)\left(\tilde{\tau} - \tau\right)'\right] = \sigma^2\Delta$$

where

$$\Delta = X_f A^{-1} W' + W A^{-1} X_f' - X_f A^{-1} X_f' + \lambda^2 \left(\Psi - V' M V\right).$$

An unbiased estimate of it is $(\upsilon/h)\,\Delta$.
Further, unbiased estimate of third term of $M\left(\tilde{\tau}_s\right)$ is

$$\left(\frac{k}{h+2}\right)^2 \frac{r^2(\eta^*)}{\eta^{*2}} \left\{W\beta^*\beta^{*'}W'\right\}. \tag{A.7}$$

Now, we have to obtain unbiased estimator of the second term of $M\left(\tilde{\tau}_s\right)$. We know that conditional distribution of u_f given u is normal $N\left(V'\Omega u, \sigma^2(\Psi - V'\Omega V)\right)$. We have

$$\begin{aligned}
& E\left[\frac{r(\eta^*)}{\eta^*}\left(\tilde{\tau}-\tau\right)\beta_S^{*'}W'\right] \\
=& E\left[\frac{r(\eta^*)}{\eta^*}\left\{X_f\left(\beta^*-\beta\right)+\lambda\left\{X_f\left(\beta^*-\beta\right)-X_f\left(\beta^*-\beta\right)\right.\right.\right. \\
&+ \left.\left.\left. V'\Omega\left(y - X\left(X'\Omega X\right)^{-1}X'\Omega y\right)-u_f\right\}\right\}\beta^{*'}W'\right] \\
=& E\left[\frac{r(\eta^*)}{\eta^*}\left\{X_f\left(\beta^*-\beta\right)+\lambda V'M\left(X\beta+u\right)-\lambda u_f\right\}\beta^{*'}W'\right] \\
=& E\left[\frac{r(\eta^*)}{\eta^*}\left\{X_f\left(\beta^*-\beta\right)+\lambda V'Mu-\lambda V'\Omega u\right\}\beta^{*'}W'\right] \\
=& WE\left[\frac{r(\eta^*)}{\eta^*}\left(\beta^*-\beta\right)\beta^*\right]W'.
\end{aligned} \tag{A.8}$$

Let $Z = (X'\Omega X)^{1/2}\beta^*$, $\theta = (X'\Omega X)^{1/2}\beta$ and $r^*(\eta^*) = \frac{r(\eta^*)}{\eta^*}$. Notice that $Z \sim N(\theta, \sigma^2 I_{p\times p})$, $\upsilon/\sigma^2 \sim \chi^2(h)$ and Z and υ are independently distributed. Thus we have

$$\begin{aligned}
& E\left[\frac{r(\eta^*)}{\eta^*}\left(\beta^*-\beta\right)\beta^{*'}\right] \\
&= \left(X'\Omega X\right)^{-1/2} E\left[\left(Z-\theta\right)Z'r^*\left(\frac{Z'Z}{\upsilon}\right)\right]\left(X'\Omega X\right)^{-1/2} \\
&= \left(X'\Omega X\right)^{-1/2}\sigma^2 E\left[\frac{\partial}{\partial Z}\left(Z'r^*\left(\frac{Z'Z}{\upsilon}\right)\right)\right]\left(X'\Omega X\right)^{-1/2} \\
&= \left(X'\Omega X\right)^{-1/2}\sigma^2 E\left[r^*\left(\frac{Z'Z}{\upsilon}\right)+2\frac{ZZ'}{\upsilon}r^*\left(\frac{Z'Z}{\upsilon}\right)\right]\left(X'\Omega X\right)^{-1/2} \\
&= \left(X'\Omega X\right)^{-1}\sigma^2 E\left[r^*\left(\eta^*\right)\right]+2E\left[\beta^*\beta^{*'}\frac{\sigma^2}{\upsilon}r^*\left(\eta^*\right)\right].
\end{aligned} \tag{A.9}$$

Hence, the second term in (A.6) becomes

$$
-\frac{k}{h+2}W\left[2\left(X'\Omega X\right)^{-1}\sigma^2 E\left\{r^*\left(\eta^*\right)\right\} + 4E\left\{\beta^*\beta^{*\prime}\frac{\sigma^2}{\upsilon}r^*\left(\eta^*\right)\right\}\right]W'
$$

$$
= -\left\{\frac{2k}{h+2}W\left(X'\Omega X\right)^{-1}E_{\beta^*}\left[E_\upsilon\left(\sigma^2 r^*\left(\eta^*\right)\right)\right]W'\right.
$$

$$
\left. +\frac{4k}{h+2}W E_{\beta^*}\left[E_\upsilon\left(\beta^*\beta^{*\prime}\frac{\sigma^2}{\upsilon}r^*\left(\eta^*\right)\right)\right]W'\right\}. \tag{A.10}
$$

For a continuous differentiable function $H(\upsilon)$ of υ, it can be shown that

$$
\frac{1}{\sigma^2}E_\upsilon\left[\left(\upsilon - h\sigma^2\right)H(\upsilon)\right] = 2E_\upsilon\left[\upsilon\frac{\partial}{\partial\upsilon}H(\upsilon)\right]
$$

or

$$
E_\upsilon\left[\upsilon H(\upsilon)\right] = \sigma^2 E_\upsilon\left[\upsilon H(\upsilon) + 2\upsilon\frac{\partial}{\partial\upsilon}H(\upsilon)\right]. \tag{A.11}
$$

Equation (A.11) implies that if we set

$$
\upsilon H(\upsilon) + 2\upsilon\frac{\partial}{\partial\upsilon}H(\upsilon) = r^*\left(\eta^*\right). \tag{A.12}
$$

Then an unbiased estimate of $\sigma^2 E_\upsilon\left(r^*\left(\eta^*\right)\right)$ is $\upsilon H(\upsilon)$.

Equation (A.12) is a first order differential equation and its solution is given by

$$
H(\upsilon) = \frac{1}{2}\int_0^1\left\{t^{h/2}\frac{r(\eta^*/t)}{\eta^*}\right\}dt.
$$

Therefore an unbiased estimator of $2E_{\beta^*}\left[E_\upsilon\left(\sigma^2 r^*(\eta^*)\right)\right]$ is

$$
\left[\upsilon\int_0^1\left\{t^{h/2-1}\frac{r(\eta^*/t)}{\eta^*}\right\}dt\right].
$$

Utilizing the above expression, we observe that an unbiased estimator of

$$
2E_{\beta^*}\left[E_\upsilon\left(\frac{\sigma^2}{\upsilon}\frac{d}{d\eta^*}r^*\left(\eta^*\right)\right)\right]
$$

is given by

$$
\int_0^1 t^{\frac{h}{2}-1}\frac{d}{d\eta^*}r^*(\eta^*/t)dt
$$

$$
= \int_0^1 t^{\frac{h}{2}-1}\frac{d}{d(\eta^*/t)}r^*(\eta^*/t)\frac{d}{d\eta^*}(\eta^*/t)dt
$$

$$
= \int_0^1 t^{\frac{h}{2}-1}\frac{r'(\eta^*/t)}{\eta^*}dt - \int_0^1 t^{h/2}\frac{r(\eta^*/t)}{\eta^{*2}}dt.
$$

Hence an unbiased estimator of

$$4E_{\beta^*}\left[\beta^*\beta^{*'}E_\upsilon\left(\frac{\sigma^2}{\upsilon}\frac{d}{d\eta^*}r^{*'}(\eta^*)\right)\right]$$

is obtained as

$$2\left[\int_0^1 t^{h/2-1}\frac{r'(\eta^*/t)}{\eta^*}dt - \int_0^1 t^{h/2}\frac{r(\eta^*/t)}{\eta^{*2}t}dt\right]\beta^*\beta^{*'}.$$

Putting these values in (A.9), we have the following unbiased estimator of second term of (A.6) as

$$\frac{k}{h+2}W\left[\upsilon\int_0^1 t^{h/2}\frac{r(\eta^*/t)}{\eta^*}dt\right]\left(X'\Omega X\right)^{-1}W'$$
$$+\frac{2k}{h+2}W\left[\int_0^1 t^{\frac{h}{2}-1}\frac{r'(\eta^*/t)}{\eta^*}dt - \int_0^1 t^{h/2}\frac{r(\eta^*/t)}{\eta^{*2}}dt\right]\beta^*\beta^{*'}W'. \quad (A.13)$$

Putting (A.7), (A.8) and (A.13) in (A.6), we have

$$\begin{aligned}\hat{M}(\tilde{\tau}_s) = {} & \frac{\upsilon}{h}\Delta - \frac{k}{h+2}\frac{\upsilon^2}{\eta^*}W\left[\int_0^1 t^{h/2}r\left(\frac{\eta^*}{t}\right)dt\right]\left(X'\Omega X\right)^{-1}W' \\ & -\frac{2k}{h+2}W\left[\int_0^1 t^{h/2-2}\left[\frac{t}{\eta^*}r'\left(\frac{\eta^*}{t}\right)\right.\right. \\ & \left.\left.-\left(\frac{t}{\eta^*}\right)^2 r\left(\frac{\eta^*}{t}\right)\right]dt\right]\beta^*\beta^{*'}W' \\ & +\left(\frac{k}{h+2}\right)^2\left[\frac{r^2(\eta^*)}{\eta^{*2}}\left\{W\beta^*\beta^{*'}W'\right\}\right]W\beta^*\beta^{*'}W'\end{aligned}$$

which leads to (17).

Proof of Theorem 3:

Let $\hat{M}(\sigma^{-1}\tilde{\tau}_s)$ be an unbiased estimator of $M(\sigma^{-1}\tilde{\tau}_s)$. Then

$$\begin{aligned}M(\sigma^{-1}\tilde{\tau}_s) &= E\left[\tilde{M}(\sigma^{-1}\tilde{\tau}_s)\right] \\ &= \Delta - \frac{1}{\sigma^2}\frac{k}{h+2}E\left[\frac{r(\eta^*)}{\eta^*}\left\{W\beta^*(\tilde{\tau}-\tau)' + (\tilde{\tau}-\tau)\beta^{*'}W'\right\}\right] \\ &\quad + \left(\frac{k}{h+2}\right)^2\frac{1}{\sigma^2}E\left[\frac{r^2(\eta^*)}{\eta^{*2}}\left\{W\beta^*\beta^{*'}W'\right\}\right]. \quad (A.14)\end{aligned}$$

From (A.10), an unbiased estimator of the second term of (A.14) is given by

$$-\left\{\frac{2k}{h+2}W\left(X'\Omega X\right)^{-1}r^*(\eta^*)W'+\frac{4k}{h+2}W\beta^*\beta^{*'}\frac{1}{\upsilon}r^*(\eta^*)W'\right\}.$$

In addition, by virtue of equation (A.11), we obtain an unbiased estimator of the third term of equation (A.14) as

$$\left[\left(\frac{k}{h+2}\right)^2 W\beta^*\beta^{*'}\left\{(h-2)\frac{r^{*2}(\eta^*)}{\upsilon}+4r^*(\eta^*)\frac{\partial}{\partial\upsilon}r^*(\eta^*)\right\}W'\right]. \tag{A.15}$$

Putting these in equation (A.14), we have

$$\begin{aligned}\tilde{M}(\sigma^{-1}\hat{\tau}_s) = \Delta - \frac{2k}{h+2}W\left\{\left(X'\Omega X\right)^{-1}r^*(\eta^*)+2\beta^*\beta^{*'}\frac{1}{\upsilon}r^{*'}(\eta^*)\right\}W \\ +\left[\left(\frac{k}{h+2}\right)^2 W\beta^*\beta^{*'}\left\{(h-2)\frac{r^{*2}(\eta^*)}{\upsilon}\right.\right. \\ \left.\left.+\,4r^*(\eta^*)\frac{\partial}{\partial\upsilon}r^*(\eta^*)\right\}W'\right].\end{aligned} \tag{A.16}$$

Substituting the values of $r^*(\eta^*)$ and $r^{*'}(\eta^*)$ in (A.16) leads to the result of the theorem.

References

Chaturvedi A, Singh SP (2000) Stein rule prediction of the composite target function in a general linear regression model. Statistical Papers 41:359–367

Chaturvedi A, Wan ATK, Singh SP (2002) Improved multivariate prediction in a general linear model with an unknown error covariance matrix. Journal of Multivariate Analysis 83:166–182

Copas JB (1983) Regression, prediction and shrinkage (with discussion). Journal of the Royal Statistical Society (B) 45:311–354

Copas JB, Jones MC (1987) Regression shrinkage methods and autoregressive time series prediction. Australian Journal of Statistics 29:264–277

Dube M, Singh V (1986) Mixed and Improved Estimation of Coefficients in Regression Models A Predictive Perspective. Journal of Applied Statistical Science 2/3, 5:161–171

Gotway CA, Cressie N (1993) Improved multivariate prediction under a general linear model. Journal of Multivariate Analysis 45:56–72

Hendersion CR (1972) General flexibility of linear model techniques for sire evaluation. Journal of Dairy Sciences 57:963–972

Hill RC, Fomby TB (1983) The Effects of Extrapolation on Mini-Max Stein-Rule Prediction Hill. In: WE Griffiths, H Lütkepohl, and ME Bock (eds) Readings in Econometrics Theory and Practice: A volume in Honor of George Judge, North Holland, Amsterdam 3–32

Judge GG , Griffiths WE , Hill RC, Lütkepohl H, Lee TC (1985) The Theory and Practice of Econometrics. Wiley, New York

Kariya T, Toyooka Y (1985) Non-linear version of the Gauss-Markov Theorem and its application to SUR model. Annals of the Institute of Statistical Mathematics 34:281–297

Kariya T, Toyooka Y (1985) Bounds for normal approximations to the distributions of generalized least squares predictors and estimators. Journal of Statistical Planning and Inference 30:213–221

Khan S, Bhatti MI (1998) Predictive inference on equicorrelated linear regression models. Applied Mathematics and Computation 95:205–217

Rao CR, Toutenburg H, Shalabh, Heumann, C. (2008) Linear Models and Generalizations: Least Squares and Alternatives (3rd edition). Springer, Berlin Heidelberg New York

Shalabh (1995) Performance of Stein-rule procedure for simultaneous prediction of actual and average values of study variable in linear regression models. Bulletin of the International Statistical Institute 56:1375–1390

Shalabh (1998) Unbiased Prediction in Linear Regression Models with Equi-Correlated Responses. Statistical Papers 39:237–244

Srivastava AK, Shalabh (1996) A composite target function for prediction in econometric models. Indian Journal of Applied Econometrics 5:251–257

Theil H (1971) Principle of Econometrics. Wiley, New York

Toutenburg H (1982) Prior Information in Linear Models. Wiley, New York

Toutenburg H, Shalabh (1996) Predictive performance of the methods of restricted and mixed regression estimators. Biometrical Journal 38:951–959

Toyooka Y (1982) Prediction error in a linear model with estimated parameters. Biometrica 69:453–459

Tuchscheres A , Herrendorfer G and Tuchscheres M (1998) Evaluation of the Best Linear Unbiased Prediction in Mixed Linear Models with

Estimated Variance Components by Means of the MSE of Prediction and the Genetic Selection Differential. Biometrical Journal 40:949–962

Zellner A, Hong C (1989) Forecasting International growth rates using Bayesian Shrinkage and Other Procedures. Journal of Econometrics 40:183–202

Finite Mixtures of Generalized Linear Regression Models

Bettina Grün[1] and Friedrich Leisch[2]

[1] Institut für Statistik und Wahrscheinlichkeitstheorie, Technische Universität Wien, Wiedner Hauptstrasse 8-10/1071, A-1040 Vienna, Austria Bettina.Gruen@ci.tuwien.ac.at
[2] Department of Statistics, University of Munich, Ludwigstrasse 33, 80539 Munich, Germany Friedrich.Leisch@stat.uni-muenchen.de

1 Introduction

Finite mixture models have now been used for more than hundred years (Newcomb (1886), Pearson (1894)). They are a very popular statistical modeling technique given that they constitute a flexible and easily extensible model class for (1) approximating general distribution functions in a semi-parametric way and (2) accounting for unobserved heterogeneity. The number of applications has tremendously increased in the last decades as model estimation in a frequentist as well as a Bayesian framework has become feasible with the nowadays easily available computing power.

The simplest finite mixture models are finite mixtures of distributions which are used for model-based clustering. In this case the model is given by a convex combination of a finite number of different distributions where each of the distributions is referred to as component. More complicated mixtures have been developed by inserting different kinds of models for each component. An obvious extension is to estimate a generalized linear model (McCullagh and Nelder (1989)) for each component. Finite mixtures of GLMs allow to relax the assumption that the regression coefficients and dispersion parameters are the same for all observations. In contrast to mixed effects models, where it is assumed that the distribution of the parameters over the observations is known, finite mixture models do not require to specify this distribution a-priori but allow to approximate it in a data-driven way.

In a regression setting unobserved heterogeneity for example occurs if important covariates have been omitted in the data collection and hence their influence is not accounted for in the data analysis. In addition in some areas of application the modeling aim is to find groups of observations with similar regression coefficients. In market segmentation (Wedel and Kamakura (2001)) one kind of application among others of finite mixtures of GLMs aims for example at determining groups of consumers with similar price elasticities in order to develop an optimal pricing policy for a market segment.

Other areas of application are biology or medicine, see Aitkin (1999), Follmann and Lambert (1989), Wang et al. (1996), Wang and Puterman (1998). An example for a biological application is illustrated by the "Aphids" data set in Boiteau et al. (1998). The data contains the results of 51 independent experiments in which varying numbers of aphids were released in a flight chamber containing 12 infected and 69 healthy tobacco plants. After 24 hours, the flight chamber was fumigated to kill the aphids, and the previously healthy plants were moved to a greenhouse and monitored to detect symptoms of infection. The number of plants displaying such symptoms was recorded. The relationship between the proportion of infected plants given the number of released aphids is depicted in Figure 1.

Clearly the proportion of infected plants in dependence of the number of released aphids does not cluster around a single regression line, but around two different regression lines. For one regression line no infection takes place while for the other the proportion of infected plants increases with the number of aphids. Fitting a finite mixture of binomial logit models allows to determine the expected number of infected plants given the number of released aphids for each of the components and the proportion of times where no infection takes place.

In Section 2 the finite mixture model of GLMs is specified starting with the standard GLM formulation. The general mixture model class is presented and several special cases which are included in this model class are discussed. In Section 3 the identifiability of finite mixtures of GLMs is analyzed and sufficient conditions to guarantee "generic" identifiability are given. As additional problems to the case of finite mixtures of distributions can occur in the regression setting, model identification has to be again investigated and results obtained for mixtures of distributions can not be directly transferred without further consideration. After an outline of model estimation using the EM algorithm and a brief overview on Bayesian methods in Section 4 the application of the model class is illustrated in a cluster-wise regression setting as well as in a situation where overdispersion in a Poisson standard GLM is observed and a random intercept model is fitted to account for this

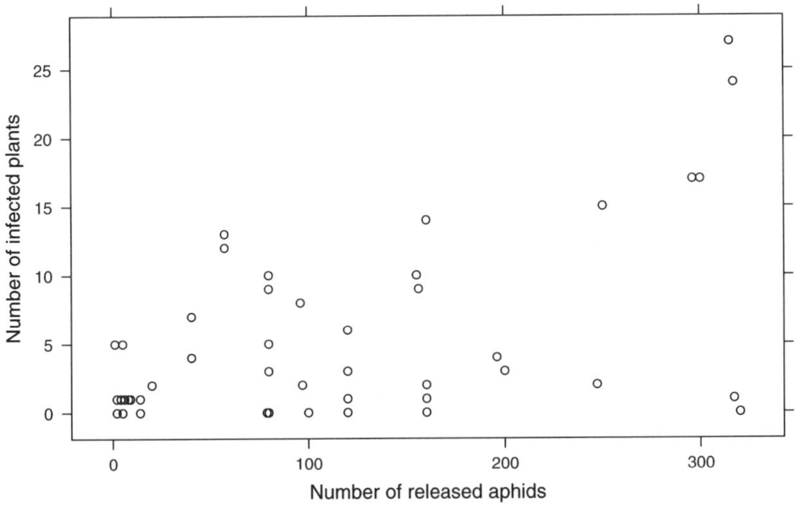

Fig. 1. "Aphids" data set.

overdispersion. An outlook on several possible extensions is given in the last section. All computations and graphics in this paper have been done using package **flexmix** (Leisch (2004b), Grün and Leisch (2006), Grün and Leisch (2007)) in R, an environment for statistical computing and graphics (R Development Core Team (2007)).

2 Model Specification

In the standard linear model the dependent variable y is assumed to follow a Gaussian distribution where the mean value is determined through a linear relationship given the covariates x:

$$\mathbb{E}[y|x] = x'\beta,$$

where β are the regression coefficients. This signifies that $y|x \sim N(x'\beta, \sigma^2)$.

The assumption that the dependent variable follows a Gaussian distribution is relaxed in the generalized linear model framework. The distribution of the dependent variable is assumed to be from the exponential family of distributions (e.g. Gaussian, binomial, Poisson or gamma). This allows to take certain data characteristics into account such as that the dependent variable y is for example a counting variable with values in $\mathbb{N}$ which is then in general assumed to follow a Poisson distribution.

The density of a distribution from the exponential family is given by

$$f(y|\theta,\phi) = \exp\left\{\frac{y\theta - b(\theta)}{a(\phi) + c(y,\theta)}\right\}$$

for some specific functions $a(\cdot)$, $b(\cdot)$ and $c(\cdot)$. For the Gaussian distribution $N(\mu, \sigma^2)$ with mean μ and variance σ^2 and the assumption that $\theta = \mu$ and $\phi = \sigma^2$ these functions are for example given by

$$a(\phi) = \phi, \qquad b(\theta) = \frac{\theta^2}{2}, \qquad c(y,\phi) = -\frac{1}{2}\left\{\frac{y^2}{\phi} + \log(2\pi\phi)\right\}.$$

The relationship between the linear predictor η and the expected value μ of the dependent variable y is modeled via a link function

$$\eta = g\left(\mathbb{E}[y|x]\right) = x'\beta,$$

where η is the linear predictor and $g(\cdot)$ the link function. Different link functions are possible. A special link function is the canonical link which is given by

$$\eta = x'\beta = \theta.$$

For the Gaussian distribution the identity function is the canonical link, for the Poisson the log function, for the binomial the logit function and for the gamma distribution the reciprocal function.

The GLM framework is embedded in the finite mixture framework by inserting GLMs into the components. The resulting models are also referred to as GLIMMIX models (Wedel and Kamakura (2001)). A finite mixture density of GLMs with K components is given by

$$h(y|x,\Theta) = \sum_{k=1}^{K} \pi_k f_k(y|x,\theta_k)$$

where Θ denotes the vector of all parameters for the mixture density $h()$. The dependent variable is y and the independent variables are x. f_k is the component specific density function which is assumed to be univariate and from the exponential family of distributions. The component specific parameters are given by $\theta_k = (\beta_k', \phi_k)$ where β_k are the regression coefficients and ϕ_k is the dispersion parameter. The mean of each component is given by

$$\mu_k(x) = g_k^{-1}(x'\beta_k),$$

where $g_k()$ is the component specific link function.

For the component weights π_k it holds

$$\sum_{k=1}^{K} \pi_k = 1 \quad \text{and} \quad \pi_k > 0, \forall k. \tag{1}$$

Several special cases and extensions of this model class exist. Often it is assumed that the component specific densities are from the same parametric family for each component, i.e., $f_k \equiv f$ for notational simplicity, and that the link function is also the same for all components ($g_k \equiv g$). In a cluster-wise regression setting this will be an obvious model choice as no a-priori knowledge about differences in distributional families of the components is available. Another popular extension is to have a so-called concomitant variable model for the prior class probabilities, such that the π_i also depend on a set of explanatory variables (e.g., using a multinomial logit model).

A special case where different component specific distributions are used is a model where only a single component is specified to follow a different distribution in order to allow this component to capture outlying observations (Dasgupta and Raftery (1998)). This approach is similar to the specification of zero-inflated models (Böhning et al. (1999)). Even though the component specific densities are in general assumed to be from the same parametric family (e.g. Poisson or binomial), the parameters are fixed a-priori for one component such that this component absorbs all excess zeros in the zero-inflated model.

In order to decrease the number of parameters equality constraints can be imposed over the components for a subset of the component specific parameters θ_k. A special case are random intercept models where only the intercept follows a finite mixture distribution while all other regression coefficients are constant over the components, see Follmann and Lambert (1989). These models are often used if overdispersion is encountered in Poisson or binomial GLMs in order to determine a model which describes the data in an appropriate way.

3 Identification

Statistical models are in general represented by parameter vectors. For finite mixture models the parameter vector Θ which consists of the component weights and the component specific parameters determines a mixture distribution, i.e., there is a mapping from the parameter space to the model space. For identifiability this mapping has to be injective, i.e., for each model in the model space there is a unique parameter vector in the parameter space which is mapped to the model. Lack of identifiability can be a problem for model estimation or if parameters are interpreted.

In the following let Ω denote the space of admissible parameters for K-component mixtures where the following conditions are fulfilled

- $\pi_k > 0 \; \forall k = 1, \ldots, K$, and
- $\forall k, l \in \{1, \ldots, K\}$: $k \neq l \Rightarrow \theta_k \neq \theta_l$.

These two conditions prevent overfitting and identifiability problems which occur due to empty components where θ_k cannot be uniquely determined and due to components with equal component parameter vectors where different values for π_k are possible.

Let $\mathscr{A}_K = \mathscr{A}_K(f, \Omega)$ be the set of all finite mixture models with K components, component specific density function f and mixture densities of form $h(\cdot|\cdot, \Theta)$, $\Theta \in \Omega$. Each parameter vector $\Theta \in \Omega$ corresponds to one model $a \in \mathscr{A}_K$, but each model a has at least $K!$ parameterizations Θ due to all possible permutations of the components, also known as *label switching* (Redner and Walker (1984)).

$\mathscr{A}_K$ induces a system of equivalence classes Ξ on Ω where two elements of Ω are in the same equivalence class if they correspond to the same model a:

$$\Theta, \tilde{\Theta} \in \Xi \Leftrightarrow h(\cdot|\cdot, \Theta) \equiv h(\cdot|\cdot, \tilde{\Theta}).$$

The usual definition of model identifiability is that either all equivalence classes contain only one element (which is trivially not true for mixture models), or that at least a unique representative for each equivalence class can be selected.

Let $\text{ident}(\Omega) \subset \Omega$ be the subset of parameterizations which contain only one permutation of each possible set of component parameters. $\text{ident}(\Omega)$ can be obtained from Ω by imposing an ordering constraint on the components with respect to a certain parameter (or a combination of several parameters). We refer to any identifiability problems which are present for $\text{ident}(\Omega)$ as *generic* (Frühwirth-Schnatter (2006)).

3.1 Generic Identifiability

Generic identifiability problems have already been analyzed for finite mixtures of distributions by Titterington et al. (1985). In nearly all cases only mixtures where the component distribution is from the same distributional family have been considered. General results for certain kinds of distribution as well as specific results for a given component specific distribution function have been derived. Generic identifiability is guaranteed for important continuous distributions such as the Gaussian, gamma and Poisson distribution. A special case are finite mixtures of binomial distributions which are only identifiable if the number of components is limited. For the model class of finite mixtures of binomial distributions $\text{Bi}(\pi, T)$ with success probability π and repetition parameter T a necessary and sufficient condition for identifiability is $T \geq 2K - 1$.

The analysis of identifiability of mixtures of Gaussian regression models revealed that requiring a covariate matrix of full rank – as postulated previously for example by Wang and Puterman (1998) – is not sufficient (Hennig (2000)). Contrarily, it is necessary to check a coverage condition in order to ensure identifiability. With respect to generic identifiability of finite mixtures of regression models three influencing factors can therefore be distinguished:

- component distribution f,
- covariate matrix and
- repeated observations/labelled observations.

Repeated observations where the class membership is fixed are necessary for mixtures of binomial distributions to be identifiable. In a regression setting repetitions over different covariate points can help in making a mixture identifiable as it changes the set of feasible hyperplanes for the coverage condition. Labels for some observations indicating that they belong to the same component have the same influence.

In order to present a theorem on sufficient conditions for identifiability of finite mixtures of GLMs a data representation is necessary which takes repeated observations of the same individual where the component membership is fixed into account. The observations for an individual t are combined and given by:

$$(y_t, x_t) = (y_i, x_i)_{i \in I_t},$$

where I_t contains the set of indices corresponding to the observations of individual t. In the following X and Y denote the matrix of all x and y observations of all N individuals.

Theorem 1. *The model defined by*

$$h(Y \mid X, \Theta) = \prod_{t=1}^{N} \left[\sum_{k=1}^{K} \pi_k \prod_{i \in I_t} f(y_i | \mu_i^k, \phi_k) \right]$$

and

$$g^{-1}\left(\mu_i^k\right) = x_i' \beta_k$$

is identifiable if the following conditions are fulfilled:

1. (a) $\exists \tilde{I} \neq \emptyset$: $\tilde{I} \subseteq \bigcup_{t=1}^{N} I_t$: *The mixture of distributions given by*

$$\sum_{k=1}^{K} \pi_k f(y_i | \mu_i^k, \phi_k)$$

is identifiable $\forall i \in \tilde{I}$.

(b) $q^* > K$ *with*

$$q^* := \min\left\{ q : \forall i^* \in \tilde{I} : \exists H_j \in \{H_1, \ldots, H_q\} : \right.$$
$$\left. \{x_i : i \in I_{t(i^*)} \cap \tilde{I}\} \subseteq H_j \wedge H_j \in \mathcal{H}_U \right\}$$

where $\mathcal{H}_U$ *is the set of* $H(\alpha) := \{x \in \mathbb{R}^U : \alpha' x = \mathbf{0}\}$ *with* $\alpha \neq \mathbf{0}$.

2. *The matrix* X *has full column rank.*

The proof is straight-forward given the previous results for finite mixtures of standard linear regression models by Hennig (2000) and finite mixtures of GLMs and multinomial logit models with varying and fixed effects in the regression coefficients by Grün (2006), Grün and Leisch (2007).

For Condition (1a) the generic identifiability of finite mixtures with the given component specific distribution is essential. If the component specific distribution is either Gaussian, Poisson or gamma this condition is no restriction as mixtures of these distributions are generically identifiable, i.e., $\tilde{I} = \bigcup_{t=1}^{N} I_t$. In the case of the binomial distribution the repetition parameter has to be checked for each observation in order to determine if it can be included in $\tilde{I}$. Condition (1b) indicates that for each individual t there has to be one of the q hyperplanes through the origin H_j which covers all identifiable observations of this individual. The rank condition (2) ensures that the regression coefficients can be uniquely determined given the linear predictor.

These conditions indicate that identifiability problems can especially occur if the covariate matrix contains categorical variables. We refer to identifiability problems due to the violation of the coverage condition as

Intra-component label switching: If the labels are fixed in one covariate point according to some ordering constraint, then labels may switch in other covariate points for different parameterizations of the model.

For mixtures where the component distributions are identifiable this means that the component weights and possible dispersion parameters are unique, but the regression coefficients vary because they depend on the combination of the components between the covariate points. This identifiability problem is also of concern for prediction, because given the class membership the predicted value for new data depends on the chosen solution.

Unidentified mixture models with several isolated non-trivial (global) modes in the likelihood are to some extent more of a theoretical problem, because, e.g., minimal changes of the component weights π_k often make the model identified by breaking symmetry. However, models "close" to an unidentified model will have multiple local modes.

The following example presents a simple mixture of regression models with intra-component label switching. The model is unidentified (with two non-trivial modes) only if both components have exactly the same probability.

Example 1. Assume we have a mixture of standard linear regression models with one measurement per object and a single categorical regressor with two levels. The usual design matrix for a model with intercept uses the two covariate points $x_1 = (1,0)'$ and $x_2 = (1,1)'$. Furthermore, let the mixture consist of two components with equal component weights. The mixture regression is given by

$$h(y|x,\Theta) = \frac{1}{2} f_N(\mu_1(x), 0.1) + \frac{1}{2} f_N(y|\mu_2(x), 0.1)$$

where $\mu_k(x) = x'\beta_k$ and $f_N(y|\mu,\sigma^2)$ is the normal distribution with mean μ and variance σ^2.

Now let $\mu_1(x_1) = 1$, $\mu_2(x_1) = 2$, $\mu_1(x_2) = -1$ and $\mu_2(x_2) = 4$. As Gaussian mixture distributions are generically identifiable the means, variances and component weights are uniquely determined in each covariate point given the mixture distribution. However, as the coverage condition is not fulfilled, the two possible solutions for β are:

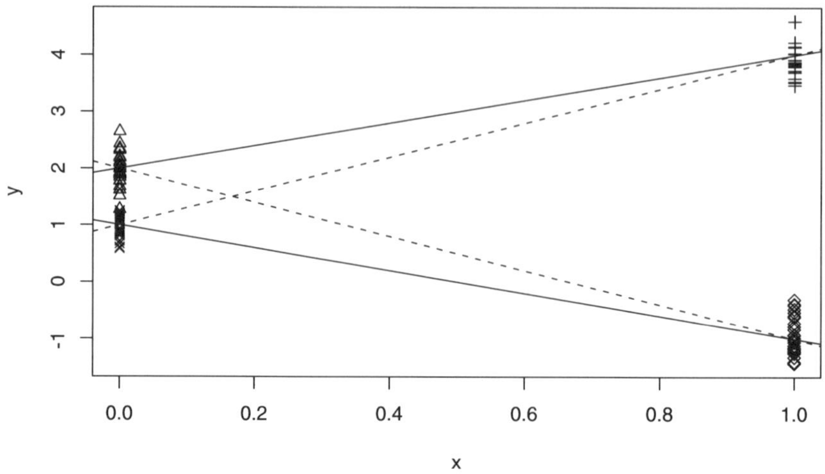

Fig. 2. Balanced sample from the artificial example with the two theoretical solutions. The solid lines correspond to solution 1 and the dashed lines to solution 2.

Solution 1: $\beta_1^{(1)} = (2, \quad 2)'$, $\beta_2^{(1)} = (1, -2)'$
Solution 2: $\beta_1^{(2)} = (2, -3)'$, $\beta_2^{(2)} = (1, \quad 3)'$.

In Figure 2 a balanced sample with 50 observations in each covariate point is plotted together with the two solutions for combining x_1 and x_2.

This mixture model would be identifiable if either

1. three different covariate points were available, or
2. observations for both covariate points for the same object were available, or
3. the component weights were unequal, e.g. $\pi_1 = 0.6$.

Condition 1 is not an option, for instance, for a single 2-level categorical regressor. Condition 2 is not possible if the categorical regressor cannot change for repeated observations of the same subject like, for instance, the gender of a person. However, when developing a suitable measurement design, the possibility of these problems to occur should be considered in order to develop a suitable design matrix.

The identifiability conditions given in Theorem 1 have the drawback that they are only sufficient conditions for a certain model class. The conditions can therefore only indicate if the model class contains at least one single model which is not identifiable. In addition they are hard to verify in practice as it is an NP hard problem (Hennig (2000)). In general it will be of interest if a fitted model suffers from identifiability problems. This means that it has to be checked if there exist several modes of the likelihood in the parameter space ident(Ω) given data sets sampled from the fitted mixture model. In a frequentist estimation setting bootstrap methods can be used to investigate potential identifiability problems of a fitted finite mixture model, see Grün and Leisch (2004, 2007).

4 Estimation

Finite mixture models can be either estimated within a frequentist framework, within a Bayesian framework, with moment estimators (Lindsay (1989)) or by applying graphical tools (Titterington et al. (1985)). An important characteristic of the estimation method is if the number of components has to be fixed a-priori or is simultaneously estimated. In the following maximum likelihood estimation with the EM algorithm is described and a short overview on Bayesian estimation using MCMC samplers is given.

4.1 Frequentist Maximum Likelihood with the EM Algorithm

There exist different methods for frequentist estimation of finite mixture models. The most popular is the EM algorithm (Dempster et al. (1977), McLachlan and Krishnan (1997)) which aims at determining the ML estimator for a finite mixture model with a given number of components K. The EM algorithm has the advantage that it provides a general framework for estimating different kinds of mixture models as often only the M-step has to be modified if different component specific models are used. In addition, already available tools for weighted maximum likelihood estimation can be applied. Nevertheless, there are also some known disadvantages such as slow convergence or that one might get stuck in local optima, i.e., it is in general difficult to ensure that the root corresponding to the maximum likelihood estimator was detected.

The EM algorithm uses a data augmentation scheme and is a general estimation method in the presence of missing data. In the case of finite mixture models the missing data is the latent variable $D_t \in \{0, 1\}^K$

for each individual t which indicates the component membership. This means that D_{tk} equals 1 if individual t is from component k and 0 otherwise. The data is therefore augmented with estimates of the component memberships, i.e., the estimated a-posteriori probabilities $\hat{p}_{tk}$.

For simplicity of notation it is in the following assumed that the component density function $f(\cdot|\cdot)$ takes all observations from each individual as arguments. For a sample of N individuals $\{(y_1, x_1), \dots, (y_N, x_N)\}$ the EM-algorithm is given by:

E-step: Given the current parameter estimates $\Theta^{(j)}$ in the jth iteration, replace the missing data D_{tk} by the estimated a-posteriori probabilities

$$\hat{p}_{tk} = \frac{\pi_k^{(j)} f(y_t | x_t, \theta_k^{(j)})}{\sum_{l=1}^{K} \pi_l^{(j)} f(y_t | x_t, \theta_l^{(j)})}.$$

M-step: Given the estimates for the a-posteriori probabilities $\hat{p}_{tk}$ (which are functions of $\Theta^{(j)}$), obtain new estimates $\Theta^{(j+1)}$ of the parameters by maximizing

$$Q(\Theta^{(j+1)} | \Theta^{(j)}) = Q_1(\theta^{(j+1)} | \Theta^{(j)}) + Q_2(\pi^{(j+1)} | \Theta^{(j)}),$$

under the restriction for the component weights given in Equation (1) and where

$$Q_1(\theta^{(j+1)} | \Theta^{(j)}) = \sum_{t=1}^{N} \sum_{k=1}^{K} \hat{p}_{tk} \log(f(y_t | x_t, \theta_k^{(j+1)}))$$

and

$$Q_2(\pi^{(j+1)} | \Theta^{(j)}) = \sum_{t=1}^{N} \sum_{k=1}^{K} \hat{p}_{tk} \log(\pi_k^{(j+1)}).$$

Q_1 and Q_2 can be maximized separately. The maximization of Q_1 gives new estimates $\theta^{(j+1)}$ and the maximization of Q_2 gives $(\pi_k^{(j+1)})_{k=1,\dots,K}$. Q_1 is maximized using weighted ML estimation of GLMs and the parameter estimates $\pi_k^{(j+1)}$ which maximize Q_2 are given by

$$\pi_k^{(j+1)} = \frac{1}{N} \sum_{t=1}^{N} \hat{p}_{tk} \quad \forall k = 1, \dots, K.$$

Before each M-step the average component sizes (over the given data points) are checked and components which are smaller than a given (relatively) small size are omitted in order to avoid too small components where fitting problems might arise. This strategy has also been recommended for the a variant of the EM algorithm, the stochastic EM (Celeux and Diebolt (1988)), in order to determine the number of components. For the SEM algorithm an additional step between the E- and M-step is performed where estimates for D_{kt} are determined by drawing a sample from the multinomial distribution implied by the posteriors for each observations and these estimates are then used as weights in the M-step. If the algorithm is started with too many components they will be omitted during the estimation process. The algorithm is stopped if the relative change in the likelihood is smaller than a pre-specified ϵ or the maximum number of iterations is reached.

It has been shown that the values of the likelihood are monotonically increased during the EM algorithm. This ensures the convergence of the EM algorithm if the likelihood is bounded. Unboundedness of the likelihood, however, might occur at the edge of the parameter space, e.g., if the variance of one component tends to zero for mixtures of Gaussian distributions. As even in the case of boundedness only the detection of a local maximum can be guaranteed, it is in general recommended to repeat the EM algorithm with different initializations and to choose as final solution the one with the maximum likelihood. Different initialization strategies for the EM algorithm have been proposed, as its convergence to the optimal solution depends on the initialization.

4.2 Bayesian MCMC Sampling

Estimation within a Bayesian framework has become popular with the advent of MCMC methods, an overview on the different sampling approaches is given in Frühwirth-Schnatter (2006, chap. 3). Gibbs sampling is the most commonly used approach and it is done by augmenting the data with the unobservable variable of class membership similar to the EM algorithm. A drawback of the Gibbs sampler is that it might fail to escape the attraction area of one mode and therefore does not explore the entire parameter space. It was therefore suggested to use Metropolis-Hastings sampling schemes. Alternatively, the permutation sampler may be used.

5 Application

Three different applications of finite mixtures of regressions are presented. As the main purpose is to illustrate the application of the model class data sets are chosen which can be easily visualized in order to facilitate the understanding of the fitted models. In two cases ("Aphids" and "Movies" data set) the presence of latent groups is assumed and clustering the observations is one of the modeling aim. The difference between the two application however is that for the "Aphids" data set the presence of two separate groups with different regression coefficients can already be visually distinguished while for the "Movies" data set no separate groups can be observed even though considerable heterogeneity in the regression coefficients is present between the observations. If a mixed-effects model was fitted to the "Movies" data set this heterogeneity would be modeled through an a-priori specified distribution. The advantage of finite mixtures in this application are that (1) it is not required to specify the distribution for modeling heterogeneity in regression coefficients a-priori and (2) the components allow to easily inspect the range of heterogeneity present in the data. For the third data set ("Fabric faults") a random intercept model is assumed in order to account for overdispersion in the data.

5.1 Infection of Tobacco Plants

A finite mixture of binomial logit models is fitted to the "Aphids" data set from Section 1. The model is given by

$$h(\texttt{n.inf}|\texttt{n.aphids}, \Theta) = \sum_{k=1}^{K} \pi_k f_{\mathrm{Bi}}(\texttt{n.inf}|\pi_k(\texttt{n.aphids}), 69),$$

where $f_{\mathrm{Bi}}(\cdot|\pi, T)$ denotes the binomial distribution with success probability π and repetition parameter T which is in this application given by 69. `n.inf` is the number of infected plants and `n.aphids` the number of released aphids. The component specific mean value is given by

$$\mathrm{logit}(\pi_k(\texttt{n.aphids})) = \beta_{k1} + \texttt{n.aphids}\beta_{k2}.$$

Figure 1 suggests that the number of components $K = 2$. In addition to the visual inspection the number of components can be selected by fitting mixtures with different number of components to the data and determine the model with the minimum BIC. The BIC values for the mixtures with components 1 to 5 are 424.04, 274.92, 284.18, 295.5 and

305.83 where each of the mixtures is the best result of 5 different runs with random initialization to avoid local optima. This criterion hence confirms the results of the visual inspection. The fitted regression lines for each of the components together with the data are given in Figure 3. The relative sizes π_k of the 2 components are 0.54 and 0.46.

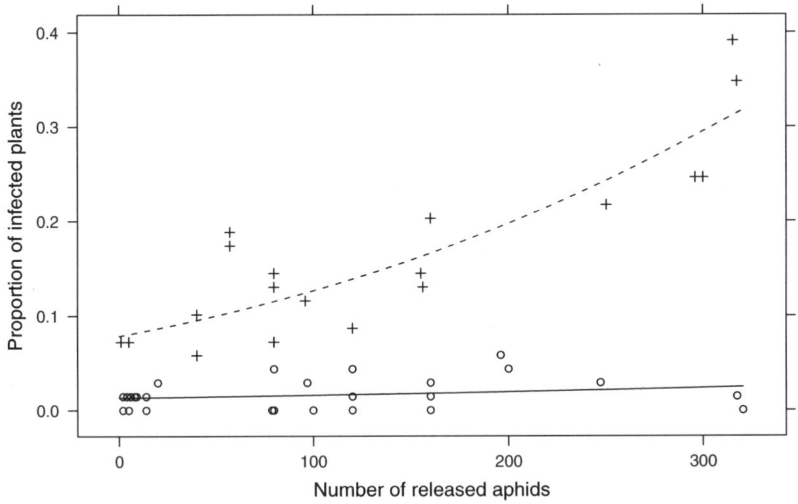

Fig. 3. "Aphids" data set with fitted regression lines for each component. The observations are plotted in different symbols according to the assignment to the component with the maximum a-posteriori probability.

As the repetition parameter T is equal to 69 the mixtures of binomial distributions are identifiable in each observation point for mixtures with up to 35 components as induced by the constraint $T \geq 2K - 1$. Given that observations are available for a range of different `n.aphids` values generic identifiability is guaranteed for the fitted mixtures with up to 5 components.

The suitability of the fitted mixture to induce a clustering of the data can be assessed by investigating the a-posteriori probabilities. If for each observation the maximum a-posteriori probability over all components is high the observations can be with a high confidence assigned to one of the components and hence a partitioning of the observations into K groups can be reasonably done using the fitted mixture model. For the "Aphids" data set the maximum a-posteriori probabilities have a mean of 0.98 with a standard deviation of 0.05 and a median of

1.00. This indicates that for each observation (n.inf,n.aphids) it can be with high confidence decided to which component it belongs. This also means that the two components are strongly separated and in fact constitute two different regimes.

From a practitioner's point of view further investigations are important to identify reasons why and when the two different regimes emerge. One possible explanation is that some batches of aphids consisted of insects that had passed their "maiden" phase. Low or zero levels of transmission of the virus are observed in this case because after the maiden phase the aphids tend to settle on the first plant they encounter.

5.2 Market Share Patterns of Movies

Finite mixtures of Gaussian regression models have been previously fitted to market share data of movies at the box office and theaters in the USA to investigate different patterns of decay (Jedidi et al. (1998)). The box office and theaters data for 407 movies playing between May 5, 2000 and December 7, 2001 were collected from a popular website of movie records (www.the-number.com), see Krider et al. (2005). The gross box-office takings for the 40 most popular movies for each weekend in the time period are recorded and transformed into market shares to account for the difference in volume between weekends. The market share is used as dependent variable and the number of weeks since release as covariate. For the data analysis the data is restricted to the first 20 weeks after release of a movie. This reduces the number of movies in the data set to 394. On average 8 observations are available for each movie which gives a total of 3149 observations.

The data is given in Figure 4. Each line represents a movie and its development of market share over the weeks after release. Most of the movies have a decline in market share over the weeks, but there also some films where an increase in market share over the first weekends can be observed. Due to this opposite trends and also due to the differences in decay for the movies loosing market shares the overlap in market shares between the movies is high which renders it impossible to discern different patterns of decay.

As most movies exhibit an exponential decay in market share the following mixture model is used to describe the data

$$h(\texttt{share}|\texttt{week}, \Theta) = \sum_{k=1}^{K} \pi_k f_N(\log(\texttt{share})|\mu_k(\texttt{week}), \sigma_k^2),$$

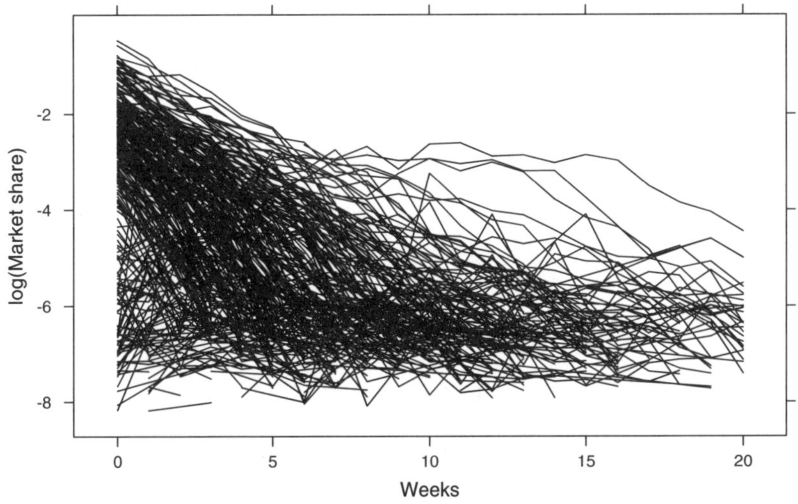

Fig. 4. Market share patterns of the "Movies" data set.

with the mean given by

$$\mu_k = \beta_{1k} + \texttt{week}\beta_{2k}.$$

As it is assumed that the component membership is fixed over the weeks for the movies, the information which observations are from the same movie is included in the estimation process.

Using an exponential decay model signifies that movies with a rise in market share at the beginning and a decline afterwards can only be approximated through a straight line which is still reasonable considering the small recorded time interval of 20 weeks. In addition we restrict the feasible mixtures to those where all component weights are at least 0.1, i.e., each component represents 39 movies or more.

Finite mixtures with 1 to 10 components are fitted and for each number of components the EM algorithm is repeated 10 times with random initialization in order to insure that the global optimum is detected. The BIC criterion is again used to determine the optimal number of components. The BIC suggests 5 components. However, it has to be noted that even though mixtures with up to 10 components are initially specified the EM algorithm did not converge to a mixture with more than 5 components as components with a weight of less than 0.1 are omitted during the run of the algorithm.

The parameter estimates are given in Table 1. C_k indicates that the parameters in this column belong to the kth component. The components are sorted in decreasing order with respect to parameter β_{1k}. The predicted mean values of market share for each component are depicted in Figure 5. The numbers indicate the component to which the line corresponds. For ease of comparison of the fitted parameters between the components they are plotted together with approximate 95% confidence intervals in Figure 6.

Parameter	C_1	C_2	C_3	C_4	C_5
π	0.15	0.17	0.23	0.13	0.32
β_{1k}	-1.99	-2.39	-2.95	-4.73	-6.49
β_{2k}	-0.29	-0.42	-0.61	-0.03	-0.01
σ	0.80	0.66	0.74	1.21	0.62

Table 1. Estimated parameters for the mixture with 5 components fitted to the "Movies" data set.

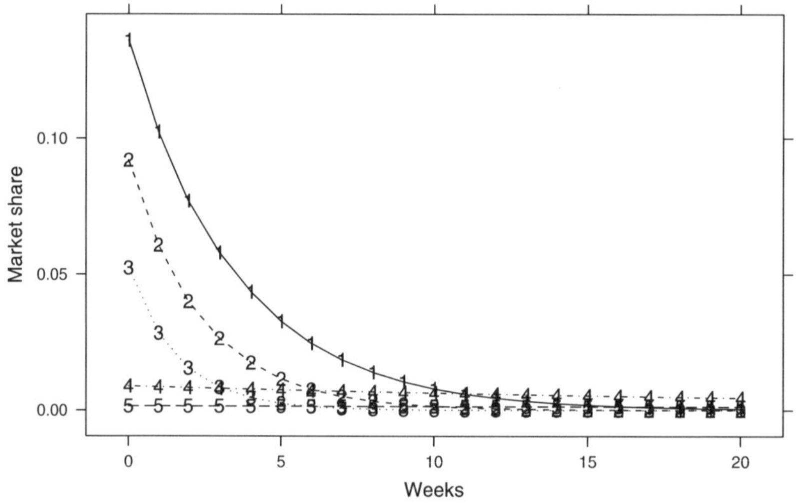

Fig. 5. Mean market share patterns of the finite mixture fitted to the "Movies" data set.

Comparing the intercepts given by β_{1k} indicates that there are three components with higher market shares at the release weekend. Components 1, 2, and 3 start with market shares of around 8.7%. The other two components achieve only market shares of 0.9% and 0.2% respectively on their release weekend. With respect to β_{2k}, which indicates the long-term success of a movie, component 3 has the strongest decline over the weeks indicating that in contrast to component 1 and 2 it is not able to stay on a high market share level for a longer time period. Component 1 seems to consist of the successful films which are also highly promoted leading to high market shares at the beginning and a slow decay over the weeks. Component 4 and 5 both have insignificant decay coefficients which indicates that they stay at about the same low level of market share during the first 20 weeks after release.

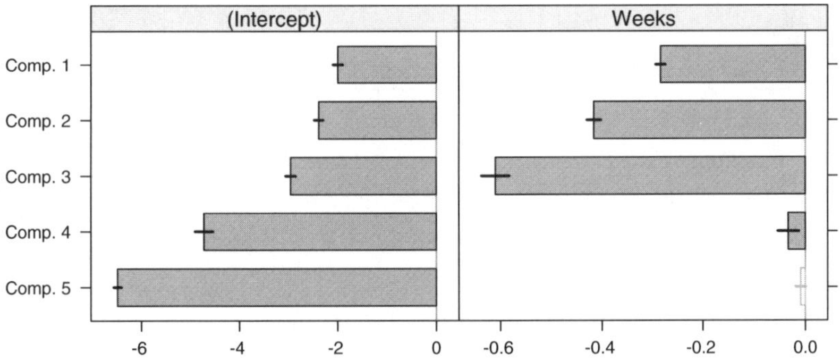

Fig. 6. Fitted regression coefficients and their approximate 95% confidence intervals for the "Movies" data set.

The a-posteriori probabilities are determined for each movie and used to assign them to the different components. Most of the films can be with high confidence assigned to one of the components. The mean of the maximum a-posteriori probabilities is 0.97 with a standard deviation of 0.08 and the median is 1.00. Rootograms of the posteriors of each component are given in Figure 7 (Leisch (2004a)). A rootogram is a modified version of a histogram where the square roots of the frequencies instead of the frequencies are used as heights for each bar. Please note that posteriors of less than 10^{-4} are omitted in order to ensure that the bar at zero does not dominate the plot.

The overlap of the components can be investigated by plotting the posteriors which correspond to observations assigned to a given component in a different color. If the posteriors for component 5 are highlighted it can be observed that the overlap with component 1 which consists of the most successful films is surprisingly high.

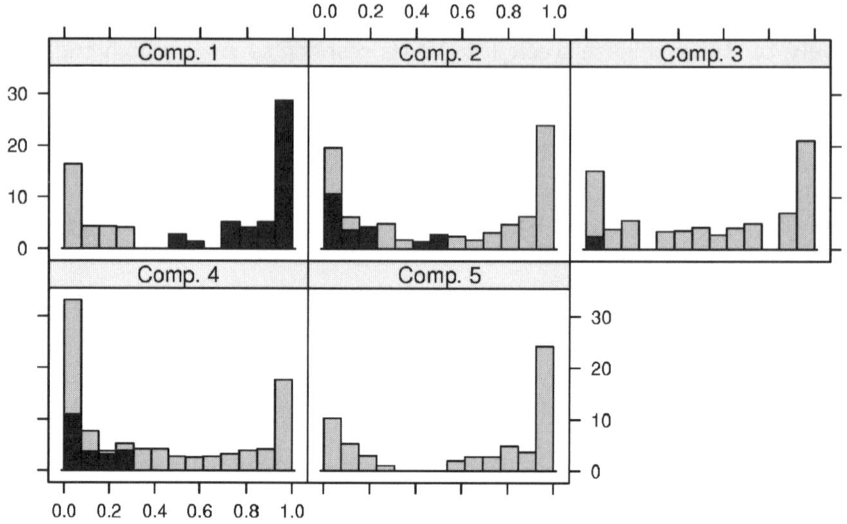

Fig. 7. Rootograms of the a-posteriori probabilities of the fitted mixture to the "Movies" data set. The posteriors of observations which are assigned to component 5 are shaded in dark grey.

The proportions of movies assigned to each component using the maximum a-posteriori probabilities are 0.34, 0.16, 0.11, 0.23 and 0.15. The quality of the partition of the data achieved by using the fitted finite mixture model can be investigated in Figure 8 where the market share patterns of the are plotted in different panels for each cluster.

5.3 Fabric Faults

The "Fabric faults" data set consists of 32 observations of number of faults in rolls of fabric of different length (Aitkin (1996)). The dependent variable is the number of faults (`n.fault`) and the covariate is the length of role in meters (`length`). The data is given in Figure 9.

As the dependent variable is a counting variable in a first step a standard GLM with Poisson distribution is fitted to the data where the logarithm of the lengths is used as independent variable. The fitted regression line is given in the left panel in Figure 10. An analysis of the model fit indicates that substantial overdispersion is present with a residual deviance of 64.54 on 30 degrees of freedom. To account for this overdispersion a random intercept model is fitted which is given by

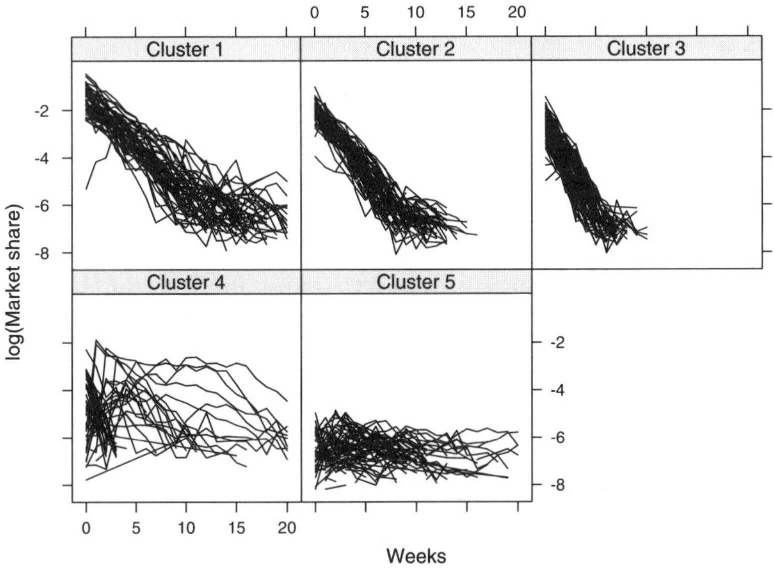

Fig. 8. Clustered market share patterns of the "Movies" data set.

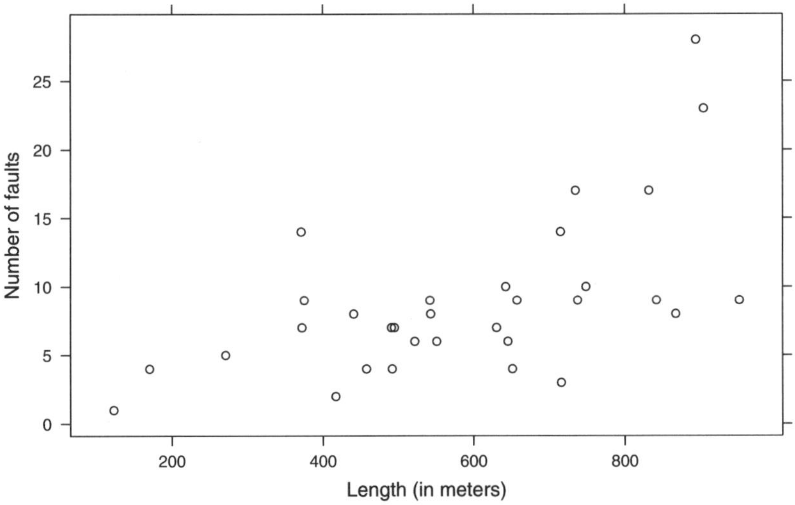

Fig. 9. "Fabric faults" data set.

$$h(\texttt{n.fault}|\texttt{length}, \Theta) = \sum_{k=1}^{K} \pi_k f_{\text{Poi}}(\texttt{n.fault}|\lambda_k(\texttt{length}))$$

where $f_{\mathrm{Poi}}(\cdot|\lambda)$ denotes the Poisson distribution with mean λ. The mean λ_k is in the random intercept model given by

$$\log(\lambda_k) = \beta_{1k} + \log(\texttt{length})\beta_2.$$

Please note that the coefficient of the covariate does not have an index k which means that it is constant over the components.

Again the optimal number of components is selected using the BIC criterion after fitting the model with the EM algorithm for different number of components ranging from 1 to 5 and 5 repeated fittings with random initialization and the number of components fixed. The BIC values are 194.77, 186.53, 193.46, 200.39 and 207.32 and consequently the mixture with 2 components is selected. The resulting regression lines for each of the components separately are the dashed lines in the right panel of Figure 10. The full line represents the fitted regression line of the random intercept model to the complete data set. The plotting symbols of the observations in the right panel are according to an assignment of the observations to the two components given the maximum a-posteriori probabilities.

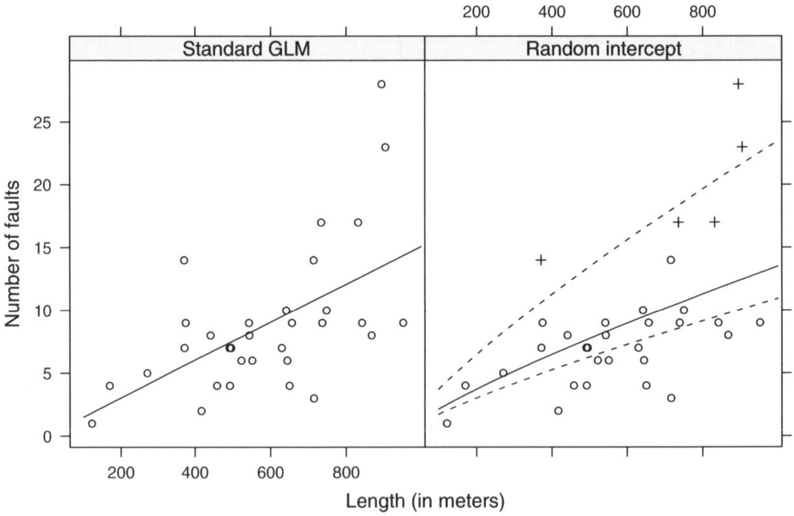

Fig. 10. Fitted regression lines to the "Fabric faults" data set for the standard GLM and a random intercept model with 2 components.

6 Conclusion and Outlook

Finite mixtures of GLMs are an important statistical modeling technique which is an obvious extension of standard GLMs. They relax the assumption of homogeneity of parameters, but do not require to a-priori specify and fix the distribution which accounts for the heterogeneity in parameters as in mixed-effects models. In addition this flexible model class contains important special cases such as zero-inflated or random intercepts models.

The model class of finite mixtures of GLMs can be easily specified within the finite mixture model framework and the modification of existing estimation methods is often straight-forward in order to be able to fit the models. For the EM algorithm it is only necessary to adapt the M-step by determining the weighted ML estimator for the component specific model. Different problems in model fitting and diagnostics than in standard mixtures of distributions however might be encountered due to trivial and generic identifiability problems.

Further extensions of finite mixtures are possible for the regression case. Instead of using GLMs as component specific models generalized additive models can be used which allow to relax the assumption that the functional relationship between covariates and dependent variable is a-priori known. Another possibility is to relax the assumption of homogeneity within the components and fit a mixed-effects model in each component.

In the future model identification and diagnostics need further investigation in the regression case for finite mixtures. The performance of newly proposed methods such as a new model selection criterion for mixtures of regression models (Naik et al. (2007)) needs for example to be validated in real applications on different empirical data sets. In addition new visualization techniques which enable the researcher to easily explore the characteristics of a fitted model and compare competing models would be a valuable enhancement of the finite mixture modeling toolbox.

Acknowledgement

This research was supported by the Austrian Science Foundation (FWF) under grant P17382.

References

Aitkin M (1996) A general maximum likelihood analysis of overdispersion in generalized linear models. Statistics and Computing 6:251–262

Aitkin M (1999) Meta-analysis by random effect modelling in generalized linear models. Statistics in Medicine 18(17–18):2343–2351

Böhning D, Dietz E, Schlattmann P, Mendonça L, Kirchner U (1999) The zero-inflated Poisson model and the decayed, missing and filled teeth index in dental epidemiology. Journal of the Royal Statistical Society A 162(2):195–209

Boiteau G, Singh M, Singh RP, Tai GCC, Turner TR (1998) Rate of spread of pvy-n by alate myzus persicae (sulzer) from infected to healthy plants under laboratory conditions. Potato Research 41(4):335–344

Celeux G, Diebolt J (1988) A random imputation principle: The stochastic EM algorithm. Rapports de Recherche 901, INRIA

Dasgupta A, Raftery AE (1998) Detecting features in spatial point processes with clutter via model-based clustering. Journal of the American Statistical Association 93(441):294–302

Dempster AP, Laird NM, Rubin DB (1977) Maximum likelihood from incomplete data via the EM-algorithm. Journal of the Royal Statistical Society B 39:1–38

Follmann DA, Lambert D (1989) Generalizing logistic regression by non-parametric mixing. Journal of the American Statistical Association 84(405):295–300

Frühwirth-Schnatter S (2006) Finite Mixture and Markov Switching Models. Springer Series in Statistics, Springer, New York

Grün B (2006) Identification and estimation of finite mixture models. PhD thesis, Institut für Statistik und Wahrscheinlichkeitstheorie, Technische Universität Wien, Friedrich Leisch, advisor

Grün B, Leisch F (2004) Bootstrapping finite mixture models. In: Antoch J (ed) Compstat 2004 — Proceedings in Computational Statistics, Physica Verlag, Heidelberg, pp 1115–1122

Grün B, Leisch F (2006) Fitting finite mixtures of linear regression models with varying & fixed effects in R. In: Rizzi A, Vichi M (eds) Compstat 2006—Proceedings in Computational Statistics, Physica Verlag, Heidelberg, Germany, pp 853–860

Grün B, Leisch F (2007) Flexmix 2.0: Finite mixtures with concomitant variables and varying and fixed effects. Submitted for publication

Grün B, Leisch F (2007) Identifiability of finite mixtures of multinomial logit models with varying and fixed effects, unpublished manuscript

Grün B, Leisch F (2007) Testing for genuine multimodality in finite mixture models: Application to linear regression models. In: Decker R, Lenz HJ (eds) Advances in Data Analysis, Proceedings of the 30th Annual Conference of the Gesellschaft für Klassifikation, Springer-Verlag, Studies in Classification, Data Analysis, and Knowledge Organization, vol 33, pp 209–216

Hennig C (2000) Identifiability of models for clusterwise linear regression. Journal of Classification 17(2):273–296

Jedidi K, Krider RE, Weinberg CB (1998) Clustering at the movies. Marketing Letters 9(4):393–405

Krider RE, Li T, Liu Y, Weinberg CB (2005) The lead-lag puzzle of demand and distribution: A graphical method applied to movies. Marketing Science 24(4):635–645

Leisch F (2004a) Exploring the structure of mixture model components. In: Antoch J (ed) Compstat 2004 — Proceedings in Computational Statistics, Physica Verlag, Heidelberg, pp 1405–1412

Leisch F (2004b) FlexMix: A general framework for finite mixture models and latent class regression in R. Journal of Statistical Software 11(8), URL `http://www.jstatsoft.org/v11/i08/`

Lindsay BG (1989) Moment matrices: Applications in mixtures. The Annals of Statistics 17(2):722–740

McCullagh P, Nelder JA (1989) Generalized Linear Models (2nd edition). Chapman and Hall

McLachlan GJ, Krishnan T (1997) The EM Algorithm and Extensions, 1st edn. John Wiley and Sons

Naik PA, Shi P, Tsai CL (2007) Extending the Akaike information criterion to mixture regression models. Journal of the American Statistical Association 102(477):244–254

Newcomb S (1886) A generalized theory of the combination of observations so as to obtain the best result. American Journal of Mathematics 8:343–366

Pearson K (1894) Contributions to the mathematical theory of evolution. Philosophical Transactions of the Royal Society A 185:71–110

R Development Core Team (2007) R: A language and environment for statistical computing. R Foundation for Statistical Computing, Vienna, Austria, URL `http://www.R-project.org`

Redner RA, Walker HF (1984) Mixture densities, maximum likelihood and the EM algorithm. SIAM Review 26(2):195–239

Titterington DM, Smith AFM, Makov UE (1985) Statistical Analysis of Finite Mixture Distributions. Wiley

Wang P, Puterman ML (1998) Mixed logistic regression models. Journal of Agricultural, Biological, and Environmental Statistics 3(2):175–200

Wang P, Puterman ML, Cockburn IM, Le ND (1996) Mixed Poisson regression models with covariate dependent rates. Biometrics 52:381–400

Wedel M, Kamakura WA (2001) Market Segmentation — Conceptual and Methodological Foundations (2nd edition). Kluwer Academic Publishers

Higher-order Dependence in the General Power ARCH Process and the Role of Power Parameter

Changli He[1], Hans Malmsten[2] and Timo Teräsvirta[3]

[1] Dalarna University, Sweden, and Tianjin University of Finance and Economics, China chh@du.se
[2] Länsförsäkringar, Stockholm, Sweden
Hans.Malmsten@lansforsakringar.se
[3] CREATES, School of Economics and Management, University of Aarhus, Denmark, and Stockholm School of Economics, Sweden
tterasvirta@econ.au.dk

1 Introduction

In a recent paper, Ding, Granger and Engle (1993) introduced a class of autoregressive conditional heteroskedastic models called Asymmetric Power Autoregressive Conditional Heteroskedastic (A-PARCH) models. The authors showed that this class contains as special cases a large number of well-known ARCH and GARCH models. The A-PARCH model contains a particular power parameter that makes the conditional variance equation nonlinear in parameters. Among other things, Ding, Granger and Engle showed that by letting the power parameter approach zero, the A-PARCH family of models also includes the logarithmic GARCH model as a special case. Hentschel (1995) defined a slightly extended A-PARCH model and showed that after this extension, the A-PARCH model also contains the exponential GARCH (EGARCH) model of Nelson (1991) as a special case as the power parameter approaches zero. Allowing this to happen in a general A-PARCH model forms a starting-point for our investigation.

A notable feature of the A-PARCH model is that, due to its parameterization, it is only possible to find analytically certain fractional moments of the absolute values of the original process related to the power parameter. Expressions for such moments were derived in He and Teräsvirta (1999d). In this paper we first define a slight generalization of the class of EGARCH models. Then we derive the autocorrelation function of squared and logarithmed observations for this class of models. For Nelson's EGARCH model it is possible to reconcile our results with those in Breidt, Crato and de Lima (1998). Furthermore, we show that this autocorrelation function follows as a limiting case from the autocorrelation function of some fractional powers of the absolute values of the original observations.

On the other hand, if we want to derive the autocorrelation function of squares of the original observations and not their logarithms for the EGARCH model then the techniques applied in this paper do not apply. The solution to that problem can be found in He, Teräsvirta and Malmsten (2002). The autocorrelation functions can be used for evaluating an estimated model by checking how well the model is able to reproduce stylized facts; see Malmsten and Teräsvirta (2004) for an example. This means estimating the autocorrelation function from the data and comparing it with the corresponding autocorrelations obtained by plugging in the parameter estimates from the PARCH model into the theoretical expressions of the autocorrelations.

This approach cannot be applied if the autocorrelations compared are, say, autocorrelations of squared observations instead of autocorrelations of suitable fractional moments of their absolute values. This is the case for example when one wants to compare autocorrelations of squares implied by two different models, for instance a standard GARCH and a symmetric PARCH model , with each other. The only possibility is to estimate the autocorrelations of squared observations for the PARCH model by simulation. This becomes an issue in this paper, for the role of the power parameter in the PARCH model will be an object of our investigation.

Applications of the A-PARCH model to return series of stocks and exchange rates have revealed some regularities in the estimated values of the power parameter; see Ding, Granger and Engle (1993), Brooks, Faff, McKenzie and Mitchell (2000) and McKenzie and Mitchell (2002). We add to these results by fitting symmetric first-order PARCH models to return series of 30 most actively traded stocks of the Stockholm Stock Index. Our results agree with the previous ones and suggest that the power parameter lowers the autocorrelations of squared observations compared to the corresponding autocorrelations implied, other things equal, by the standard first-order GARCH model. In the present situation this means estimating the autocorrelation function of the squared observations from the data and comparing that with the corresponding values obtained by plugging the parameter estimates into the theoretical expressions of the autocorrelations. Another example can be found in He and Teräsvirta (1999d).

The plan of the paper is as follows. Section 2 defines the class of models of interest and introduces notation. The main theoretical results appear in Section 3. Section 4 contains a comparison of autocorrelation functions of squared observations for different models and Section 5 a discussion of empirical examples. Finally, conclusions appear in Section 6. All proofs can be found in Appendix.

2 The Model

Let $\{\varepsilon_t\}$ be a real-valued discrete time stochastic process generated by

$$\varepsilon_t = z_t h_t \tag{1}$$

where $\{z_t\}$ is a sequence of independent identically distributed random variables with mean zero and unit variance, and h_t is a $\mathcal{F}_{t-1}$-measurable function, where $\mathcal{F}_{t-1}$ is the sigma-algebra generated by $\{z_{t-1}, z_{t-2}, z_{t-3}, ...\}$, and positive with probability one. Let

$$h_t^{2\delta} = \alpha_0 + c_\delta(z_{t-1})h_{t-1}^{2\delta}, \ \delta > 0 \tag{2}$$

where α_0 is a positive scalar and $c_{\delta t} = c_\delta(z_t)$ is a well-defined function of z_t. The sequence $\{c_{\delta t}\}$ is a sequence of independent identically distributed random variables such that each $c_{\delta t}$ is stochastically independent of $h_t^{2\delta}$. Function $c_{\delta t}$ contains parameters that determine the moment structure of $\{\varepsilon_t\}$. Constrains on these parameters are necessary to guarantee that $h_t^{2\delta}$ remains positive with probability 1. We call (1) and (2) a general power ARCH (GPARCH$(\delta, 1, 1)$) model. This model appeared in He and Teräsvirta (1999d) in a slightly more general form with $\alpha_0 = g(z_t)$ being a stochastic variable.

Setting $c_\delta(z_{t-1}) = \alpha(|z_t| - \phi z_t)^{2\delta} + \beta$ in equation (2), defines, together with equation (1), the Asymmetric Power ARCH (A-PARCH) (1,1) model of Ding, Granger and Engle (1993). Note that these authors use δ in place of 2δ in equation (2) but that does not affect the results. Hentschel (1995) also defined a parametric family of GARCH models similar to (1) and (2) in order to highlight relationships between different GARCH models and their treatment of asymmetry.

In this paper we are interested in the limiting case $\delta \to 0$. Taking logarithms of (1) yields

$$\ln \varepsilon_t^2 = \ln z_t^2 + \ln h_t^2. \tag{3}$$

On the other hand, equation (1) can be modified such that it relates the Box-Cox transformed ε_t^2, that is, $\varphi_\delta(\varepsilon_t^2) = (\varepsilon_t^{2\delta} - 1)/\delta$, to $\{(z_t h_t)^{2\delta} - 1\}/\delta$. Then by applying l'Hôpital's rule it can be shown that letting $\delta \to 0$ in the modified equation also leads to (3). This entitles us to consider certain exponential GARCH models as limiting cases of the power ARCH model (1) and (2). In order to see that, rewrite (2) in terms of $(h_t^{2\delta} - 1)/\delta$ and define $c_\delta(z_t) = \delta g(z_t) + \beta$. It can be shown that under certain conditions, as $\delta \to 0$, equation (2) becomes

$$\ln h_t^2 = \alpha_0 + g(z_{t-1}) + \beta \ln h_{t-1}^2 \tag{4}$$

where $g(z_t)$ is a well-defined function of z_t. Equation (4) is thus a limiting case of (2). We call the models defined by equations (1) and (4) or (3) and (4) a limiting class of GPARCH(1,1) models. They contain certain well-known models as special cases. For example, setting $g(z_t) = \phi z_t + \psi(|z_t| - \mathsf{E}\,|z_t|)$ in (4) yields

$$\ln h_t^2 = \alpha_0 + \phi z_{t-1} + \psi(|z_{t-1}| - \mathsf{E}\,|z_{t-1}|) + \beta \ln h_{t-1}^2. \tag{5}$$

This equation, jointly with (1), defines the EGARCH(1,1) model of Nelson (1991). Similarly, we may set $c_\delta(z_t) = \alpha g_1^\delta(z_t) + \beta$ where $g_1(z_t) > 0$ for all t with probability one. Then, by l'Hôpital's rule, (2) converges to

$$\ln h_t^2 = \alpha_0 + \alpha \ln g_1(z_{t-1}) + (\alpha + \beta) \ln h_{t-1}^2 \tag{6}$$

as $\delta \to 0$. Equations (1) and (6) define a class of logarithmic GARCH (LGARCH(1,1)) models. Setting $g_1(z_t) = z_t^2$ in (6) yields

$$\ln h_t^2 = \alpha_0 + \alpha \ln \varepsilon_{t-1}^2 + \beta \ln h_{t-1}^2 \tag{7}$$

which is the LGARCH(1,1) model of Geweke (1986) and Pantula (1986). Since (4) and (6) have a similar structure, we mainly consider results for the limiting GPARCH(1,1) model (1) and (4). They can be easily modified to apply to the class of LGARCH(1,1) models.

3 The Limiting Results

In this section we derive the asymptotic moment structure of the GPARCH(1,1) model (1) and (2) as $\delta \to 0$ under the Box-Cox transformation. We first give the moment structure of (1) and (2) for $\delta > 0$. Having done that we derive the moment structure of model (3) with (4). Finally, we show that this result may be also obtained as a limiting case of model (1) with (2) as $\delta \to 0$.

To formulate our first result let $\gamma_\delta = \mathsf{E}c_{\delta t}$ and $\gamma_{2\delta} = \mathsf{E}c_{\delta t}^2$. We have

Lemma 1. *For the GPARCH(δ,1,1) model (1) with (2), a necessary and sufficient condition for the existence of the 4δ-th unconditional moment $\mu_{4\delta} = \mathsf{E}\,|\varepsilon_t|^{4\delta}$ of $\{\varepsilon_t\}$ is*

$$\gamma_{2\delta} < 1. \tag{8}$$

If (8) holds, then

$$\mu_{4\delta} = \alpha_0^2 \nu_{4\delta}(1+\gamma_\delta)/\{(1-\gamma_\delta)(1-\gamma_{2\delta})\} \tag{9}$$

where $\nu_{2\psi} = \mathsf{E}\,|z_t|^{2\psi}, \psi > 0$. The autocorrelation function $\rho_n(\delta) = \rho(|\varepsilon_t|^{2\delta}, |\varepsilon_{t-n}|^{2\delta})$, $n \geq 1$, of $\{|\varepsilon_t|^{2\delta}\}$ has the form

$$\rho_1(\delta) = \frac{\nu_{2\delta}[\overline{\gamma}_\delta(1-\gamma_\delta^2) - \nu_{2\delta}\gamma_\delta(1-\gamma_{2\delta})]}{\nu_{4\delta}(1-\gamma_\delta^2) - \nu_{2\delta}^2(1-\gamma_{2\delta})} \tag{10}$$

where $\overline{\gamma}_\delta = \mathsf{E}(|z_t|^{2\delta} c_{\delta t})$, and $\rho_n(\delta) = \gamma_\delta \rho_{n-1}(\delta)$, $n \geq 2$.

Proof:

See Appendix.

Let $\mathcal{M}_\delta(\mu_{4\delta}, \rho_n(\delta))$ denote the analytic second moment structure defined by Lemma 1 for the GPARCH(δ,1,1) model (1) and (2). It consists of $\mu_{4\delta}$ and the autocorrelations $\rho_n(\delta)$, $n \geq 1$. It is seen that $\mathcal{M}_\delta(\cdot)$ is a function of power parameter δ. Note that the autocorrelation function of $\{|\varepsilon_t|^{2\delta}\}$ is decaying exponentially with the discount factor γ_δ. In particular, setting $\delta = 1$ in equations (8) and (10) yields the existence condition of the fourth moment and the autocorrelation function of the squared observations of the standard GARCH(1,1) model (Bollerslev (1986)) with non-normal errors.

It is customary to also consider the kurtosis of any given GARCH process, see, for example, Bollerslev (1986) or He and Teräsvirta (1999b). In this case, the kurtosis of $|\varepsilon_t|^\delta$ or $\varphi_\delta(|\varepsilon_t|) = (|\varepsilon_t|^\delta - 1)/\delta$ may be defined as

$$\begin{aligned}\kappa_4(\delta) &= \frac{\mathsf{E}(|\varepsilon_t|^\delta - \mathsf{E}\,|\varepsilon_t|^\delta)^4}{\{\mathsf{E}(|\varepsilon_t|^\delta - \mathsf{E}\,|\varepsilon_t|^\delta)^2\}^2} \\ &= \frac{\mathsf{E}(\varphi_\delta(|\varepsilon_t|) - \mathsf{E}\varphi_\delta(|\varepsilon_t|))^4}{\{\mathsf{E}(\varphi_\delta(|\varepsilon_t|) - \mathsf{E}\varphi_\delta(|\varepsilon_t|))^2\}^2}\end{aligned}$$

so that the limiting case

$$\lim_{\delta\to 0}\kappa_4(\delta) = \frac{\mathsf{E}(\ln|\varepsilon_t| - \mathsf{E}\ln|\varepsilon_t|)^4}{\{\mathsf{E}(\ln|\varepsilon_t| - \mathsf{E}\ln|\varepsilon_t|)^2\}^2}. \tag{11}$$

The kurtosis (11) is thus the limiting case of the kurtosis of the absolute-valued process $\{|\varepsilon_t|^\delta\}$. Computing it would require the expectations $\mathsf{E}(\ln|\varepsilon_t|)^4$ and $\mathsf{E}(\ln|\varepsilon_t|)^3$ or, alternatively, $\mathsf{E}(\ln\varepsilon_t^2)^4$ and $\mathsf{E}(\ln\varepsilon_t^2)^3$ for which no analytical expressions have been derived above. The kurtosis of $\ln|\varepsilon_t|$ is a concept quite different from that of ε_t, and for this reason it is not considered any further here.

For the limiting GPARCH(1,1) process we obtain the following result:

Lemma 2. *For the limiting GPARCH(1,1) process (3) and (4), assume that variances of* $(\ln z_t^2)^2$ *and* $(g(z_t))^2$ *are finite for any* t. *Then the second unconditional moment of* $\ln\varepsilon_t^2$ *exists if and only if*

$$|\beta| < 1. \tag{12}$$

When (12) holds, this second moment can be expressed as

$$\mu_0 = \mathsf{E}(\ln\varepsilon_t^2)^2 = \frac{\Delta}{(1-\beta)(1-\beta^2)} \tag{13}$$

where $\Delta = \gamma_{(\ln z^2)^2}(1-\beta)(1-\beta^2) + 2\gamma_{\ln z^2}(\alpha_0+\gamma_g)(1-\beta^2) + [\alpha_0^2(1+\beta) + 2\alpha_0(1+\beta)\gamma_g + 2\beta\gamma_g^2 + (1-\beta)\gamma_{g^2}]$ *and* $\gamma_{(\ln z^2)^2} = \mathsf{E}(\ln z_t^2)^2$, $\gamma_{\ln z^2} = \mathsf{E}\ln z_t^2$, $\gamma_g = \mathsf{E}g(z_t)$ *and* $\gamma_{g^2} = \mathsf{E}(g(z_t))^2$. *Furthermore, the autocorrelation function* $\rho_n^0 = \rho(\ln\varepsilon_t^2, \ln\varepsilon_{t-n}^2)$, $n \geq 1$, *of* $\{\ln\varepsilon_t^2\}$ *has the form*

$$\rho_1^0 = \frac{(1-\beta^2)(\gamma_{g\ln z^2} - \gamma_g\gamma_{\ln z^2}) + \beta(\gamma_{g^2} - \gamma_g^2)}{(1-\beta^2)(\gamma_{(\ln z^2)^2} - \gamma_{\ln z^2}^2) + (\gamma_{g^2} - \gamma_g^2)},$$
$$\rho_n^0 = \rho_1^0\beta^{n-1}, n \geq 2, \tag{14}$$

where $\gamma_{g\ln z^2} = \mathsf{E}(g(z_t)\ln z_t^2)$.

Proof:

See Appendix.

Nelson (1991) derived the autocovariance function of the logarithm of the conditional variance of the EGARCH process. Breidt, Crato and de Lima (1998) obtained the autocorrelation function of $\{\ln\varepsilon_t^2\}$ for the EGARCH model. In both articles the authors made use of the infinite moving average representation of the logarithm of the conditional variance. Lemma 2 gives the corresponding result for the first-order process directly in terms of the parameters of the original model, which is practical for model evaluation purposes.

Let $\mathcal{M}_0(\mu_0, \rho_n^0)$ denote the second moment structure defined by Lemma 2 for the limiting GPARCH(1,1) process (3) and (4) and assume that $\delta < \delta_0$ such that $\gamma_{2\delta_0} < 1$. We have

Theorem *Assume that $\mathcal{M}_\delta(\cdot)$ is defined for $\gamma_{2\delta_0} < 1$ and the functions defining $\mathcal{M}_\delta(\cdot)$ are continuous and twice differentiable with respect to δ. Then, under the transformation $\varphi_\delta(\varepsilon_t^2) = (\varepsilon_t^{2\delta} - 1)/\delta$,*

$$\mathcal{M}_\delta(\mu_{4\delta}, \rho_n(\delta)) \to \mathcal{M}_0(\mu_0, \rho_n^0) \tag{15}$$

that is, $\mu_{4\delta} \to \mu_0$ and $\rho_n(\delta) \to \rho_n^0$, $n \geq 1$, as $\delta \to 0$.

Proof:

See Appendix.

Remark. It has been pointed out above that, under the Box-Cox transformation $\varphi_\delta(\varepsilon_t^2) = (\varepsilon_t^{2\delta} - 1)/\delta$, equation (1), when appropriately modified, converges to equation (3) as $\delta \to 0$. The theorem then says that under this transformation the moment structure of the GPARCH(δ,1,1) model (1) and (2) approaches the moment structure of the limiting GPARCH(1,1) model as $\delta \to 0$: $\mu_{4\delta} \to \mu_0$ and $\rho_n(\delta) \to \rho_n^0$. This convergence shows that the moment structure $\mathcal{M}_0(\cdot)$ belongs to the class of structures $\mathcal{M}_\delta(\cdot)$ as a boundary case. Besides, the parameter δ in the GPARCH(δ,1,1) process defines a value for which the autocorrelation function $\rho(|\varepsilon_t|^\delta, |\varepsilon_{t-k}|^\delta)$, $k \geq 1$, decays exponentially with k.

In order to consider the practical value of these results suppose, for example, that $\gamma_4 < 1$. Then we have a class of GPARCH$(\delta, 1, 1)$ models with the same parameter values such that the available $\mathcal{M}_\delta(\cdot)$ is defined on $[0, 1]$, that is, $\gamma_{2\delta} < 1$, $\delta \leq 1$. Practitioners may want to use these results to see what kind of moment implications GPARCH models they estimate may have. Results in $\mathcal{M}_\delta(\cdot)$ defined on $[0, 1]$ may also be useful in checking how well different GPARCH models represent the reality, which is done by comparing parametric moment estimates from a GPARCH$(\delta, 1, 1)$ model with corresponding nonparametric ones obtained directly from the data. First-order LGARCH and EGARCH models may thus be compared with, say, a standard GARCH(1,1) model in this respect if both are estimated using the same data.

4 Autocorrelation Functions of Squared Observations

In this section we show how the autocorrelation function of ε_t^2 varies with δ across GPARCH(δ,1,1) models with normal errors. We demonstrate how the power parameter increases the flexibility of the specification compared to the GARCH model. We also include the symmetric first-order EGARCH model and LGARCH models in our comparison. The three parameters in the GARCH model are selected such that the unconditional variance equals unity, the kurtosis equals 12, and the decay rate of the autocorrelations of ε_t^2 equals 0.95. For the GARCH(1,1) model, this decay rate is obtained by setting $\alpha + \beta = 0.95$. The parameter values for the EGARCH model and the LGARCH are chosen such as to make the models as comparable with the GARCH model as possible. Thus, $\beta = 0.95$ in the EGARCH model and $\alpha + \beta = 0.95$ in the LGARCH model correspond to $\alpha + \beta = 0.95$ in the GARCH model, because β and $\alpha + \beta$, respectively, control the decay of the autocorrelation function of the squared observations in these two models. Note, however, that while the decay rate of the autocorrelation of ε_t^2 in the GARCH(1,1) model equals $\alpha + \beta$, it only approaches β from below with increasing lag length in the EGARCH(1,1) model and $\alpha + \beta$ from below in the LGARCH(1,1) model. The individual parameters are chosen such that the unconditional variance and the kurtosis are the same in all three models as well. This can be done using the analytic expressions for the relevant moments of the EGARCH(1,1) model in He, Teräsvirta and Malmsten (2002) and the LGARCH(1,1) model in the Appendix (Lemma 3).

In order to illustrate the role of δ, we consider the GPARCH(1,1) model with $\delta = 1.5$ and $\delta = 1$ under the assumption that the other parameters are the same as in the GARCH(1,1) model. For $\delta = 1.5$ the autocorrelations of ε_t^2 cannot be obtained analytically, and we have computed them by simulation from 1,000 series of 100,000 observations each. For $\delta = 1$, they are available from He and Teräsvirta (1999b) where this special case is considered under the name absolute-valued GARCH model. It can be seen from Figure 1 that $\delta < 2$ reduces the autocorrelations of ε_t^2 (other things equal) compared to $\delta = 2$ (the GARCH model). The difference in autocorrelations between $\delta = 1.5$ and $\delta = 1$ is smaller than the corresponding difference between $\delta = 2$ and $\delta = 1.5$ which is quite large. The autocorrelations of ε_t^2 for the EGARCH and the LGARCH model are different from the ones for the GARCH model. As already mentioned, the decay is exponential for the autocorrelations of the GARCH model but faster than exponential and exponential only asymptotically (as a function of the lag length), both for the EGARCH model and the LGARCH model. This is also in fact the case for the GPARCH models for which $\delta < 2$.

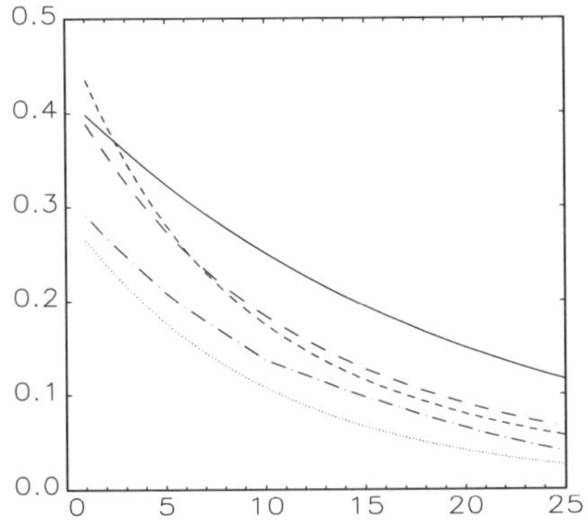

Fig. 1. Autocorrelation functions of squares for five first-order GARCH models, EGARCH (dashed line, short dashes), GARCH (solid line), LGARCH (dashed line, long dashes), PGARCH $\delta = 1.5$ (dashed-dotted line) and PGARCH $\delta = 1$ (dotted line).

5 Empirical Examples

Ding, Granger and Engle (1993) demonstrated the potential of the GPARCH model by fitting the model with normal errors to the long S&P 500 daily stock return series from January 3, 1928, to August 30, 1991, 17055 observations in all. The estimate of the power parameter δ was equal to 1.43 and significantly different from two (the GARCH model). Brooks, Faff, McKenzie and Mitchell (2000) applied the GPARCH(1,1) model with t-distributed errors to national stock market returns for 10 countries plus a world index for the period February 1989 to December 1996, which amounted to a total of 2062 daily observations. Except for three extremes cases, the power parameter estimates were between 1.17 and 1.45, with most values close to the mean value 1.36. The authors concluded that in the absence of leverage effects there is moderate evidence supporting the need for the power parameter. In the case of six countries plus the world index, the standard GARCH model could not be rejected in favour of the symmetric GPARCH model at the 5% significance level. The evidence against the standard GARCH model was, however, much stronger in case of a combination of leverage and power effects. More specifically, with the exception of one national return series, the GARCH model was strongly rejected in favour of the asymmetric GPARCH model.

McKenzie and Mitchell (2002) applied the GPARCH model to daily return series of 17 heavily traded bilateral exchange rates and found the estimated power parameter equal to 1.37 on average. All power parameter estimates were between one and two. For seven of the estimated models the power parameter was significantly different from two. The results were thus quite similar to the ones Brooks, Faff, McKenzie and Mitchell (2000) reported.

Tse and Tsui (1997) fitted the A-PARCH model to two exchange rate return series, the Malaysian ringgit and the Singaporean dollar. Their results do not fit the aforementioned pattern. The most notable fact was the large change in the estimated value of δ when the t-distributed errors are substituted for the normal ones. The estimated degrees of freedom of the t-distribution were in both cases so low that they alone practically excluded the existence of the finite fourth moment for the underlying GARCH process.

In order to further explore the role of the power parameter in practice we consider daily return series of the 30 most actively traded stocks in Stockholm Stock Exchange and estimate a symmetric GPARCH(δ,1,1) model (with normal errors), a standard GARCH model and an EGARCH model for these series. The names of the stocks can be found in Table 1 together with information about the length of the series. The period investigated ends April 24, 2001. The return series have been obtained from Datastream.

In Table 1 we report the maximum likelihood estimates of the power parameter δ. The estimates of δ lie between 1.21 and 1.49, most of them close to the mean value of 1.40. We find that the estimates are remarkably similar (around a mean value of 1.40) to the ones Brooks, Faff, McKenzie and Mitchell (2000) obtained for their return series. The estimates of δ are significantly different from two in a majority of cases, see Table 1, where the p-value of the test is less than 0.01 in 15 cases out of 29. It should be noted, however, that some of the estimated autocorrelations may not actually have a theoretic counterpart because the moment condition $\gamma_{2\delta} < 1$ appearing in Lemma 1 is not satisfied. This does not mean that the 2δth moment of ε_t cannot exist, because $\widehat{\gamma}_{2\delta}$ is an estimate, but empirical support for the existence of this moment cannot be argued to be strong. If we merely compare standard GARCH and EGARCH models using tests of non-nested hypotheses, the results reported in Malmsten (2004) indicate that both models fit the 30 series more or less equally well.

Table 1. The stocks, the estimates of the power parameter, length of the series, p-values of the likelihood ratio test of GARCH against GPARCH, and the estimated left-hand side of the moment condition of Lemma 1.

y	δ	T	p	$\widehat{\gamma}_{2\delta}$
ABB	1.41	3717	0.008	1.013
Assa A.	1.37	1617	0.382	0.733
Assi D.	1.40	1769	0.003	1.015
Astra	1.37	3591	3×10^{-7}	1.029
Atlas C.	1.40	2915	0.001	0.976
Autoliv	1.30	1690	9×10^{-5}	1.055
Electrolux	1.42	4577	0.1648	0.959
Ericsson	1.42	4576	1×10^{-6}	0.962
FSB	1.49	1470	0.1915	0.622
Gambro	1.41	2454	0.037	1.004
Holmen	1.42	4568	0.022	0.959
Industriv.	1.43	2061	0.200	0.905
Investor	1.42	4146	0.009	0.944
Nokia	1.41	2907	5×10^{-5}	0.993
OMG	1.43	2084	9×10^{-6}	0.962
Pharmacia	1.44	1370	0.339	0.826
Sandvik	1.38	4576	9×10^{-6}	1.037
Scania	1.38	1268	2×10^{-14}	1.173
Securitas	1.33	2461	9×10^{-5}	1.025
Skandia	1.42	4566	0.314	0.959
SEB	1.43	2984	0.003	0.957
Skanska	1.41	4337	0.173	0.984
SKF	1.43	4578	0.012	0.939
SSAB	1.45	2963	0.719	0.784
Stora	1.42	3263	0.197	0.964
SCA	1.39	4576	0.290	1.019
SHB	1.43	2612	5×10^{-5}	0.920
Sw. Match	1.21	1239	0.045	1.039
VOLVO	1.37	5324	2×10^{-8}	1.033

As a detailed example we consider the return series of SEB which is plotted in Figure 2. For this series we estimate the autocorrelation function of the squared observations from the data and compare them with the autocorrelations obtained by plugging the parameter estimates for the three estimated models into the theoretical expressions of the autocorrelations. Note that from the GARCH(1,1) model estimated for this series one obtains $\widehat{\gamma}_4 < 1$, so the fourth-moment condition is satisfied and we can discuss the autocorrelation function of squares of the GPARCH(1,1) model with some confidence. It is seen from Figure 3 that for all models the discrepancy between the autocorrelation functions and the autocorrelations estimated directly from data is large at small lags. For long lags, the gap between the two is much smaller for the GPARCH(δ,1,1) model than for the two other models. As already noted, augmenting the GARCH model by the power parameter δ, other things equal, reduces autocorrelations of squared observations compared to the two other models. This probably explains the results obtained by Brooks, Faff, McKenzie and Mitchell (2000) and McKenzie and Mitchell (2002).

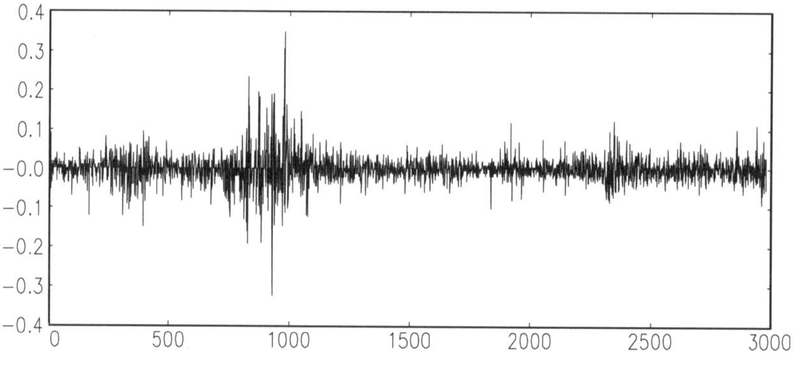

Fig. 2. Daily returns of the stock SEB, from 1989 to April 2001.

The present example shows that the estimated power parameter considerably improves the correspondence between the estimated autocorrelations on the one hand and the autocorrelation estimates from the model on the other. But then, the rapid decrease of the autocorrelations at first lags is not accounted for by any of the models; a higher-order model is required for the purpose. He and Teräsvirta (1999c) showed how a second-order GARCH model already can have an autocorrelation function of squared observations that is much more flexible

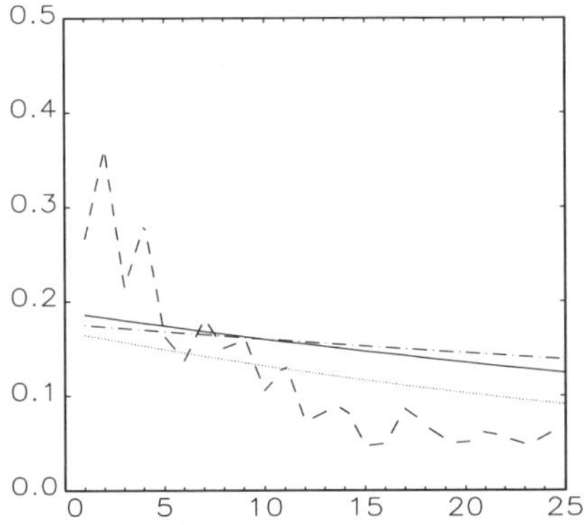

Fig. 3. Autocorrelation functions of squared observations, estimated for the SEB (dashed line), and computed from three estimated first-order GARCH models; GARCH (solid line), EGARCH (dashed-dotted line), and GPARCH (dotted line).

than the corresponding autocorrelation function for the GARCH(1,1) model.

6 Final Remarks

In this chapter we derive the autocorrelation structure of the logarithms of squared observations of a class of power ARCH processes and show that it may be obtained as a limiting case of a general power ARCH model. An interesting thing to notice is that the autocorrelation structure of the δth power of absolute-valued observations of this first-order GPARCH process is exponential for all GPARCH($\delta, 1, 1$) processes such that the 4δth fractional moment exists. This property is retained at the limit as the power parameter approaches zero, which means that the autocorrelation function of the process of logarithms of squared observations also decay exponentially. While this is true for the logarithmed squared observations of an LGARCH(1,1) or EGARCH(1,1) process it cannot simultaneously be true for the untransformed observations defined by these processes as shown in He, Teräsvirta and Malmsten (1999) for the EGARCH(1,1) case.

Conversely, if we have the original GARCH(1,1) [GPARCH(1,1,1)] process of Bollerslev (1986) with the autocorrelations of $\{\varepsilon_t^2\}$ decaying exponentially, the autocorrelation function of $\{\ln \varepsilon_t^2\}$ does not have this property. The practical value of these facts when discriminating between GARCH(1,1) and EGARCH(1,1) models is not clear, but they illustrate the theoretical differences in the higher-order dynamics between these two classes of models. Note that possible asymmetry is not an issue here. Nelson's EGARCH(1,1) model is a member of the limiting GPARCH(1,1) family independent of the value of the asymmetry parameter. Likewise, if the standard GARCH(1,1) process is generalized to an asymmetric GJR-GARCH(1,1) (Glosten, Jagannathan and Runkle (1993)) process the argument remains the same. This is because the GJR-GARCH model is still a member of the GPARCH(1,1,1) class; see Ding, Granger and Engle (1993) and He and Teräsvirta (1999b) for more discussion.

In order to explain the role of the power parameter we present a detailed analysis of how the autocorrelation function of ε_t^2 differ across members of the GPARCH(δ,1,1) models. We demonstrate that $\delta < 2$ reduces the autocorrelations of ε_t^2 (other things equal) compared to $\delta = 2$ (the GARCH model). This fact may explain the regularities in estimation results in papers in which GPARCH models have been fitted to stock return series. In an empirical example we show that the estimated power parameter considerably improves the correspondence between the estimated autocorrelations on the one hand and the autocorrelation estimates from the model on the other.

Proofs

A.1 Lemma 1

(i) We shall show that $\{\varepsilon_t\}$ defined in (1) and (2) is strictly stationary if $\gamma_\delta < 1$. Note that under $\gamma_\delta < 1$ (2) has a representation

$$h_t^{2\delta} = \alpha_0 + \sum_{i=1}^{\infty} c_\delta(z_{t-i}).$$

Since $\{c_\delta(z_t)\}$ is a sequence of iid and $Var(c_\delta(z_t))$ is finite, $\sum_{i=1}^{\infty} Var(c_\delta(z_{t-i})) < \infty$. It follows from Billingsley (1986) (Theorem 22.6) that $\{h_t^{2\delta}\}$ is finite almost surely. This, combined with Theorem 2.1 in Nelson (1991) and $\gamma_{2\delta} < 1$ implies that $\{\varepsilon_t\}$ in (1) and (2) is strictly stationary.

(ii) That $\{h_t^{4\delta}\}$ is finite almost surely follows by the fact that $\gamma_{2\delta} < 1$ and (2) is strictly stationary. Thus, the results in He and Teräsvirta (1999b) and He and Teräsvirta (1999c) apply and thus (9) and (10) hold.

(iii) It follows from Theorems 22.3 and 22.8 in Billingsley (1986) that if $\sum_{i=1}^{\infty} Var(c_\delta(z_{t-i})) = \infty$, then $\{h_t^{4\delta}\} = \infty$ almost surely. Thus the necessary condition (8) holds.

This completes the proof.

A.2 Lemma 2

(i) Similarly to (i) of Lemma 1, strict stationarity of $\{\ln \varepsilon_t^2\}$ in (3) and (4) follows from the fact that $|\beta| < 1$, and $Var(\ln z_t^2)^2$ and $Var(g(z_t))^2$ are finite.

(ii) As (ii) and (iii) in Lemma 1, under the assumptions of Lemma 2, $\{(\ln \varepsilon_t^2)^2\}$ is finite almost surely if and only $|\beta| < 1$.

(iii) To compute μ_0 under (12), we repeatedly apply (4) to $\ln h_t^2$, which yields

$$\ln h_t^2 = \alpha_0 \sum_{i=1}^{k+1} \beta^{i-1} + \sum_{i=1}^{k+1} \beta^{i-1} g(z_{t-1}) + \beta^{k+1} \ln h_{t-(k+1)}^2. \tag{A.1}$$

Taking expectations of both sides of (A.1) and letting $k \to \infty$ yield

$$\mathsf{E}(\ln h_t^2) = (\alpha_0 + \gamma_g)/(1-\beta). \tag{A.2}$$

Similarly, repeated application of

$$(\ln h_t^2)^2 = (\alpha_0 + g(z_{t-1}))^2 + 2\beta(\alpha_0 + g(z_{t-1})) \ln h_{t-1}^2 + \beta^2 (\ln h_{t-1}^2)^2$$

yields

$$(\ln h_t^2)^2 = \sum_{i=1}^{k} [\beta^{2(i-1)} (\alpha_0 + g(z_{t-i}))^2] + 2\beta \sum_{i=1}^{k} \beta^{2(i-1)} (\alpha_0 + g(z_{t-i}))(\ln h_{t-i}^2) + \beta^{2k} (\ln h_{t-k}^2)^2.$$

Thus, under (12) by letting $k \to \infty$ and taking expectations

$$\mathsf{E}(\ln h_t^2)^2 = [\alpha_0^2(1+\beta) + 2\alpha_0(1+\beta)\gamma_g + 2\beta\gamma_g^2 + (1-\beta)\gamma_{g^2}]/[(1-\beta)(1-\beta^2)]. \tag{A.3}$$

It follows from formulas (3), (A.2) and (A.3) that expression (13) is valid.

Next, consider the n-th order autocorrelation of $\{\ln \varepsilon_t^2\}$

$$\rho_n^0 = \frac{\mathsf{E}(\ln \varepsilon_t^2 \ln \varepsilon_{t-n}^2) - (\mathsf{E}(\ln \varepsilon_t^2))^2}{\mathsf{E}(\ln \varepsilon_t^2)^2 - (\mathsf{E}(\ln \varepsilon_t^2))^2}. \tag{A.4}$$

We have

$$\ln \varepsilon_t^2 \ln \varepsilon_{t-n}^2 = \ln z_t^2 \ln z_{t-n}^2 + \ln z_t^2 \ln h_{t-n}^2 + \ln h_t^2 \ln z_{t-n}^2 + \ln h_t^2 \ln h_{t-n}^2. \tag{A.5}$$

It follows from (A.1) that

$$\begin{aligned}(\ln h_t^2)(\ln h_{t-n}^2) = {} & \alpha_0 \sum_{i=1}^{n} \beta^{i-1} \ln h_{t-n}^2 + (\sum_{i=1}^{n} \beta^{i-1} g(z_{t-1})) \ln h_{t-n}^2 \\ & + \beta^n (\ln h_{t-n}^2)^2 \end{aligned} \tag{A.6}$$

and

$$\begin{aligned}\ln h_t^2 \ln z_{t-n}^2 = {} & \alpha_0 \sum_{i=1}^{n} \beta^{i-1} \ln z_{t-n}^2 + (\sum_{i=1}^{n-1} \beta^{i-1} g(z_{t-1})) \ln z_{t-n}^2 \\ & + \beta^{n-1} g(z_{t-n}) \ln z_{t-n}^2 + \beta^n \ln z_{t-n}^2 \ln h_{t-n}^2. \end{aligned} \tag{A.7}$$

The expectation of (A.5) is obtained by taking expectations of both sides of (A.6) and (A.7) and inserting them to (A.5). Applying this expectation to (A.4) yields (14).
This completes the proof.

A.3 Theorem

For the ease of exposition, write (2) as

$$h_t^{2\delta} = \alpha_0^* + c_\delta(z_{t-1}) h_{t-1}^{2\delta} \tag{A.8}$$

where $c_\delta(z_{t-1}) = \delta g(z_{t-1}) + \beta$. Following Ding, Granger and Engle (1993), decompose α_0^* as

$$\alpha_0^* = (1 - \gamma_\delta)\omega^\delta \tag{A.9}$$

where $\gamma_\delta = \delta \gamma_g + \beta$ and $\omega^\delta = \mathsf{E} h_t^{2\delta}$, $\omega > 0$. Rewrite (A.8) as

$$\begin{aligned}(h_t^{2\delta} - 1)/\delta = {} & (\alpha_0^* + \beta - 1)/\delta + g(z_{t-1}) h_{t-1}^{2\delta} \\ & + \beta(h_{t-1}^{2\delta} - 1)/\delta. \end{aligned} \tag{A.10}$$

Insert (A.9) into (A.10) and let $\delta \to 0$ on both sides of (A.10). Then, by l'Hôpital's rule (A.10) converges to (4). In particular,

$$(\alpha_0^* + \beta - 1)/\delta \rightarrow \alpha_0, \tag{A.11}$$

where $\alpha_0 = (1 - \beta)(\mathsf{E} \ln h_t^2) - \gamma_g$ is the constant term in (4). Besides, from (A.9) we have, as $\delta \rightarrow 0$,

$$\alpha_0^* \rightarrow 1 - \beta. \tag{A.12}$$

The convergence results (A.11) and (A.12) are used to prove the following results.

(i) We shall show that $\mu_{4\delta} \rightarrow \mu_0$ as $\delta \rightarrow 0$ under the Box-Cox transformation. From Lemma 1 we obtain

$$\mu_{2\delta} = \mathsf{E}\varepsilon_t^{2\delta} = \frac{\alpha_0^* \nu_{2\delta}}{1 - \gamma_\delta}. \tag{A.13}$$

From (A.13) it follows for the Box-Cox transformed $\varepsilon_t^{2\delta}$ that

$$\mathsf{E}\varphi_\delta(\varepsilon_t^2) = \frac{[\alpha_0^*(\nu_{2\delta} - 1) + (\alpha_0^* + \beta - 1)]/\delta + \gamma_g}{1 - \gamma_\delta}. \tag{A.14}$$

Letting $\delta \rightarrow 0$ on both sides of (A.14) and applying (A.11) and (A.12) to the right-hand side of (A.14) gives

$$\mu_2^0 = \mathsf{E} \ln \varepsilon_t^2 = \frac{\gamma_{\ln z^2}(1 - \beta) + (\alpha_0 + \gamma_g)}{1 - \beta}. \tag{A.15}$$

From (9) it follows that

$$\mathsf{E}(\varphi_\delta(\varepsilon_t^2))^2 = \left[\frac{\alpha_0^{*2}\nu_{4\delta}(1 + \gamma_\delta)}{(1 - \gamma_\delta)(1 - \gamma_{2\delta})} - 2\mathsf{E}\varepsilon_t^{2\delta} + 1\right]/\delta^2. \tag{A.16}$$

Applying (A.13) to the right-hand side of expression (A.16) it is seen that (A.16) is equivalent to

$$\begin{aligned}\mathsf{E}(\varphi_\delta(\varepsilon_t^2))^2 = \frac{1}{(1 - \gamma_\delta)(1 - \gamma_{2\delta})} & \left\{\frac{1}{\delta^2}\left[\alpha_0^{*2}\nu_{4\delta}(1 + \beta) - 2\alpha_0^*\nu_{4\delta}(1 - \beta^2)\right.\right. \\ & \left. +1 - \beta - \beta^2 + \beta^3\right] + \frac{1}{\delta}\left[\alpha_0^{*2}\nu_{4\delta}\gamma_g + 4\alpha_0^*\beta\nu_{2\delta}\gamma_g\right. \\ & \left. -\gamma_g(1 + 2\beta - 3\beta^2)\right] + \left[2\alpha_0^*\beta\nu_{2\delta}\gamma_{g^2} - \gamma_{g^2}(1 - \beta)\right. \\ & \left.\left. +2\beta\gamma_g^2\right] + \delta\gamma_g\gamma_{g^2}\right\}. \end{aligned} \tag{A.17}$$

Note that, as $\delta \rightarrow 0$, $\mathsf{E}(\varphi_\delta(\varepsilon_t^2))^2 \rightarrow \mathsf{E}(\ln \varepsilon_t^2)^2$, $(\nu_{4\delta} - 2\nu_{2\delta} + 1)/\delta^2 \rightarrow \mathsf{E}(\ln z_t^2)^2$ and $(\nu_{2\delta} - 1)/\delta \rightarrow \mathsf{E}(\ln z_t^2)$. Apply those facts and (A.11) and (A.12) to the right-hand side of (A.17) while letting $\delta \to 0$ on both sides of (A.17). It follows from l'Hôpital's rule that (A.17) converges to

$$\mu_0 = \mathsf{E}(\ln \varepsilon_t^2)^2 = \frac{\Delta}{(1-\beta)(1-\beta^2)}. \tag{A.18}$$

Then $\mu_{4\delta} \to \mu_0$ holds in (15).

(ii) We shall now prove that $\lim_{\delta\to 0} \rho_n(\delta) = \rho_n^0$. Since $\lim_{\delta\to 0} \rho_n(\delta) = \lim_{\delta\to 0} \rho_1(\delta)\gamma_\delta^{n-1} = \beta^{n-1} \lim_{\delta\to 0} \rho_1(\delta)$, we have to prove that $\lim_{\delta\to 0} \rho_1(\delta) \to \rho_1^0$.

Let $\rho_1(\delta) = u/v$ in (10) where $u = \nu_{2\delta}[\overline{\gamma}_\delta(1-\gamma_\delta^2) - \nu_{2\delta}\gamma_\delta(1-\gamma_{2\delta})]$ and $v = \nu_{4\delta}(1-\gamma_\delta^2) - \nu_{2\delta}^2(1-\gamma_{2\delta})$. Since $\lim_{\delta\to 0} u = 0$ and $\lim_{\delta\to 0} v = 0$ we need to apply l'Hôpital's rule in order to obtain $\lim_{\delta\to 0} \rho_1(\delta)$. Note that

$$\begin{aligned}\frac{\partial}{\partial\delta} u = {} & (\nu_{2\delta} - \beta^2\nu_{2\delta} - \nu_{2\delta}^2 + \beta^2\nu_{2\delta}^2)\overline{\gamma}_g \\ & + \delta\frac{\partial}{\partial\delta}(\nu_{2\delta} - \beta^2\nu_{2\delta} - \nu_{2\delta}^2 + \beta^2\nu_{2\delta}^2)\overline{\gamma}_g \\ & + \frac{\partial}{\partial\delta}(-\delta^3\nu_{2\delta}\overline{\gamma}_g\gamma_g^2 - \delta^2\beta\nu_{2\delta}^2\gamma_g^2 - 2\delta^2\beta\nu_{2\delta}\overline{\gamma}_g\gamma_g \\ & + \delta^3\nu_{2\delta}^2\gamma_g^2\gamma_{g^2} + \delta^2\beta\nu_{2\delta}^2\gamma_{g^2} + 2\delta^2\beta\gamma_g^2)\end{aligned}$$

and

$$\begin{aligned}\frac{\partial}{\partial\delta} v = {} & \frac{\partial}{\partial\delta}(\nu_{4\delta}(1 - \delta^2\gamma_g^2 - 2\delta\beta\gamma_g - \beta^2) \\ & - \nu_{2\delta}^2(1 - \delta^2\gamma_{g^2} - 2\delta\beta\gamma_g - \beta^2))\end{aligned}$$

imply that $\lim_{\delta\to 0} \frac{\partial}{\partial\delta} u = 0$ and $\lim_{\delta\to 0} \frac{\partial}{\partial\delta} v = 0$. Thus we have to calculate $\frac{\partial^2}{\partial\delta^2} u$ and $\frac{\partial^2}{\partial\delta^2} v$. We obtain

$$\begin{aligned}\frac{\partial^2}{\partial\delta^2} u = {} & \frac{\partial}{\partial\delta}[(\nu_{2\delta} - \beta^2\nu_{2\delta} - \nu_{2\delta}^2 + \beta^2\nu_{2\delta}^2)\overline{\gamma}_g] \\ & + \frac{\partial}{\partial\delta}(\nu_{2\delta}\overline{\gamma}_g - \beta^2\nu_{2\delta}\overline{\gamma}_g - \nu_{2\delta}^2\overline{\gamma}_g + \beta^2\nu_{2\delta}^2\overline{\gamma}_g) \\ & + \delta\frac{\partial^2}{\partial\delta^2}(\nu_{2\delta}\overline{\gamma}_g - \beta^2\nu_{2\delta}\overline{\gamma}_g - \nu_{2\delta}^2\overline{\gamma}_g + \beta^2\nu_{2\delta}^2\overline{\gamma}_g) \\ & + \frac{\partial^2}{\partial\delta^2}(-\delta^3\nu_{2\delta}\overline{\gamma}_g\gamma_g^2 - \delta^2\beta\nu_{2\delta}^2\gamma_g^2 - 2\delta^2\beta\nu_{2\delta}\overline{\gamma}_g\gamma_g \\ & + \delta^3\nu_{2\delta}^2\gamma_g^2\gamma_{g^2} + \delta^2\beta\nu_{2\delta}^2\gamma_{g^2} + 2\delta^2\beta\gamma_g^2)\end{aligned}$$

and

$$\frac{\partial^2}{\partial\delta^2}v = (1-\delta^2\gamma_g^2 - 2\delta\beta\gamma_g - \beta^2)(\frac{\partial^2}{\partial\delta^2}\nu_{4\delta})$$
$$-4(\delta\gamma_g^2 - \beta\gamma_g)(\frac{\partial}{\partial\delta}\nu_{4\delta}) - 2\gamma_g^2\nu_{4\delta}$$
$$-2\nu_\delta(1-\delta^2\gamma_{g^2} - 2\delta\beta\gamma_g - \beta^2)(\frac{\partial^2}{\partial\delta^2}\nu_{2\delta})$$
$$-2(1-\delta^2\gamma_{g^2} - 2\delta\beta\gamma_g - \beta^2)(\frac{\partial}{\partial\delta}\nu_{2\delta})^2$$
$$+4\nu_\delta(\delta\gamma_{g^2} - \beta\gamma_g)(\frac{\partial}{\partial\delta}\nu_{2\delta}) + 2\nu_{2\delta}^2\gamma_{g^2}.$$

Note that

$$\lim_{\delta\to 0}\frac{\partial}{\partial\delta}\nu_{2\delta} = \lim_{\delta\to 0}(\frac{\partial}{\partial\delta}\mathsf{E}z_t^{2\delta}) = \lim_{\delta\to 0}\int\frac{\partial}{\partial\delta}x^{2\delta}f(x)dx$$
$$= \int\lim_{\delta\to 0}x^{2\delta}(\ln x^2)f(x)dx = \mathsf{E}(\ln z_t^2). \qquad (A.19)$$
$$\lim_{\delta\to 0}\frac{\partial^2}{\partial\delta^2}\nu_{2\delta} = \lim_{\delta\to 0}(\frac{\partial^2}{\partial\delta^2}\mathsf{E}z_t^{2\delta}) = \lim_{\delta\to 0}\int\frac{\partial^2}{\partial\delta^2}x^{2\delta}f(x)dx$$
$$= \int\lim_{\delta\to 0}x^{2\delta}(\ln x^2)^2f(x)dx = \mathsf{E}(\ln z_t^2)^2. \qquad (A.20)$$

$$\lim_{\delta\to 0}\frac{\partial}{\partial\delta}\overline{\gamma}_g = \lim_{\delta\to 0}\frac{\partial}{\partial\delta}\mathsf{E}(z_t^{2\delta}g(z_t)) = \lim_{\delta\to 0}\int\frac{\partial}{\partial\delta}(x^{2\delta}g(x))f(x)dx$$
$$= \int\lim_{\delta\to 0}(x^{2\delta}g(x)\ln x^2)f(x)dx = \mathsf{E}(g(z_t)\ln z_t^2). \qquad (A.21)$$

Applying (A.19) - (A.21) gives $\lim_{\delta\to 0}\frac{\partial^2}{\partial\delta^2}u$ and $\lim_{\delta\to 0}\frac{\partial^2}{\partial\delta^2}v$, respectively. We see that $\rho_1(\delta)\to\rho_1^0$ as $\delta\to 0$.

This completes the proof.

A.4 Lemma 3

Consider the LGARCH (1,1) model (1) and (7) and assume z_t's to be identically distributed following $N(0,1)$. Then the autocorrelation function of squared observations has the form

$$\rho_n = \frac{\frac{2\Gamma(1+0.5\alpha(\alpha+\beta)^{n-1})}{B(1+0.5\alpha(\alpha+\beta)^{n-1},0.5)} \prod_{i=1}^{n-1} \frac{\Gamma_{1i}(.)}{B_{1i}(.)} \prod_{i=1}^{\infty} \frac{\Gamma_{2i}(.)}{B_{2i}(.)} - \prod_{i=1}^{\infty} (\frac{\Gamma_{1i}(.)}{B_{1i}(.)})^2}{3\prod_{i=1}^{\infty} \frac{\Gamma_{3i}(.)}{B_{3i}(.)} - \prod_{i=1}^{\infty} (\frac{\Gamma_{1i}(.)}{B_{1i}(.)})^2} \tag{A.22}$$

where

$$\frac{\Gamma_{1i}(.)}{B_{1i}(.)} = \frac{\Gamma(0.5\alpha(\alpha+\beta)^{i-1})}{B(0.5\alpha(\alpha+\beta)^{i-1},0.5)}$$

$$\frac{\Gamma_{2i}(.)}{B_{2i}(.)} = \frac{\Gamma(0.5\alpha(1+(\alpha+\beta)^{n})(\alpha+\beta)^{i-1})}{B(0.5\alpha(1+(\alpha+\beta)^{n}(\alpha+\beta)^{i-1},0.5)}$$

$$\frac{\Gamma_{3i}(.)}{B_{3i}(.)} = \frac{\Gamma(\alpha(\alpha+\beta)^{i-1})}{B(\alpha(\alpha+\beta)^{i-1},0.5)}$$

and $\Gamma(\cdot)$ and $B(\cdot)$ are the Gamma function and the Beta function, respectively.

Proof:

See He, Teräsvirta and Malmsten (2002).
This completes the proof.

References

Billingsley P (1986) Probability and Measure. Wiley, New York

Bollerslev T (1986) Generalized Autoregressive Conditional Heteroskedasticity. Journal of Econometrics 31:307–327

Breidt FJ, Crato N, de Lima P (1998) The Detection and Estimation of Long Memory in Stochastic Volatility. Journal of Econometrics 83:325–348

Brooks R, Faff RW, McKenzie MD, Mitchell H (2000) A Multi-Country Study of Power ARCH Models and National Stock Market Returns. Journal of International Money and Finance 19:377–397

Ding Z, Granger CWJ, Engle RF (1993) A Long Memory Property of Stock Market Returns and a New Model. Journal of Empirical Finance 1:83–106

Geweke J (1986) Modelling Persistence of Conditional Variances: Comment. Econometric Reviews 5:57–61

Glosten L, Jagannathan R, Runkle D (1993) On the Relation Between Expected Value and the Volatility of the Nominal Excess Return on Stocks. Journal of Finance 48:1779–1801

He C, Teräsvirta T (1999a) Fourth Moment Structure of the GARCH(p,q) Process. Econometric Theory 15:824–846

He C, Teräsvirta T (1999b) Properties of Moments of a Family of GARCH Processes. Journal of Econometrics 92:173–192

He C, Teräsvirta T (1999c) Properties of the Autocorrelation Function of Squared Obeservations for Second-Order GARCH Processes under Two Sets of Parameter Constraints. Journal of Time Series Analysis 20:23–30

He C, Teräsvirta T (1999d) Statistical Properties of the Asymmetric Power ARCH Process. In: Engle RF and White H (eds) Cointegration, Causality, and Forecasting. Festschrift in Honour of Clive W J Granger. Oxford University Press, 462–474

He C, Teräsvirta T, Malmsten H (1999) Fourth moment structure of a family of first-order exponential GARCH models. Working Paper Series in Economics and Finance, No. 345, Stockholm School of Economics

He C, Teräsvirta T, Malmsten H (2002) Moment Structure of a Family of First-Order Exponential GARCH Models. Econometric Theory 18:868–885

Hentschel L (1995) All in the Family. Nesting Symmetric and Asymmetric GARCH Models. Journal of Financial Economics 39:71–104

Malmsten H, Teräsvirta (2004) Stylized Facts of Financial Time Series and Three Popular Models of Volatility. Working Paper Series in Economics and Finance, No. 563, Stockholm School of Economics

Malmsten H (2004) Evaluating Exponential GARCH Models. Working Paper Series in Economics and Finance, No. 564, Stockholm School of Economics

McKenzie MD, Mitchell H (2002) Generalized Asymmetric Power ARCH Modelling of Exchange Rate Volatility. Applied Financial Economics 12:555–564

Nelson DB (1991) Conditional heteroskedasticy in asset returns: a new approach. Econometrica 59:347–370

Pantula SG (1986) Modelling Persistence of Conditional Variances: Comment. Econometric Reviews 5:71–74

Tse YK, Tsui AKC (1997) Conditional Volatility in Foreign Exchange Rates: Evidence from the Malaysian Ringgit and Singapore Dollar. Pacific-Basin Finance Journal 5:345–356

Regression Calibration for Cox Regression Under Heteroscedastic Measurement Error — Determining Risk Factors of Cardiovascular Diseases from Error-prone Nutritional Replication Data

Thomas Augustin[1], Angela Döring[2] and David Rummel[3]

[1] Department of Statistics, University of Munich, Ludwigstrasse 33, 80539 Munich, Germany thomas@stat.uni-muenchen.de
[2] GSF-National Research Center for Environment and Health, Ingolästter Landstrasse 1, 85764 Neuherberg, Germany doering@gsf.de
[3] emnos GmbH, Theresienhöhe 12, 80339 Munich, Germany d_rummel@web.de

1 Introduction

A widespread problem in applying regression analysis is the presence of data deficiency. In most surveys a not negligible proportion of data is missing, and sophisticated methods are needed to avoid severely biased estimation. Reviews on this important topic are provided, in particular, by Rao et al. (2008, Chapter 8), Little and Rubin (2002), Toutenburg et al. (2002) and Toutenburg, Fieger and Heumann (2000). Recent developments include, for instance, Toutenburg and Srivastava (1999) and Toutenburg and Srivastava (2004), who discuss corrected estimation of population characteristics from partially incomplete survey data. Toutenburg and Shalabh (2001), Heumann (2004), Shalabh and Toutenburg (2005), Toutenburg and Shalabh (2005), Toutenburg et al. (2006), Toutenburg et al. (2005) and Toutenburg, Srivastava and Shalabh (2006) provide neat methods for handling missing data in linear and nonlinear regression models, while, among others, Strobl, Boulesteix and Augustin (2007) and Svejdar et al. (2007) are concerned with classification under missing data.[1]

[1] A completely different paradigm to handle missing data has been developed, among others, by Horowitz and Manski (2000), Manski and Tamer (2002) Zaffalon (2002), Manski (2003), Zaffalon and de Cooman (2004), Zaffalon M (2005)

But even if data is recorded completely, or correction procedures for missing data have been applied, a similar problem remains, since the data at hand do often not exactly convey the information in which one is interested: Frequently, the variables of material interest, called *ideal variables* or *gold standard*, cannot be observed directly or measured correctly, and one has to be satisfied with *surrogates* (often also named *indicators* or *proxies*), i.e., with somehow related, but different variables (see also Figure 1). If one ignores the difference between the ideal variables in the model and the observable variables and just plugs in the surrogates instead of the variables ('naive estimation'), then all the estimators must be suspected to be severely biased. *Error-in-variables* modeling, also called *measurement error* modeling, provides a methodology.[2] In such cases it develops procedures to adjust for the measurement error based on an error model describing the relation between ideal variables and surrogates. Recent surveys, also containing many examples from different fields of application, include Cheng and Van Ness (1999) and Wansbeek and Meijer (2000), who mainly concentrate on linear models, and Stefanski (2000), Caroll et al. (2006), Van Huffel et al. (2002), Schneeweiß and Augustin (2006), who are concerned with non-linear models.

Some recent developments on measurement error correction are concerned with linear models (see, e.g., Shalabh (2001a), Shalabh (2003), Shalabh (2001b)), polynomial regression (like Cheng and Schneeweiß (1998)), Cheng et al. (2000), Huang and Huwang (2001), Kuha and Temple (2003), Kukush et al. (2004), Kukush and Schneeweiss (2005a), Shklyar, Schneeweiss and Kukush (2007) and generalized linear models (cp., for instance, Kukush, Schneeweiss and Wolf (2004), Kukush and Schneeweiss (2005b), Shklyar and Schneeweiss (2005),

and Utkin and Augustin (2007). They advocate in favor of a cautious, but reliable, handling of missing data by considering a *set* of models, namely the set of all models being compatible with potential observations of the missing values. Then the theory of interval probability or imprecise probability (Kuznetsov (1991),Walley (1991), Weichselberger (2001)) is used for statistical analysis and decision making.

[2] Typically the terms 'measurement error' and 'error-in-variables' are applied to continuous data only. The corresponding problem for categorical data is usually termed 'misclassification', see, in particular, Küchenhoff et al. (2006) and Küchenhoff et al. (2007), for a recently developed method and further references on this topic. For handling deliberately contaminated data for purposes of data disclosure and anonymity see, e.g., Schmid (2006); Schmid et al. (2007), Ronning (2005), and the references therein; for investigations on measurement error arising from heaping and rounding data see, e.g., Ahmad (2006), Augustin and Wolff (2004), Wolff and Augustin (2003).

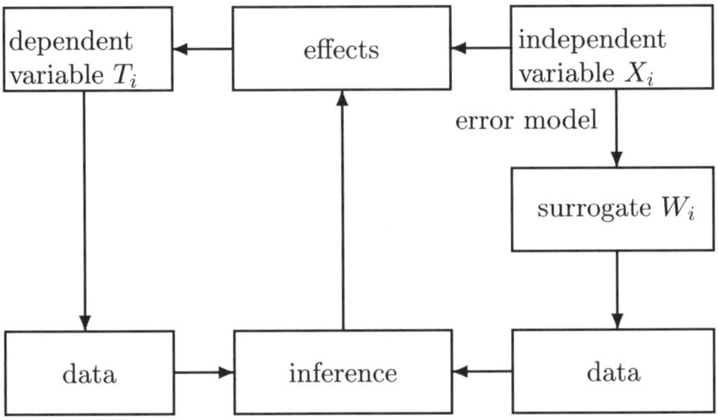

Fig. 1. Regression under covariate measurement error

Schneeweiss and Cheng (2006), Heid et al. (2006)). Also in the Cox model, which is considered here, covariate measurement error correction has become an area of intensive and fruitful research in the last decade: Next to the regression calibration approaches listed below, so-called functional approaches are developed and extended, e.g., in Buzas (1998), Kong (1999), Kong and Gu (1999), Huang and Wang (2000), Li and Lin (2003), Hu et al. (2002), Augustin (2004), Hu and Lin (2004), Huang and Wang (2006), Yi and Lawless (2007) and Martin-Magniette and Taupin (2006), while, for instance, Hu et al. (1998), Pepe, Self and Prentice (1989), Bender, Augustin, Bletter (2005), Dupuy (2005) and Rummel et al. (2007) consider the so-called structural model and the Berkson error; see also the survey and comparison of basic approaches, including the classical work by Prentice (1982) and Nakamura (1992), in Augustin and Schwarz (2002) and Liu et al. (2004).

From the practical point of view, it should be stressed explicitly that the topic of measurement error is not simply a matter of sloppy research; quite often the 'true value' is unascertainable *eo ipso*. A typical example, which also motivates the present contribution, is the recording of protein intakes in surveys on the influence of eating habits on certain diseases. Though much attention is paid to the high quality of the questionnaire and the subsequent procedures, a considerable random distortion in the data cannot be avoided. Below we reanalyze data from the WHO MONICA Augsburg substudy on dietary intake, see Döring and Kußmaul (1997), Winkler et al. (1991).[3] This study, which is embedded into the WHO MONICA project (MONItoring of trends and determinants in Cardiovascular disease), is concerned with the question whether changes in dietary intake can explain trends in the incidence and mortality of cardiac infarctions.[4]

Since severe error is present in the measurements of animal and plant protein intake from a seven day food diary, naively applying Cox regression to answer this question could lead to wrong conclusions. As a first approach to adjust for the measurement error, we rely on a variant of the regression calibration approach, which is one of the most universal methods to correct for measurement error (see Caroll et al. (2006, Chapter 4) for a general description). Its basic idea is to run a standard analysis where the unobservable variables are replaced by values predicted from the observable ones. For Cox regression, regression calibration type methods were introduced by Prentice (1982) and were studied and developed further in Pepe, Self and Prentice (1989), Clayton (1991), Hughes (1993), Wang et al. (1997) and Xie et al. (2001).

Here we adapt and extend this method taking into account three general methodological issues, which also deserve special attention in the data analyzed below:

- *Heteroscedastic* measurement error. Recent research in nutritional epidemiology strongly suggests that the measurement error must be expected to vary considerably among the different study participants (cf., e.g., Willett (1998, pp. 33-48)).
- The presence of *replication data.* The protein intake measurements are based on diaries, where all food intake had to be recorded in great detail for seven days. Taking for every individual the errors in these measurements as independently and identically distributed gives us the opportunity to estimate the error variances.
- The *non-linearity* of the influence. Previous studies showed that the effect of protein intake on morbidity and mortality could be nonlinear: both types of extreme intakes, very high as well as very

[3] The MONICA Augsburg study is currently continued as the KORA study (Cooperative health research in the area of Augsburg).

[4] The quality of Swedish nutrition data was investigated in Johansson et al. (2002), where the reproducibility of food frequency measurements of a sample of respondents to the Swedish MONICA study was considered. It may be mentioned that, if such local studies are combined and compared, additional measurement error arises: It is quite important to take into account the variation in these aggregated observations (cf. Kulathinal et al. (2002)).

low intakes, could be detrimental, and so it is of great importance to work with quadratic predictors. While introducing non-linearity in the covariates does not encounter much difficulty in the error-free situation, under measurement error it is often very difficult to handle non-linear terms. (For the problems already arising in the linear polynomial model see, e.g., Cheng and Schneeweiß (1998). For some models a general result (Stefanski (1989, Theorem 1)) can be used to prove even the non-existence of a so-called corrected score function.)

As shown below, the convenience of regression calibration is maintained in this extended setting; still the core parts of the estimation can be done by standard software packages. Applying this correction method shows a complex relationship between naive and corrected estimates. After having adjusted for measurement error, some of the estimates change substantially, others do not. Sometimes there is a high deattenuation, sometimes the absolute values even get smaller. Since, however, regression and calibration is known to be only an approximative correction method, reducing the bias but not necessarily producing consistent estimators, we understand our analysis more as an illustration of methodological issue and as a motivation for further research than as the last word on the topic.

The paper is organized as follows: The next section describes our modeling of the replication data. Section 3 adapts the idea of regression calibration to replication data and to quadratic predictors. The application to the MONICA data is reported in Section 4, while Section 5 concludes by sketching some topics for further research.

2 Survival Data with Replicated Covariate Measurements

The basic setup is described in this section , followed by a brief detour on systematic measurement error.

2.1 The Main Setting

Let n be the sample size and $T_1, \ldots, T_n$ the lifetimes, which may be subject to noninformative independent censorship in the sense of, e.g., Kalbfleisch and Prentice (2002). For every $i = 1, \ldots, n$ we split the vector of covariates into a vector X_i and a vector Z_i. All error-prone variables are collected in X_i, while Z_i consists of the correctly measured variables. Let all elements of X_i be measured on a metrical scale, Z_i may contain metrical and categorical covariates in 0/1-coding. Both types of covariates should be not time-varying. With the application below in mind, we additionally consider another vector, denoted by $X_i^{②}$, which contains the squared elements of X_i.

We assume that Cox's (Cox (1972)) proportional hazard model describes the relationship between the lifetimes and the covariates; the individual hazard rate $\lambda(t|X_i, Z_i)$ has the form

$$\lambda(t|X_i, Z_i) = \lambda_0(t) \exp\left(\beta_1' X_i + \beta_2' X_i^{②} + \beta_Z' Z_i\right), \tag{1}$$

with the unspecified baseline hazard rate $\lambda_0(t)$ and the regression parameter vector $\beta = (\beta_1', \beta_2', \beta_Z')'$.

For X_i, i.e., plant and animal protein in the application discussed below, replicated measurements $W_{i1}, \ldots, W_{ik}$, $k > 1$ (later on, k=7) are available for every unit i. We assume them to follow the additive error model

$$W_{ij} = X_i + U_{ij}, \qquad j = 1, \ldots, k, \quad i = 1, \ldots, n, \tag{2}$$

and make the usual assumptions. The errors (U_{ij}), $j = 1, \ldots, k$, $i = 1, \ldots, n$, have zero mean (see Section 2.2) and are independent among each other as well as of $X_1, \ldots, X_n$ and $T_1, \ldots, T_n$. It will prove important to allow for heteroscedasticity of the errors, where, for i fixed, $U_{i1}, \ldots, U_{ik}$ are i.i.d., but the covariance matrix Σ_i may vary among the units $i = 1, \ldots, n$. The common covariance matrix in the homoscedastic case will be denoted by Σ.

In a naive analysis, for every unit i, the individual average

$$\overline{W}_i := \frac{1}{k} \sum_{j=1}^{k} W_{ij} \tag{3}$$

would function as the surrogate for X_i. Additionally defining

$$\overline{U}_i := \frac{1}{k} \sum_{j=1}^{k} U_{ij} \tag{4}$$

leads us back to the classical error model

$$\overline{W}_i = X_i + \overline{U}_i, \quad i = 1, \ldots, n, \tag{5}$$

with $\mathbb{E}(\overline{U}_i) = 0$ and $\mathbb{V}(\overline{U}_i) = \frac{1}{k}\Sigma_i$. The particular attractiveness of replication data is based on the fact that the measurement error variances can be estimated from the data.[5] Therefore, in contrast to most cases relying on the classical error model, it is possible here to avoid additional assumptions, which are quite often difficult to justify.

2.2 A Note on Systematic Measurement Error

Before addressing the main topic, the assumption $\mathbb{E}(U_{ij}) = 0$ deserves some attention. If it is violated, i.e., if systematic measurement error with $\mathbb{E}(U_{ij}) = a \neq 0$ of unknown size a is present, then it becomes important to distinguish whether the covariates act merely linearly or also in a nonlinear way. In order to bring out this point most clearly, concentrate on the following special case: X_i is one-dimensional, there are no error-free covariates Z_i, and there is only a deterministic error a so that (5) reads as

$$\overline{W}_i = X_i + a, \quad i = 1, \dots, n.$$

In the case of no quadratic influence, where in (1) the parameter β_2 in (1) is set to zero and then (1) can be written as

$$\lambda_0(t)\exp(\beta_1\overline{W}_i) = \lambda_0(t)\exp(\beta_1 a + \beta_1 X_i) =: \lambda_0^*(t)\exp(\beta_1 X_i)\,. \qquad (6)$$

Therefore, the naive partial likelihood estimator based on replacing X_i by $\overline{W}_i$ still estimates β_1 consistently, and a bias only occurs in the estimation of $\lambda_0(t)$, where the naive standard methods estimate $\lambda_0^*(t) = \lambda_0(t)\exp(\beta_1 a)$ instead of $\lambda_0(t)$ itself. If, however, quadratic terms are taken into account, then we have to consider

$$\begin{aligned} &\lambda_0(t)\exp(\beta_1\overline{W}_i + \beta_2\overline{W}_i^2) \\ &= \lambda_0(t)\exp(\beta_1 a + \beta_1 X_i + \beta_2 X_i^2 + 2\beta_2 a X_i + \beta_2 a^2) \qquad (7) \\ &=: \lambda_0^{**}(t)\exp(\beta_1 X_i + \beta_2 X_i^2 + 2\beta_2 a X_i)\,, \end{aligned}$$

and also inconsistencies in the estimation of the regression parameters must be expected.

3 Regression Calibration Under Replication Data

In this section, the regression calibration approach is elaborated in the context of the intended application.

[5] See Shalabh (2003) for a sophisticated method where replication data are directly used to construct consistent estimator in linear models under measurement error.

3.1 The Basic Concept

Regression calibration (cf., in particular, Caroll et al. (2006, Chapter 4)) is an universally applicable, easy-to-handle method to adjust for measurement error. The main idea is to utilize the surrogate $\overline{W}_i$ from (3), together with the error-free variable Z_i, to predict the corresponding value of the unobservable variable X_i, and then to proceed with a standard analysis where X_i is replaced by its prediction $\widehat{X_i}$.

Applying this concept, the vector $(X_i', Z_i')'$ of covariates is assumed to be i.i.d., with unknown mean vector $(\mu_X', \mu_Z')'$ and unknown covariance matrix

$$\begin{pmatrix} \Sigma_{X,X} & \Sigma_{X,Z} \\ \Sigma'_{X,Z} & \Sigma_{Z,Z} \end{pmatrix}.$$

Based on Relation (5), the best linear prediction of X_i given $\overline{W_i}$ and Z_i is (cf., e.g., Rao et al. (2008))

$$\widehat{X_i} = \mu_X + (\Sigma_{X,X}\ \Sigma_{X,Z}) \begin{pmatrix} \Sigma_{X,X} + \frac{1}{k}\Sigma_i & \Sigma_{X,Z} \\ \Sigma'_{X,Z} & \Sigma_{Z,Z} \end{pmatrix}^{-1} \begin{pmatrix} \overline{W}_i - \mu_X \\ Z_i - \mu_Z \end{pmatrix}. \tag{8}$$

If additionally X_i, U_i and Z_i are Gaussian then (8) is exactly the conditional expectation of X_i given W_i and Z_i.

As mentioned above, replication data play an important role in measurement error modeling. Here they are used to estimate all nuisance parameters in (8), i.e., the parameters μ_X, μ_Z, Σ_X and Σ_Z of the distribution of $(X_i', Z_i')'$ as well as the measurement error variances Σ_i, from the data. We firstly adopt the procedure for the homoscedastic case ($\Sigma_i \equiv \Sigma$), taken from Caroll et al. (2006, Chapter 4.4.2), and then discuss the generalization to the heteroscedastic case.

3.2 The Case of Homoscedastic Measurement Error

Following (5) the overall mean is suggested as

$$\overline{W} := \frac{1}{n}\sum_{i=1}^{n} \overline{W}_i \tag{9}$$

which is an unbiased estimator for μ_X; analogously μ_Z is estimated by $\overline{Z} := \frac{1}{n}\sum_{i=1}^{n} Z_i$.

In order to derive the estimators for other parameters, it is illuminating to embed the situation under homoscedastic measurement error into the theory of design of experiments. Then (2) is reinterpreted as a one-factorial model with a random effect (e.g., Toutenburg (2002, pp. 147-150),), yielding the estimators

$$\widehat{\Sigma} = \frac{1}{n(k-1)} \sum_{i=1}^{n} \sum_{j=1}^{k} \left(W_{ij} - \overline{W}_i\right) \left(W_{ij} - \overline{W}_i\right)', \tag{10}$$

$$\widehat{\Sigma}_{X,X} = \left(\frac{1}{n-1} \sum_{i=1}^{n} \left(\overline{W}_i - \overline{W}\right) \left(\overline{W}_i - \overline{W}\right)' \right) - \frac{1}{k} \widehat{\Sigma}, \tag{11}$$

$$\widehat{\Sigma}_{X,Z} = \frac{1}{n-1} \sum_{i=1}^{n} \left(\overline{W}_i - \overline{W}\right) \left(Z_i - \overline{Z}\right)', \tag{12}$$

$$\widehat{\Sigma}_{Z,Z} = \frac{1}{n-1} \sum_{i=1}^{n} \left(Z_i - \overline{Z}\right) \left(Z_i - \overline{Z}\right)'. \tag{13}$$

3.3 The Case of Heteroscedastic Measurement Error

Under heteroscedastic measurement error, the relation

$$\begin{aligned} \mathbb{V}(U_{ij}) &= \mathbb{V}(U_{ij}|X_i) = \mathbb{V}(X_i|X_i) + \mathbb{V}(U_{ij}|X_i) = \mathbb{V}(X_i + U_{ij}|X_i) \\ &= \mathbb{V}(W_{ij}|X_i) \end{aligned}$$

plays a central role. It provides

$$\widehat{\Sigma}_i = \frac{1}{(k-1)} \sum_{j=1}^{k} \left(W_{ij} - \overline{W}_i\right) \left(W_{ij} - \overline{W}_i\right)', \quad i = 1, \ldots, n, \tag{14}$$

as an estimator for the error covariance matrices Σ_i at the individual level. $\Sigma_{X,Z}$ and $\Sigma_{Z,Z}$ are estimated in the same way as in (12) and in (13). To get an idea how to estimate $\Sigma_{X,X}$, it is helpful to apply the covariance decomposition formula

$$\begin{aligned} \operatorname{Cov}(\overline{W}_i[l_1], \overline{W}_i[l_2]) = &\operatorname{Cov}\left(\mathbb{E}\left(\overline{W}_i[l_1]\,|X_i\right), \mathbb{E}\left(\overline{W}_i[l_2]\,\middle|\,X_i\right)\right) \\ &+\mathbb{E}\left(\operatorname{Cov}\left(\overline{W}_i[l_1], \overline{W}_i[l_2]\,\middle|\,X_i\right)\right) \end{aligned}$$

to every pair $(\overline{W}_i[l_1], \overline{W}_i[l_2])$ of components of $\overline{W}_i$. For the covariance matrices this finally yields, in somewhat informal notation, the relation

$$\mathbb{V}(\overline{W}_i) = \mathbb{V}\left(\mathbb{E}(\overline{W}_i|X_i)\right) + \mathbb{E}\left(\mathbb{V}(\overline{W}_i|X_i)\right) = \mathbb{V}(X_i) + \mathbb{E}\left(\mathbb{V}(\overline{U}_i|X_i)\right),$$

which suggests to generalize (11) by using the pooled version

$$\widehat{\Sigma}_{X,X} = \frac{1}{n} \sum_{i=1}^{n} \left(\left(\overline{W}_i - \overline{W}\right) \left(\overline{W}_i - \overline{W}\right)' - \frac{1}{k} \widehat{\Sigma}_i \right). \tag{15}$$

3.4 Calibrating the Quadratic Part

One natural way to deal with the quadratic part $X_i^{②}$ is to replace every component $(X_i[l])^2$ of $X_i^{②}$ by the square $\left(\widehat{X_i[l]}\right)^2$ of the corresponding component $\widehat{X_i}[l]$ of $\widehat{X_i}$. Alternatively to that procedure, which is also pursued in the analysis below, one could prefer to calibrate $(X_i[l])^2$ 'directly' by an appropriate approximation to $\mathbb{E}((X_i[l])^2|W_i, Z_i)$. By means of the relation

$$\begin{aligned}\mathbb{E}\Big((X_i[l])^2|W_i, Z_i\Big) &= \Big(\mathbb{E}(X_i[l]|W_i, Z_i)\Big)^2 + \mathbb{V}(X_i[l]|W_i, Z_i)\\ &\approx \left(\widehat{X_i[l]}\right)^2 + \mathbb{V}(X_i[l]|W_i, Z_i)\end{aligned}$$

and arguments very similar to (6), both approaches lead to the same estimator for β as long as $\mathbb{V}(X_i[l]|W_i, Z_i)$ does not depend on i. This is the case, for instance, if under homoscedastic Gaussian measurement error $(X_i', Z_i')'$ are Gaussian, too.

4 Application to the MONICA Data

In this section the methods just developed are applied to the MONICA data.

4.1 The Data

Within the WHO MONICA project (MONItoring of trends and determinants in CArdiovascular disease) also the influence of nutrition was considered. We reanalyze data from a panel of the WHO MONICA substudy on dietary intake, conducted in 1984/1985 in Southern Germany, which is currently continued as the KORA study (Cooperative health research in the area of Augsburg), see Döring and Kußmaul (1997), Winkler et al. (1991). A subpopulation of 899 male respondents, aged from 45 to 65, filled in a comprehensive diary. For seven consecutive days all meals had been listed in detail. By using a nutritional data base also containing standard recipes, nutritional variables were derived from the raw data given in every-day units like ladle or gram of certain ingredients. Among other questions the role of the amount of plant protein intake (PLANT in the table below) and animal protein intake (ANIMAL) was investigated. Though high attention has been paid to the exactness of the measurement procedure, substantial error in the calculation of protein intake is unavoidable, and so we applied the correction methods developed above to adjust for it.

By a mortality and morbidity follow-up for more than 10 years, the respondents' first cardiac infarctions (total number 71 of 858 observations) and deaths (114 cases of 892 observations[6]) had been registered.[7] The main interest focused on the influence protein intake had on the response variable which was defined as age at the event. In the analysis also confounders were incorporated, namely cholesterol (mg/dl) (CHOL), daily alcohol consumption (g/day) (ALC) as continuous variables, as well as hypertension (HYPER) and smoking[8] (SMOKER) as categorical variables (1=yes, 0=no). The measurement error in these variables may be expected to be quite low compared to that in the protein intakes, and so the confounders were treated as error free. The estimated regression coefficients are written in the form $\hat{\beta}[VARIABLE]$, i.e., $\hat{\beta}[PLANT]$, $\hat{\beta}[ANIMAL]$, etc.

4.2 The Results

Table 1 summarizes the results of naive and corrected proportional hazards regression. The first two columns belong to the naive analysis, which used the seven-days averages of calculated animal protein intake and of calculated plant protein intake as surrogates for the true corresponding intake. They contain the naive estimates and the p-values based on them.[9] Column 3 and 5 report the corrected estimates after having adjusted for homoscedastic measurement error by the methods of Subsection 3.2, and for heteroscedastic measurement error along the lines of Subsection 3.3, respectively. In Column 4 and 6 also "approximative p-values" are given, which, however, have to be used with particular reservation here. They are based on the standard errors which usual software calculates after every X_i was replaced by the corresponding $\widehat{X_i}$; they are only meant to give a very rough impression and should not be taken literally. Correct estimators for the standard error

[6] The number of overall observations slightly differs for the two events, because for some units there was no information about morbidity, but it could be found out whether they died or survived the follow-up period.

[7] The median of the follow-up times with respect to the occurrence of infarction was 2302 days for the cases and 3996 days for the censored observations. The median of the follow-up times concerning the death event was 2598.5 days for the cases and 4006 days for the censored observations, respectively.

[8] In this analysis persons who are currently smoking or are ex-smokers were summarized into the smoker category.

[9] It may be noted explicitly that not only the naive estimators of the regression parameters are inconsistent, but also the estimators of the standard error.

of regression calibration estimators are not straightforwardly found (cf. Caroll et al. (2006, Chapter 4.6)), and so we used those easy available values as a rule of thumb to judge the significance. Though they are not correct, they still should give an impression of the correct magnitude.

In order to illustrate the overall influence of animal and plant protein intake on morbidity and mortality, it is helpful to look at the functions

$$f(x) = \hat{\beta}[ANIMAL]x + \hat{\beta}[(ANIMAL)^2]x^2 \tag{16}$$

$$g(y) = \hat{\beta}[PLANT]y + \hat{\beta}[(PLANT)^2]y^2 \,. \tag{17}$$

They describe the estimated effect of the animal protein intake x, and of the plant protein intake y, respectively, on the predictor in the hazard function in (1). The domains of x and y are chosen such that they cover approximately the whole range of the observed values. These functions are plotted in Figure 4, where the dotted and dashed line corresponds to the naive estimation. The results, after having adjusted for homoscedastic or heteroscedastic measurement error, are plotted by thin and thick solid lines, respectively.[10]

The Naive Analysis

For the naive analysis, the seven-days averages of calculated animal protein intake and of calculated plant protein intake were used as surrogates for the true corresponding intake in a proportional hazards regression. The naive analysis judges the linear and quadratic terms for animal protein to be significant at the five percent level, and cholesterol to have a highly significant influence on morbidity. For mortality the estimates $\hat{\beta}[PLANT]$ and $\hat{\beta}[(PLANT)^2]$ are significant at least at the ten percent level, and hypertension becomes highly significant.

The decisive question following the naive analysis now is: are these result still valid if one takes into account the substantial measurement error which is naturally inherent in the protein intake?

[10] As discussed in Section 5, we understand our analysis mainly as an illustration of the methods, and therefore only as a first step motivating further investigations, before neat material conclusion may be drawn.

Table 1. Estimates for the influence on morbidity and mortality

	naive estimation		homoscedastic error		heteroscedastic error	
	estimate	p-value	estimate	approx. p-value	estimate	approx. p-value
Morbidity						
$ANIMAL$	$-5.62 \cdot 10^{-5}$	0.04	$-1.07 \cdot 10^{-4}$	0.04	$-5.49 \cdot 10^{-5}$	0.28
$PLANT$	$-2.77 \cdot 10^{-5}$	0.79	$-1.32 \cdot 10^{-5}$	0.93	$-6.60 \cdot 10^{-5}$	0.64
$(ANIMAL)^2$	$4.68 \cdot 10^{-10}$	0.01	$9.09 \cdot 10^{-10}$	0.02	$5.27 \cdot 10^{-10}$	0.17
$(PLANT)^2$	$1.47 \cdot 10^{-11}$	0.99	$-4.10 \cdot 10^{-10}$	0.88	$4.51 \cdot 10^{-10}$	0.87
$CHOL$	$8.28 \cdot 10^{-3}$	0.00	$8.14 \cdot 10^{-3}$	0.00	$7.81 \cdot 10^{-3}$	0.00
$HYPER$	$4.60 \cdot 10^{-1}$	0.06	$4.70 \cdot 10^{-1}$	0.06	$4.71 \cdot 10^{-1}$	0.06
$SMOKER$	$8.79 \cdot 10^{-1}$	0.03	$8.68 \cdot 10^{-1}$	0.03	$8.35 \cdot 10^{-1}$	0.04
ALC	$8.00 \cdot 10^{-5}$	0.98	$5.01 \cdot 10^{-5}$	0.99	$-2.03 \cdot 10^{-5}$	1.00
Mortality						
$ANIMAL$	$-1.01 \cdot 10^{-5}$	0.73	$-1.54 \cdot 10^{-5}$	0.79	$-6.43 \cdot 10^{-6}$	0.89
$PLANT$	$-1.15 \cdot 10^{-4}$	0.06	$-1.61 \cdot 10^{-4}$	0.07	$-1.72 \cdot 10^{-4}$	0.03
$(ANIMAL)^2$	$4.55 \cdot 10^{-11}$	0.84	$8.23 \cdot 10^{-11}$	0.85	$3.41 \cdot 10^{-11}$	0.93
$(PLANT)^2$	$2.07 \cdot 10^{-9}$	0.04	$2.94 \cdot 10^{-9}$	0.05	$3.16 \cdot 10^{-9}$	0.03
$CHOL$	$7.34 \cdot 10^{-4}$	0.74	$6.77 \cdot 10^{-4}$	0.76	$5.78 \cdot 10^{-4}$	0.79
$HYPER$	$5.43 \cdot 10^{-1}$	0.00	$5.41 \cdot 10^{-1}$	0.00	$5.42 \cdot 10^{-1}$	0.00
$SMOKER$	$6.79 \cdot 10^{-1}$	0.02	$6.77 \cdot 10^{-1}$	0.02	$6.97 \cdot 10^{-1}$	0.02
ALC	$3.00 \cdot 10^{-3}$	0.29	$3.06 \cdot 10^{-3}$	0.28	$2.86 \cdot 10^{-3}$	0.32

Adjusting for Homoscedastic Measurement Error

First the homoscedastic error model is considered. In order to obtain corrected estimates the regression calibration method based on (8) and the estimators from (10) to (13) are applied. Column 3 and 4 of Table 1 report the corrected estimates for the influence on morbidity. In comparison to the naive estimates the effects of animal protein are estimated about twice as high; this results in the thin solid line in Figure 4 (top left). The point of minimal risk (x=59038) is about the same as in the naive analysis (x=60079), and also the zeros are equal in essence, but the curve is much steeper. $\hat{\beta}[PLANT]$ is half as high as the naive estimate. Now $\hat{\beta}[(PLANT)^2]$ has a negative sign, too. The corresponding function $g(y)$, which is depicted as the thin solid line in Figure 4 (top right), is concave and decreasing in the amount of intake in a monotone way: the higher the plant intake the higher is the reduction of the risk by an additional unit of intake.

The role of the confounders is more or less the same. The estimated strong influence of hypertension and smoking is confirmed. The regression parameter for alcohol intake changes its sign, but it remains insignificant.

Turning to mortality (cf. Table 1, the lower part), the absolute values of the regression parameters of the linear and the quadratic terms in the protein variables become higher by factors between 1.4 and 1.8, the effects of the confounders remain unchanged in essence. The figures in the lower part of Figure 4 show the corresponding curves, which are of the same shape as those from the naive analysis, but run steeper again.

Adjusting for Heteroscedastic Measurement Error

As discussed above, the presence of replication data also allows, for every unit i, $i = 1, \ldots, n$, to estimate the covariance matrix Σ_i of the error variable in animal protein intake and in plant protein intake at the individual level (cf. Equation (14)). Even if one takes into account that only seven observations are available to estimate Σ_i, the variation in the estimated variances (Fig. 2 and Fig. 3) is high enough that a detailed study of heteroscedastic measurement error appears promising.

The last two columns of the upper part in Table 1 refer to the corrected estimates for morbidity, the corresponding curves are shown by the thick solid lines in the upper parts of Fig. 4. Compared to the analysis assuming homoscedastic measurement error, the absolute

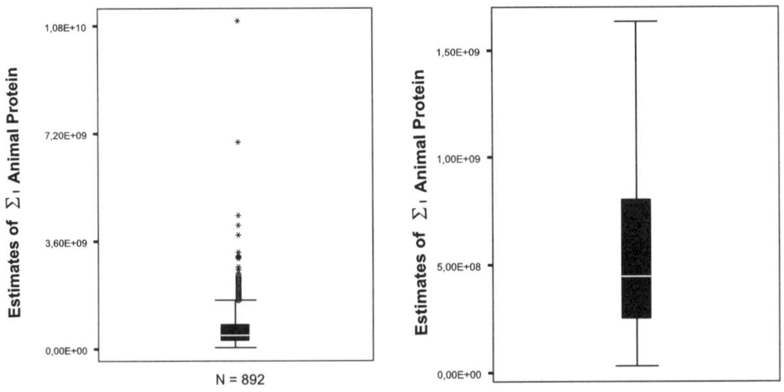

Fig. 2. Estimated individual error variances for animal protein intake: overall and detail figure

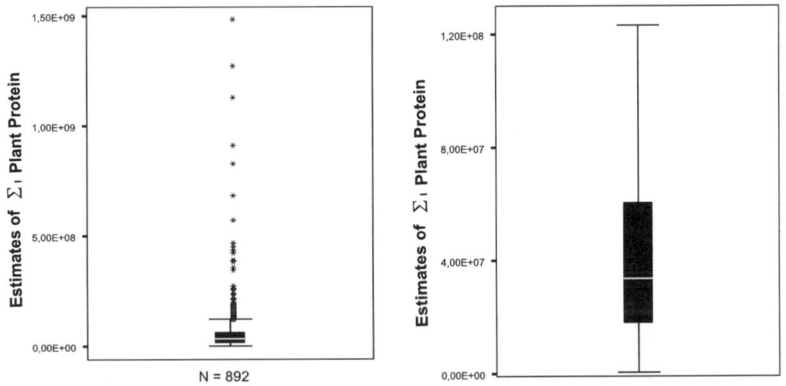

Fig. 3. Estimated individual error variances for plant protein intake: overall and detail figure

values of the estimates of the regression coefficients for the linear and the quadratic terms in animal protein intake are attenuated, indeed they are even closer to the results from the naive analysis. The curve grows flatter (cf. Fig. 4, top left), the point of minimal risk and the second zero are shifted to the left: from about $x = 60000$ to $x = 52408$, and from about $x = 120000$ to $x = 104098$, respectively. In contrast to this, the quadratic nature of the influence of plant protein becomes much clearer. The regression coefficient for the quadratic term now again has a positive sign, its value is about 30 times as high as in the naive analysis. As can also be seen in Figure 4 (top right), the risk is still decreasing with increasing plant protein intake, but now the curve is clearly convex: the relative gain in risk reduction becomes the smaller the higher the intake is, and there would be a border value (outside the domain of the data, at y=73241), where further intake would increase the risk again.

Correcting for heteroscedastic measurement error in the estimation of mortality confirms the results obtained from the homoscedastic error model for plant protein intake (cf. also Fig. 4, down right). The absolute values of the estimated coefficients of animal protein intake are lower by the factor 2.4, which results in a much flatter curve in Fig. 4 (down left).

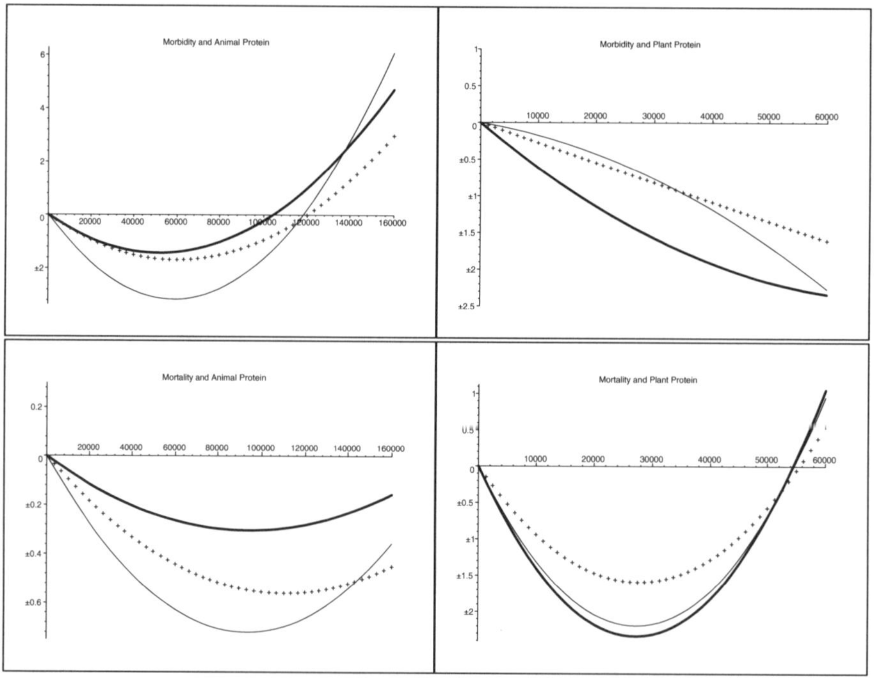

Fig. 4. Estimated overall influence of animal protein intake (*left figures*) and plant protein intake on morbidity (*upper part*) and mortality (cf. (16) and (17)), calculated from the naive estimates (*dotted line*), from the estimates after having corrected for homoscedastic measurement error (*thin solid line*), and from the estimates after having corrected for heteroscedastic measurement error (*thick solid line*), respectively

It is also worth mentioning that – in our analysis – morbidity and mortality differ with respect to the consequences a certain amount of protein intake has. Very high plant protein intake considerably reduces the risk of cardiac infarction, but increases the risk of death. In the case of animal protein the intake which minimizes the risk of death (x=94263 for the heteroscedastic error model) has already a rather high risk for cardiac infarction.

5 Concluding Remarks

We discussed an extended version of regression calibration to correct for possibly heteroscedastic measurement error in Cox regression with quadratic error-prone predictors when replication data are available. This method was applied to a part of the MONICA Augsburg survey to study the influence of eating habits on cardiovascular diseases.

It has become clear how important it is to take into account measurement error carefully. In particular under heteroscedastic measurement error there is a complex relationship between naive and corrected estimation, which may alter the estimates substantially. Nevertheless, the results reported here must be taken only as a first step towards a more comprehensive analysis, suggesting and motivating further research in several directions. Three areas of further investigation should be mentioned explicitly, namely limitations of the measurement error modeling and correction applied here, the use of alternative models, and the dynamic nature of the problem:

First of all, it must not be forgotten that the regression calibration method is only an approximate method, reducing the bias of naive analysis but not necessarily producing consistent estimators. Furthermore, the parameter estimates have to be interpreted in relative terms because correct estimators for their standard errors are missing, and therefore also no confidence regions were given. To derive such appropriate estimators is demanding (cf. Caroll et al. (2006, Chapter 4.6)), an interesting alternative would be bootstrapping.

Also alternative correction methods should be applied, in order to justify, or to correct, the preliminary results obtained here. Of special interest here is a so-to-say dynamic regression calibration procedure, developed by Xie et al. (2001), where at every failure time only those units are taken into account which still are under risk (cf. also Wang et al. (2001)). Another powerful method to correct for homoscedastic measurement error in the Cox model was developed by Nakamura

(1992) and extended to heteroscedastic error by Augustin (2004). However, prior to applying this method, further theoretic development is needed, in order to be able to model the quadratic influence of the covariates. The inherent restriction to linear predictors is also the main hurdle for an application of the nonparametric functional correction method from Huang and Wang (2000), which would provide an appealing alternative to utilize replicated measurements. In addition the error model itself leaves space for further improvement. In particular, there are good reasons to doubt the assumption made above that the measurement error should be independent of the true protein intake and other covariates, and so more complex error models deserve special attention (cf., e.g., Heitmann et al. (1995), Prentice (1996), Caroll et al. (1998)). Also most probable the errors in the measurements of the same units are not independent of each other.

The second issue to keep in mind is that valuable insights in the data may be gained by applying different models, and so a study of accelerated failure time models or additive hazard models seems attractive. Techniques for measurement error correction in accelerated failure time models have not yet received much attention. One of the very rare exceptions is Nakamura (1990), where Nakmura illustrates his general method of corrected score functions with members of the exponential family. His approach is generalized to possibly censored Weibull distributed lifetimes in Giménez et al. (1999), see also Giménez et al. (2006); He et al. (2007) applies the SIMEX method to accelerate failure time models. Song and Huang (2006) consider covariate measurement error in the additive hazards model, and Cheng and Wang (2001) propose a method to correct for measurement error the semi-parametric transformation model. A procedure handling error-prone covariates in the nonparametric log-linear lifetime model is suggested by Wang (2000), while Augustin (2002, Chapter 5f) proposes two methods for corrected quasi-likelihood estimation in arbitrary parametric accelerated failure time models. As discussed there, the latter approaches need some non-standard treatment of censored observations, but have, on the other hand, the advantage of being able to take also error-prone lifetimes into account.Another promising variation in modeling concerns the effects themselves, and deeper insight may also be obtained by going beyond the quadratic approach pursued here and allowing for flexible modeling of the effects (see Rummel (2006, Chapter 4.3) for analyzing the data by flexible binary regression, see also Rummel et al. (2007)).

The last item to be mentioned here may be the most difficult one: Eating habits may change! Even if the X_i to be measured by the diary could be determined exactly, this measurement would only stem from a cursory glance at a process developing over time. Morbidity and mortality is also affected by the intake before as well as after the recording. This leads to the superposition of the heteroscedastic measurement error treated here with a complex kind of measurement error where a time-dependent covariate is only observed at a certain time point. (Compare for this also de Bruijne et al. (2001) and Andersen and Liestol (2003), who consider Cox models where a time dependent covariate is only observed irregularly.)

Acknowledgements

We are grateful to Professor Helmut Küchenhoff and Professor Hans Schneeweiss for many helpful comments.

References

Ahmad A (2006) Statistical analysis of heaping and rounding effects. Dr Hut. Munich

Andersen PK, Liestol K (2003) Attenuation caused by infrequently updated covariates in survival analysis. Biostatistics 4: 633–649

Augustin T (2002) Survival analysis under measurement error. Habilitation (post-doctorial) thesis, Faculty of Mathematics, Informatics and Statistics, University of Munich, Munich

Augustin T (2004) An exact corrected log-likelihood function for Cox's proportional hazards model under measurement error and some extensions. Scandinavian Journal of Statistics 31: 43–50

Augustin T, Schwarz R (2002) Cox's proportional hazards model under covariate measurement error — A review and comparison of methods. In: Van Huffel S,Lemmerling P (eds) Total least squares and errors-in-variables modeling: analysis, algorithms and applications. Kluwer, Dordrecht: 175–184

Augustin T, Wolff J (2004) A bias analysis of Weibull models under heaped data. Statistical Papers 45: 211–229

Bender R, Augustin T, Bletter M (2005) Simulating survival times for Cox regression models. Statistics in Medicine 24: 1713–1723

de Bruijne MHJ, le Cessie S, Kluin-Neemans HC, van Houwelingen HC (2001) On the use of Cox regression in the presence of an irregularly observed time-dependent covariate. Statistics in Medicine 20: 3817–3829

Buzas JS (1998) Unbiased scores in proportional hazards regression with covariate measurement error. Journal of Statistical Planning and Inference 67: 247–257

Carroll RJ, Freedman LS, Kipnis V, Li L (1998) A new class of measurement error models, with applications to dietary data. Canadian Journal of Statistics 26: 467–477

Carroll RJ, Ruppert D, Stefanski LA, Crainiceanu CM (2006) Measurement error in nonlinear models: a modern perspective, 2nd edition. Chapman and Hall/CRC. Boca Raton, FL, USA

Cheng SC, Wang NY (2001) Linear transformation models for failure time data with covariate measurement error. Journal of the American Statistical Association 96: 706–716

Cheng C-L, Schneeweiß H (1998) The polynomial regression with errors in the variables. Journal of the Royal Statistical Society Series B 60: 189–199

Cheng C-L, Schneeweiss H, Thamerus M (2000) A small sample estimator for a polynomial regression with errors in the variables. Journal of the Royal Statistical Society B 62: 699–709

Cheng C-L, Van Ness JW (1999) Statistical regression with measurement error. Arnold, London

Clayton DG (1991) Models for the analysis of cohort and case-control studies with inaccurately measured exposures. In: Dwyer JH, Feinleib M, Lipsert P et al. (eds) Statistical Models for Longitudinal Studies of Health. Oxford University Press, New York: 301–331

Cox DR (1972) Regression models and life tables (with discussion). Journal of the Royal Statistical Society Series B 34: 187–220

Döring A, Kußmaul B. (1997) Ernährungsdeterminanten des Herzinfarktrisikos. Report GSF-Fe-7629. GSF — National Research Center for Environment and Health, Neuherberg

Dupuy JF (2005) The proportional hazards model with covariate measurement error. Journal of Statistical Planning and Inference 135: 260–275

Giménez P, Bolfarine H, Colosimo EA (1999) Estimation in Weibull regression model with measurement error. Communications in Statistics — Theory and Methods 28: 495–510

Giménez P, Bolfarine H, Colosimo EA (2006) Asymptotic relative efficiency of score tests in Weibull models with measurement errors, Statistical Papers 47: 461–470.

He W, Yi G, Xiong J. (2007) Accelerated failure time models with covariates subject to measurement error. Statistics in Medicine (Appearing)

Heid IM, Kchenhoff H, Rosario AS, Kreienbrock L and Wichmann HE (2006) Impact of measurement error in exposures in German radon studies. Journal of Toxicology and Environmental Health A 69: 701–721

Heitmann BL, Lissner L (1995) Dietary underreporting by obese individuals — is it specific or non-specific? British Medical Journal 311: 986–989

Heumann C (2004) Monte Carlo methods for missing data in generalized linear and generalized linear mixed models. Habilitation thesis, Faculty of Mathematics, Informatics and Statistics, University of Munich, Munich

Horowitz J, Manski CF (2000) Nonparametric analysis of randomized experiments with missing covariate and outcome data. Journal of the American Statistical Association, 95: 77–84

Hu C, Lin DY (2002) Cox regression with covariate measurement error. Scandinavian Journal of Statistics 29: 637–655

Hu CC, Lin DY (2004) Semiparametric failure time regression with replicates of mismeasured covariates. Journal of the American Statistical Association 99: 105–118

Hu P, Tsiatis A, Davidian M (1998) Estimating the parameters in the Cox model when covariate variables are measured with error. Biometrics 54: 1407–1419

Huang HS, Huwang L (2001) On the polynomial structural relationship. The Canadian Journal of Statistics 29: 493–511

Huang Y, Wang CY (2000) Cox regression with accurate covariates unascertainable: a nonparametric-correction approach. Journal of the American Statistical Association 95: 1209–1219

Huang YJ, Wang CY (2006) Errors-in-covariates effect on estimating functions: Additivity in limit and nonparametric correction. Statistica Sinica 16: 861–881

Hughes MD (1993) Regression dilution in the proportional hazards model. Biometrics 49: 1056–1066

Johansson I, Hallmans G, Wikman A, Biessy C, Riboli E, Kaaks R (2002) Validation and calibration of food-frequency questionnaire measurement in the Northern Sweden health and disease cohort. Public Health Nutrition 5: 487–496

Kalbfleisch JD, Prentice RL (2002) The Statistical analysis of failure time data. 2nd edition. Wiley, New York

Kong FH (1999) Adjusting regression attenuation in the Cox proportional hazards model. Journal of Statistical Planning and Inference 79: 31–44

Kong FH, Gu M (1999) Consistent estimation in Cox proportional hazards model with covariate measurement errors. Statistica Sinica 9: 953–969

Küchenhoff H, Lederer W, Lesaffre E (2007) Asymptotic variance estimation for the misclassification SIMEX. Computational Statistics & Data Analysis 51: 6197–6211

Küchenhoff H, Mwalili SM, Lesaffre E (2006) A general method for dealing with misclassification in regression: The misclassification SIMEX. Biometrics 62: 85–96

Kuha JT, Temple J (2003) Covariate measurement error in quadratic regression. International Statistical Review 71: 131–150

Kukush A, Markovsky I, Van Huffel S (2004) Consistent estimation in an implicit quadratic measurement error model. Computational Statistics & Data Analysis 47: 123–147

Kukush A, Schneeweiss H, Wolf R (2004) Three estimators for the Poisson regression model with measurement errors. Statistical Papers 45: 351–368

Kukush A, Schneeweiss H (2005) Relative efficiency of three estimators in a polynomial regression with measurement errors. Journal of Statistical Planning and Inference 127: 179–203

Kukush A, Schneeweiss H (2005) Comparing different estimators in a nonlinear measurement error model. I and II. Mathematical Methods of Statistics 14: 53–79 and 203–223

Kulathinal SB, Kuulasmaa K, Gasbarra D (2002) Estimation of an errors-in-variables regression model when the variances of the measurement errors vary between the observations. Statistics in Medicine 21: 1089–1101

Kuznetsov VP (1991) Interval statistical models (in Russian), Radio and Communication, Moscow

Li Y, Lin XH (2003) Functional inference in frailty measurement error models for clustered survival data using the SIMEX approach. Journal of the American Statistical Association 98: 191–203

Little RJA, Rubin DB (2002) Statistical analysis with missing data, 2nd edition. Wiley, New York

Liu K, Stone RA, Mazumdar S, Houck PR, Reynolds CF (2004) Covariate measurement error in the Cox model: A simulation study. Communications in Statistics – Simulation and Computation 33: 1077–1093

Manski CF (2003) Partial identification of probability distributions. Springer, New York

Manski CF, Tamer E (2002) Inference on regressions with interval dara on a regressor or outcome. Econometrica 70: 519–546

Martin-Magniette M-L, Taupin M-L (2006) Semi-parametric estimation of the hazard function in a model with covariate measurement error. Arxiv preprint math.ST/0606192.

Nakamura T (1990) Corrected score functions for errors-in-variables models: Methodology and application to generalized linear models. Biometrika 77: 127–137

Nakamura T (1992) Proportional hazards model with covariates subject to measurement error. Biometrics 48: 829–838

Pepe MS, Self SG, Prentice RL (1989) Further results in covariate measurement errors in cohort studies with time to response data. Statistics in Medicine 8: 1167–1178

Prentice RL (1982) Covariate measurement errors and parameter estimation in a failure time regression model. Biometrika 69: 331–342

Prentice RL (1996) Measurement error and results from analytic epidemiology: dietary fat and breast cancer. Journal of the National Cancer Institute 88: 1738–1747

Rao CR, Toutenburg H, Shalabh, Heumann C (2008) Linear Models: Least Squares and Generalizations (3rd edition). Springer, Berlin Heidelberg New York

Ronning G (2005) Randomized response and the binary probit model. Economics Letters 86: 221–228

Rummel D (2006) Correction for covariate measurement error in nonparametric regression. Dr Hut. Munich

Rummel D, Augustin T, Küchenhoff H (2007) Correction for covariate measurement error in nonparametric longitudinal regression. SFB-Discussion Paper 513. Department of Statistics. University of Munich

Schmid M (2006) Estimation of a linear regression model with microaggregated data. Dr Hut, Munich

Schmid M, Schneeweiß H, Küchenhoff H (2007) Estimation of a linear regression under microaggregation with the response variable as a sorting variable. Statistica Neerlandica (Appearing)

Schneeweiß H, Augustin T (2006) Some recent advanced in measurement error models and methods. Allgemeines Statistisches Archiv – Journal of the German Statistical Association 90: 183–197 (also printed In: Hübler O, Frohn J (eds, 2006) Modern econometric analysis – survey on recent developments, 183–198)

Schneeweiss H, Cheng C-L (2006) Bias of the structural quasi-score estimator of a measure-ment error model under misspecification of the regressor distribution. Journal of Multivariate Analysis 97: 455–473

Shalabh (2001) Consistent estimation through weighted harmonic mean of inconsistent estimators in replicated measurement error models. Econometric Reviews. 20: 507–510

Shalabh (2001) Least squares estimators in measurement error models under the balanced loss function. TEST 10: 301–308

Shalabh (2003) Consistent estimation of coefficients in measurement error models with replicated observations. Journal of Multivariate Analysis. 86: 227–241

Shalabh, Toutenburg H (2005) Estimation of linear regression model with missing data: the role of stochastic linear constraints. Communications in Statistics – Theory and Methods 34: 375–387

Shklyar S, Schneeweiss H (2005) A comparison of asymptotic covariance matrices of three consistent estimators in the Poisson regression model with measurement errors. Journal of Multivariate Analysis 94: 250–270

Shklyar S, Schneeweiss H, Kukush A (2007) Quasi score is more efficient than corrected score in a polynomial measurement error model. Metrika 65: 275–295

Song X, Huang YJ (2006) A corrected pseudo-score approach for additive hazards model with longitudinal covariates measured with error. Lifetime Data Analysis 12: 97–110

Stefanski LA (1989) Unbiased estimation of a nonlinear function of a normal mean with application to measurement error models. Communications in Statistics — Theory and Methods 18: 4335–4358

Stefanski LA (2000) Measurement error models. Journal of the American Statistical Association 95: 1353–1358

Strobl C, Boulesteix A-L , Augustin T (2007) Unbiased split selection for classification trees based on the Gini index. Computational Statistics & Data Analysis 52: 483–501

Svejdar V, Augustin T and Strobl C (2007) Variablenselektion in Klassifikationsbämen unter spezieller Bercksichtigung von fehlenden Werten. Technical report (`http://www.statistik.lmu.de/~carolin/research.html`)

Toutenburg H (2002) Statistical Analysis of Designed Experiments (2nd edition). Springer, New York

Toutenburg H, Fieger A, Heumann C (2000) Regression modelling with fixed effects: missing values and related problems. In: Rao CR, Székely GJ (eds) Statistics for the 21st century. Marcel Dekker, New York

Toutenburg H, Heumann C, Nittner T, Scheid S (2002) Parametric and nonparametric regression with missing X's - a review. Journal of the Iranian Statistical Society 1: 79–110

Toutenburg H, Shalabh, Heumann C (2006) Use of prior information in the form of interval constraints fir the improved estimation of linear regression models with some missing responses. Journal of Statistical Planning and Inference 136: 2430–2445

Toutenburg H, Shalabh (2005) Estimation of regression coefficients subject to exact linear restrictions when some observations are missing and quadratic error balanced loss function is used. TEST 14: 385–396

Toutenburg H, Shalabh (2001) Use of minimum risk approach in the estimation of regression models with missing observations. Metrika 54: 247–259

Toutenburg H, Srivastava VK (1999) Estimation of ratio of population means in survey sampling when some observations are missing. Metrika 48: 177–187

Toutenburg H, Srivastava VK (2004) Efficient estimation of population mean using incomplete survey data on study and auxiliary characteristics. Statistica (Bologna) 63: 223–236

Toutenburg H, Srivastava VK, Shalabh, Heumann C (2005) Estimation of parameters in multiple linear regression with missing covariates using a modified first order regression proce-dure. Annals of Economics and Finance 6: 289–301

Toutenburg H, Srivastava VK, Shalabh (2006) Estimation of linear regression models with missing observations on both the explanatory and study variables. Quality Technology & Quantitative Management 3: 179–189

Utkin LV, Augustin T (2007) Decision making under imperfect measurement using the imprecise Dirichlet model. International Journal of Approximate Reasoning 44, 332–338

Van Huffel S, Lemmerling P (eds, 2002) Total least squares and errors-in-variables modeling: analysis, algorithms and applications. Kluwer, Dordrecht

Walley P (1991) Statistical reasoning with imprecise probabilities, Chapman and Hall, London

Wang CY, Hsu L, Feng ZD, Prentice RL (1997) Regression calibration in failure time regression. Biometrics 53: 131–145

Wang CY, Xie SX, Prentice RL (2001) Recalibration based on an approximate relative risk estimator in cox regression with missing covariates. Statistica Sinica 1: 1081–1104

Wang Q (2000) Estimation of linear error-in-covariables models with validation data under random censorship. Journal of Multivariate Analysis 74: 245–266

Wansbeek T, Meijer E (2000) Measurement error and latent variables in econometrics. Elsevier, Amsterdam

Weichselberger K (2001) Elementare Grundbegriffe einer allgemeineren Wahrscheinlichkeitsrechnung, Volume I: Intervallwahrscheinlichkeit als umfassendes Konzept, Physika, Heidelberg

Willett W (1998) Nutritional Epidemiology (2nd edition). Oxford University Press, New York

Winkler G, Döring A, Keil U (1991) Selected nutrient intakes of middle-aged men in Southern Germany: Results from the WHO MONICA Augsburg Dietary Survey of 1984/1985. Annals of Nutrition and Metabolism 35: 284–291

Wolff J, Augustin T (2003) Heaping and its consequences for duration analysis - a simulation study. Allgemeines Statistisches Archiv - Journal of the German Statistical Association 87: 1–28

Xie SX, Wang CY, Prentice RL (2001) A risk set calibration method for failure time regression by using a covariate reliability sample. Journal of the Royal Statistical Society Series B 63: 855–870

Yi GY, Lawless JF (2007) A corrected likelihood method for the proportional hazards model with covariates subject to measurement error. Journal of Statistical Planning and Inference 137(6): 1816–1828

Zaffalon M (2002) Exact credal treatment of missing data, Journal of Statistical Planning and Inference. 105: 105–122

Zaffalon M (2005) Conservative rules for predictive inference with incomplete data, In: F. G. Cozman, R. Nau, T. Seidenfeld (eds.), ISIPTA '05, Proceedings of the fourth international symposium on imprecise probabilities and their applications, Pittsburgh, PA, USA, SIPTA, Manno (CH), 406–415

Zaffalon M, de Cooman G (2004) Updating beliefs with incomplete observations. Artificial Intelligence 159: 75–125

Homoscedastic Balanced Two-fold Nested Model when the Number of Sub-classes is Large

Shu-Min Liao[1] and Michael Akritas[2]

[1] Department of Statistics, Penn State University, 325, Thomas Bldg., University Park, PA16802, U.S.A. sx1340@psu.edu
[2] Department of Statistics, Penn State University, 325 Thomas Bldg., University Park, PA16802, U.S.A. mga@stat.psu.edu

1 Introduction

Analysis of variance (ANOVA) is a corner stone of statistical applications. The classical ANOVA model assumes that the error terms are i.i.d. normal, in which case F-statistics have certain optimality properties (cf. Arnold (1981, Chapter 7)). Arnold (1980) showed that the classical F-test is robust to the normality if the sample sizes tend to infinity while the number of levels stays fixed. The past decade has witnessed the generation of large data sets, involving a multitude of factor levels, in several areas of scientific investigation. For example, in agricultural trials it is not uncommon to see a large number of treatments but limited replication per treatment. See Brownie and Boos (1994) and Wang and Akritas (2006). Another application arises in certain type of microarray data in which the nested factor corresponds to a large number of genes. In addition to the aforementioned papers by Brownie and Boos and Wang and Akritas, other relevant literature includes Akritas and Arnold (2000), Bathke (2002) and Akritas and Papadatos (2004).

The above papers deal only with crossed designs. In this article we consider the two-fold nested design and establish the asymptotic theory, both under the null and alternative hypotheses, for the usual F-test statistics of sub-class effects when the number of sub-classes goes to infinity but the number of classes and the number of observations in each sub-class remain fixed. The fixed, random and mixed effects models are all considered. The main finding of the paper is that the classical, normality-based, test procedure is asymptotically robust to departures from the normality assumption.

The rest of this manuscript is organized as follows. The next section contains a review of the statistical models, and states three results that are useful for the asymptotic derivations. In Sections 3 we present the asymptotic theory for the fixed-effects model, while Section 4 presents the asymptotic theory for both the random and the mixed-effects model. Some simulation results are shown in Section 5, and finally Section 6 states conclusions.

2 The Statistical Models and Auxiliary Results

2.1 The Fixed-effects Model

In the balanced two-fold fixed-effects model , we observe

$$Y_{ijk} = \mu_{ij} + \sigma e_{ijk}, \quad i = 1, \ldots, r; \quad j = 1, \ldots, c; \quad k = 1, \ldots, n, \tag{1}$$

where the μ_{ij} are bounded and e_{ijk} are independent with

$$E(e_{ijk}) = 0,\ Var(e_{ijk}) = 1,\ E(e_{ijk}^4) = \kappa. \tag{2}$$

The means μ_{ij} are typically decomposed as

$$\mu_{ij} = \mu + \alpha_i + \delta_{ij}, \tag{3}$$

where

$$\sum_{i=1}^{r} \alpha_i = 0 \text{ and } \delta_{ic} \equiv \sum_{j=1}^{c} \delta_{ij} = 0,\ \forall i, \text{ for any chosen } c.$$

In this paper, we are mainly interested in testing H_0: $\delta_{ij} = 0$ (no subclass effect). Let

$$\begin{aligned} MS\delta &= \frac{n\sum_{i=1}^{r}\sum_{j=1}^{c}(\bar{Y}_{ij\cdot} - \bar{Y}_{i\cdot\cdot})^2}{r(c-1)}, \\ MSE &= \frac{\sum_{i=1}^{r}\sum_{j=1}^{c}\sum_{k=1}^{n}(Y_{ijk} - \bar{Y}_{ij\cdot})^2}{rc(n-1)}, \\ \mathsf{M}_c^{\delta} &\equiv \frac{1}{\sigma^2}\begin{pmatrix} MS\delta \\ MSE \end{pmatrix}, \end{aligned} \tag{4}$$

then the usual F-test statistic for testing H_0: $\delta_{ij} = 0$ is

$$F_c^\delta \equiv \frac{MS\delta}{MSE}. \tag{5}$$

If the e_{ijk} are i.i.d. $N(0,1)$, then

$$F_c^\delta \sim F_{r(c-1),\, rc(n-1)}, \text{ under } H_0 : \delta_{ij} = 0. \tag{6}$$

2.2 The Random-effects Model

In the balanced two-fold nested random-effects model, we observe

$$Y_{ijk} = \mu + \sigma_a a_i + \sigma_d d_{ij} + \sigma e_{ijk}, \tag{7}$$

$i = 1, \ldots, r$; $j = 1, \ldots, c$; $k = 1, \ldots, n$, where a_i, d_{ij}, e_{ijk} are random variables independent of each other, and

$$\begin{aligned} E(a_i) = E(d_{ij}) &= E(e_{ijk}) = 0, \\ Var(a_i) = Var(d_{ij}) &= Var(e_{ijk}) = 1, \\ E(a_i^4) = \kappa_a,\ E(d_{ij}^4) &= \kappa_d,\ E(e_{ijk}^4) = \kappa. \end{aligned} \tag{8}$$

Let

$$\gamma_d = \frac{n\sigma_d^2}{\sigma^2}, \quad \bar{d}_{i\cdot} = \frac{1}{c}\sum_{j=1}^{c} d_{ij}. \tag{9}$$

Define MSE, MSd, and M_c^d as MSE, $MS\delta$, and M_c^δ in (4). In the random-effects model, the usual null hypothesis for testing the sub-class effect is $H_0 : \sigma_d^2 = 0$ (no sub-class effect) $\Leftrightarrow \gamma_d = 0$, and the corresponding F-test statistic could be defined as

$$F_c^d = \frac{MSd}{MSE}. \tag{10}$$

If the a_i, d_{ij} and e_{ijk} are i.i.d. $N(0,1)$, then

$$F_c^d \sim F_{r(c-1),\, rc(n-1)}, \text{ under } H_0 : \sigma_d^2 = 0. \tag{11}$$

2.3 The Mixed-effects Model

In the balanced two-fold nested mixed-effects model , we observe

$$Y_{ijk} = \mu + \alpha_i + \sigma_d d_{ij} + \sigma e_{ijk}, \tag{12}$$

$i = 1, \ldots, r$; $j = 1, \ldots, c$; $k = 1, \ldots, n$, where the d_{ij}, e_{ijk} are random variables independent of each other, and

$$E(d_{ij}) = E(e_{ijk}) = 0,\ Var(d_{ij}) = Var(e_{ijk}) = 1,$$
$$E(d_{ij}^4) = \kappa_d,\ E(e_{ijk}^4) = \kappa. \tag{13}$$

Further assume

$$\sum_{i=1}^{r} \alpha_i = 0, \tag{14}$$

and let γ_d and $\bar{d}_{i\cdot}$ be defined as (9). For testing $H_0 : \sigma_d^2 = 0$ in the mixed-effects model, we define MSE, MSd, M_c^d and F_c^d as in the random-effects model. Under the normality assumption, (11) is true also for the mixed-effects model.

2.4 Auxiliary Results and Notations

In this section we state two theorems from Akritas and Arnold (2000) which are useful for our asymptotic derivations, state a simple asymptotic result concerning the F-distribution, and introduce some notation.

Following the notation in Akritas and Arnold (2000), we use E and Cov for the expected value and covariance of vectors under the *general* model, while E_N and Cov_N are used when the *normal* model is assumed.

Theorem 1. *Suppose that* $\mathsf{e} = (e_1, \ldots, e_q)'$ *is a* q*-dimensional random vector such that the* e_i *are independent with mean 0, variance 1, and* $E(e_i^4) = \kappa$. *Let* $\boldsymbol{\alpha}_1, \ldots, \boldsymbol{\alpha}_p$ *be* q*-dimensional constant vectors,* $\boldsymbol{\alpha} = (\boldsymbol{\alpha}_1, \ldots, \boldsymbol{\alpha}_p)$*, and* $\mathsf{A}_1, \ldots, \mathsf{A}_p$ *be* $q \times q$ *symmetric matrices. Define*

$$Q_j(\mathsf{e}, \boldsymbol{\alpha}_j) = (\mathsf{e} + \boldsymbol{\alpha}_j)'\mathsf{A}_j(\mathsf{e} + \boldsymbol{\alpha}_j),$$

$$\mathsf{Q}(\mathsf{e}, \boldsymbol{\alpha}) = \begin{pmatrix} Q_1(\mathsf{e}, \boldsymbol{\alpha}_1) \\ \vdots \\ Q_p(\mathsf{e}, \boldsymbol{\alpha}_p) \end{pmatrix}.$$

Then,

$$E(\mathsf{Q}(\mathsf{e},\,\boldsymbol{\alpha})) = E_N(\mathsf{Q}(\mathsf{e},\,\boldsymbol{\alpha})).$$

In addition, if each A_i *has constant diagonals and either* $\boldsymbol{\alpha} = \mathsf{0}$ *or* $E(e_i^3) = 0$, *then*

$$Cov(\mathsf{Q}(\mathsf{e},\,\boldsymbol{\alpha})) = Cov_N\,(\mathsf{Q}(\mathsf{e},\,\boldsymbol{\alpha}))) + \frac{\kappa - 3}{q}(E_N(\mathsf{Q}(\mathsf{e},\,\mathsf{0})))(E_N(\mathsf{Q}(\mathsf{e},\,\mathsf{0})))'.$$

Theorem 2. *Consider the setting and notation of Theorem 1, and assume further that* $\mathsf{a} = (a_1, \ldots, a_s)'$ *is a s-dimensional random vector such that the* a_i *are independent with mean 0, variance 1, and* $E(a_i^4) = \kappa_a$. *Let* $\mathsf{L}_1, \ldots, \mathsf{L}_p$ *be* $q \times s$ *fixed matrices and let* $\mathsf{C}_j = \mathsf{L}_j'\mathsf{A}_j\mathsf{L}_j$. *Define*

$$Q_j(\mathsf{e},\,\mathsf{a},\,\boldsymbol{\alpha}_j) = (\mathsf{e} + \boldsymbol{\alpha}_j + \mathsf{L}_j\mathsf{a})'\mathsf{A}_j(\mathsf{e} + \boldsymbol{\alpha}_j + \mathsf{L}_j\mathsf{a}),$$

$$\mathsf{Q}(\mathsf{e},\,\mathsf{a},\,\boldsymbol{\alpha}) = \begin{pmatrix} Q_1(\mathsf{e},\,\mathsf{a},\,\boldsymbol{\alpha}_1) \\ \vdots \\ Q_p(\mathsf{e},\,\mathsf{a},\,\boldsymbol{\alpha}_p) \end{pmatrix}.$$

Then,

$$E(\mathsf{Q}(\mathsf{e},\,\mathsf{a},\,\boldsymbol{\alpha})) = E_N(\mathsf{Q}(\mathsf{e},\,\mathsf{a},\,\boldsymbol{\alpha})).$$

In addition, if each A_i *and each* C_i *has constant diagonals and either* $\boldsymbol{\alpha} = \mathsf{0}$ *or* $E(e_i^3) = E(a_i^3) = 0$, *then*

$$\begin{aligned} Cov(\mathsf{Q}(\mathsf{e},\,\mathsf{a},\,\boldsymbol{\alpha})) = {} & Cov_N\,(\mathsf{Q}(\mathsf{e},\,\mathsf{a},\,\boldsymbol{\alpha}))) \\ & + \frac{\kappa - 3}{q}(E_N(\mathsf{Q}(\mathsf{e},\,\mathsf{0},\,\mathsf{0})))(E_N(\mathsf{Q}(\mathsf{e},\,\mathsf{0},\,\mathsf{0})))' \\ & + \frac{\kappa_a - 3}{s}(E_N(\mathsf{Q}(\mathsf{0},\,\mathsf{a},\,\mathsf{0})))(E_N(\mathsf{Q}(\mathsf{0},\,\mathsf{a},\,\mathsf{0})))'. \end{aligned}$$

Theorem 3. *Let the random variable* X_c *have the* $F_{(c-k_1)\ell_1,(c-k_2)\ell_2}$ *distribution, where* k_1, k_2, ℓ_1, *and* ℓ_2 *are constants. Then, as* $c \to \infty$,

$$\sqrt{c}\,(X_c - 1) \xrightarrow{d} N\left(0, 2\left(\frac{1}{\ell_1} + \frac{1}{\ell_2}\right)\right).$$

The proof of the proposition is straight forward and is omitted.

We close this section by giving the following additional notation.

$$\mathsf{U}_c \approx \mathsf{V}_c \quad \Leftrightarrow \quad \sqrt{c}(\mathsf{U}_c - \mathsf{V}_c) \xrightarrow{P} 0, \quad \text{as } c \to \infty,$$
$$\mathsf{a}_c \approx \mathsf{b}_c \quad \Leftrightarrow \quad \sqrt{c}(\mathsf{a}_c - \mathsf{b}_c) \to 0, \quad \text{as } c \to \infty,$$

where U_c and V_c are two sequences of random vectors, while a_c and b_c are two sequences of constant vectors.

3 Fixed Effects Model

In this section, we consider the fixed-effects model and derive the asymptotic null distribution of F_c^δ, defined in (5), as $c \to \infty$ but r, n remain fixed. The asymptotic distribution under alternatives is also derived. Define

$$U_{ij}^\delta = n(\bar{e}_{ij\cdot} + \frac{\delta_{ij}}{\sigma})^2, \quad \bar{U}_{ic}^\delta = \frac{1}{c}\sum_{j=1}^{c} U_{ij}^\delta, \quad \bar{U}_{\cdot c}^\delta = \frac{1}{r}\sum_{i=1}^{r} \bar{U}_{ic}^\delta, \tag{15}$$

$$W_{ij} = \frac{\sum_{k=1}^{n}(e_{ijk} - \bar{e}_{ij\cdot})^2}{n-1}, \quad \bar{W}_{ic} = \frac{1}{c}\sum_{j=1}^{c} W_{ij}, \quad \bar{W}_{\cdot c} = \frac{1}{r}\sum_{i=1}^{r} \bar{W}_{ic}, \tag{16}$$

$$\mathsf{V}_{ij}^\delta = \begin{pmatrix} U_{ij}^\delta \\ W_{ij} \end{pmatrix}, \quad \bar{\mathsf{V}}_{ic}^\delta = \begin{pmatrix} \bar{U}_{ic}^\delta \\ \bar{W}_{ic} \end{pmatrix}, \quad \bar{\mathsf{V}}_{\cdot c}^\delta = \begin{pmatrix} \bar{U}_{\cdot c}^\delta \\ \bar{W}_{\cdot c} \end{pmatrix}. \tag{17}$$

Note that $\bar{U}_{\cdot c}^\delta$, $\bar{W}_{\cdot c}$ are related to $MS\delta$, MSE respectively. In particular,

$$\bar{W}_{\cdot c} = \frac{1}{\sigma^2} MSE, \quad \text{and} \tag{18}$$

$$\bar{U}_{\cdot c}^\delta = \frac{1}{\sigma^2} MS\delta + \left[\frac{nc}{r(c-1)}\sum_{i=1}^{r} \bar{e}_{i\cdot\cdot}^2 - \frac{1}{c-1}\bar{U}_{\cdot c}^\delta\right].$$

It is straightforward to see that, as $c \to \infty$ and r, n remain fixed,

$$\sqrt{c}\frac{nc}{r(c-1)}\sum_{i=1}^{r} \bar{e}_{i\cdot\cdot}^2 \xrightarrow{P} 0, \quad \text{and} \quad \sqrt{c}\frac{1}{c-1}\bar{U}_{\cdot c}^\delta \xrightarrow{P} 0.$$

Combining the above we have that, as $c \to \infty$ and r, n remain fixed,

$$\bar{\mathsf{V}}_{\cdot c}^\delta \approx \mathsf{M}_c^\delta \tag{19}$$

where M_c^δ is defined in (4).

Hence, the asymptotic joint distribution of $MS\delta/\sigma^2$ and MSE/σ^2 is the same as the asymptotic joint distribution of $\bar{U}_{\cdot c}^\delta$ and $\bar{W}_{\cdot c}$.

3.1 Null Distribution

Theorem 4. *Consider the model and assumptions given in (1), (2), and the decomposition of the means given in (3). Under the null hypothesis $H_0 : \delta_{ij} = 0$, and as $c \to \infty$ while r, n stay fixed,*

$$\sqrt{c}(F_c^\delta - 1) \xrightarrow{d} N\left(0, \frac{2n}{r(n-1)}\right). \tag{20}$$

Corollary 1. *Under the model and assumptions of Theorem 4, the classical, normality-based, F-test for the hypothesis $H_0 : \delta_{ij} = 0$ is asymptotically valid.*

Proof of Corollary 1:

The proof follows easily from Theorem 3 and Theorem 4.

Proof of Theorem 4:

Under H_0, we have

$$U_{ij}^{\delta} = n(\bar{e}_{ij\cdot})^2, \quad W_{ij} = \frac{\sum_{k=1}^{n}(e_{ijk} - \bar{e}_{ij\cdot})^2}{n-1}.$$

We will use Theorem 1 in order to find the expected value of $\mathbf{V}_{ij}$ and then we will use the multivariate CLT to find the asymptotic distribution of $\bar{\mathsf{V}}_{\cdot c}^{\delta}$. Because under the normal model, U_{ij}^{δ} and W_{ij} are independent, $U_{ij}^{\delta} \sim \chi_1^2$ and $(n-1)W_{ij} \sim \chi_{n-1}^2$, application of Theorem 1 with $\boldsymbol{\alpha} = \mathsf{0}$ yields

$$E(\mathsf{V}_{ij}^{\delta}) = \begin{pmatrix} E_N(U_{ij}^{\delta}) \\ E_N(W_{ij}) \end{pmatrix} = \begin{pmatrix} 1 \\ 1 \end{pmatrix} \triangleq \boldsymbol{\mu}$$

$$Cov(\mathsf{V}_{ij}^{\delta}) = \begin{pmatrix} 2 & 0 \\ 0 & \frac{2}{n-1} \end{pmatrix} + \frac{\kappa - 3}{n}\begin{pmatrix} 1 & 1 \\ 1 & 1 \end{pmatrix} \triangleq \Sigma.$$

Since V_{ij}^{δ} are i.i.d., by the multivariate CLT,

$$\sqrt{c}(\bar{\mathsf{V}}_{ic}^{\delta} - \boldsymbol{\mu}) \xrightarrow{d} N_2(\mathsf{0}, \Sigma),$$

and since $\bar{\mathsf{V}}_{ic}^{\delta}$ are independent it follows that

$$\sqrt{c}(\bar{\mathsf{V}}_{\cdot c}^{\delta} - \boldsymbol{\mu}) \xrightarrow{d} N_2(\mathsf{0}, \Sigma/r), \text{ as } c \to \infty.$$

Thus, by (19)

$$\sqrt{c}(\mathsf{M}_{c}^{\delta} - \boldsymbol{\mu}) \xrightarrow{d} N_2(\mathsf{0}, \Sigma/r), \text{ as } c \to \infty. \tag{21}$$

Relation (21) and the Δ-method imply

$$\sqrt{c}(F_c^{\delta} - 1) \xrightarrow{d} N\left(\mathsf{0}, \mathsf{s}'\Sigma\mathsf{s}/r\right), \text{ where } \mathsf{s}' = (1, -1).$$

Since $\mathsf{s}'\Sigma\mathsf{s}/r = 2n/r(n-1)$, the result follows.

3.2 Alternative Distribution

Theorem 5. *Consider the model and assumptions given in Theorem 4. In addition assume that*

$$E(e_{ijk}^3) = 0, \quad \text{and} \quad E|e_{ijk}|^{4+2\epsilon} < \infty \quad \text{for some } \epsilon > 0.$$

Then, under alternatives δ_{ij} *which satisfy*

$$\theta_{ic}^{\delta} = \frac{n \sum_{j=1}^{c} \delta_{ij}^2}{c\sigma^2} \approx \theta_i,$$

as $c \to \infty$ *while* r, n *stay fixed,*

$$\sqrt{c}\left(F_c^{\delta} - (1+\theta)\right) \xrightarrow{d} N\left(0, \frac{2n + 4n\theta + 2\theta^2}{r(n-1)} + \frac{\kappa - 3}{rn}\theta^2\right).$$

Proof:

Let U_{ij}^{δ} and W_{ij} be as defined in (15) and (16). We will use Theorem 1 again to find the expected value and covariance matrix of V_{ij}^{δ}, defined in (17). Under the normal model, U_{ij}^{δ} and W_{ij} are independent, and

$$U_{ij}^{\delta} \sim \chi_1^2\left(\frac{n\delta_{ij}^2}{\sigma^2}\right), \quad (n-1)W_{ij} \sim \chi_{n-1}^2.$$

Thus, by Theorem 1,

$$E(\mathsf{V}_{ij}^{\delta}) = \begin{pmatrix} E_N(U_{ij}^{\delta}) \\ E_N(W_{ij}) \end{pmatrix} = \begin{pmatrix} 1 + \frac{n\delta_{ij}^2}{\sigma^2} \\ 1 \end{pmatrix},$$

$$Cov(\mathsf{V}_{ij}^{\delta}) = \begin{pmatrix} 2 + 4\frac{n\delta_{ij}^2}{\sigma^2} & 0 \\ 0 & \frac{2}{n-1} \end{pmatrix} + \frac{\kappa - 3}{n}\begin{pmatrix} 1 & 1 \\ 1 & 1 \end{pmatrix},$$

where we used the fact that, if $U \sim \chi_1^2(\gamma)$ then

$$E(U) = 1 + \gamma, \quad Var(U) = 2 + 4\gamma.$$

Hence,

$$E(\bar{\mathsf{V}}_{ic}^{\delta}) = \begin{pmatrix} 1 + \theta_{ic}^{\delta} \\ 1 \end{pmatrix} \approx \begin{pmatrix} 1 + \theta_i \\ 1 \end{pmatrix} \triangleq \boldsymbol{\mu}_i$$

and

$$c \cdot Cov(\bar{\mathsf{V}}_{ic}^{\delta}) = \begin{pmatrix} 2 + 4\theta_{ic}^{\delta} & 0 \\ 0 & \frac{2}{n-1} \end{pmatrix} + \frac{\kappa - 3}{n} \begin{pmatrix} 1\,1 \\ 1\,1 \end{pmatrix}$$
$$\approx \begin{pmatrix} 2 + 4\theta_i & 0 \\ 0 & \frac{2}{n-1} \end{pmatrix} + \frac{\kappa - 3}{n} \begin{pmatrix} 1\,1 \\ 1\,1 \end{pmatrix} \triangleq \Sigma_i.$$

Note that now V_{ij}^{δ} are *not* i.i.d. Under the assumption that $E|e_{ijk}|^{4+2\epsilon} < \infty$ for some $\epsilon > 0$, Lyapanov's theorem and the Cramér-Wold theorem yield,

$$\sqrt{c}(\bar{\mathsf{V}}_{ic}^{\delta} - \boldsymbol{\mu}_i) \xrightarrow{d} N_2(0, \Sigma_i), \text{ and } \sqrt{c}(\bar{\mathsf{V}}_{\cdot c}^{\delta} - \boldsymbol{\mu}) \xrightarrow{d} N_2(0, \Sigma),$$

where

$$\theta = \frac{1}{r}\sum_{i=1}^{r} \theta_i, \quad \boldsymbol{\mu} = \begin{pmatrix} 1 + \theta \\ 1 \end{pmatrix}, \quad \Sigma = \begin{pmatrix} \frac{2}{r} + \frac{4\theta}{r} & 0 \\ 0 & \frac{2}{r(n-1)} \end{pmatrix} + \frac{\kappa - 3}{rn} \begin{pmatrix} 1\,1 \\ 1\,1 \end{pmatrix}.$$

By the asymptotic equivalence between $\bar{\mathsf{V}}_{\cdot c}$ and M_c^{δ} shown in (19),

$$\sqrt{c}(\mathsf{M}_c^{\delta} - \boldsymbol{\mu}) \xrightarrow{d} N_2(0, \Sigma), \text{ as } c \to \infty.$$

Note that if $\mathsf{s}' = (1, -(1 + \theta))$, $\sqrt{c}\,\mathsf{s}'(\mathsf{M}_c^{\delta} - \boldsymbol{\mu}) = [MS\delta - (1 + \theta)MSE]/\sigma^2$ which, by Slutsky's theorem, is asymptotically equivalent to $\sqrt{c}\left(F_c^{\delta} - (1 + \theta)\right)$. Thus, by the Δ-method, we have that as $c \to \infty$,

$$\sqrt{c}\left(F_c^{\delta} - (1 + \theta)\right) \xrightarrow{d} N\left(0, \mathsf{s}'\Sigma\mathsf{s}\right)$$
$$= N\left(0, \frac{2n + 4n\theta + 2\theta^2}{r(n-1)} + \frac{\kappa - 3}{rn}\theta^2\right). \quad (22)$$

Note that when $\theta = 0$, equation (22) is in fact equivalent to equation (20).

4 Random and Mixed Effects Models

4.1 Random-effects Models

The following theorem presents the asymptotic distribution of the F-statistic under both the null and alternative hypotheses.

Theorem 6. *Consider the model and assumptions given in (7) and (8), and the notation in (9). Then as* $c \to \infty$*, while* r*,* n *remain fixed,*

$$\sqrt{c}\left(F_c^d - (1+\gamma_d)\right) \xrightarrow{d} N\left(0,\ \tfrac{2n(1+\gamma_d)^2}{r(n-1)} + \tfrac{\gamma_d^2}{r}\left(\kappa_d - 3 + \tfrac{\kappa-3}{n}\right)\right), \quad (23)$$

where F_c^d *is defined in (10). Under the null hypothesis* $H_0 : \sigma_d^2 = 0$ *of no sub-class effect, which is equivalently stated as* $\gamma_d = 0$*, we have*

$$\sqrt{c}\left(F_c^d - 1\right) \xrightarrow{d} N\left(0,\ \frac{2n}{r(n-1)}\right), \quad \text{as } c \to \infty. \quad (24)$$

Remarks:

1. The asymptotic null distribution in this case is the same as in the fixed-effects case; see equation (20).
2. Arguing as in Corollary 1, it is shown that under the model and assumptions of Theorem 6, the classical, normality-based, F-test for the hypothesis $H_0 : \sigma_d^2 = 0$ is asymptotically valid.

Proof:

Define W_{ij}, $\bar{W}_{ic}$, and $\bar{W}_{\cdot c}$ as in equation (16), and define

$$U_{ij}^d = n\left(\bar{e}_{ij\cdot} + \frac{\sigma_d}{\sigma} d_{ij}\right)^2, \quad \bar{U}_{ic}^d = \frac{1}{c}\sum_{j=1}^{c} U_{ij}^d, \quad \bar{U}_{\cdot c}^d = \frac{1}{r}\sum_{i=1}^{r} \bar{U}_{ic}^d,$$

$$\mathsf{V}_{ij}^d = \mathsf{V}_{ij}^d(\mathsf{e}_{ij}, d_{ij}) = \begin{pmatrix} U_{ij}^d \\ W_{ij} \end{pmatrix}, \quad \bar{\mathsf{V}}_{ic}^d = \begin{pmatrix} \bar{U}_{ic}^d \\ \bar{W}_{ic} \end{pmatrix}, \quad \mathsf{V}_{\cdot c}^d = \begin{pmatrix} \bar{U}_{\cdot c}^d \\ \bar{W}_{\cdot c} \end{pmatrix},$$

where

$$\mathsf{e}_{ij}' = (e_{ij1}, \ldots, e_{ijn})_{n \times 1}.$$

Note that the relationship between $\bar{W}_{\cdot c}$ and MSE remains the same as (18). In addition,

$$\bar{U}_{\cdot c}^d = \frac{1}{\sigma^2} MSd + \left[\frac{nc}{r(c-1)}\sum_i \left(\bar{e}_{i\cdot\cdot} + \frac{\sigma_d}{\sigma}\bar{d}_{i\cdot}\right)^2 - \frac{1}{c-1}\bar{U}_{\cdot c}^d\right].$$

It can be shown that, as $c \to \infty$ and r, n remain fixed,

$$\sqrt{c}\frac{nc}{r(c-1)}\sum_i \left(\bar{e}_{i\cdot\cdot} + \frac{\sigma_d}{\sigma}\bar{d}_{i\cdot}\right)^2 \xrightarrow{P} 0, \quad \text{and} \quad \sqrt{c}\frac{1}{c-1}\bar{U}_{\cdot c}^d \xrightarrow{P} 0.$$

Combining the above we have that, as $c \to \infty$ and r, n remain fixed,

$$\bar{\mathsf{V}}^d_{\cdot c} \approx \mathsf{M}^d_c \tag{25}$$

where M^d_c is defined in Section (2.2). Hence, the asymptotic joint distribution of MSd/σ^2 and MSE/σ^2 is the same as the asymptotic joint distribution of $\bar{U}^d_{\cdot c}$ and $\bar{W}_{\cdot c}$. To find the expectation and covariance matrix of V^d_{ij}, we will use Theorem 2. Under the normal model, U^d_{ij} and W_{ij} are independent, and

$$\frac{U^d_{ij}}{1+\gamma_d} \sim \chi^2_1, \quad (n-1)W_{ij} \sim \chi^2_{n-1}.$$

Thus, by Theorem 2, with $\boldsymbol{\alpha} = \mathsf{0}$,

$$E(\mathsf{V}^d_{ij}) = \begin{pmatrix} E_N(U^d_{ij}) \\ E_N(W_{ij}) \end{pmatrix} = \begin{pmatrix} 1+\gamma_d \\ 1 \end{pmatrix} \triangleq \boldsymbol{\mu}.$$

In addition, under the normality,

$$E_N\left(\mathsf{V}^d_{ij}(\mathsf{e}_{ij}, 0)\right) = \begin{pmatrix} 1 \\ 1 \end{pmatrix}, \quad \text{and} \quad E_N\left(\mathsf{V}^d_{ij}(0, d_{ij})\right) = \begin{pmatrix} \gamma_d \\ 1 \end{pmatrix},$$

where

$$\mathsf{V}^d_{ij}(\mathsf{e}_{ij}, 0) = \begin{pmatrix} n\,\bar{e}^2_{ij\cdot} \\ \frac{\Sigma_k (e_{ijk}-\bar{e}_{ij\cdot})^2}{n-1} \end{pmatrix} \quad \text{and} \quad \mathsf{V}^d_{ij}(0, d_{ij}) = \begin{pmatrix} \gamma_d\, d^2_{ij} \\ 0 \end{pmatrix}.$$

Thus, by Theorem 2, with $\boldsymbol{\alpha} = \mathsf{0}$, again

$$Cov(\mathsf{V}^d_{ij}) = \begin{pmatrix} 2(1+\gamma_d)^2 & 0 \\ 0 & \frac{2}{n-1} \end{pmatrix} + (\kappa_d - 3)\gamma^2_d \begin{pmatrix} 1\,0 \\ 0\,0 \end{pmatrix} + \frac{\kappa-3}{n} \begin{pmatrix} 1\,1 \\ 1\,1 \end{pmatrix} \triangleq \Sigma.$$

Since V^d_{ij}, $j = 1, \ldots, c$, are i.i.d., by the multivariate CLT and the Cramér-Wold theorem, we have

$$\sqrt{c}(\bar{\mathsf{V}}^d_{ic} - \boldsymbol{\mu}) \xrightarrow{d} N_2(\mathsf{0}, \Sigma), \text{ and } \sqrt{c}(\bar{\mathsf{V}}^d_{\cdot c} - \boldsymbol{\mu}) \xrightarrow{d} N_2(\mathsf{0}, \Sigma/r).$$

Thus, by (25),

$$\sqrt{c}(\mathsf{M}^d_c - \boldsymbol{\mu}) \xrightarrow{d} N_2(\mathsf{0}, \Sigma/r), \text{ as } c \to \infty.$$

The Δ-method can now be used to obtain

$$\sqrt{c}(F^d_c - (1+\gamma_d)) \xrightarrow{d} N\left(\mathsf{0}, \mathsf{s}'\Sigma\mathsf{s}/r\right), \text{ where } \mathsf{s}' = (1, -(1+\gamma_d)),$$

which implies the theorem.

4.2 Mixed-effects Model

Theorem 7. *Consider the model and assumptions given in (12)–(14), and the notation in (9). Then as $c \to \infty$, while r, n remain fixed, the asymptotic distribution of the F-statistic for testing the sub-class effect in the mixed-effects model is the same as the one in the random-effects model.*

Remark: Arguing as in Corollary 1 it is shown that under the model and assumptions of Theorem 7, the classical, normality-based, F-test for the hypothesis $H_0 : \sigma_d^2 = 0$ is asymptotically valid.

Proof:

Since the F-statistic for testing the sub-class effect in the mixed-effects model is defined in the same way as the one in the random-effects model (see section 2.3), the proof follows by the same arguments used to show Theorem 6.

5 Simulations

In this section, simulations under the fixed-effects model are used to compare the achieved sizes of two procedures. The first procedure rejects at level α if

$$\sqrt{c}(F_c^\delta - 1) > \sqrt{\frac{2n}{r(n-1)}} Z_\alpha, \tag{26}$$

where F_c^δ is defined in (5) and Z_α is the $(1-\alpha)100$th percentile of the standard normal distribution. The second procedure rejects at level α if

$$F_c^\delta > F^\alpha_{r(c-1),rc(n-1)}, \tag{27}$$

where F_c^δ is as before and $F^\alpha_{r(c-1),rc(n-1)}$ is the $(1-\alpha)100$th percentile of the $F_{r(c-1),rc(n-1)}$ distribution. Thus the first procedure uses the asymptotic null distribution (note that the null hypothesis is only rejected for large values of the test statistic), while the second procedure can be thought of as a finite-sample correction to the asymptotic distribution.

The simulation uses 5 classes ($r = 5$) while the numbers of sub-classes used are $c = 5, 30, 100$, and 500. The other parameters of the decomposition in (3) are chosen as follows: $\mu = 0$ and $\boldsymbol{\alpha} = (\alpha_1, \alpha_2, \alpha_3, \alpha_4\, \alpha_5)' = (-3, -2, -1, 2, 4)'$. There are 3 observations ($n = 3$) for each sub-class with randomly-generated errors e_{ijk} from one of four distributions: (1) the standard normal; (2) the exponential distribution with $\lambda = 1$; (3) the log-normal distribution whose logarithm has mean 0 and standard deviation 2; and (4) the mixture distribution defined as $UX + (1-U)Y$, where $U \sim$ Bernoulli($p = 0.9$), $X \sim N(-1.11, 1)$ and $Y \sim N(10, 1)$. All e_{ijk} are standardized to have mean 0 and standard deviation 1. The simulated sizes are shown in Table 1, in which "F-dist" represents the procedure with rejection rule (27), while "N-dist" represents the procedure with rejection rule (26). As we can see in Table 1, except in the case of the log-normal distrib-

	$c = 5$		$c = 30$		$c = 100$		$c = 500$	
$\alpha = 0.01$	F-dist	N-dist	F-dist	N-dist	F-dist	N-dist	F-dist	N-dist
Normal	0.0090	0.0454	0.0118	0.0238	0.0106	0.0165	0.0103	0.0125
Exponential	0.0126	0.0500	0.0119	0.0238	0.0105	0.0179	0.0098	0.0127
LNorm	0.0158	0.0348	0.0217	0.0285	0.0223	0.0253	0.0181	0.0201
Mixture	0.0189	0.0494	0.0151	0.0254	0.0124	0.0206	0.0106	0.0127
$\alpha = 0.05$	F-dist	N-dist	F-dist	N-dist	F-dist	N-dist	F-dist	N-dist
Normal	0.0481	0.0972	0.0523	0.0706	0.0499	0.0601	0.0492	0.0522
Exponential	0.0535	0.0988	0.0523	0.0696	0.0486	0.0582	0.0495	0.0536
LNorm	0.0359	0.0587	0.0407	0.0487	0.0396	0.0433	0.0350	0.0371
Mixture	0.0521	0.0905	0.0570	0.0753	0.0552	0.0630	0.0510	0.0540
$\alpha = 0.10$	F-dist	N-dist	F-dist	N-dist	F-dist	N-dist	F-dist	N-dist
Normal	0.0977	0.1413	0.1016	0.1193	0.0993	0.1072	0.0985	0.1019
Exponential	0.0994	0.1446	0.1000	0.1166	0.0992	0.1079	0.0981	0.1006
LNorm	0.0589	0.0807	0.0600	0.0666	0.0578	0.0609	0.0514	0.0529
Mixture	0.0908	0.1303	0.1024	0.1199	0.1032	0.1114	0.1020	0.1048
$\alpha = 0.20$	F-dist	N-dist	F-dist	N-dist	F-dist	N-dist	F-dist	N-dist
Normal	0.2014	0.2280	0.2010	0.2109	0.1970	0.2018	0.1995	0.2019
Exponential	0.1957	0.2231	0.1994	0.2108	0.1964	0.2019	0.1957	0.1969
LNorm	0.1050	0.1220	0.1008	0.1047	0.0987	0.1004	0.0919	0.0925
Mixture	0.1775	0.1998	0.1929	0.2016	0.1974	0.2012	0.1997	0.2012

Table 1. Sizes under over 10,000 simulation runs ($r = 5$, $n = 3$).

ution, the size achieved by procedure (27) is closer to the nominal size than that achieved by procedure (26). In fact, the sizes achieved by procedure (27) are surprisingly accurate and robust against the normal assumption in most cases. In addition, the number of sub-classes (c) seems to have small effect on the performance of procedure (27). On the other hand, we can see that procedure (26) tends to be liberal but becomes less so as $c \to \infty$.

6 Concluding Comments

We have established that the asymptotic null distribution of the usual F-test statistic for the sub-class effect in the balanced two-folded nested homoscedastic model is independent of the normal assumption as $c \to \infty$, but r, n remain fixed. Moreover, simulations indicate that the traditional test procedure based on the F-distribution serves as a successful finite sample correction to the asymptotic distribution of the test statistic.

References

Akritas MG, Papadatos N (2004) Heteroscedastic One-Way ANOVA and Lack-of-Fit tests. Journal of the American Statistical Association 99:368–382

Akritas MG, Arnold SF (2000) Asymptotics for Analysis of Variance when the Number of Levels is Large. Journal of the American Statistical Association 95:212–226

Arnold SF (1980) Asymptotic Validity of F Tests for the Ordinary Linear Model and the Multiple Correlation Model. Journal of the American Statistical Association 75:890–894

Arnold SF (1981) The Theory of Linear Models and Multivariate Analysis. Wiley, New York

Bathke A (2002) ANOVA for a Large Number of Treatments. Mathematical Methods of Statistics 11:118–132

Brownie C, Boos DD (1994) Type I Error Robustness of ANOVA and ANOVA on Ranks when the Number of Treatments is Large. Biometrics 50:542–549

Wang L and Akritas MG (2006) Two-way Heteroscedastic ANOVA when the Number of Levels is Large. Statistica Sinica 16:1387-1408

QR-Decomposition from the Statistical Point of View

Hilmar Drygas

Department of Mathematics, University of Kassel, 34109 Kassel, Germany
drygas@mathematik.uni.kassel.de

1 Introduction.

In this paper we will deal with the linear model

$$\mathrm{E}(y) = X\beta \;, \quad \mathrm{Cov}(y) = \sigma^2 I \tag{1}$$

where y is a $n \times 1$ vector of observations on response variable, X is a $n \times k$ matrix of n observations on each of the k explanatory variables and β is a $k \times 1$ vector of associated regression coefficients. We denote by $\mathrm{span}\{z_1, \ldots, z_r\}$ the linear subspace spanned by the vectors $z_1, \ldots, z_r$ of the vector-space V. For a matrix A we denote the range of the matrix A by $\mathrm{im}(A)$ and its rank by $\mathrm{r}(A)$. A g-inverse of A is denoted A^- and the Moore-Penrose of A is denoted by A^+. P_L denotes the orthogonal projection into the linear subspace L of the vector-space V.

It should be noted that the model (1) is the most general linear model. Further, (1) is understood with respect to a general inner product, i.e.,

$$\mathrm{E}(y, a) = (X\beta, a), \quad \mathrm{Var}(y, a) = \sigma^2 (a, a) \tag{2}$$

with respect to a given inner product. If with respect to the classical inner product $(x, y)_o = x'y$ and the relation $\mathrm{Cov}(y) = Q$, (Q is nonnegative definite matrix) is correct, then if $\mathrm{im}(X) \subseteq \mathrm{im}(Q)$ with respect to the inner product

$$(x, z) = (x, Q^+ z)_0 = x' Q^+ z, \tag{3}$$

then the relation $\mathrm{Var}(y, a) = \sigma^2(a, a)$ holds. (x, z) is also equal to $(x, \rho)_o$ where ρ is any solution of $Q\rho = z$. This definition does not depend on the selection ρ. If $\mathrm{im}(X) \subseteq \mathrm{im}(Q)$ does not hold, then with respect to the estimation of $\mathrm{E}(y)$, we can "regularize" the model by replacing Q by $W = Q + cXX', c > 0$. Then $\mathrm{im}(X) \subseteq \mathrm{im}(W)$ and for the inner product

$$(x, z) = (x, W^{+}z)_o = x'W^{+}z, \tag{4}$$

again $\mathrm{Var}(y, a) = \sigma^2(a, a)$. This is just what $\mathrm{Cov}\, y = \sigma^2 I$ means.

In the model (1) the BLUE (Best Linear Unbiased Estimator) or GME (Gauss-Markov estimator) of $\mathrm{E}y$ is

$$\mathrm{G}y = P_{\mathrm{im}(X)}y, \tag{5}$$

which is the orthogonal projection of y onto $\mathrm{im}(X)$. If $X = (x_1, \dots, x_k)$, $x_i \in \mathbb{R}_{k\times 1}$ and $\beta = (\beta_1, \dots, \beta_k)'$ then

$$X\beta = \beta_1 x_1 + \ldots + \beta_k x_k.$$

Gy is easy to compute if x_i's are pairwise orthogonal, i.e., $(x_i, x_j) = 0$ if $i \neq j$. In this case

$$\mathrm{G}y = \sum_{i=1}^{k} \hat{\beta}_i x_i, \quad \hat{\beta}_i = \begin{cases} \dfrac{(x_i, y)}{(x_i, x_i)} & \text{if } x_i \neq 0 \\ \text{arbitrary} & \text{if } x_i = 0. \end{cases} \tag{6}$$

Also the measure of determination $R^2 = ||\mathrm{G}y||^2/||y||^2$ can be easily computed. Indeed,

$$||\mathrm{G}y||^2 = \sum_{i:x_i\neq 0} \frac{(x_i, y)^2}{(x_i, x_i)} = \sum_{i=1}^{k} \hat{\beta}_i^2 (x_i, x_i)$$

and

$$R^2 = \sum_{i:x_i\neq 0} \frac{(x_i, y)^2}{(x_i, x_i)(y, y)} = \sum_{i:x_i\neq 0} \mathrm{Corr}^2(x_i, y)$$

where $\mathrm{Corr}(x, y)$ denotes the (empirical) correlation coefficient between x and y.

When x_i's are not pairwise orthogonal, then by the orthogonalization method due to Erhard Schmidt, late professor at "Humboldt-University at Berlin", it is possible to represent $X\beta$ as follows:

$$X\beta = \sum_{i=1}^{s} \alpha_i q_i, \tag{7}$$

where $s = \mathrm{r}(X)$ and q_i's are non-zero vectors which are pairwise orthogonal. They are determined by the orthogonalization algorithm which in the literature is mostly called Gram-Schmidt orthogonalization method. If

$$x_j = \sum_{l=1}^{s} r_{lj} q_l,\ j = 1, \ldots, k\,,$$

then it follows that $(x_j, q_l) = r_{lj}(q_l, q_l)$ and

$$r_{lj} = \frac{(x_j, q_l)}{(q_l, q_l)}\,. \tag{8}$$

Without loss of generality, we can assume that $x_1, \ldots, x_s$ are linearly independent. If we then let $q_1 = x_1$, then $r_{11} = 1$ and if also $r_{ll} = 1$, $l = 2, \ldots, s$, then

$$q_l = x_l - \sum_{j=1}^{l-1} \frac{(x_l, q_j)}{(q_j, q_j)} q_j. \tag{9}$$

Since $(x_j, q_l) = (q_j, q_l)$ if $l \geq j$, it follows that $r_{ll} = 1$ and $r_{lj} = 0$ if $j < l$. Let

$$Q_1 = (q_1, \ldots, q_s), \quad R_1 = (r_{lj};\, l = 1, \ldots, s,\ j = 1, \ldots, k).$$

Then

$$X = Q_1 R_1 \tag{10}$$

where R_1 is an upper triangular matrix. This representation is called QR-decomposition in numerical analysis (See Björck-Dahlquist (1972) or Lawson and Hanson (1974)). Since $X\beta = Q_1\alpha$, $\alpha = (\alpha_1, \ldots, \alpha_s)'$, it follows that $\hat{\alpha} = (Q_1'Q_1)^{-1}Q_1'y$ is the least squares estimator (BLUE, GME) of α. From

$$Q_1\alpha = \sum_{l=1}^{s} \alpha_l q_l = \sum_{j=1}^{k} \beta_j x_j = \sum_{j=1}^{k}(\sum_{l=1}^{s} r_{lj} q_l) = \sum_{l=1}^{s}(\sum_{j=1}^{k} r_{lj}\beta_j) q_l = X\beta,$$

it follows that

$$\alpha_l = \sum_{j=1}^{k} r_{lj}\beta_j = (R_1\beta)_l, \quad l = 1, \ldots, r,$$

implying $R_1\beta = \alpha$.

Theorem 1. *Let $\hat{\alpha} = (Q_1'Q_1)^{-1}Q_1'y$ and $\hat{\beta}$ be any solution of the equation $R_1\hat{\beta} = \hat{\alpha}$. Then $(l, \hat{\beta})$ is BLUE of (l, β) whenever (l, β) is estimable. There exists at least one solution of $R_1\hat{\beta} = \hat{\alpha}$.*

Proof:

a) We have the representation $X = Q_1R_1$ and (l, β) is estimable iff $l \in \text{im}(X')$, i.e., $l \in \text{im}(R_1'Q_1')$ or $l = R_1'Q_1'z$ for some z. Then

$$(l, \hat{\beta}) = (R_1'Q_1'z, \hat{\beta}) = (Q_1'z, R_1\hat{\beta}) = (z, Q_1\hat{\alpha}) \tag{11}$$

and this is BLUE of $(z, Q_1\alpha) = (z, X\beta) = (X'z, \beta) = (l, \beta)$.

b) The equation $R_1\beta = \alpha$ is solvable. Let $R = (R_{11}' \vdots R_{12})$ where R_{11} is of order $s \times s$ and R_{12} of order $s \times k - s$. Since R_{11} is upper triangular with diagonal elements equal to one, it follows that the determinant of R_{11} equals one. Hence R_{11} is regular. If we split up $\hat{\beta}$ as $\hat{\beta}' = (\hat{\beta}_1', \hat{\beta}_2')$ where β_1 of order $s \times 1$ and β_2 of order $k - s \times 1$, then $\hat{\beta}_1 = R_{11}^{-1}\alpha$, $\hat{\beta}_2 = 0$, form a solution of $R_1\beta = \hat{\alpha}$.

Since R_1 and R_{11} are upper triangular the system $R_1\hat{\beta} = \hat{\alpha}$ and $R_{11}\hat{\beta}_1 = \hat{\alpha}$ can be easily solved successively as follows:

$$\hat{\beta}_s = \hat{\alpha}_s = \frac{(q_s, y)}{(q_s, q_s)} \tag{12}$$

$$\begin{aligned} \hat{\beta}_l &= \hat{\alpha}_l - \sum_{j=l+1}^{s} r_{lj}\hat{\beta}_j \\ &= \frac{(q_l, y)}{(q_l, q_l)} - \sum_{j=l+1}^{s} r_{lj}\hat{\beta}_j, \; l = s-1, s-2, \ldots, 1. \end{aligned} \tag{13}$$

2 Estimable Functions and their Estimation

In the model $\mathrm{E}(y) = X\beta$, $\mathrm{Cov}(y) = \sigma^2 I$, the least squares estimator of β is

$$\hat{\beta} = (X'X)^{-1}X'y,$$

if X has full column-rank and

$$\mathrm{Cov}(\hat{\beta}) = \sigma^2(X'X)^{-1}.$$

Let $X = (X_0 \vdots x_k)$, where X_0 is of order $n \times (k-1)$ and x_k of order $n \times 1$. Then

$$X'X = \begin{pmatrix} X_0'X_0 & x_k'X_0 \\ X_0'x_k & x_k'x_k \end{pmatrix}$$

and

$$(X'X)^{-1} = \begin{pmatrix} A & B \\ C & D \end{pmatrix},$$

where D is the inverse of the Schur-complement as

$$\begin{aligned} x_k' x_k - x_k' X_0 (X_0' X_0)^{-1} X_0' x_k &= x_k'(I - X_0(X_0'X_0)^{-1}X_0')x_k \\ &= x_k'(I - P_{\mathrm{im}(X_0)})x_k \\ &= ||(I - P_{\mathrm{im}(X_0)})x_k||^2. \end{aligned} \tag{14}$$

Let $q = (I - P_{\mathrm{im}(X_0)})x_k$. Then if $(a, b) = b'a$ is the classical inner product used in this model, then

$$\frac{q'y}{q'q}$$

is an estimator with variance $\sigma^2/(q'q)$ and $q'q$ is the Schur-complement of $x_k' x_k$. Since

$$\mathrm{E}(q'y) = q'X\beta = q'(X_0\beta_0 + \beta_k x_k) = q'\beta_k x_k = \beta_k q' x_k = \beta_k q' q, \tag{15}$$

it follows that

$$\frac{q'y}{q'q} \tag{16}$$

is the best linear unbiased estimator (BLUE) of β_k. This result is also correct in more general situations.

Theorem 2. *Let the linear model* $\mathrm{E}(y) = X\beta$, $\mathrm{Cov}(y) = \sigma^2 I$ *be given and* $l \in \mathbb{R}^k$. *Then for* $l \neq 0$

(i) (l, β) *is estimable iff* $Xl \notin X(l)^{\perp}$,

(ii) $(l, \hat{\beta}) = ||l||^2 \dfrac{(q, y)}{(q, q)}$ *is BLUE of* (l, β), *where* $q = (I - P_{X(l)^{\perp}})Xl$,

(iii) $\mathrm{Var}(l, \hat{\beta}) = \sigma^2 ||l||^4 ||q||^{-2}$.

Proof:

The proof can already be found in Drygas (1976), but for the sake of completeness, it is repeated here.

(i) If $Xl \in X(l)^{\perp}$ then $Xl = Xl_1$ for some $l_1 \perp l$. Then $X(l - l_1) = 0$, $(l, l - l_1) = (l, l) - (l, l_1) = (l, l) \neq 0$. Hence $l \notin \mathrm{im}(X') = (X^{-1}(0))^{\perp}$.

(ii) Let $q = (I - P_{X(l)^{\perp}})Xl$. Then

$$\mathrm{E}(l, \hat{\beta}) = ||l||^2 (X'(I - P_{X(l)^{\perp}})Xl, \beta)/||q||^2.$$

Now

$$X'(I - P_{X(l)^{\perp}})Xl = X'P_{X'^{-1}(\mathrm{span}(l))}Xl = \lambda l$$

for some $\lambda \in \mathbb{R}$ and

$$\begin{aligned} \lambda ||l||^2 &= \lambda(l, l) = \lambda(l, X'(I - P_{X(l)^{\perp}})Xl) \\ &= \lambda(Xl, (I - P_{X(l)^{\perp}})Xl) = ||q||^2, \end{aligned}$$

i.e., $\lambda = ||q||^2 ||l||^{-2}$ and consequently

$$\mathrm{E}(l, \hat{\beta}) = (l, \beta).$$

(iii) (c, y) is BLUE of its expectation $(X'c, \beta) = (c, X\beta)$ iff $c \in \mathrm{im}(X)$. This follows from the linear version of the Lehmann-Scheffé Theorem which says that (c, y) is BLUE iff it is uncorrelated with every linear unbiased estimator of 0. (b, y) is an unbiased estimator of zero iff $X'b = 0$. Thus (c, y) is BLUE iff $(c, b) = 0$ for every $b \in X^{-1'}(0)$, i.e., $c \in (X'^{-1}(0))^{\perp} = \mathrm{im}(X)$. Since $q \in \mathrm{im}(X)$, the proof is complete.

Corollary 3 *Let x_i be the i-th column of X and X_i be the matrix obtained from X by deleting the i-th column. Then*

$$\hat{\beta}_i = \frac{(y, (I - X_i(X_i'X_i)^- X_i')x_i}{(x_i, (I - X_i(X_i'X_i)^- X_i')x_i)} \tag{17}$$

is BLUE of β_i.

Still the problem of an actual computation of the least squares estimator $(l, \hat{\beta})$ remains. To solve this we need an appropriate form of the Gram-Schmidt orthogonalization method.

Theorem 4. *Let $x_1, \ldots, x_k$ be the arbitrary vectors of the inner product vector-space V and let P_i be the orthogonal projection onto span$\{x_1, \ldots, x_i\}$. If $q_1 = x_1$ and*

$$q_i = x_i - P_{i-1}x_i, \; i = 2, \ldots, k$$

then $\{q_1, \ldots, q_k\}$ form a system of orthogonal vectors and span$\{x_1, \ldots, x_k\}$ = span$\{q_1, \ldots, q_k\}$.

Proof:

Firstly, we show that $\text{span}\{x_1, \ldots, x_i\} = \text{span}\{q_1, \ldots, q_i\}$, $i = 1, \ldots, k$. This assertion is true for $i = 1$ since $x_1 = q_1$. Since by induction assumption $P_{i-1}\ x_i \in \text{span}\{x_1, \ldots, x_{i-1}\} = \text{span}\{q_1, \ldots, q_{i-1}\}$ it follows that $x_i = q_i + P_{i-1}x_i \in \text{span}\{q_1, \ldots, q_i\}$. But $q_i \in x_i - P_{i-1}x_i \in \text{span}\{x_1, \ldots, x_i\}$. Thus $\text{span}\{x_1, \ldots, x_i\}$ and $\text{span}\{q_1, \ldots, q_i\}$ coincide. Moreover, if $i < j$ then from $q_i \in \text{span}\{q_1, \ldots, q_{j-1}\} = \text{span}\{x_1, \ldots, x_{j-1}\}$ it follows that

$$\begin{aligned}(q_i, q_j) &= (q_i, x_j) - (q_i, P_{j-1}x_j) = (q_i, x_j) - (P_{j-1}q_i, x_j) \\ &= (q_i, q_j) - (q_i, q_j) = 0.\end{aligned}$$

Clearly, $P_i y = \sum_{j \le i : q_j \ne 0} \dfrac{(q_j, y)q_j}{(q_j, q_j)}$ for both sides are equal to q_i if $y = q_i$ and vanish if $y \perp q_1, \ldots, q_i$. From the properties of projections it also follows that

$$||q_i||^2 = ||x_i||^2 - ||P_{i-1}x_i||^2 = ||x_i||^2 - \sum_{j \le i : q_j \ne 0} \frac{(x_i, q_j)^2}{||q_j||^2}.$$

This is an important formula for making computations.

Now we are in a position to solve the computation-problem for $(l, \hat{\beta})$. Let $l_1, \ldots, l_{k-1}$ be an basis of $(l)^\perp$ and perform the Gram-Schmidt orthogonalization procedure for

$$Xl_1, \ldots, Xl_{k-1}, Xl.$$

Then

$$\begin{aligned}q_k &= Xl - P_{\text{span}\{Xl_1, \ldots, Xl_{k-1}\}}Xl \\ &= Xl - (P_{X(l)^\perp})Xl = (I - P_{X(l)^\perp})Xl.\end{aligned}$$

Thus

$$\frac{||l||^2(q_k, y)}{||q_k||^2} \tag{18}$$

is the BLUE of (l, β).

In order to find the BLUE of $\beta_i = (e_i, \beta) = e_i'\beta$, e_i the i-th unit-vector, we have to apply the orthogonalization process to the sequence

$$x_1, \dots, x_{i-1}, x_k, x_{i+1}, \dots, x_{k-1}, x_i.$$

Having found one formula, the other formulae are just obtained by interchanging index k and i.

As an example let us at first consider the case $k = 2$. Then

$$q_1 = x_1,\ q_2 = x_2 - \frac{(x_2, x_1)}{(x_1, x_1)} x_1,\ \text{if}\, x_1 \neq 0,$$

(if $x_1 = 0$, then $q_2 = x_2$ and we have indeed the case $k = 1$),

$$||q_2||^2 = (x_2, x_2) - \frac{(x_1, x_2)^2}{(x_1, x_1)}$$

and

$$\begin{aligned} \hat{\beta}_2 &= \frac{(q_2, y)}{||q_2||^2} \\ &= \frac{(x_2, y) - \dfrac{(x_2, x_1)}{(x_1, x_1)}(x_1, y)}{(x_2, x_2) - \dfrac{(x_1, x_2)^2}{(x_1, x_1)}} \\ &= \frac{(x_1, x_1)(x_2, y) - (x_2, x_1)(x_1, y)}{(x_2, x_2)(x_1, x_1) - (x_1, x_2)^2}. \end{aligned} \tag{19}$$

(if $q_2 = 0$, then β_2 is not estimable and $\hat{\beta}_1 = \dfrac{(x_1, y)}{(x_1, x_1)}$).

By interchanging index 1 and 2 we get $\hat{\beta}_1$:

$$\hat{\beta}_1 = \frac{(x_2, x_2)(x_1, y) - (x_2, x_1)(x_2, y)}{(x_2, x_2)(x_1, x_1) - (x_1, x_2)^2}. \tag{20}$$

From $R_1\hat{\beta} = \hat{\alpha}$, we get an alternative representation of $\hat{\beta}_1$, namely

$$\hat{\beta}_1 = \frac{(x_1, y)}{(x_1, x_1)} - \frac{(x_2, x_1)}{(x_1, x_1)}\hat{\beta}_2. \tag{21}$$

This gives a possibility to check the computation of $\hat{\beta}_1$ and $\hat{\beta}_2$ computationally and numerically.

Now let us also consider the example $k = 3$

$$q_1 = x_1,\ q_2 = x_2 - \frac{(x_2, x_1)}{(x_1, x_1)} x_1,\ q_3 = x_3 - \frac{(x_3, x_1)}{(x_1, x_1)} x_1 - \frac{(x_3, q_2)}{(q_2, q_2)} q_2, \tag{22}$$

$$||q_3||^2 = (x_3, x_3) - \frac{(x_3, x_1)^2}{(x_1, x_1)} - \frac{(x_3, q_2)^2 (x_1, x_1)}{(x_2, x_2)(x_1, x_1) - (x_2, x_1)^2}. \tag{23}$$

After some rearrangement the formula

$$\begin{aligned} q_3 = x_3 &- \frac{(x_3, x_2)(x_1, x_1) - (x_2, x_1)(x_3, x_1)}{(x_2, x_2)(x_1, x_1) - (x_2, x_1)^2} x_2 \\ &+ \left(\frac{(x_2, x_1)}{(x_1, x_1)} \frac{(x_3, x_2)(x_1, x_1) - (x_2, x_1)(x_3, x_1)}{(x_2, x_2)(x_1, x_1) - (x_2, x_1)^2} - \frac{(x_3, x_1)}{(x_1, x_1)} \right) x_1 \end{aligned} \tag{24}$$

is obtained.

By interchanging Index 3 and 1 we get:

$$\begin{aligned} q_3^{(1)} = x_1 &- \frac{(x_1, x_2)(x_3, x_3) - (x_2, x_3)(x_3, x_3)}{(x_3, x_3)(x_2, x_2) - (x_2, x_3)^2} x_2 \\ &+ \left(\frac{(x_2, x_3)}{(x_3, x_3)} \frac{(x_1, x_2)(x_3, x_3) - (x_2, x_3)(x_3, x_1)}{(x_2, x_2)(x_3, x_3) - (x_2, x_3)^2} - \frac{(x_3, x_1)}{(x_3, x_3)} \right) x_3 \end{aligned} \tag{25}$$

and by interchanging Index 3 and 2:

$$\begin{aligned} q_3^{(2)} = x_2 &- \frac{(x_3, x_2)(x_1, x_1) - (x_3, x_1)(x_2, x_1)}{(x_3, x_3)(x_1, x_1) - (x_3, x_1)^2} x_3 \\ &+ \left(\frac{(x_2, x_1)}{(x_1, x_1)} \frac{(x_3, x_2)(x_1, x_1) - (x_2, x_3)(x_2, x_1)}{(x_3, x_3)(x_1, x_1) - (x_3, x_1)^2} - \frac{(x_2, x_1)}{(x_3, x_3)} \right) x_1. \end{aligned} \tag{26}$$

Of course, $\hat{\beta}_1$ and $\hat{\beta}_2$ can be obtained from $\hat{\beta}_3$ and $\hat{\alpha}_i = \dfrac{(q_i, y)}{(q_i, q_i)}$, $i = 1, 2$ by solving $R_1 \hat{\beta} = \hat{\alpha}$.

A final remark on the computation. There are, of course, computer-programs such as R, SPSS etc. by which estimates of regression coefficients can easily and efficiently be calculated. But even with a pocket-calculator or similar equipment it is possible to calculate the formula derived in this section.

A pocket-calculator can calculate from a data-array x (an n-dimensional vector x) the (empirical) mean and the (empirical) standard-deviation (SD), if the numbers are plugged in. Moreover, for two arrays x and y the (empirical) linear regression $y = \alpha + \beta x$ can be computed by just plugging in all numbers. The above formulae show that from the inner products (x_i, x_j) and (x_i, y) the regression coefficients estimators can be obtained by some very few elementary calculations. The inner product (x, x) can be obtained from mean and variance. Now the estimator of regression coefficient β in the regression $y = \alpha + \beta x$ is given by

$$\hat{\beta} = \frac{(x, y) - n\bar{x}\bar{y}}{\sum_{i=1}^{n}(x_i - \bar{x})^2}. \tag{27}$$

Thus the inner product (x, y) can be obtained from mean, variance and regression coefficient. Another method is to use the formula

$$(x, y) = \frac{1}{2}((x + y, x + y) - (x, x) - (y, y))$$

and computing the standard deviation of $x + y$. An alternative to (28) is the Jordan v.Neumann formula

$$(x, y) = \frac{1}{4}(||x + y||^2 - ||x - y||^2) = \frac{1}{4}((x + y, x + y) - (x - y, x - y)).$$

Here mean and variance of $x + y$ and $x - y$ are needed.

3 Linear Sufficiency

Baksalary and Kala (1981) and Drygas (1983) have introduced the concept of a linearly sufficient statistic $z = Ty$ in the linear model $\mathrm{E}y = X\beta$, $\mathrm{Cov}(y) = \sigma^2 Q$. Ty is called linear sufficient if the BLUE of $\mathrm{E}(y)$ can be computed from Ty alone. The following theorem was proved:

Theorem 5. *Ty is linearly sufficient iff*

$$\operatorname{im}(X) \subseteq \operatorname{im}(WT'), \tag{28}$$

where $W = Q + cXX'$ is such that $c \geq 0$ and $\operatorname{im}(X) \subseteq \operatorname{im}(W)$.

If we introduce the inner product $(x, y)_{W^+} = (x, W^+y) = x'W^+y$, then WT is just the adjoint T^* of T with respect to this inner product. Indeed, for $x, y \in \operatorname{im}(W)$

$$\begin{aligned}(Tx, y) = (x, T'y) &= (WW^+x, T'y) = (W^+x, WT'y) \\ &= (x, WT'y)_{W^+} = (x, T^*y)_{W^+}.\end{aligned} \tag{29}$$

Thus $\operatorname{im}(X) \subseteq \operatorname{im}(T^*)$ is the more transparent formulation of linear sufficiency.

Let us now return to the model $\operatorname{E}(y) = X\beta$, $\operatorname{Cov}(y) = \sigma^2 I$ and let $q_1, \ldots, q_s, q_{s+1}, \ldots, q_k$ be the vectors obtained from the columns of X by applying the Gram-Schmidt orthogonalization procedure to them. Without restricting generality it can be assumed that $q_{s+1} = q_{s+2} = \ldots = q_k = 0$. Then let

$$Q = (q_1, \ldots, q_s, 0, \ldots, 0) = (q_1, \ldots, q_k)$$

and

$$Q_1 = (q_1, \ldots, q_s).$$

Theorem 6. *$Q_1'y$ and $Q'y$ are linearly sufficient statistics.*

Proof:

Since $\operatorname{im}(X) = \operatorname{im}(Q) = \operatorname{im}(Q_1)$ it follows from theorem 5 that $Q'y$ and $Q_1'y$ are linearly sufficient.

Linear sufficiency means that the BLUE of $X\beta$ can be obtained from $Q'y$ and $Q_1'y$, respectively. Since $Q(Q'Q)^-Q'y$ and $Q_1(Q_1'Q_1)^{-1}Q_1'y$, respectively are the orthogonal projections onto $\operatorname{im}(X)$, this property is evident.

The model $\operatorname{E}(y) = X\beta$ can be split up into $X = (X_1 \vdots X_2)$, where $X_2 = X_1 A$. Thus $\operatorname{E}(y) = X_1(\beta_1 + A\beta_2)$, where $\beta = (\beta_1', \beta_2')$. Then $X_1 = Q_1 R_{11}$ and $X_2 = Q_1 R_{12}$ and

$$Q_1(Q_1'Q_1)^{-1}Q_1' = Q_1 R_{11} R_{11}^{-1}(Q_1'Q_1)^{-1}Q_1'y = X_1 R_{11}^{-1}(Q_1'Q_1)^{-1}Q_1'y.$$

Thus $R_{11}^{-1}(Q_1'Q_1)^{-1}Q_1'y$ is the BLUE of $\beta_1 + A\beta_2$ in the model $\mathrm{E}(y) = X_1(\beta_1 + A\beta_2)$, $\mathrm{Cov}(y) = \sigma^2 I$.

A similar representation is more complicated for Q. At first

$$X = QR,\ R = \begin{pmatrix} R_1 \\ \cdots \\ 0 \end{pmatrix} = \begin{pmatrix} R_{11} & \vdots & R_{12} \\ \cdots & \cdots & \cdots \\ 0 & \vdots & 0 \end{pmatrix}. \tag{30}$$

By Searle (1971, p. 4) generalized inverse of R and $Q'Q$ can be found as follows:

$$R^- = \begin{pmatrix} R_{11}^{-1} & \vdots & 0 \\ \cdots & \cdots\cdots \\ 0 & \vdots & 0 \end{pmatrix},\ (Q'Q)^- = \begin{pmatrix} (Q_1'Q_1)^{-1} & \vdots & 0 \\ \cdots & & \cdots\cdots \\ 0 & & \vdots\ 0 \end{pmatrix}. \tag{31}$$

Then $RR^-(Q'Q)^- = (Q'Q)^-$ and

$$Q(Q'Q)^-Q' = QRR^-(Q'Q)^-Q' = XR^-(Q'Q)^-Q'. \tag{32}$$

Thus $R^-(Q'Q)^-Q'y = \hat{\beta}$ can be considered as an estimator of β in the sense that $(l, \hat{\beta})$ is BLUE of (l, β) whenever (l, β) is estimable.

It is remarkable to note that in the model $\mathrm{E}(Q'y) = Q'X\beta = (Q'Q)R\beta$, $\mathrm{Cov}(Q'y) = \sigma^2(Q'Q)$ there is a linear unbiased estimator of β if and only if $Q'Q$ is regular, i.e., $s = k$. In this case

$$R^{-1}(Q'Q)^{-1}y = \hat{\beta}$$

is the only unbiased estimator. It is at the same time the Aitken-estimator

$$\begin{aligned} &((Q'Q)R)'(Q'Q)^{-1}((Q'Q)R)^{-1}((Q'Q)R)'(Q'Q)^{-1}y \\ &= (R'(Q'Q)R)^{-1}R'y \\ &= R^{-1}(Q'Q)^{-1}(R')^{-1}R'y \\ &= R^{-1}(Q'Q)^{-1}y. \end{aligned}$$

4 Application: Diabetes Mellitus

Diabetes mellitus is a disease where the autoregulation of metabolism is disturbed. Normally after eating, the content of glucosis in the blood decreases very rapidly after perhaps one hour to a value below 100 mg/dl (5.55 mmol/l). Under diabetes mellitus it takes perhaps 4 hours to reach such a value - even under medicaments. However, by physical training it is possible to get a lower value perhaps already after two hours.

We will here discuss the behavior of glucosis during the night and the early morning. Though there is in general no intensive consumption of food in the late evening and during the night an acceptable value in the morning seems to pose a serious problem.

Some years ago it was said that it can not be recommended to use antidiabetic oral medicaments in the late evening unless you eat regularly during the night. To avoid hypoglycemia during the night (or counter-reactions), it was recommended to eat one bread-unit (12 g carbon-hydrats) just before bedtime. In later years the opinion about this procedure was changed and oral antidiabetic medicaments of an appropriate dose were also recommended before bedtime. There was, however, no recommendation of change of the amount of food that should be consumed at bedtime. The patients continued to eat one bread-unit (BU) just before bedtime. Thus simultaneously measures against too high and too low glucosis-values were taken. This seems to be a rather contradictory proposal. In my opinion with fixed medicaments, the amount of food just before bedtime should depend on the value of glucosis at this time. I finally decided that 100 mg/dl (5.55 mmol/l) and 150 mg/dl (8.32 mmol/l) should be the critical values for this decision. If the glucosis is equal or above 150 mg/dl (8.32 mmol/l) you should not eat anything. If however, the value is equal or below 100 mg/dl, then you should eat one bread-unit (1 BU). If the value is between 100 and 150, then the last meal of day is less accordingly. For example, if the value is 120, then 0.6 BU are eaten.

If you awake during the night between perhaps 1 A.M. and 2 A.M. you repeat the procedure of the evening. If the value is equal or below 100mg/dl, you eat 1 BU and you eat no BU for values equal or above 150 mg/dl. For values between 100 and 150, you eat a fraction of a BU.

I have pursued this method since some time and I have constructed and computed a regression model describing the behavior of the changes of the glucosis-values. We have the following Variables:

- y, change of the glucosis-values, i.e., the difference of the value during the night and the value in the evening or the difference of the values in the morning and value during the night
- x, the amount of food, measured in BU, taken in the evening and during the night, respectively and
- t, time between two measurements.

The model which will be studied in the sequel is as follows:

$$y = \alpha(t - D) + \beta x + \epsilon. \tag{33}$$

Here

$$D = I_{\{x>0\}} = \begin{cases} 1 & x > 0 \\ 0 & x = 0\,. \end{cases}$$

The idea behind this modelation is as follows: If you eat an amount x then the glucosis will increase and it will reach its highest point after approximately one hour.

Since the model (33) is difficult to treat by a pocket-calculator, we have changed it to

$$\hat{y} = \frac{y}{(t-D)} = \alpha + \beta\hat{x} + \delta\ ,\ \hat{x} = \frac{x}{(t-D)}.$$

Table 1. Night/morning measurements (transformed)

No.	Date	y	t	x	$\hat{y} = y/(t-D)$	$\hat{x} = x/(t-D)$
1	26.3.07	-23	5.22	0.4	-5.4502	0.0948
2	28.3.07	-42	5.82	0.2	-8.7137	0.0415
3	29.3.07	30	3.12	1.0	14.1509	0.4717
4	30.3.07	-41	5.18	0.4	-9.1086	0.0957
5	31.3.07	-28	6.22	0	-4.5016	0
6	2.4.07	-29	5.28	0.6	-6.7757	0.1402
7	4.4.07	-13	3.53	0.6	-5.1383	0.2372
8	5.4.07	9	5.05	1.0	2.2222	0.2469
9	6.4.07	-48	5.17	0.4	-10.0719	0.0959
10	7.4.07	-18	3.42	0.2	-7.4380	0.0826
m=Mean		-19.7		0.48	-4.1525	0.15065
σ=Standard Deviation		23.4665		0.3293	7.3329	0.1368

Regression $\hat{y} = \hat{\alpha} + \hat{\beta}\hat{x}$, $\hat{y} = -11.2354 + 47.0156\hat{x}$, r=0.8768090
$1 - r^2 = 0.2312059$, $\sum_{i=1}^{10}(\hat{y}_i - \bar{\hat{y}})^2 = 9\sigma_0^2 = 483.9428$,
$\hat{\sigma}^2 = \frac{1}{8} \times 0.2312059 \times 483.9428 = 13.986304 = (3.7398)^2$.

Table 2. Evening/night measurements (transformed)

No.	Date	y	t	x	$\hat{y} = y/(t-D)$	$\hat{x} = x/(t-D)$
1	25./26.3.07	-49	2.98	0	-1.6443	0
2	26./27.3.07	21	3.6	0	5.8333	0
3	27./28.3.07	-39	3.85	0	-10.1299	0
4	28./29.3.07	-85	3.73	0	-22.7882	0
5	29./30.3.07	0	3.57	0.2	0	0.0778
6	30./31.3.07	81	3.55	1	31.7647	0.3922
7	31.3/1.4.07	79	4.77	1	20.9549	0.2653
8	1.4./2.4.07	-38	3.73	0	-10.1877	0
9	2.4./3.4.07	-26	3.8	0	-6.8421	0
10	3.4./4.4.07	-3	5.03	0.6	-0.7444	0.1489
m=Mean		-5.9		0.28	0.62163	0.08842
σ=Standard Deviation		53.9247		0.4237	15.8256	0.1391
Regression $\hat{y} = \hat{\alpha} + \hat{\beta}\hat{x}$, $\hat{y} = -8.1859 + 99.6107\hat{x}$, $r = 0.8758071$ $1 - r^2 = 0.23296$, $\sum_{i=1}^{10}(\hat{y}_i - \bar{\hat{y}})^2 = 9\sigma_0^2 = 2254.05$ $\hat{\sigma}^2 = \frac{(1-r^2)}{8} \times \sum_{i=1}^{10}(\hat{y}_i - \bar{\hat{y}})^2 = 65.6378 = (8.1017)^2$						

The variation of y is very high, so is the coefficient $\hat{\beta}$.

Since the value for the time t, as we will see later, is almost identical for most values, so it could be hoped that there will be no large difference in the estimated regression coefficients. We have to see whether this is actually the case.

We begin with the Night/morning values. We now proceed to Table 3 with the original data.

In order to find the measure of determination we must compute (q_1, y) and (q_2, y). Here

$$q_1 = x,$$
$$q_2 = (t - D) - \frac{(t - D, x)}{(x, x)} x$$

Table 3. Night/morning measurements (original)

No.	Date	y	t	x	$t-D$	$y+x$	$y+t-D$	$x+t-D$
1	26.3.07	-23	5.22	0.4	4.22	-22.6	-18.78	4.62
2	28.3.07	-42	5.82	0.2	4.82	-41.8	-37.18	5.02
3	29.3.07	30	3.12	1.0	2.12	31.0	32.12	3.12
4	30.3.07	-41	5.18	0.4	4.18	-40.6	-36.82	4.58
5	31.3.07	-28	6.22	0	6.22	-28	-21.78	6.22
6	2.4.07	-29	5.28	0.6	4.28	-28.4	-24.72	4.88
7	4.4.07	-13	3.53	0.6	2.53	-12.4	-10.47	3.13
8	5.4.07	9	5.05	1.0	4.05	10.0	13.05	5.05
9	6.4.07	-42	5.17	0.4	4.17	-41.6	-37.83	4.57
10	7.4.07	-18	3.42	0.2	2.42	-17.8	-15.58	2.62
m=Mean		-19.7		0.48	3.901	-19.22	-15.791	4.381
s=Standard Deviation		23.4665		0.3293	1.2421	23.7229	22.7683	1.0993
$SQ = 9s^2 + 10m^2$		8837		3.28	166.0631	8759.08	7161.6633	202.8071

$$(y,x) = \tfrac{1}{2}((y+x,y+x)-(y,y)-(x,x)) = -40.6$$

$$(y,t-D) = \tfrac{1}{2}((y+t-D,y+t-D)-(y,y)-(t-D,t-D)) = -920.699$$

$$(x,t-D) = \tfrac{1}{2}((x+t-D,x+t-D)-(x,x)-(t-D,t-D)) = 16.732$$

$$\hat{\alpha} = \frac{(y,t-D)(x,x)-(y,x)(t-D,x)}{(x,x)(t-D,t-D)-(x,t-D)^2} = \frac{-2340.5765}{264.7272} = -8.8415$$

$$\hat{\beta} = \frac{-(t-D,x)(y,t-D)-(t-D,t-D)(y,x)}{(x,x)(t-D,t-D)-(x,t-D)^2} = \frac{-8662.9888}{264.7272} = 32.7242$$

$$= \frac{(y,x)-(t-D,x)\hat{\alpha}}{(x,x)} = \frac{147.936-40.6}{3.28} = \frac{107.336}{3.28} = 32.7244$$

Thus the regression function equals $y = -8.8415(t-D) + 32.724x$.

$$\begin{aligned}
(q_1,y) &= (x,y) = -40.6,\\
(q_2,y) &= (t-D,y) - \frac{(t-D,x)}{(x,x)}(x,y)\\
&= -920.6999 + \frac{16.732}{3.28}40.6 = -713.5904,\\
(y,y) &= 8837,
\end{aligned}$$

$$
\begin{aligned}
(q_1, q_1) &= (x, x) = 3.28, \\
(q_2, q_2) &= (t - D, t - D) - \frac{(t - D, x)^2}{(x, x)} \\
&= 166.0631 - \frac{16.732^2}{3.28} = 80.7095, \\
(q_1, y)^2 &= 1648.36, \\
(q_2, y)^2 &= (713.5904)^2 = 509211.259, \\
R^2 &= 0.05687 + 0.71395 = 0.77082.
\end{aligned}
$$

The estimator $\hat{\sigma}^2$ of σ^2 is given by

$$
\hat{\sigma}^2 = \frac{1}{8}(1 - R^2)(y, y) = 253.1580 = (15.9109)^2.
$$

Finally we analyze the Evening/Night measurements in their original form, see Table 4 for this. Thus the (estimated) regression function is

$$
y = -10.3167(t - D) + 105.68752x \,.
$$

For the computation of the measure of determination let

$$
\begin{aligned}
q_1 &= (t - D), \\
q_2 &= x - \frac{(x, t - D)}{(t - D, t - D)}(t - D).
\end{aligned}
$$

Then

$$
\begin{aligned}
(q_1, y) &= -285.86998, \\
(q_1, q_1) &= 122.49, \\
(q_2, y) &= 158.2 - \frac{(9.252)(-285.86998)}{122.49} = 179.79523, \\
(q_2, q_2) &= (x, x) - \frac{(x, t - D)^2}{(t - D, t - D)} = 2,4 - \frac{9.252^2}{122.49} = 1.70117.
\end{aligned}
$$

Thus

$$
\begin{aligned}
R^2 &= \frac{1}{(y, y)} \left(\frac{(q_1, y)^2}{(q_1, q_1)} + \frac{(q_2, y)^2}{(q_2, q_2)} \right) \\
&= 0.025181 + 0.07165366 = 0.7417176.
\end{aligned}
$$

In the evening/night-tables, the high variance of y and the high coefficients of the x-variables is very remarkable.

Table 4. Evening/night measurements (original)

No.	Date	y	x	t	$t-D$
1	25./26.3.07	-49	0	2.98	2.98
2	26./27.3.07	21	0	3.6	3.6
3	27./28.3.07	-39	0	3.85	3.85
4	28./29.3.07	-85	0	3.73	3.73
5	29./30.3.07	0	0.2	3.57	2.57
6	30./31.3.07	81	1	3.55	2.55
7	31.3/1.4.07	79	1	4.77	3.77
8	1.4./2.4.07	-38	0	3.73	3.73
9	2.4./3.4.07	-26	0	3.8	3.8
10	3.4./4.4.07	-3	0.6	5.03	4.03
m=Mean		-5.9	0.28		3.461
s=Standard Deviation		53.1247	0.4237		0.5482
$9s^2 = \sum_{i=1}^{n}(z_i - \bar{z})^2$		26170.90	1.6160		2.4707
$SQ = 9s^2 + 10m^2$		26519	2.4		122.49

$y = a_1 + b_1 x$, $y = -36.1733 + 108.1188x$, $r = 0.8496$

$(y,x) - 10\bar{x}\bar{y} = 174.7199$, $(y,x) = 158.2$

$y = a_2 + b_2(t-D) = 98.6086 - 30.1961(t-D)$, $r = -0.3070$

$(y, t-D) - 10\bar{y}(\bar{t} - \bar{D}) = -81.67098$, $(t-D) = -285.86998$

$t - D = a_3 + b_3 x = -3.537 - 0.2715x$, $r = -0.20589$

$(t-D,x) - 10(\bar{t} - \bar{D})\bar{x} = -0.4388$, $(t-D,x) = 9.252$, $(t-D,x)^2 = 85.5995$

$\text{Det} = (x,x)(t-D,t-D) - (x,t-d)^2 = 208.3762$

$A_1 = (t-D,t-D)(y,x) - (t-D,x)(y,t-D) = 22022.76414$

$A_2 = -(t-D,x)(y,x) + (x,x)(y,t-D) = -2149.7544$

$\hat{\beta} = \frac{A_1}{\det} = 105.68752$

$\hat{\alpha} = \frac{A_2}{\det} = \frac{[(y,t-D) - \hat{\beta}(x,t-D)]}{(t-D,t-D)} = -10.3167.$

$$\hat{\sigma}^2 = \frac{1}{8}(1 - R^2)(y,y) = 854.2538 = (29.22762)^2.$$

It may perhaps be interesting how large the fastening actual values in the morning have been. These were as follows:

Date	26.3	27.3	28.3	29.3	30.3	31.3	1.4	2.4	3.4	4.4	5.4	6.4	7.4
mg/dl	104	99	97	103	98	124	94	98	98	104	108	86	123

This is very a good result because these values are normal or close to normality. The relative high values on 31th of March and 7th of April can be explained by an additional medicament taken on that day (Saturday).

References

Baksalary JK, Kala R (1981) Linear transformation preserving the best linear unbiased estimator in a general Gauss-Markov model. Annals of Statistics 9:913-916

Björck-Dahlquist (1972) Numerische Methoden. R.Oldenbourg Verlag München Wien

Drygas H (1976) Weak and Strong Consistency of the Least Squares Estimator in Regression Models. Zeitschr. f. Wahrsch.-Th. u. verwandte Gebiete, 34:119-127

Drygas H (1983) Sufficiency and Completeness in the general Gauss-Markov model, Sankhya (A), Vol. 43, Part 1, 88-98

Lawson CL, Hanson RJ (1974) Solving Least Squares Problems. Prentice-Hall Inc., New Jersey

Searle SR (1971) Linear Models. Wiley, New York

On Penalized Least-Squares: Its Mean Squared Error and a Quasi-Optimal Weight Ratio

Burkhard Schaffrin

School of Earth Sciences, Ohio State University, Columbus, Ohio, U.S.A. and Geodetic Institute, University of Karlsruhe, Germany schaffrin.1@osu.edu

1 Introduction

It is well known in a Random Effects Model, that the Best inhomogeneously LInear Prediction (inhomBLIP) of the random effects vector is equivalently generated by the standard Least-Squares (LS) approach. This LS solution is based on an objective function that consists of two parts, the first related to the observations and the second to the prior information on the random effects; for more details, we refer to the book by Rao, Toutenburg, Shalabh and Heumann (2008). We emphasize that, in this context, the second part cannot be interpreted as "penalization term".

A very similar objective function, however, could be applied in the Gauss-Markov model where no prior information is available for the unknown parameters. In this case, the additional term would serve as "penalization" indeed as it forces the Penalized Least-Squares (PLS) solution into a chosen neighborhood, not specialized through the model. This idea goes, at least, back to Tykhonov (1963) and Phillips (1962) and has since become known as (a special case of) "Tykhonov regularization" for which the weight ratio between the first and the second term in the objective function determines the degree of smoothing to which the estimated parameters are subjected to. This weight ratio is widely known as "Tykhonov regularization parameter"; for more details, we refer to Grafarend and Schaffrin (1993) or Engl et al. (1996), for instance.

It is now of interest how the MSE-matrices in the two abovementioned cases differ even though the numerical results for the parameters may coincide. Moreover, we shall explore the possibility of variance component estimation in the random effects model to find a new "quasi-optimal regularization parameter" in the Gauss-Markov model, independent of the L-curve approach by Hansen and O'Leary (1993) and the principle of cross-validation favored by Golub et al. (1979). Such an idea was first proposed by Schaffrin (1995) and later sketched out by Schaffrin (2005).

2 A Brief Review of Standard Least-Squares in the Random Effects Model

Let us introduce the following Random Effects Model (REM)

$$y = Ax + e, \;\; \beta_0 = x + e_0 \quad (\beta_0 \text{ is given}), \tag{1}$$
$$e \sim (0, \Sigma), \; e_0 \sim (0, \Sigma_0), \; C\{e, e_0\} = 0,$$

where y denotes a $n \times 1$ vector of observations (increments), A is a known $n \times m$ coefficient matrix with $\text{rank}(A) \leq m < n$, x is a $m \times 1$ vector of unknown random effects (increments), β_0 is the $m \times 1$ vector of "prior information" on x, e is a $n \times 1$ random error vector with zero expectation and positive-definite dispersion matrix Σ, e_0 is a $m \times 1$ random error vector with zero expectation and positive-definite dispersion matrix Σ_0, and C denotes the "covariance". Note that e and e_0 are uncorrelated.

Now the standard (weighted) least-squares solution for x is based on the minimization of the objective function

$$\Phi(x) := (y - Ax)^T P (y - Ax) + \lambda(\beta_0 - x)^T P_0 (\beta_0 - x)$$

where the weight matrices may be chosen as $P := \sigma_0^2 \Sigma^{-1}$ and $P_0 := (\sigma_0^2/\lambda)\Sigma_0^{-1}$, respectively. Obviously, σ_0^2 denotes the observational variance component, and λ denotes the appropriate variance ratio between observational and prior information. Now we obtain

$$\tilde{x} = (N + \lambda P_0)^{-1}(c + \lambda P_0 \beta_0) \text{ for } [N, c] := A^T P\,[A, y] \tag{2}$$

as standard LS-solution, which incidentally yields the Best inhomogeneously LInear Prediction (inhomBLIP) of x following Goldberger (1962), Schaffrin (1985), or Rao, Toutenburg, Shalabh and Heumann (2008). The latter is defined as a linear prediction

$$\tilde{x} = Ly + \gamma_0 \tag{3}$$

for which the $m \times n$ matrix L and the $m \times 1$ vector γ_0 are to be determined in such a way that the trace of the mean squared error matrix

$$
\begin{aligned}
MSE(\tilde{x}) &= D(\tilde{x}-x) + E(\tilde{x}-x)E(\tilde{x}-x)^T \\
&= D\left[(LA - I_m)x + Le\right] \\
&\quad + E\left[(LA - I_m)x + \gamma_0\right] E\left[(LA - I_m)x + \gamma_0\right]^T \\
&= L(\Sigma + A\Sigma_0 A^T)L^T - LA\Sigma_0 - \Sigma_0 A^T L^T + \Sigma_0 \\
&\quad + \left[\gamma_0 - (I_m - LA)\beta_0\right]\left[\gamma_0 - (I_m - LA)\beta_0\right]^T
\end{aligned} \tag{4}
$$

is being minimized where E denotes "expectation" and D "dispersion". This objective function immediately leads to

$$\gamma_0 = (I_m - LA)\beta_0$$

and

$$(\Sigma + A\Sigma_0 A^T)L^T - A\Sigma_0 = 0$$

from which we first obtain the matrix L as

$$
\begin{aligned}
L &= \Sigma_0 A^T(\Sigma + A\Sigma_0 A^T)^{-1} = \Sigma_0 A^T(I_n + \Sigma^{-1}A\Sigma_0 A^T)^{-1}\Sigma^{-1} \\
&= \Sigma_0(I_m + \sigma_0^{-2}N\Sigma_0)^{-1}A^T\Sigma^{-1} = (N + \lambda P_0)^{-1}A^T P,
\end{aligned}
$$

and finally the predicted vector

$$\tilde{x} = \beta_0 + L(y - A\beta_0) = \beta_0 + (N + \lambda P_0)^{-1}(c - N\beta_0)$$

in full agreement with (2). Apparently, this solution of type inhomBLIP is automatically *weakly unbiased* in the sense of

$$E(\tilde{x}) = \beta_0 = E(x) \text{ for the given } \beta_0 \tag{5}$$

and, therefore, represents the Best inhomogeneously LInear (weakly) Unbiased Prediction (inhomBLUP) as well.

Consequently, the mean squared error matrix from (4) results in

$$
\begin{aligned}
MSE(\tilde{x}) &= D(\tilde{x}-x) = L\Sigma L^T + (LA - I_m)\Sigma_0(LA - I_m)^T \\
&= \sigma_0^2(N + \lambda P_0)^{-1}N(N + \lambda P_0)^{-1} \\
&\quad + (N + \lambda P_0)^{-1}(N - N - \lambda P_0)\Sigma_0 \\
&\quad \times (N - N - \lambda P_0)(N + \lambda P_0)^{-1} \\
&= (N + \lambda P_0)^{-1}(\sigma_0^2 N + \lambda^2 P_0\Sigma_0 P_0)(N + \lambda P_0)^{-1} \\
&= \sigma_0^2(N + \lambda P_0)^{-1}
\end{aligned} \tag{6}
$$

which differs from the dispersion matrix

$$D(\tilde{x}) = L\Sigma L^T = \sigma_0^2(N + \lambda P_0)^{-1}N(N + \lambda P_0)^{-1}. \tag{7}$$

We again emphasize that the second term in the objective function from (2) should not be misinterpreted as "penalization" in this model.

3 Penalized Least-Squares in the Gauss-Markov Model

In order for the second term in the objective function from (2) to be interpreted as "penalization", only the first term ought to refer to random errors in the model. Consequently, we have to restrict ourselves to the observation equations of a standard Gauss-Markov model, namely

$$y = A\xi + e, \quad e \sim (0, \sigma_0^2 P^{-1}), \tag{8}$$

where the random effects vector x is now replaced by the $m \times 1$ vector ξ of fixed (i.e., nonrandom) parameters.

This automatically entails that $R := \lambda P_0$ can no longer be related to a cofactor matrix (such as $\sigma_0^{-2}\Sigma_0$ in section 2). Instead, the symmetric positive-(semi)definite matrix P_0 must be chosen independently to define the degree of penalization relative to the components of the vector $\beta_0 - \xi$ of deviations from an "educated guess" (such as β_0).

In contrast, we shall leave the coefficient λ unspecified at this point, but emphasize that it no longer describes a variance ratio. An interpretation as "weight ratio", however, is still possible as it regulates the impact of the penalization term on the final estimates of ξ, oftentimes felt as (over-)smoothing; for more details, see section 4 below.

Now the application of (R-weighted) Penalized Least-Squares (PLS) in minimizing the objective function

$$\Phi(\xi) := (y - A\xi)^T P(y - A\xi) + (\beta_0 - \xi)^T R(\beta_0 - \xi)$$

leads to "Tykhonov regularization" with

$$\hat{\xi}_\lambda = (N + R)^{-1}(c + R\beta_0) = \beta_0 + (N + R)^{-1}(c - N\beta_0)$$

as a linear estimate of ξ which, however, turns out to be biased in the GM-Model from (8) according to:

$$\begin{aligned} E(\hat{\xi}_\lambda) - \xi &= (N + R)^{-1}(E(c) + R\beta_0) - \xi \\ &= (N + R)^{-1}\left[(N\xi + R\beta_0) - (N + R)\xi\right] \\ &= (N + R)^{-1}R(\beta_0 - \xi) \neq 0, \end{aligned} \tag{9}$$

unless the "guess" β_0 happens to coincide with the unknown parameter vector ξ. (If we knew this right away, we would not bother to estimate ξ.)

As a consequence, the mean squared error matrix for the PLS-solution is computed as follows:

$$\begin{aligned} MSE(\hat{\xi}_\lambda) &= D[\hat{\xi}_\lambda] + [E(\hat{\xi}_\lambda) - \xi][E(\hat{\xi}_\lambda) - \xi]^T \\ &= (N+R)^{-1}\left[D(c) + R(\beta_0 - \xi)(\beta_0 - \xi)^T R\right](N+R)^{-1} \\ &= \sigma_0^2 (N+R)^{-1}\left[N + R(\beta_0 - \xi)\sigma_0^{-2}(\beta_0 - \xi)^T R\right](N+R)^{-1}. \end{aligned} \tag{10}$$

This matrix cannot coincide with that in (6) unless the rank-1 matrix

$$(\beta_0 - \xi)\sigma_0^{-2}(\beta_0 - \xi)^T \in \left\{R^- \,|RR^- R = R\right\} \tag{11}$$

belongs to the g-inverses of $R := \lambda P_0$ for which the matrix R needs to have the rank 1 itself. Obviously, this would rather be the exception from the rule.

Therefore, it makes sense to ask the question under which condition the mean squared error matrix in (10) of the penalized least-squares solution within the GM-model turns out to be better than the mean squared error matrix in (6) of the inhomBLIP within the random effects model, which is also known as "Bayesian estimate"; cf. Rao (1976).

Theorem 1. *The penalized least-squares solution $\hat{\xi}_\lambda$ with $R := \lambda P_0$ is superior to the "Bayesian estimate" $\tilde{x}$ with $P_0 := (\sigma_0^2/\lambda)\Sigma_0^{-1}$ whenever the difference between their mean squared error matrices is, at least, positive-(semi)definite:*

$$MSE(\tilde{x}) - MSE(\hat{\xi}_\lambda) \geq_L 0$$

where $\geq_L$ denotes Löwner's partial ordering of matrices; see, e.g., Marshall and Olkin (1979). This is the case if and only if the inequality

$$(\beta_0 - \xi)^T P_0 (\beta_0 - \xi) \leq \sigma_0^2/\lambda$$

holds true.

Proof:

A direct comparison of (6) with (10) yields the following equivalent statements

$$MSE(\tilde{x}) \geq_L MSE(\hat{\xi}) \Leftrightarrow R \geq_L R(\beta_0 - \xi)\sigma_0^{-2}(\beta_0 - \xi)^T R,$$

implying that

$$(\beta_0 - \xi)^T R(\beta_0 - \xi) \geq \left[(\beta_0 - \xi)^T R(\beta_0 - \xi)\right]^2 / \sigma_0^2\,. \tag{12}$$

If $R(\beta_0 - \xi) = 0$, then the PLS-solution $\hat{\xi}$ automatically becomes superior to the "Bayesian estimate" $\tilde{x}$.

If $R(\beta_0 - \xi) \neq 0$, then the inequality in (12) further implies

$$(\beta_0 - \xi)^T P_0(\beta_0 - \xi) \leq \sigma_0^2/\lambda$$

as in (1). This inequality, however, is also sufficient for (1) to hold, by virtue of a result by Baksalary and Kala (1983).
This completes the proof.

Of course, the inequality in (1) is impractical as long as the quantities ξ, σ_0^2 and λ are unknown. In an empirical version of (1), these quantities may be replaced by their estimates which would include $\hat{\xi}$ from (3), as well as suitable estimates for σ_0^2 and λ. We note that λ represents Tykhonov's "regularization parameter".

4 A Quasi-optimal Choice for the Weight Ratio λ

When operating in the Random Effects Model from (1) and (2) that may be rewritten as

$$y_{ext} := \begin{bmatrix} y \\ \beta_0 \end{bmatrix} = \begin{bmatrix} A \\ I_m \end{bmatrix} x + \begin{bmatrix} e \\ e_0 \end{bmatrix} =: A_{ext}x + e_{ext}, \tag{13}$$

$$\begin{bmatrix} e \\ e_0 \end{bmatrix} =: e_{ext} \sim \left(\begin{bmatrix} 0 \\ 0 \end{bmatrix}, \begin{bmatrix} \Sigma & 0 \\ 0 & \Sigma_0 \end{bmatrix} \right) \tag{14}$$

where

$$\begin{bmatrix} \Sigma & 0 \\ 0 & \Sigma_0 \end{bmatrix} = \sigma_0^2 \begin{bmatrix} P^{-1} & 0 \\ 0 & 0 \end{bmatrix} + (\sigma_0^2/\lambda) \begin{bmatrix} 0 & 0 \\ 0 & P_0^{-1} \end{bmatrix},$$

we observe that λ could be estimated as the ratio of two variance components, for instance on the basis of the *reproBIQUUE* principle (reproducing Best Invariant Quadratic Uniformly Unbiased Estimate), yielding

$$\hat{\lambda} = \hat{\sigma}_0^2/\hat{\sigma}_{00}^2 \text{ for } \sigma_{00}^2 := \sigma_0^2/\lambda. \tag{15}$$

As was shown in all detail by Schaffrin (1983), the corresponding nonlinear system of equations would read:

$$\begin{bmatrix} tr(\hat{W}Q\,\hat{W}Q) & |tr(\hat{W}Q\,\hat{W}Q_0) \\ tr(\hat{W}Q\,\hat{W}Q_0) & |tr(\hat{W}Q_0\,\hat{W}Q_0) \end{bmatrix} \begin{bmatrix} \hat{\sigma}_0^2 \\ \hat{\sigma}_{00}^2 \end{bmatrix} = \begin{bmatrix} y_{ext}^T \hat{W} Q \hat{W} y_{ext} \\ y_{ext}^T \hat{W} Q_0 \hat{W} y_{ext} \end{bmatrix} \tag{16}$$

with

$$\hat{W} = W(\hat{\sigma}_0^2, \hat{\sigma}_{00}^2) := \hat{\Sigma}_{ext}^{-1} - \hat{\Sigma}_{ext}^{-1} A_{ext}(A_{ext}^T \hat{\Sigma}_{ext}^{-1} A_{ext})^{-1} A_{ext}^T \hat{\Sigma}_{ext}^{-1}$$

and

$$\hat{\Sigma}_{ext} = \hat{\sigma}_0^2 Q + \hat{\sigma}_{00}^2 Q_0 := \hat{\sigma}_0^2 \begin{bmatrix} P^{-1} & 0 \\ 0 & 0 \end{bmatrix} + \hat{\sigma}_{00}^2 \begin{bmatrix} 0 & 0 \\ 0 & P_0^{-1} \end{bmatrix};$$

see also Rao and Kleffe (1988), or Searle et al. (1992). Specifically we would find

$$\begin{aligned} \sigma_0^2 W &= \begin{bmatrix} P & 0 \\ 0 & \lambda P_0 \end{bmatrix} - \begin{bmatrix} P & 0 \\ 0 & \lambda P_0 \end{bmatrix} \begin{bmatrix} A \\ I_m \end{bmatrix} (N + \lambda P_0)^{-1} \left[A^T, I_m\right] \begin{bmatrix} P & 0 \\ 0 & \lambda P_0 \end{bmatrix} \\ &= \left[\begin{array}{c|c} P - PA(N+\lambda P_0)^{-1} A^T P & -\lambda PA(N+\lambda P_0)^{-1} P \\ -\lambda P_0 (N+\lambda P_0)^{-1} A^T P & \lambda P_0 - \lambda^2 P_0 (N+\lambda P_0)^{-1} P_0 \end{array}\right] \end{aligned}$$

and, consequently,

$$\sigma_0^2 (WQ) = \left[\begin{array}{c|c} I_n - PA(N+\lambda P_0)^{-1} A^T & 0 \\ -\lambda P_0 (N+\lambda P_0)^{-1} A^T & 0 \end{array}\right]$$

as well as

$$\sigma_0^2 (WQ_0) = \left[\begin{array}{c|c} 0 & -\lambda PA(N+\lambda P_0)^{-1} \\ 0 & \lambda I_m - \lambda^2 P_0 (N+\lambda P_0)^{-1} \end{array}\right]$$

which leads to the matrix elements in (16), namely

$$\begin{aligned} m_{11} &:= tr(WQWQ) = tr(\left[I_n - PA(N+\lambda P_0)^{-1} A^T\right]^2) / \sigma_0^4 , \\ m_{12} &= m_{21} := tr(WQWQ_0) \\ &= \lambda^2 tr \left[P_0 (N+\lambda P_0)^{-1} N (N+\lambda P_0)^{-1}\right] / \sigma_0^4 , \\ m_{22} &:= tr(WQ_0WQ_0) = \lambda^2 tr(\left[I_m - \lambda P_0 (N+\lambda P_0)^{-1}\right]^2 / \sigma_0^4 . \end{aligned}$$

Alternative expressions include

$$\begin{aligned} \sigma_0^4 m_{11} &= n - tr\left[N(N+\lambda P_0)^{-1}\right] \\ &\quad -\lambda tr\left[P_0 (N+\lambda P_0)^{-1} N (N+\lambda P_0)^{-1}\right], \\ \sigma_0^4 m_{22} &= \lambda^2 tr\left[N(N+\lambda P_0)^{-1} N (N+\lambda P_0)^{-1}\right] \end{aligned}$$

which first imply the relationships

$$\begin{aligned} \sigma_0^4 (\lambda m_{11} + m_{12}) &= \lambda \left(n - tr\left[N(N+\lambda P_0)^{-1}\right]\right), \\ \sigma_0^4 (\lambda m_{12} + m_{22}) &= \lambda^2 tr\left[N(N+\lambda P_0)^{-1}\right], \end{aligned}$$

and finally the "control formula"

$$\lambda^2 m_{11} + 2\lambda m_{12} + m_{22} = n(\lambda^2/\sigma_0^4) = n/\sigma_{00}^4.$$

For the PLS-solution, however, the model in (13) and (14) is not applicable and needs to be replaced by the Gauss-Markov model from (8) which, in turn, can be rewritten as

$$y_{ext} - A_{ext}\xi = e_{mod} := \left[e^T,\, \delta^T\right]^T \tag{17}$$

with

$$e \sim (0,\ \sigma_0^2 P^{-1}) \ \text{ and } \ \delta := \beta_0 - \xi.$$

Nevertheless, in order to find estimates for σ_0^2 and λ in this GM-Model, the equation system from (16) may formally be adapted even though, in a rigorous sense, the optimal properties ("reproBIQUUE") will be lost. Apparently, this approach would first result in the auxiliary vector

$$\begin{aligned} \sigma_0^2(W y_{ext}) &= \begin{bmatrix} Py - PA(N+\lambda P_0)^{-1}(c+\lambda P_0\beta_0) \\ \lambda P_0\beta_0 - \lambda P_0(N+\lambda P_0)^{-1}(c+\lambda P_0\beta_0 \end{bmatrix} \\ &= \begin{bmatrix} P(y - A\hat{\xi}_\lambda) \\ \lambda P_0(\beta_0 - \hat{\xi}_\lambda) \end{bmatrix} =: \begin{bmatrix} P\tilde{e}_\lambda \\ \lambda P_0 \hat{\delta}_\lambda \end{bmatrix} \end{aligned} \tag{18}$$

and eventually in the nonlinear equations system

$$\begin{bmatrix} \hat{m}_{11} & \hat{m}_{12} \\ \hat{m}_{12} & \hat{m}_{22} \end{bmatrix} \begin{bmatrix} \hat{\sigma}_0^2 \\ \hat{\sigma}_0^2/\hat{\lambda} \end{bmatrix} = \begin{bmatrix} (\tilde{e}_\lambda^T P \tilde{e}_\lambda)/\hat{\sigma}_0^4 \\ \hat{\lambda}^2(\hat{\delta}_\lambda^T P_0 \hat{\delta}_\lambda)/\hat{\sigma}_0^4 \end{bmatrix} \tag{19}$$

after using (15). Separately, when applying (17) and (17), the two equations become:

$$\tilde{e}_\lambda^T P \tilde{e}_\lambda = \hat{\sigma}_0^6(\hat{\lambda}\hat{m}_{11} + \hat{m}_{12}) \big/ \hat{\lambda} = \hat{\sigma}_0^2\left(n - tr\left[N(N+\hat{\lambda}P_0)^{-1}\right]\right),$$

$$\hat{\lambda}(\hat{\delta}_\lambda^T P_0 \hat{\delta}_\lambda) = \hat{\sigma}_0^6(\hat{\lambda} m_{12} + \hat{m}_{22}) \big/ \hat{\lambda}^2 = \hat{\sigma}_0^2 tr\left[N(N+\hat{\lambda}P_0)^{-1}\right]. \tag{20}$$

Obviously, $\hat{\sigma}_0^2$ can now be eliminated, and we arrive at the nonlinear equation

$$\hat{\lambda} = \left(\frac{\tilde{e}_\lambda^T P \tilde{e}_\lambda}{\hat{\delta}_\lambda^T P_0 \hat{\delta}_\lambda}\right)\left(\frac{\hat{t}}{n - \hat{t}}\right), \ \hat{t} = tr\left[(N+\hat{\lambda}P_0)^{-1}N\right],$$

that needs to be solved iteratively since $\hat{t} = t(\hat{\lambda})$. Finally, the estimated variance component is obtained through

$$\hat{\sigma}_0^2 = (\tilde{e}_\lambda^T P \tilde{e}_\lambda) / (n - \hat{t}) = (\tilde{e}_\lambda^T P \tilde{e}_\lambda + \hat{\lambda} \hat{\delta}_\lambda^T P_0 \hat{\delta}_\lambda) / n \tag{21}$$

which cannot be expected to be unbiased since both $\tilde{e}_\lambda$ and $\hat{\delta}_\lambda$ are themselves biased (as was $\hat{\xi}_\lambda$).

Theorem 2. *For the penalized least-squares solution in (3) which is a special case of "Tykhonov regularization", a quasi-optimal weight ratio, respectively a quasi-optimal "regularization parameter", has been defined through (4), along with a suitable variance component estimate via (21).*

On this basis an empirical version of the mean squared error matrix in (10) can be computed, and an empirical decision can be made via (1) whether the Penalized Least-Squares solution will be superior to the "Bayesian estimate" in (2), i.e. inhomBLIP, whose empirical mean squared error matrix would result from (6) in full analogy.

5 Conclusions

The distinction between the Penalized Least-Squares solution (within a Gauss-Markov model) and the so-called "Bayesian estimate" (within a Random Effects Model) has been carefully analyzed, particularly in terms of the respective mean squared error matrices. An inequality has been established to decide when one estimate is superior to the other.

In addition, a quasi-optimal estimate has been found for the weight ratio between the "penalization" and "best fit" terms, incidentally leading to a novel approach to determine Tykhonov's "regularization parameter". It is still an open question, however, what exactly its statistical characteristics are.

References

Baksalary JK, Kala R (1983) Partial ordering between matrices one of which is of rank one. Bulletin of Polish Academy of Sciences: Mathematics 31: 5–7

Engl H, Hanke M, Neubauer A (1996) Regularization of Inverse Problems. Kluwer: Dordrecht/NL

Goldberger AS (1962) Best linear unbiased prediction in the genelarized linear regression model. Journal of American Statistical Association. 57: 369–375

Golub GH, Heath M, Wahba G (1979) Generalized cross–validation as a method for choosing a good ridge parameter. Technometrics 21: 215–223

Grafarend E, Schaffrin B (1993) Adjustment Computations in Linear Models (in German), Bibliograph. Institute Mannheim

Hansen PC, O'Leary DP (1993) The use of the L-curve in the regularization of discrete ill-posed problems, SIAM Journal of Science and Computation. 14: 1487–1503

Marshall AW, Olkin I (1979) Inequalities. Theory of Majorization and its Applications, Academic Press, New York

Phillips DL (1962) A technique for the numerical solution of certain integral equations of the first kind, Journal of the Association for Computing Machinery. 9: 84–96

Rao CR (1976) Estimation of parameters in a linear model, Annals of Statistics. 4: 1023–1037

Rao CR, Kleffe J (1988) Estimation of Variance Components and Applications, North Holland: Amsterdam/NL

Rao CR, Toutenburg H, Shalabh, Heumann C (2008) Linear Models and Generalizations. Least Squares and Alternatives (3rd edition) Springer, Berlin Heidelberg New York

Schaffrin B (1983) Estimation of variance–covariance components for heterogenous replicated measurements (in German), German Geodetic Community, Publication C-282, Munich/Germany

Schaffrin B (1985) The geodetic datum with stochastic prior information (in German), German Geodetic Community, Publication C-313, Munich/ Germany

Schaffrin B (1995) A comparison of inverse techniques: Regularization, weight estimation and homBLUP, IUGG General Assembly, IAG Scientific Meeting U7, Boulder/CO

Schaffrin B (2005) On the optimal choice of the regularization parameter through variance ratio estimation, 14th International Workshop on Matrices and Statistics, Auckland/NZ

Searle SR, Casella G, McCulloch CE (1992) Variance Components, Wiley, New York

Tykhonov AN (1963) The regularization of incorrectly posed problems, Soviet Mathematics Doklady 4: 1624–1627

Optimal Central Composite Designs for Fitting Second Order Response Surface Linear Regression Models

Sung Hyun Park[1], Hyuk Joo Kim[2] and Jae-Il Cho[3]

[1] Department of Statistics, Seoul National University, Seoul 151-747, Korea parksh@plaza.snu.ac.kr
[2] Division of Mathematics and Informational Statistics and Institute of Basic Natural Sciences, Wonkwang University, Iksan, Jeonbuk 570-749, Korea hjkim@wonkwang.ac.kr
[3] Junior Manager, Management Innovation Part, Dongbu Electronics, Buchon, Gyeonggi 420-712, Korea jaeil.cho@dsemi.com

1 Introduction

The central composite design (CCD) is a design widely used for estimating second order response surfaces. It is perhaps the most popular class of second order designs. Since introduced by Box and Wilson (1951), the CCD has been studied and used by many researchers.

Let $x_1, x_2, \ldots, x_k$ denote the explanatory variables being considered. Much of the motivation of the CCD evolves from its use in sequential experimentation. It involves the use of a two-level factorial or fraction (resolution V) combined with the following $2k$ axial points:

x_1	x_2	$\cdots$	x_k
$-\alpha$	0	$\cdots$	0
α	0	$\cdots$	0
0	$-\alpha$	$\cdots$	0
0	α	$\cdots$	0
$\vdots$	$\vdots$	$\ddots$	$\vdots$
0	0	$\cdots$	$-\alpha$
0	0	$\cdots$	α

As a result, the design involves, say, $F = 2^k$ factorial points (or $F = 2^{k-p}$ fractional factorial points), $2k$ axial points, and n_0 center points.

The sequential nature of the design is quite obvious. The factorial points represent a variance optimal design for a first-order model or a first-order + two-factor interaction type model. The center points clearly provide information about the existence of curvature in the system. If curvature is found in the system, then addition of axial points allow for efficient estimation of the pure quadratic terms.

Among many statisticians who have studied the CCD in response surface methodology, Myers and Montgomery (2002) discussed the efficiency of experimental designs, and compared the CCD with other designs under $D-$, $A-$ and $E-$ optimality criterion. Box and Draper (1963) suggested several criteria which can be used in the selection of design. Myers (1976) suggested optimal CCDs under several design criteria and Hader and Park (1978) discussed about slope-rotatable CCDs.

This paper deals with optimal CCDs under several design criteria for fitting second order response surface regression models. In Section 2, results on optimal CCDs under the criteria of orthogonality, rotatability and slope rotatability are reviewed. In Section 3, we discuss optimal CCDs under alphabetic design optimality criteria. The appropriate values of α which minimize the squared bias when the true model is of third order are suggested in Section 4. Finally, in Section 5, considering all possible design criteria, suitable values of α for the practical design purpose are recommended.

2 Orthogonality, Rotatability and Slope Rotatability

Let us consider the model represented by

$$y_u = \beta_0 + \sum_{i=1}^{k} \beta_i x_{iu} + \sum_{i=1}^{k} \beta_{ii} x_{iu}^2 + \sum_{i<j}^{k} \beta_{ij} x_{iu} x_{ju} + \varepsilon_u \ (u = 1, 2, \ldots, N) \tag{1}$$

where x_{iu} is the value of the variable x_i at the uth experimental point, and ε_u's are uncorrelated random errors with mean zero and variance σ^2. This is the second order response surface model.

2.1 Orthogonality

In this subsection, we consider the model with the pure quadratic terms corrected for their means, that is,

$$y_u = \beta_0' + \sum_{i=1}^{k} \beta_i x_{iu} + \sum_{i=1}^{k} \beta_{ii}(x_{iu}^2 - \overline{x_i^2}) + \sum_{i<j}^{k} \beta_{ij} x_{iu} x_{ju} + \varepsilon_u \qquad (2)$$

where $\beta_0' = \sum_{i=1}^{k} \beta_{ii} \overline{x_i^2}$ and $\overline{x_i^2} = \sum_{u=1}^{N} x_{iu}^2/N$. In regard to orthogonality, this model is often used for the sake of simplicity in calculation.

Let $b_0', b_i, b_{ii}, b_{ij}$ denote the least squares estimators of $\beta_0', \beta_i, \beta_{ii}, \beta_{ij}$, respectively. In the CCD, all the covariances between the estimated regression coefficients except $Cov(b_{ii}, b_{jj})$ are zero. But if the $(X'X)^{-1}$ matrix is a diagonal matrix, then $Cov(b_{ii}, b_{jj})$ also becomes zero. This property is called orthogonality. Here the detailed contents of the X matrix are as follows:

$$X = \begin{bmatrix} 1 & x_{11} & \dots & x_{k1} & x_{11}^2 - \overline{x_1^2} & \dots & x_{k1}^2 - \overline{x_k^2} & x_{11}x_{21} & \dots & x_{k-1,1}x_{k1} \\ 1 & x_{12} & \dots & x_{k2} & x_{12}^2 - \overline{x_1^2} & \dots & x_{k2}^2 - \overline{x_k^2} & x_{12}x_{22} & \dots & x_{k-1,2}x_{k2} \\ \vdots & \vdots & \ddots & \vdots & \vdots & \ddots & \vdots & \vdots & \ddots & \vdots \\ 1 & x_{1N} & \dots & x_{kN} & x_{1N}^2 - \overline{x_1^2} & \dots & x_{kN}^2 - \overline{x_k^2} & x_{1N}x_{2N} & \dots & x_{k-1,N}x_{kN} \end{bmatrix}.$$

It is well-known (see Myers (1976, p. 134) and Khuri and Cornell (1996, p. 122)) that the condition for a CCD to be an orthogonal design is that

$$\alpha = \left\{ \frac{\sqrt{F(F + 2k + n_0)} - F}{2} \right\}^{1/2}.$$

The orthogonal CCD will provide an ease in computations and uncorrelated estimates of the response model coefficients. Table 1 shows the values of α for various k and n_0 which make the CCD orthogonal. Note that for $k = 5$ and 6, we also consider the case in which a fractional factorial is used instead of a complete factorial.

2.2 Rotatability

It is important for a second order design to possess a reasonably stable distribution of $NVar[\hat{y}(x)]/\sigma^2$ throughout the experimental design region. Here $\hat{y}(x)$ is the estimated response at the point $x = (x_1, x_2, \dots, x_k)'$. It must be clearly understood that the experimenter does not know at the outset where in the design space he or she might wish to predict, or where in the design space the optimum may lie. Thus, a reasonably stable $NVar[\hat{y}(x)]/\sigma^2$ provides values which are

Table 1. Values of α for orthogonal CCDs

k	F	α					
		$n_0 = 1$	$n_0 = 2$	$n_0 = 3$	$n_0 = 4$	$n_0 = 5$	$n_0 = 6$
2	4	1.000	1.078	1.147	1.210	1.267	1.320
3	8	1.215	1.287	1.353	1.414	1.471	1.525
4	16	1.414	1.483	1.547	1.607	1.664	1.719
5	32	1.596	1.662	1.724	1.784	1.841	1.896
5(1/2 rep)	16	1.547	1.607	1.664	1.719	1.771	1.820
6	64	1.761	1.824	1.885	1.943	2.000	2.055
6(1/2 rep)	32	1.724	1.784	1.841	1.896	1.949	2.000

roughly the same throughout the region of interest. To this purpose, Box and Hunter (1957) developed the notion of design rotatability.

A rotatable design is one for which $NVar[\hat{y}(x)]/\sigma^2$ has the same value at any two locations that have the same distance from the design center. In other words, $NVar[\hat{y}(x)]/\sigma^2$ is constant on spheres.

It is well-known that the condition for a CCD to be rotatable is that

$$\alpha = F^{1/4}.$$

This means that the value of α for a rotatable CCD does not depend on the number of center points.

Table 2. Values of α for rotatable CCDs

k	F	T	N	α
2	4	5	9	1.414
3	8	7	15	1.682
4	16	9	25	2.000
5	32	11	43	2.378
5 (1/2 rep)	16	11	27	2.000
6	64	13	77	2.828
6 (1/2 rep)	32	13	45	2.378

Table 2 gives the values of α for rotatable CCDs for various k. Note that for $k = 5$ and 6, a CCD is also suggested in which a fractional factorial is used instead of a complete factorial. Also tabulated are F and T, where $T = 2k + 1$.

The designs considered in the table 2 contain a single center point. This by no means implies that one would always use only one center point.

2.3 Slope Rotatability

Suppose that estimation of the first derivative of η is of interest (η is the expected value of the response variable y). For the second order model,

$$\frac{\partial \hat{y}(x)}{\partial x_i} = b_i + 2b_{ii}x_i + \sum_{j \neq i} b_{ij}x_j.$$

The variance of this derivative is a function of the point x at which the derivative is estimated and also a function of the design.

Hader and Park (1978) proposed an analog of the Box-Hunter rotatability criterion, which requires that the variance of $\partial \hat{y}(x)/\partial x_i$ be constant on circles ($k = 2$), spheres ($k = 3$), or hyperspheres ($k \geq 4$) centered at the design origin.

Estimates of the derivative over axial directions would then be equally reliable for all points x equidistant from the design origin. They referred to this property as slope rotatability, and showed that the condition for a CCD to be a slope-rotatable design is as follows:

$$[2(F + n_0)]\alpha^8 - [4kF]\alpha^6 - F[N(4-k) + kF - 8(k-1)]\alpha^4$$
$$+[8(k-1)F^2]\alpha^2 - 2F^2(k-1)(N-F) = 0.$$

Table 3 gives slope-rotatable values of α for $2 \leq k \leq 6$. For $k = 5$ and 6, CCDs involving fractional factorials are also considered.

3 The Alphabetic Design Optimality

3.1 D-Optimality

The best known and most often used criterion is D-optimality. D-optimality is based on the notion that the experimental design should be chosen so as to achieve certain properties in the matrix $X'X$.

Table 3. Values of α for slope-rotatable CCDs

$k = 2$

N	n_0	α
9	1	2.0903
10	2	1.9836
11	3	1.9106
12	4	1.8586
13	5	1.8203
14	6	1.7912
15	7	1.7684
16	8	1.7501
17	9	1.7352

$k = 3$

N	n_0	α
15	1	2.4324
16	2	2.3387
17	3	2.2675
18	4	2.2133
19	5	2.1716
20	6	2.1390
21	7	2.1132
22	8	2.0924
23	9	2.0753

$k = 4$

N	n_0	α
25	1	2.7988
26	2	2.7303
27	3	2.6732
28	4	2.6259
29	5	2.5869
30	6	2.5547
31	7	2.5280
32	8	2.5057
33	9	2.4869

$k = 5$

N	n_0	α
43	1	3.2034
44	2	3.1607
45	3	3.1228
46	4	3.0892
47	5	3.0597
48	6	3.0337
49	7	3.0108
50	8	2.9907
51	9	2.9729

$k = 5$ (1/2 rep)

N	n_0	α
27	1	2.8722
28	2	2.7750
29	3	2.6954
30	4	2.6321
31	5	2.5828
32	6	2.5444
33	7	2.5143
34	8	2.4903
35	9	2.4709

$k = 6$

N	n_0	α
77	1	3.6732
78	2	3.6500
79	3	3.6284
80	4	3.6085
81	5	3.5901
82	6	3.5732
83	7	3.5575
84	8	3.5431
85	9	3.5298

$k = 6$ (1/2 rep)

N	n_0	α
45	1	3.2650
46	2	3.2066
47	3	3.1551
48	4	3.1103
49	5	3.0719
50	6	3.0391
51	7	3.0112
52	8	2.9874
53	9	2.9671

Here X is the following matrix:

$$X = \begin{bmatrix} 1 & x_{11} & \dots & x_{k1} & x_{11}^2 & \dots & x_{k1}^2 & x_{11}x_{21} & \dots & x_{k-1,1}x_{k1} \\ 1 & x_{12} & \dots & x_{k2} & x_{12}^2 & \dots & x_{k2}^2 & x_{12}x_{22} & \dots & x_{k-1,2}x_{k2} \\ \vdots & \vdots & \ddots & \vdots & \vdots & \ddots & \vdots & \vdots & \ddots & \vdots \\ 1 & x_{1N} & \dots & x_{kN} & x_{1N}^2 & \dots & x_{kN}^2 & x_{1N}x_{2N} & \dots & x_{k-1,N}x_{kN} \end{bmatrix}$$

Also, the inverse of $X'X$ contains variances and covariances of the regression coefficients, scaled by $1/\sigma^2$. As a result, control of the moment matrix by design implies control of the variances and covariances.

Suppose the maximum, arithmetic mean, and geometric mean of the eigenvalues $\lambda_1, \lambda_2, \dots, \lambda_p$ of $(X'X)^{-1}$ are indicated by $\lambda_{\max}$, $\overline{\lambda}$, and $\tilde{\lambda}$. It turns out that an important norm on the moment matrix is the determinant; that is,

$$D = |X'X| = \prod_{i=1}^{p} \lambda_i^{-1} = \tilde{\lambda}^{-p}$$

where p is the number of parameters in the model. Under the assumption of independent normal errors with constant variance, the determinant of $X'X$ is inversely proportional to the square of the volume of the confidence region for the regression coefficients. The volume of the confidence region is relevant because it reflects how well the set of coefficients are estimated. A D-optimal design is one in which $|X'X|$ is maximized; that is,

$$Max_\zeta|X'X(\zeta)|$$

where Max_ζ implies that the maximum is taken over all design ζ's.

3.2 A-Optimality

The concept of A-optimality deals with the individual of the regression coefficients. Unlike D-optimality, it does not make use of covariances among coefficients; that is,

$$\begin{aligned} A &= \sum_{i=1}^{p} \lambda_i = tr(X'X)^{-1} \\ &= \sigma^{-2}\left[\mathrm{var}(b_0) + \sum_{i=1}^{k}\mathrm{var}(b_i) + \sum_{i=1}^{k}\mathrm{var}(b_{ii}) + \sum_{i<j}^{k}\mathrm{var}(b_{ij})\right]. \end{aligned}$$

The A-optimal design is defined as

$$Min_\zeta tr(X'X(\zeta))^{-1}$$

where tr represents trace, that is, the sum of the variances of the coefficients (weighted by N).

3.3 E-Optimality

The criterion E, evaluation of the smallest eigenvalue, also gains in understanding by a passage to variances. It is the same as minimizing the largest eigenvalue of the dispersion matrix; that is,

$$E = Max_i\lambda_i$$

where $i = 1, 2, \ldots, p$.

In terms of variance, it is a minimax approach. Thus the E-optimal design is defined as

$$Min_\zeta Max_i\{\lambda_i\}.$$

1	x_1	x_2	x_1^2	x_1^2	x_1x_2
1	-1	-1	1	1	1
1	-1	1	1	1	-1
1	1	-1	1	1	-1
1	1	1	1	1	1
1	$-\alpha$	0	α^2	0	0
1	α	0	α^2	0	0
1	0	$-\alpha$	0	α^2	0
1	0	α	0	α^2	0
1	0	0	0	0	0

3.4 Application to the CCD

For fitting the two factor second order model, we can consider the following CCD. It consists of (i) a 2^2 factorial, at levels ± 1, (ii) a one-factor-at-a-time array and (iii)n_0 center points. That is, the matrix X is given by

Then the matrix $X'X$ is given by

$$\begin{bmatrix} N & 0 & 0 & a & a & 0 \\ 0 & a & 0 & 0 & 0 & 0 \\ 0 & 0 & a & 0 & 0 & 0 \\ a & 0 & 0 & b & F & 0 \\ a & 0 & 0 & F & b & 0 \\ 0 & 0 & 0 & 0 & 0 & F \end{bmatrix}$$

where N is the number of experimental points, F is the number of factorial points, $a = F + 2\alpha^2$ and $b = F + 2\alpha^4$.

For the two factor CCD, for example, the value of D is

$$D = n_0 F a^2 [N(b^2 - F^2) - 2a^2(b - F)]$$

where n_0 is the number of center points.

Figure 1 shows a plot of D versus α for the indicated values of n_0 for a CCD in $k = 2$ factors.

In CCDs, the determinant of moment matrix has a tendency of increase as α increases. That is, a larger value of α is recommendable for D-optimal sense. But in a practical experiment, the region of interest

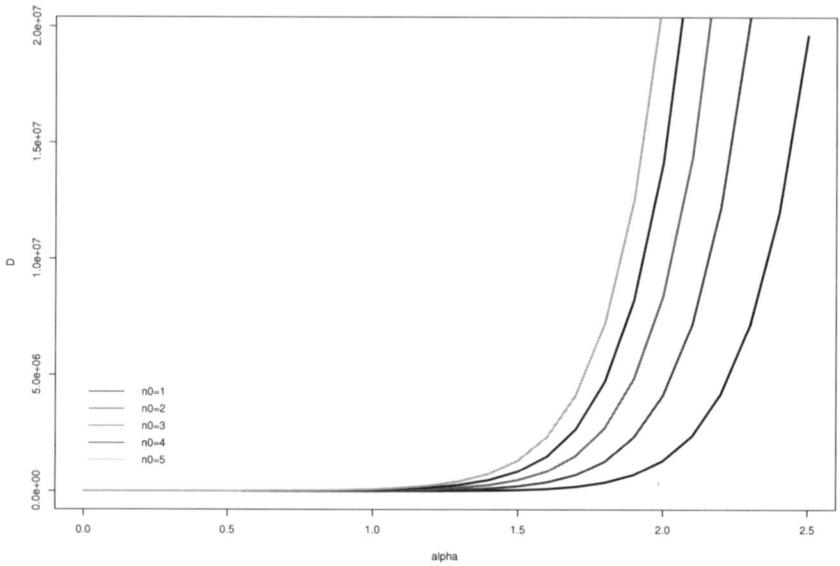

Fig. 1. Plots of D versus α for $k = 2$ factors

is usually restricted and the conditions of experiment cannot be set for a large α. So it is necessary for the experimenter to choose α as large as possible within the controllable region of interest.

On the other hand, for the two factor CCD, the value of A is

$$A = \sum \lambda_i = \frac{2}{a} + \frac{1}{F} + \frac{1}{b-F} + \frac{N+b+F}{N(b+F)-2a^2}.$$

Figure 2 shows plots of A versus α for the indicated values of n_0 for CCDs in $k = 2$ factors. Table 4 shows the results of optimal α values for two factor CCDs.

4 Optimal CCDs When the True Model is of Third Order

Suppose that we fit the second order response surface model, but the true model is of third order. For this case, what value of α should be used in the CCD?

We can generally formulate the problem by supposing that the experimenter fits a model $\hat{y}(x_1, x_2, \ldots, x_k)$ of order d_1 in a region R of the explanatory variables. However, the true model is a polynomial

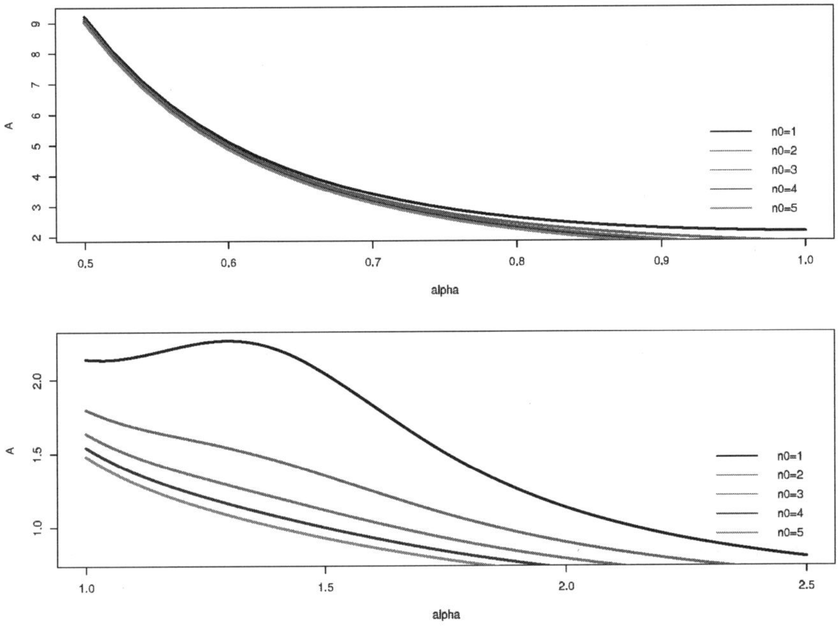

Fig. 2. Plots of A versus α for $k = 2$ factors

Table 4. Comparison of optimal α values for two factor CCDs

number of center points	orthogonality	rotatability	slope rotatability	D, A, E
$n_0 = 1$	1.000	1.414	2.0903	∞
$n_0 = 2$	1.078	1.414	1.9836	∞
$n_0 = 3$	1.147	1.414	1.9106	∞
$n_0 = 4$	1.210	1.414	1.8586	∞
$n_0 = 5$	1.267	1.414	1.8203	∞

$g(x_1, x_2, \ldots, x_k)$ of order d_2, where $d_2 > d_1$. Then, a reasonable design criterion is the minimization of

$$M = \frac{N}{\sigma^2} \int_R E[\hat{y}(x) - g(x)]^2 dx \Big/ \int_R dx. \qquad (3)$$

The multiple integral in (3) actually represents the average of the expected squared deviations of the true response from the estimated response over the region R.

Writing the integral $\int_R dx = 1/K$,

$$\begin{aligned} M &= \frac{NK}{\sigma^2} \int_R E[\hat{y}(x) - g(x)]^2 dx \\ &= \frac{NK}{\sigma^2} \int_R E[\hat{y} - E(\hat{y}) + E(\hat{y}) - g(x)]^2 dx \\ &= \frac{NK}{\sigma^2} \left[\int_R E[\hat{y} - E(\hat{y})]^2 dx + \int_R [E(\hat{y}) - g(x)]^2 dx \right]. \qquad (4) \end{aligned}$$

The first quantity in (4) is the variance of $\hat{y}$, integrated or, rather averaged over the region R, whereas the second quantity is the square of the bias, similarly averaged. Thus M is naturally divided as follows:

$$M = V + B$$

where V is the average variance of $\hat{y}$, and B is the average squared bias of $\hat{y}$.

In this section, as a reasonable choice of design we will consider the design which minimizes B. Such a design is called the all-bias design.

It is assumed here that the experimenter desires to fit a quadratic response surface in a cuboidal region R but that the true function is best described by a cubic polynomial. The actual measured variables have been transformed to $x_1, x_2, \ldots, x_k$ which are scaled so that the region of interest R is a unit cube. Also the assumption on the design is made that its center of gravity is at the origin $(0, 0, \ldots, 0)$ of the cube. The equation of the fitted model is

$$\hat{y} = x_1' \hat{\beta}_1,$$

where

$$\begin{aligned} x_1' &= [1, x_1, \ldots, x_k; x_1^2, \ldots, x_k^2; x_1 x_2, \ldots, x_{k-1} x_k] \\ \hat{\beta}_1' &= [b_0, b_1, \ldots, b_k; b_{11}, \ldots, b_{kk}; b_{12}, \ldots, b_{k-1,k}]. \end{aligned}$$

The true relationship is written as

$$E(y) = x_1' \beta_1 + x_2' \beta_2$$

where

$$x_2' = [x_1^3, x_1 x_2^2, \ldots, x_1 x_k^2; x_2^3, x_2 x_1^2, \ldots, x_2 x_k^2; \ldots; x_k^3, x_k x_1^2, \ldots, x_k x_{k-1}^2;$$

$$x_1 x_2 x_3, x_1 x_2 x_4, \ldots, x_{k-2} x_{k-1} x_k]$$

contains the cubic contribution to the actual model. The vector β_2' contains the coefficients corresponding to terms in x_2'; terms such as $\beta_{111}, \beta_{122}, \ldots$ are included. The matrix X_1 is given by

$$X_1 = \begin{bmatrix} 1 & x_{11} & \ldots & x_{k1} & x_{11}^2 & \ldots & x_{k1}^2 & x_{11}x_{21} & \ldots & x_{k-1,1}x_{k,1} \\ 1 & x_{12} & \ldots & x_{k2} & x_{12}^2 & \ldots & x_{k2}^2 & x_{12}x_{22} & \ldots & x_{k-1,2}x_{k,2} \\ \vdots & \vdots & \ddots & \vdots & \vdots & \ddots & \vdots & \vdots & \ddots & \vdots \\ 1 & x_{1N} & \ldots & x_{kN} & x_{1N}^2 & \ldots & x_{kN}^2 & x_{1N}x_{2N} & \ldots & x_{k-1,N}x_{k,N} \end{bmatrix}.$$

In this case the matrix X_2 is

$$X_2 = \begin{bmatrix} x_{11}^3 & x_{11}x_{21}^2 & \ldots & x_{11}x_{k1}^2 & x_{21}^3 & x_{21}x_{11}^2 & \ldots & x_{21}x_{k1}^2 & \ldots \\ x_{12}^3 & x_{12}x_{22}^2 & \ldots & x_{12}x_{k2}^2 & x_{22}^3 & x_{22}x_{12}^2 & \ldots & x_{22}x_{k2}^2 & \ldots \\ \vdots & \vdots & \ddots & \vdots & \vdots & \vdots & \ddots & \vdots & \ddots \\ x_{1N}^3 & x_{1N}x_{2N}^2 & \ldots & x_{1N}x_{kN}^2 & x_{2N}^3 & x_{2N}x_{1N}^2 & \ldots & x_{2N}x_{kN}^2 & \ldots \end{bmatrix}$$

$$\begin{matrix} x_{k1}^3 & x_{k1}x_{11}^2 & \ldots & x_{k1}x_{k-1,1}^2 & x_{11}x_{21}x_{31} & \ldots & x_{k-2,1}x_{k-1,1}x_{k1} \\ x_{k2}^3 & x_{k2}x_{12}^2 & \ldots & x_{k2}x_{k-1,2}^2 & x_{12}x_{22}x_{32} & \ldots & x_{k-2,2}x_{k-1,2}x_{k2} \\ \vdots & \vdots & \ddots & \vdots & \vdots & \ddots & \vdots \\ x_{kN}^3 & x_{kN}x_{1N}^2 & \ldots & x_{kN}x_{k-1,N}^2 & x_{1N}x_{2N}x_{3N} & \ldots & x_{k-2,N}x_{k-1,N}x_{kN} \end{matrix}\Bigg].$$

Let us now write

$$M_{11} = N^{-1}X_1'X_1, \qquad M_{12} = N^{-1}X_1'X_2,$$

$$\mu_{11} = K\int_R x_1 x_1' dx, \qquad \mu_{12} = K\int_R x_1 x_2' dx,$$

where $K^{-1} = \int_R dx$. One can write the bias term as

$$\begin{aligned} B = \alpha_2'[&(\mu_{22} - \mu_{12}'\mu_{11}^{-1}\mu_{12}) \\ &+(M_{11}^{-1}M_{12} - \mu_{11}^{-1}\mu_{12})'\mu_{11}(M_{11}^{-1}M_{12} - \mu_{11}^{-1}\mu_{12})]\alpha_2, \end{aligned}$$

where the vector α_2 is merely $\beta_2\sqrt{N}/\sigma$ (see Myers (1976, p. 213)).

The first term in the square brackets in (5) contains only the region moment matrices and thus is independent of the design. The bias term can be no smaller than the positive semidefinite quadratic form $\alpha_2'(\mu_{22} - \mu_{12}'\mu_{11}^{-1}\mu_{12})\alpha_2$. So the experimenter has to use designs which minimize the positive semidefinite quadratic form

$$\alpha_2'(M_{11}^{-1}M_{12} - \mu_{11}^{-1}\mu_{12})'\mu_{11}(M_{11}^{-1}M_{12} - \mu_{11}^{-1}\mu_{12})\alpha_2.$$

Now we will find out the value of α which makes the optimal design in the CCDs. In practice, it is difficult to know α_2. However, for an illustration purpose, let's assume that α_2 is a vector of ones. That is, α_2' is (1,1,1,1) for the two factor CCDs when $d_1 = 2$ and $d_2 = 3$. And, if we assume that the region of interest is $-1 \le x_i \le 1$ where $i = 1, 2, \ldots, k$, then we can obtain region moment matrices (μ_{11} and μ_{12}).

For example, let's consider the second order CCD which minimizes the squared bias from the third order terms for $k = 2$. The design consists of four factorial points, six axial points at a distance α from the origin, and two center points. Then we obtain the following design moment matrices and region moment matrices.

$$M_{11} = \frac{1}{10}\begin{bmatrix} 10 & 0 & 0 & 4+2\alpha^2 & 4+2\alpha^2 & 0 \\ 0 & 4+2\alpha^2 & 0 & 0 & 0 & 0 \\ 0 & 0 & 4+2\alpha^2 & 0 & 0 & 0 \\ 4+2\alpha^2 & 0 & 0 & 4+2\alpha^4 & 4 & 0 \\ 4+2\alpha^2 & 0 & 0 & 4 & 4+2\alpha^4 & 0 \\ 0 & 0 & 0 & 0 & 0 & 4 \end{bmatrix}$$

$$M_{12} = \frac{1}{10}\begin{bmatrix} 0 & 0 & 0 & 0 \\ \frac{4+2\alpha^4}{4+2\alpha^2} & 0 & \frac{2}{4+2\alpha^2} & 0 \\ 0 & \frac{2}{4+2\alpha^2} & 0 & \frac{4+2\alpha^4}{4+2\alpha^2} \\ 0 & 0 & 0 & 0 \\ 0 & 0 & 0 & 0 \\ 0 & 0 & 0 & 0 \end{bmatrix}$$

$$\mu_{11} = \frac{1}{4}\begin{bmatrix} 4 & 0 & 0 & \frac{4}{3} & \frac{4}{3} & 0 \\ 0 & \frac{4}{3} & 0 & 0 & 0 & 0 \\ 0 & 0 & \frac{4}{3} & 0 & 0 & 0 \\ \frac{4}{3} & 0 & 0 & \frac{4}{5} & \frac{4}{9} & 0 \\ \frac{4}{3} & 0 & 0 & \frac{4}{9} & \frac{4}{5} & 0 \\ 0 & 0 & 0 & 0 & 0 & \frac{4}{9} \end{bmatrix}$$

$$\mu_{12} = \frac{1}{4}\begin{bmatrix} 0 & 0 & 0 & 0 \\ \frac{12}{5} & 0 & \frac{4}{3} & 0 \\ 0 & \frac{4}{3} & 0 & \frac{12}{5} \\ 0 & 0 & 0 & 0 \\ 0 & 0 & 0 & 0 \\ 0 & 0 & 0 & 0 \end{bmatrix}$$

So $\alpha_2'(M_{11}^{-1}M_{12} - \mu_{11}^{-1}\mu_{12})'\mu_{11}(M_{11}^{-1}M_{12} - \mu_{11}^{-1}\mu_{12})\alpha_2$ is obtained as

$$f(\alpha) = \frac{2(32 - 14\alpha^2 + 15\alpha^4)^2}{675(2+\alpha^2)^2}.$$

The value of α which minimizes $f(\alpha)$ is found to be

$$[2(\sqrt{2}-1)]^{1/2} = 0.91018.$$

A very interesting fact is that $f(\alpha)$ has nothing to do with the number of center points. Table 5 gives the appropriate values of α for second order CCD which minimize the squared bias from the third order terms for k factors.

Table 5. Values of α for second order CCDs which minimize the squared bias from the third order

k	$f(\alpha)$	α
2	$\frac{2(32-14\alpha^2+15\alpha^4)^2}{675(2+\alpha^2)^2}$	0.91018
3	$\frac{(104-19\alpha^2+15\alpha^4)^2}{225(4+\alpha^2)^2}$	1.13644
4	$\frac{4(96-8\alpha^2+5\alpha^4)^2}{75(8+\alpha^2)^2}$	1.34088
5	$\frac{(736-29\alpha^2+15\alpha^4)^2}{135(16+\alpha^2)^2}$	1.52653
6	$\frac{2(1792-34\alpha^2+15\alpha^4)^2}{225(32+\alpha^2)^2}$	1.69446

5 Concluding Remarks

In this paper, we found out values of α which optimize CCDs for fitting second order response surface models under several criteria. Table 6 gives the value of α in Tables 1, 2 and 5.

Table 6. Values of α which optimize CCDs under several criteria

k	Minimum bias	Orthogonality			Rotatability
		$n_0 = 1$	...	$n_0 = 6$	
2	0.910	1.000	...	1.320	1.414
3	1.136	1.216	...	1.525	1.682
4	1.341	1.414	...	1.719	2.000
5	1.527	1.596	...	1.896	2.378
6	1.694	1.761	...	2.055	2.828

k	Slope rotatability					Alphabetic optimality
	$n_0 = 1$	$n_0 = 2$	$n_0 = 3$	$n_0 = 4$	$n_0 = 5$	
2	2.090	1.984	1.911	1.859	1.820	∞
3	2.432	2.339	2.268	2.213	2.172	∞
4	2.799	2.730	2.673	2.626	2.587	∞
5	3.203	3.161	3.123	3.089	3.060	∞
6	3.673	3.650	3.628	3.609	3.590	∞

From Table 6, we can find that the values of α tend to increase in the following order:

Minimum bias $<$ Orthogonality $<$ Rotatability $<$ Slope rotatability $<$ Alphabetic optimality.

Note that the optimal value of α under the minimum bias and rotatability criteria does not depend on the number of center points. Also, an interesting fact is that the optimal value of α under the minimum bias criterion is very similar to that under the orthogonality criterion with one center point.

In conclusion, we will consider reasonable choice of CCD for fitting the second order model according to the following cases:

1. When the true model is of second order ($d_2 = 2$).
2. When the true model is of third order ($d_2 = 3$).

Table 7 shows values of α recommended for the CCD considering the order d_2.

Table 7. Values of α recommended for CCD

k	$d_2 = 2$		$d_2 = 3$	
	recommended α	appropriate α	recommended α	appropriate α
2	1.320∼1.414	1.4	0.910∼1.000	1.0
3	1.525∼1.682	1.6	1.136∼1.216	1.2
4	1.719∼2.000	2.0	1.341∼1.414	1.4
5	1.896∼2.378	2.3	1.527∼1.596	1.6
6	2.055∼2.828	2.8	1.694∼1.761	1.7

Acknowledgement

This work was supported by grant No. R01-2003-000-10220-0 from the Basic Research Program of the Korea Science & Engineering Foundation.

References

Box GEP, Draper, NR (1963) The choice of a second order rotatable design Biometrika. 50:335–352.

Box GEP, Draper NR (1987) Empirical Model-Building and Response Surfaces. John Wiley.

Box GEP, Hunter JS (1957) Multifactor experimental design for exploring response surfaces. Annals of Mathematical Statistics 28:195–241.

Box GEP, Wilson KB (1951) On the experimental attainment of optimum conditions. Journal of Royal Statistical Society B13:1–38.

Hader RJ, Park SH (1978) Slope-rotatable central composite designs. Technometrics 20:413–417.

Khuri AI, Cornell JA (1996) Response Surfaces: Designs and Analyses (2nd edition). Marcel Dekker.

Myers RH (1976) Response Surface Methodology. Blacksburg, VA.

Myers RH, Montgomery DC (2002) Response Surface Methodology: Process and Product Optimization Using Designed Experiments (2nd edition). John Wiley.

Park SH (1987) A class of multifactor designs for estimating the slope of response surfaces. Technometrics 29:449–453.

Park SH, Kim HJ (1992) A measure of slope-rotatability for second order response surface experimental designs. Journal of Applied Statistics 19:391–404.

Does Convergence Really Matter?

Jörg Drechsler[1] and Susanne Rässler[2]

[1] Institute for Employment Research of the Federal Employment Agency, Department for Statistical Methods, Regensburger Strasse 104, 90478 Nürnberg, Germany joerg.drechsler@iab.de

[2] Otto-Friedrich-University Bamberg, Department of Statistics and Econometrics, Feldkirchenstrasse 21, 96045 Bamberg, Germany susanne.raessler@sowi.uni-bamberg.de

1 Introduction

For many data sets, especially for non mandatory surveys, missing data are a common problem. Deleting units that are not completely observed and using only the remaining units is a popular, easy to implement approach in this case. This can possibly introduce severe bias if the strong assumption of a missing pattern that is completely at random (MCAR) is not fulfilled (see for example Rubin (1987)). Imputing the missing values can overcome this problem. However, ad hoc methods like, e.g., mean imputation can destroy the correlation between the variables. Furthermore, imputing missing values only once (single imputation) generally doesn't account for the fact that the imputed values are only estimates for the true values. After the imputation process, they are treated like truly observed values leading to an underestimation of the variance in the data and by this to p values that are too significant.

Multiple imputation as proposed by Rubin (1978) overcomes these problems. With multiple imputation, the missing values in a data set are replaced by $m > 1$ simulated versions, generated according to a probability distribution for the missing values given the observed data. More precisely, let Y_{obs} be the observed and Y_{mis} the missing part of a data set Y, with $Y = (Y_{mis}, Y_{obs})$, then missing values are drawn from the Bayesian posterior predictive distribution of $(Y_{mis}|Y_{obs})$, or an approximation thereof.

Over the years, two different methods emerged to generate draws from the above distribution: joint modeling and fully conditional specification (FCS). The first assumes that the data follow a specific distribution, e.g. a multivariate normal distribution. Under this assumption a parametric multivariate density $P(Y|\theta)$ can be specified with θ representing parameters from the assumed underlying distribution. Within the Bayesian framework, this distribution can be used to generate draws from $(Y_{mis}|Y_{obs})$. Methods to create multivariate imputations using this approach have been described in detail by Schafer (1997), e.g., for the multivariate normal, the log-linear, and the general location model.

FCS on the other hand does not depend on an explicit assumption for the joint distribution of the data set. Instead, conditional distributions $P(Y_j|Y_{-j}, \theta_j)$ are specified for each variable separately. Thus imputations are based on univariate distributions allowing for different models for each variable. Missing values in Y_j can be imputed for example by a linear or a logistic regression of Y_j on Y_{-j}, depending on the character of Y_j, where Y_{-j} denotes all columns of Y excluding Y_j. The process of iteratively drawing from the conditional distributions can be viewed as a Gibbs sampler that will converge to draws from the theoretical joint distribution of the data.

In general, imputing missing values by joint modeling is faster and the imputation algorithms are simpler to implement. However, empirical data will seldom follow a standard multivariate distribution, especially if they consist of a mix of numerical and categorical variables. Furthermore, FCS provides a flexible tool to account for bounds, interactions, skip patterns or constraints between different variables.

Nevertheless, there is one drawback for FCS that is usually ignored: The iterative draws from the different conditional distributions will only converge to draws from the joint distribution, if this joint distribution really exists. Rubin coined the acronym PIGS (potentially incompatible Gibbs sampler) for FCS, since in practice it is often impossible to proof the existence of a joint distribution for the specified conditional distributions and it is widely unknown what happens, if the joint distribution doesn't exist. Usually convergence is assumed after a deliberate number of iterations, where the number selected depends on the complexity of the imputation model and even more problematic on the time one can afford to wait for imputation results. Measures for convergence have been proposed (Arnold, Castillo and Sarabia (1999)), but can be misleading since the monitored estimates can stay stable for hundreds of iterations before drifting off to infinity. For that reason, convergence is seldom monitored and convergence simply is assumed.

If the fraction of missing values in the data set is low, this seems a reasonable strategy. The Gibbs sampler will converge most of the times and if not, the introduced bias will not be high. But in the last years a new application of multiple imputation became more and more popular: Multiple imputation for statistical disclosure control. Here the originally observed values are replaced by multiple draws from $(Y_{syn}|Y_{obs})$, where Y_{syn} denotes synthetic variables that can be released to the public. This means that all values have to be replaced by imputed values. In this case incompatible imputation models possibly could have devastating consequences on the imputation results.

There have been some investigations on the impact of incompatible conditional distributions on the validity of the imputation results, but they are very limited in scope. This paper further investigates an example of incompatibility discussed by Van Buren et al. (2006). Their starting point is a bivariate normal distribution $Y = (Y_1, Y_2)$, where missing values are generated to resemble a design that is missing at random (MAR). An incompatible imputation model is set up by drawing new values for Y_1 from a regression on Y_2, but new values for Y_2 from a regression of Y_2 on Y_1^2. Van Buren et al. (2006) illustrate that imputations under this incompatible model can still yield valid results. However, this is only true, if all imputed values are positive. We demonstrate that even a small fraction of imputed negative values in Y can lead to biased estimates from the imputed data set. This is an important result since in general the true underlying joint distribution is only approximated and although most variables of interest will have only positive observed values by definition (e.g. the number of employees or the number of cigarettes smoked per day), the imputation model can generate some negative values. The standard procedure after imputation is to edit the data by setting negative values to zero. This seems justified if the fraction of negative values is small. Editing the data after imputation will not change descriptive statistics in a major way. However, doing so ignores the fact that the negative values are used as predictors for other variables during the imputation process. This could lead to bias if, e.g., a quadratic function of the variable is used.

The remainder of the paper is organized as follows. Section 2 recapitulates multiple imputation as a means of treating missing data problems. Section 3 introduces the two different methods to generate draws from the posterior distribution of $(Y_{mis}|Y_{obs})$ and describes the conditions necessary for compatible conditional distributions. Section 4 extends the simulation study from Van Buren et al. (2006) using different means for the bivariate normal distribution. The paper concludes with some final remarks.

2 Multiple Imputation for Missing Data

Multiple imputation, introduced by Rubin (1978) and discussed in detail in Rubin (1987), Rubin (2004), is an approach that retains the advantages of imputation while allowing the uncertainty due to imputation to be directly assessed. With multiple imputation, the missing values in a data set are replaced by $m > 1$ simulated versions, generated according to a probability distribution for the true values given the observed data. More precisely, let Y_{obs} be the observed and Y_{mis} the missing part of a data set Y, with $Y = (Y_{mis}, Y_{obs})$, then missing values are drawn from the Bayesian posterior predictive distribution of $(Y_{mis}|Y_{obs})$, or an approximation thereof. Typically, m is small, such as $m = 5$. Each of the imputed (and thus completed) data sets is first analyzed by standard methods designed for complete data; the results of the m analyses are then combined in a completely generic way to produce estimates, confidence intervals, and test statistics that reflect the missing-data uncertainty properly. In this paper, we discuss analysis with scalar parameters only, for multidimensional quantities see Little and Rubin (2002), Section 10.2. To understand the procedure of analyzing multiply imputed data sets, think of an analyst interested in an unknown scalar parameter θ, where θ could be, e.g. the mean of a variable, the correlation coefficient between two variables or a regression coefficient in a linear regression.

Inferences for this parameter for data sets with no missing values usually are based on a point estimate $\hat{\theta}$, a variance estimate $\hat{V}$, and a normal or Student's t reference distribution. For analysis of the imputed data sets, let $\hat{\theta}_i$ and $\hat{V}_i$ for $i = 1, 2, ...m$ be the point and variance estimates achieved from each of the m completed data sets. To get a final estimate over all imputations, these estimates have to be combined using the combining rules first described by Rubin (1978).

For the point estimate, the final estimate simply is the average of the m point estimates $\hat{\theta}_{MI} = \frac{1}{m}\sum_{i=1}^{m}\hat{\theta}_i$ with $i = 1, 2, ...m$. Its variance is estimated by $T = \overline{V} + (1 + m^{-1})B$, where $\overline{V} = \frac{1}{m}\sum_{i=1}^{m}\hat{V}_i$ is the "within-imputation" variance, $B = \frac{1}{m-1}\sum_{i=1}^{m}(\hat{\theta}_i - \hat{\theta}_{MI})^2$ is the "between-imputation" variance, and the factor $(1 + m^{-1})$ reflects the fact that only a finite number of completed-data estimates $\hat{\theta}_i$, $i = 1, 2, ...m$ are averaged together to obtain the final point estimate. The quantity $\hat{\gamma} = (1 + m^{-1})B/T$ estimates the fraction of information about θ that is missing due to nonresponse.

Inferences from multiply imputed data are based on $\hat{\theta}_{MI}$, T, and a Student's t reference distribution. Thus, for example, interval estimates for θ have the form $\hat{\theta}_{MI} \pm t(1-\alpha/2)\sqrt{T}$, where $t(1-\alpha/2)$ is the $(1-\alpha/2)$ quantile of the t distribution. Rubin and Schenker (1986) provide the approximate value $\nu_{RS} = (m-1)\hat{\gamma}^{-2}$ for the degrees of freedom of the t distribution, under the assumption that with complete data, a normal reference distribution would have been appropriate. Barnard and Rubin (1999) relax the assumption of Rubin and Schenker (1986) to allow for a t reference distribution with complete data, and suggest the value $\nu_{BR} = (\nu_{RS}^{-1} + \hat{\nu}_{obs}^{-1})^{-1}$ for the degrees of freedom in the multiple-imputation analysis, where $\hat{\nu_{obs}} = (1-\hat{\gamma})(\nu_{com})(\nu_{com}+1)/(\nu_{com}+3)$ and ν_{com} denotes the complete-data degrees of freedom.

3 Two Approaches to Generate Imputations for the Missing Values

As discussed in the introduction, there are two main approaches to generate draws from $P(Y_{mis}|Y_{obs})$: Joint modeling and fully conditional specification (FCS). In the following Section both methods should be described in more detail.

3.1 Joint Modeling

In general, it will not be possible to specify $P(Y_{mis}|Y_{obs})$ directly. Note however, that we can write

$$\begin{aligned} P(Y_{mis}|Y_{obs}) &= \int P(Y_{mis}, \psi|Y_{obs})d\psi \\ &= \int P(Y_{mis}|Y_{obs}, \psi)P(\psi|Y_{obs})d\psi. \end{aligned} \tag{1}$$

Given this equation, imputations can be generated in two steps:

1. Generate random draws for the parameter ψ from its observed-data posterior distribution $P(\psi|Y_{obs})$ given the observed values.
2. Generate random draws for Y_{mis} from its conditional predictive distribution $P(Y_{mis}|Y_{obs}, \psi)$ given the actual parameter ψ from step 1.

With joint modeling the second step is straight forward. The distribution of $(Y_{mis}|Y_{obs}, \psi)$ can be obtained from the underlying model. For example a multivariate normal density can be assumed for the complete data. But the first step usually requires Markov Chain Monte Carlo techniques, since the observed-data posterior distribution for $(\psi|Y_{obs})$

seldom follows standard distributions, especially if the missing pattern is not monotone. Therefore, often simple random draws from the complete-data posterior $f(\psi|Y_{obs}, Y_{mis})$ are performed. This means that even for joint modeling convergence of the Markov Chain has to be monitored and it is not guaranteed that it will ever converge. Though the probability of non-convergence might be much lower in this context than with FCS, it is still possible and Schafer (1997) provides examples where the necessary stationary distribution can never be obtained.

3.2 Fully Conditional Specification (FCS)

With FCS the problem of drawing from a k-variate distribution is replaced by drawing k times from much easier to derive univariate distributions. Every variable in the data set is treated separately using a regression model suitable for that specific variable. Thus, continuous variables can be imputed using a normal model, binary variables can be imputed with a logit model and so on. Here, we can specify $P(\psi|Y_{obs})$ directly and no iterations are necessary, because we don't have to draw from possibly awkward multivariate distributions. For example, if we want to impute a continuous variable Y, we can assume $Y|X \sim N(\mu, \sigma^2)$, where X denotes all variables that are used as explanatory variables for the imputation. The two step imputation approach described above can now be applied as follows:

Let n be the number of observations in the observed part of Y. Let k be the number of regressors to be included in the regression. Let $\hat{\sigma}^2$ and $\hat{\beta}$ be the variance and the beta-coefficient estimates obtained from regressions using only the observed data. Finally, let X_{obs} be the matrix of regressors for the observed part of Y and X_{mis} be the matrix of regressors for the fraction of the data where Y is missing. Imputed values for Y_{mis} can now be generated using the following algorithm:

Step 1: Draw new values for $\psi = (\sigma^2, \beta)$ from $P(\psi|Y_{obs}), i.e.,$

- draw $\sigma^2|X \sim (Y_{obs} - X_{obs}\hat{\beta})'(Y_{obs} - X_{obs}\hat{\beta})\chi^{-2}_{n-k}$,
- draw $\beta|\sigma^2, X \sim N(\hat{\beta}, (X'_{obs}X_{obs})^{-1}\sigma^2)$.

Step 2: Draw new values for Y_{mis} from $P(Y_{mis}|Y_{obs}, \psi)$, i.e.,

- draw $Y_{mis}|\beta, \sigma^2, X \sim N(X_{mis}\beta, \sigma^2)$.

Note that we are drawing new values for the parameters directly from the observed-data posterior distributions. This means, we don't need Markov Chain Monte Carlo techniques to obtain new values from the complete-data posterior distribution of the parameters. However, there are more variables with missing data. Thus, we generate new values for Y_{mis} by drawing from $P(Y_{mis}|\beta, \sigma^2, X)$ and the matrix of regressors X might contain imputed values from an earlier imputation step. These values have to be updated now, based on the new information in our recently imputed variable Y. Hence, we have to sample iteratively from the fully conditional distribution for every variable in the data set until the draws from the different conditional distributions converge to draws from the joint distribution.

In a more detailed notation, for multivariate Y, let $Y_j|Y_{-j}$ be the distribution of Y_j conditioned on all rows of Y except Y_j and ψ_j be the parameter specifying the distribution of $Y_j|Y_{-j}$. If Y consists of k rows, and each Y_j is univariate, then the tth iteration of the method consists of the following successive draws:

$$
\begin{aligned}
\psi_1^{(t)} &\sim P(\hat{\psi}_1|Y_1^{obs}, Y_2^{(t-1)}, ..., Y_k^{(t-1)}), \\
Y_1^{(t)} &\sim P(Y_1^{mis}|Y_2^{(t-1)}, ..., Y_k^{(t-1)}, \psi_1^{(t)}), \\
&\vdots \\
\psi_k^{(t)} &\sim P(\hat{\psi}_k|Y_k^{obs}, Y_1^{(t)}, Y_2^{(t)}, ..., Y_{k-1}^{(t)}), \\
Y_k^{(t)} &\sim P(Y_k^{mis}|Y_1^{(t)}, ..., Y_{k-1}^{(t)}, \psi_k^{(t)})
\end{aligned}
$$

The sampler will converge to the desired joint distribution of $(Y_{mis}\ |Y_{obs})$, but only if this joint distribution really exists. In practice it is often impossible to verify this (if we would know the exact joint distribution we would take samples from it directly), so its existence is implicitly assumed. This is problematic, since it will always be possible to draw from the conditional distributions and we will not get any hint that our Gibbs sampler actually never converges.

Another problem may simply occur due to overparametrization, as Van Buren et al. (2006) state: "With k incomplete variables, the vector parameters $[\psi_1, ..., \psi_k]$ will generally depend on each other, and so the sampler can be overparameterized. For example, the space spanned by $[\psi_1, ..., \psi_k]$ generally has more dimensions than appropriate. If this occurs, the implicit joint distribution does not exist."

To conclude, the consequences of taking imputations from so called incompatible Gibbs samplers as good estimates for the missing values in Y are not fully investigated yet. To further understand the meaning of incompatibility, we have to define what conditions are necessary to make a Gibbs sampler compatible.

3.3 Compatibility

Two conditional densities are considered as compatible if a joint distribution exists that has the defined distributions as its conditional densities. For example, Bhattacharryya (1943) noticed that a bivariate normal density could be modeled from two conditional normal densities with the same variance by linear regression. In 1974 Besag (1974) proofed that two conditional densities $f(x|y)$ and $g(y|x)$ are compatible in the above sense if and only if the ratio of their densities $f(x|y)/g(y|x)$ can be factorized into $u(x)v(y)$, where both u and v have to be integrable functions.

4 An Example for Incompatibility

In their paper, Van Buren et al. (2006) use the following example to evaluate the potential consequences of incompatible imputation models: Their original data consist of 1000 draws from the bivariate normal distribution $Y = (Y_1, Y_2)$ with $\mu_1 = \mu_2 = 5$, $\sigma_1^2 = \sigma_2^2 = 1$ and $\rho_{12} = 0.6$. Missing values are generated using three different missing data mechanisms MARRIGHT, MARMID, and MARTAIL. For MARRIGHT the probability to be missing in Y_1 increases with increasing values in Y_2 and vice versa, MARMID generates more missing values at the center of the distribution and for MARTAIL more missing values are generated at the tails of the distribution. The exact missing data mechanisms are given by the following formulae:

$$\begin{array}{ll}
\text{MARRIGHT:} & logit(Pr(Y_1 = missing)) = -1 + Y_2/5 \\
 & logit(Pr(Y_2 = missing)) = -1 + Y_1/5 \\
\text{MARTAIL:} & logit(Pr(Y_1 = missing)) = -1 + 0.4 * |Y_1 - mean(Y_1)| \\
 & logit(Pr(Y_2 = missing)) = -1 + 0.4 * |Y_2 - mean(Y_2)| \\
\text{MARMID:} & 1 - Pr(MARTAIL).^1
\end{array}$$

The authors note that for MARRIGHT "(t)he multivariate missing data were not entirely MAR because the cases where Y_1 or Y_2 (or

[1] In our setting we changed the parameters for MARMID to $1 - (-0.1 + 0.4 * |Y_2 - mean(Y_2)|)$ because the originally suggested parameters $1 - (-1 + 0.4 * |Y_2 - mean(Y_2)|)$ cause the probability that either Y_1 or Y_2 or both are missing to become more than 88%, leaving only a small fraction of fully observed values to estimate the imputation models. This leads to severe bias even for the compatible imputation model in some of the settings. With our parameters the above probability decreases to roughly 70%.

both) is (are) missing were more frequent for the higher values." But they argue that "(t)he regression lines are (...) not affected because the nonresponse is generated symmetrically around the regression lines" (Van Buren et al. (2006, p. 1059)). Argumentation for the other missing mechanisms follows along the same line.

Multiple imputations for the missing values are generated under three different imputation models. The compatible model imputes missing values Y_1^*, Y_2^* by alternately drawing from $Y_1^*|Y_2 \sim N(\mu_1^* + \beta_2^* Y_2, \sigma_2^{2*})$ and $Y_2^*|Y_1 \sim N(\mu_1^* + \beta_1^* Y_1, \sigma_1^{2*})$, where $*$ indicates drawn values from the appropriate observed-data posterior distributions for the different parameters. The first incompatible model replaces the second imputation step given above by draws from $Y_2^*|Y_1 \sim N(\mu_1^* + \beta_1^* Y_1^2, \sigma_2^{2*})$. Another incompatible model uses $\log(Y_1)$ instead of Y_1^2 in the second imputation step. For all simulations the number of iterations between each imputation and the number of imputations are set to five. Finally the number of replications for each setting is set to 500.

In our paper we focus on the first incompatible model. Results for the setting described by Van Buren et al. (2006) are given in Table 1. Note however that we generate 10,000 draws from the bivariate normal distribution. For $n = 1,000$ the sampling error overlays the bias introduced by the incompatible Gibbs sampler in some of the following examples. Thus, for $n = 1,000$ we would see the bias only in the point estimates and not in the 95%-confidence interval. Choosing $n = 10,000$ drives down the sampling error visualizing the bias in the inference.

The estimates of interest are the mean of Y_1 and the regression coefficient β in the linear model $Y_1 = \alpha + \beta Y_2 + \varepsilon$. For the regression coefficient the standard error and the coverage are also reported. The estimates $E(\beta_1/\sigma_1^2)$ and $E(\beta_2/\sigma_2^2)$, with β and σ^2 taken from the imputation model, are indicators for compatibility. Since, for the conditionals from a bivariate normal distribution $E(\beta_1/\sigma_1^2) = E(\beta_2/\sigma_2^2)$ must hold. Note that $E(\beta_1/\sigma_1^2) \neq E(\beta_2/\sigma_2^2)$ does not necessarily mean that the models are generally incompatible in the above sense, they are just incompatible with the bivariate normal distribution.

In the original setting ($\mu_1 = \mu_2 = 5$) the compatible Gibbs sampler provides good results, although $\mathrm{E}(Y_1)$ is slightly biased downwards for MARRIGHT and $E(\beta)$ is slightly biased upwards for MARMID. Imputations under the incompatible Gibbs sampler are only slightly biased for MARTAIL ($E(\beta)$) and MARRIGHT ($\mathrm{E}(Y_1)$), while the complete case analysis shows severe bias for $E(\beta)$ under MARMID and MARTAIL with coverage rates close to zero. Based on these results, Van Buren et al. (2006) conclude that "imputation using the Gibbs sam-

Table 1. Regression slopes, standard errors, and coverages (95% c.i.) under one compatible and one incompatible multiple imputation model (bivariate normal data, $\rho = 0.6$, $n = 10,000$, $m = 5$, three symmetric data mechanisms, 500 replications) when compared with complete case analysis

		Compatibility		Estimates				
Mechanism	*Method*	$E(\frac{\beta_1}{\sigma_1^2})$	$E(\frac{\beta_2}{\sigma_2^2})$	$E(Y_1)$	$E(\beta)$	$E(se(\beta))$	Fmi[2]	Cov
	Theoretical values			5.00	0.600			0.95
MARRIGHT	CC analysis[3]			4.94	0.596	0.0160		0.94
	MI comp. linear	0.99	0.99	4.96	0.602	0.0137	0.75	0.93
	MI incomp. quadr.	0.98	0.097	4.96	0.596	0.0134	0.75	0.92
MARMID	CC analysis			5.00	0.649	0.0137		0.06
	MI comp. linear	0.99	1.00	5.00	0.609	0.0117	0.69	0.88
	MI incomp. quadr.	1.01	0.10	5.00	0.602	0.0118	0.69	0.92
MARTAIL	CC analysis			5.00	0.555	0.0125		0.04
	MI comp. linear	0.97	0.98	5.00	0.597	0.0113	0.56	0.93
	MI incomp. quadr.	0.99	0.10	5.00	0.591	0.0113	0.56	0.89

pler seems to be robust against incompatible-specified conditionals in terms of bias and precision, thus suggesting that incompatibility may be a relatively minor problem in multivariate imputation" (Van Buren et al. (2006, p. 1061)).

This assumption only holds however, if Y_1 and Y_2 will never include any negative values, neither in the original data nor in the imputed data. According to our calculations, negative values in Y_2 will lead to negative values in Y_1 because of the linear imputation model and the positive correlation in the original data. But negative values in Y_1 will drive down the linear correlation between Y_1 and Y_2 if we use the incompatible imputation model, because imputations for Y_2 are based solely on Y_1^2. This can be illustrated easily, if we take the example in Table 1 and set $\mu_1 = \mu_2 = 0$. Table 2 provides the results for this setting. Obviously, the compatible model still provides good results for all estimates. In fact the estimates are almost similar to the estimates in Table 1. For the incompatible model, the correlation and by this the estimated regression coefficient is heavily biased. (Note, that for this setting the coefficient β_1 equals ρ_{12}) As expected, the correlation

[2] Fmi denotes the fraction of the data for which either Y_1,Y_2 or both are missing.
[3] CC analysis (complete case analysis) = Only fully observed units are used for the analysis.

is underestimated. Interestingly, the estimates for one variable alone ($E(Y_1)$) are not affected.

The second setting displayed in Table 3 with $\mu_1 = \mu_2 = 2.75$ aims to illustrate that only a very small fraction of negative values is necessary to cause bias in the incompatible model (note that for a bivariate normal distribution with $\mu = 2.75$ and $\sigma^2 = 1$ the probability for having negative values is $\approx 0.6\%$). Though the bias is small compared to the setting with $\mu_1 = \mu_2 = 0$, it is surprising that such a small fraction of negative values can cause a bias of this magnitude. This is an important result since in practical settings the true underlying distributions are unknown and the defined fully conditional distributions for the imputation model are only approximations of the true distributions. In general this is not a problem since, if the explanatory power of the covariates used is high, approximations will be very close to the true distributions. Still it can happen that these approximations lead to negative imputed values for a variable that by definition can only have positive values, especially if the variable to be imputed contains a high number of zeros. A common approach in this case is to check the data after the imputation for such inconsistencies and, if the fraction of negative values is low, set these negative values to zero. Only if this fraction is high, the imputation model is usually revised. This seems justified since these changes for only a very small fraction of the data will not change inferences in a major way. But this assumption ignores the fact that during the imputation process this variable is used as a covariate in the imputation model for other variables that need to be imputed. If only a quadratic function of the variable is used in the imputation model, the correlation between the two variables could be underestimated.

To illustrate this, we introduce a new missing data mechanism MARLEFT, where the probability to be missing decreases with an increasing value of the explanatory variable. Missing values are generated using the formula:

$$\begin{aligned}\text{MARLEFT: } & logit(Pr(Y_1) = missing) = -0.5 + 0.05 * (-Y_2)\\ & logit(Pr(Y_2) = missing) = -0.5 + 0.05 * (-Y_1).\end{aligned}$$

To mimic a setting where all observed values are positive, but the assumed underlying distribution could be negative, data sets are generated as before by drawing from a bivariate normal distribution with

Table 2. Setting similar to Table 1 with $\mu_1 = \mu_2 = 0$

		Compatibility		Estimates				
Mechanism	*Method*	$E(\frac{\beta_1}{\sigma_1^2})$	$E(\frac{\beta_2}{\sigma_2^2})$	$E(Y_1)$	$E(\beta)$	$E(se(\beta))$	*Fmi*	Cov
	Theoretical values			0.00	0.600			0.95
MARRIGHT	CC analysis			-0.03	0.596	0.0110		0.93
	MI comp. linear	0.93	0.92	-0.01	0.599	0.0102	0.47	0.93
	MI incomp. quadr.	0.54	-0.03	-0.01	0.402	0.0125	0.47	0.00
MARMID	CC analysis			0.00	0.648	0.0137		0.06
	MI comp. linear	0.93	0.92	0.00	0.609	0.0118	0.69	0.88
	MI incomp. quadr.	0.43	0.02	0.00	0.304	0.0135	0.69	0.00
MARTAIL	CC analysis			0.00	0.556	0.0126		0.05
	MI comp. linear	0.90	0.92	0.00	0.598	0.0115	0.56	0.93
	MI incomp. quadr.	0.41	-0.02	0.00	0.313	0.0139	0.56	0.00

Table 3. Setting similar to Table 1 with $\mu_1 = \mu_2 = 2.75$

		Compatibility		Estimates				
Mechanism	*Method*	$E(\frac{\beta_1}{\sigma_1^2})$	$E(\frac{\beta_2}{\sigma_2^2})$	$E(Y_1)$	$E(\beta)$	$E(se(\beta))$	*Fmi*	Cov
	Theoretical values			2.75	0.600			0.95
MARRIGHT	CC analysis			2.70	0.597	0.0131		0.93
	MI comp. linear	0.87	0.90	2.73	0.601	0.0116	0.62	0.95
	MI incomp. quadr.	0.87	0.15	2.73	0.584	0.0116	0.62	0.73
MARMID	CC analysis			2.75	0.649	0.0137		0.06
	MI comp. linear	0.93	0.94	2.75	0.610	0.0120	0.69	0.88
	MI incomp. quadr.	0.89	0.15	2.75	0.587	0.0119	0.69	0.83
MARTAIL	CC analysis			2.75	0.555	0.0126		0.08
	MI comp. linear	0.93	0.92	2.75	0.598	0.0114	0.56	0.93
	MI incomp. quadr.	0.87	0.15	2.75	0.579	0.0114	0.56	0.58

$\mu_1 = \mu_2 = 2.75$ and $\Sigma = \begin{pmatrix} 1 & 0.6 \\ 0.6 & 1 \end{pmatrix}$. But now missing values are generated under the MARLEFT missing design and the imputation process only starts if no negative values remain in the observed part of the data set after MARLEFT is applied. Otherwise a new data set is drawn and missing values are generated in the above manner. The parameters for the missing mechanism are selected to generate a fraction of missing values that roughly equals the fraction of missing values generated by the other MAR mechanisms, while on the other hand keeping the number of draws necessary to fulfill the above requirements (no negative values in the observed data) as low as possible. Drawing 10,000 times from the above distribution will make it almost impossible that no single negative value remains in the data set after the MARLEFT design is applied. So we decided to set $n = 1,000$ accepting the increase in the sampling error. Even with $n = 1,000$, the average number of draws necessary under this setting is 35.36, so the data sets selected for the imputation reflect draws from a truncated bivariate normal distribution. The results for the original data in Table 4 indicate, however, that the introduced bias is only marginal. Nevertheless, the bias introduced by the imputations under the incompatible model remains, although the fraction of imputed negative values is very low (the empirical fraction of negative values was 0.093% for Y_1 and 0.090% for Y_2 respectively). This result, although alarming at first sight, does not necessarily mean

Table 4. Estimates for a data set with no observed negative values after the application of the MARLEFT missing data mechanism, (bivariate normal data, $\rho = 0.6, n = 1,000, m = 5, \mu_1 = \mu_2 = 2.75$, 500 replications)

		Compatibility		Estimates				
Mechanism	*Method*	$E(\frac{\beta_1}{\sigma_1^2})$	$E(\frac{\beta_2}{\sigma_2^2})$	$E(Y_1)$	$E(\beta)$	$E(se(\beta))$	*Fmi*	Cov
	Theoretical values			2.75	0.600			0.95
MARLEFT	Original Data			2.758	0.594	0.0255		0.96
	CC analysis			2.77	0.592	0.0392		0.95
	MI comp. linear	0.80	0.85	2.76	0.591	0.0350	0.57	0.95
	MI incomp. quadr.	0.80	0.13	2.76	0.577	0.0356	0.57	0.90

that all imputation models using quadratic terms will have to guarantee that no negative values will be imputed. The scenario depicted here is somewhat theoretical in a sense that regression models hardly ever contain only the quadratic term of a covariate and omit the linear term.

Table 5 presents results for the different settings discussed above, but now imputations of Y_2 are based on the standard model if a quadratic function of the covariates is considered for imputation: $Y_2 = \beta_0 + \beta_1 Y_1 + \beta_2 Y_1^2 + \varepsilon$. Not surprisingly, under this specification all estimates are unbiased for all settings, since the model includes a linear term again and we could think of the quadratic term as being part of the residuals. This will lead to residuals with an awkward distribution and a mean that is definitely not zero. Nevertheless, the above equations specify two linear conditional models for which a joint distribution generally exists.

Table 5. Estimates for the mean of Y_1 and the regression coefficient β in the linear model $Y_1 = \alpha + \beta Y_2 + \varepsilon$ from the imputation model: $Y_1 = \beta_0 + \beta_1 Y_2 + \varepsilon$ and $Y_2 = \beta_0 + \beta_1 Y_1 + \beta_2 Y_1^2 + \varepsilon$ (bivariate normal data, $\rho = 0.6, n = 10,000, m = 5$, different μ, 500 replications)

mean	*Mechanism*	Compatibility		Estimates				
		$E(\frac{\beta_1}{\sigma_1^2})$	$E(\frac{\beta_2}{\sigma_2^2})$	$E(Y_1)$	$E(\beta)$	$E(se(\beta))$	Fmi	Cov
$\mu_1 = \mu_2 = 5$	MARRIGHT	0.94	-0.01	4.96	0.602	0.0137	0.75	0.93
	MARMID	0.94	-0.01	5.00	0.609	0.0117	0.69	0.85
	MARTAIL	0.93	0.00	5.00	0.598	0.0113	0.56	0.91
$\mu_1 = \mu_2 = 2.75$	MARRIGHT	0.96	-0.01	2.73	0.601	0.0115	0.63	0.92
	MARMID	0.95	-0.01	2.75	0.609	0.0117	0.69	0.86
	MARTAIL	0.93	0.00	2.75	0.598	0.0113	0.56	0.91
$\mu_1 = \mu_2 = 0$	MARRIGHT	0.90	0.00	-0.01	0.600	0.0102	0.47	0.94
	MARMID	1.00	0.01	0.00	0.609	0.0116	0.69	0.85
	MARTAIL	0.93	0.00	0.00	0.598	0.0113	0.56	0.91

5 Concluding Remarks

Fully conditional specifications are very flexible tools to generate multiple imputations. They are especially useful for settings where the desired joint distribution for multivariate data doesn't follow standard distributions. In this case, joint modeling - generally based on the assumption of a multivariate normal distribution - can lead to biased estimates from the imputed data. However, this increased flexibility comes at a price since FCS can be problematic, if the specified conditional distributions don't have a joint distribution. This means that the underlying Gibbs sampler will never converge. The consequences for the imputation results are still unknown.

In this paper we illustrate that an incompatible imputation model can lead to biased estimates from the imputed data. These findings are in contrast to the results in Van Buren et al. (2006) where the authors presume that incompatible Gibbs samplers might have only minor influences on the imputation results.

Obviously, there is not yet a best practice how to deal with this problem. In general, we would suggest to monitor convergence carefully when using FCS. The depicted example for an incompatible imputation model is somewhat theoretical of course, since imputations will never be based solely on the quadratic term of a covariate. Maybe it is possible to show that only implausible imputation models can lead to non-convergence. Then, even if FCS are PIGS, that doesn't mean they are not useful in many settings. For this reason one field of further research could be to evaluate the risk of defining an incompatible imputation model by chance.

References

Rubin DB (1978) Multiple Imputation in Sample Surveys - a Phenomenological Baysian Approach to Nonresponse. American Statistical Association Proceedings of the Section on Survey Research Methods 20–40

Van Buuren S, Brand JPL, Groothuis-Oudshoorn CGM, Rubin DB (2006) Fully conditional specification in multivariate imputation. Journal of Statistical Computation and Simulation 76:1049–1064

Rubin DB (1987) Multiple Imputation for Nonresponse in Surveys. John Wiley & Sons, New York

Schafer J (1997) Analysis of Incomplete Multivariate Data. Chapman and Hall, London

Arnold BC, Castillo E, and Sarabia JM (1999) Conditional Specification of Statistical Models. Springer-Verlag, New York

Rubin DB (2004) The Design of a General and Flexible System for Handling Nonresponse in Sample Surveys. The American Statistician 58:298–302

Little RJA, Rubin DB (2002) Statistical Analysis With Missing Data. John Wiley & Sons, Hoboken

Rubin DB, Schenker N (1986) Multiple Imputation for Interval Estimation From Simple Random Samples With Ignorable Nonresponse. Journal of the American Statistical Association 81:366–374

Barnard J, Rubin DB (1999) Small-sample Degrees of Freedom With Multiple Imputation. Biometrika 86:948–955

Bhattacharryya A (1943) On some sets of sufficient conditions leading to the normal bivariate distribution. Sankhya 6:399–406

Besag J (1974) Spatial interaction and the statistical analysis of lattice systems (with discussions). Journal of the Royal Statstical Society B36:192–236

OLS-Based Estimation of the Disturbance Variance Under Spatial Autocorrelation

Walter Krämer[1] and Christoph Hanck[2]

[1] Department of Statistics, University of Dortmund, Vogelpothsweg 78, 44221 Dortmund, Germany walterk@statistik.uni-dortmund.de
[2] Department of Statistics, University of Dortmund, Vogelpothsweg 78, 44221 Dortmund, Germany christoph.hanck@uni-dortmund.de

1 Introduction

We consider the standard linear regression model with N observations

$$y = X\beta + u,$$

where y is $(N \times 1)$ vector of observations on response variable, X is a nonstochastic $(N \times K)$ matrix of N observations on each of the K explanatory variables with rank K and β is an unknown $(K \times 1)$ vector of associated regression coefficients. The components of u have expected value $E(u) = 0$ and a common variance $E(u_i^2) = \sigma^2$ $(i = 1, 2, \ldots, N)$. The OLS estimate for β is $\hat{\beta} = (X'X)^{-1}X'y$, and the OLS-based estimate for σ^2 is

$$s^2 = \frac{1}{N-K}(y - X\hat{\beta})'(y - X\hat{\beta}) = \frac{1}{N-K}u'Mu, \tag{1}$$

where $M = I - X(X'X)^{-1}X'$. It has long been known that s^2 is in general (and contrary to $\hat{\beta}$) biased whenever $V := Cov(u)$ is no longer a multiple of the identity matrix. Krämer (1991) and Krämer and Berghoff (1991) show that this problem disappears asymptotically for certain types of temporal correlation such as stationary AR(1)-disturbances, although it is clear from Kiviet and Krämer (1992) that the relative bias of s^2 might still be substantial for any finite sample size. The present paper extends these analyses to the case of spatial correlation, where we allow the disturbance vector u to be generated by the spatial autoregressive scheme

$$u = \rho W u + \epsilon, \tag{2}$$

where ϵ is a $N \times 1$ random vector with mean zero and scalar covariance matrix $\sigma_\epsilon^2 I$ and W is some known $N \times N$-matrix of nonnegative spatial weights with $w_{ii} = 0$ $(i = 1, \ldots, N)$. Such patterns of dependence are often entertained when the objects under study are positioned in some "space," whether geographical or sociological (in some social network, say) and account for spillovers from one unit to its neighbors, whichever way "neighborhood" may be defined. They date back to Whittle (1954) and have become quite popular in econometrics recently. See Anselin and Florax (1995) or Anselin (2001) for surveys of this literature.

The coefficient ρ in (2) measures the degree of correlation, which can be both positive and negative. Below we focus on the empirically more relevant case of positive disturbance correlation, where

$$0 \leqslant \rho \leqslant \frac{1}{\lambda_{\max}}$$

and where $\lambda_{\max}$ is the Frobenius-root of W (i.e., the unique positive real eigenvalue such that $\lambda_{\max} \geqslant |\lambda_i|$ for arbitrary eigenvalues λ_i). The disturbances are then given by

$$u = (I - \rho W)^{-1} \epsilon,$$

so

$$V = Cov(u) = \sigma_\epsilon^2 [(I - \rho W)'(I - \rho W)]^{-1} \tag{3}$$

and

$$V = \sigma_\epsilon^2 I$$

whenever $\rho = 0$.

Of course, for our analysis to make sense, the main diagonal of V should be constant, i.e.,

$$V = \sigma^2 \Sigma, \tag{4}$$

where Σ is the correlation matrix of the disturbance vector.[1] It is therefore important to clarify that many, though not all, spatial autocorrelation schemes are compatible with homoscedasticity. Consider for instance the following popular specification for the weight matrix known as "one ahead and one behind:"

[1] Note that $\sigma^2 = Var(u_i)$ need not be equal to $\sigma_\epsilon^2 = Var(\epsilon_i)$, unless $\Sigma = I$. In the sequel, we keep σ_ϵ^2 fixed, so σ^2 will in general vary with W and N.

$$\tilde{W} := \begin{pmatrix} 0 & 1 & 0 & \cdots & 0 & 1 \\ 1 & 0 & \ddots & 0 & \cdots & 0 \\ 0 & \ddots & \ddots & \ddots & \ddots & \vdots \\ \vdots & \ddots & \ddots & 0 & \ddots & 0 \\ 0 & \cdots & 0 & 1 & 0 & 1 \\ 1 & 0 & \cdots & 0 & 1 & 0 \end{pmatrix}$$

and renormalize the rows such that the row sums are 1. Then it is easily seen that $E(u_i^2)$ is independent of i, and analogous results hold for the more general "j ahead and j behind" weight matrix W which has non-zero elements in the j entries before and after the main diagonal, with the non-zero entries equal to $j/2$. This specification has been considered by, for instance, Kelejian and Prucha (1999) and Krämer and Donninger (1987).

As another example, consider the equal-weight matrix (see, e.g., Kelejian and Prucha (2002), Lee (2004), Case (1992) or Kelejian et al. (2006)), defined by

$$W^{EW} = (w_{ij}^{EW}) = \begin{cases} \frac{1}{N-1} & \text{for } i \neq j \\ 0 & \text{for } i = j. \end{cases} \tag{5}$$

One easily verifies that, for $|\rho| < 1$,

$$(I - \rho W^{EW})^{-1} = \delta_1 J_N + \delta_2 I_N,$$

where

$$\delta_1 = \frac{\rho}{(N-1+\rho)(1-\rho)}, \quad \delta_2 = \frac{N-1}{N-1+\rho}$$

and J_N is an $(N \times N)$ matrix of ones. Without loss of generality, let $\sigma_\epsilon^2 = 1$. We then have, using symmetry of W,

$$\begin{aligned} V &= [(I - \rho W^{EW})'(I - \rho W^{EW})]^{-1} \\ &= (I - \rho W^{EW})^{-1}(I - \rho W^{EW})^{-1} \\ &= (\delta_1 J_N + \delta_2 I_N)^2. \end{aligned}$$

Carrying out the multiplication, it is seen that

$$E(u_i^2) = (\delta_1^2 + \delta_2^2)^2 + (N-1)\delta_1^2 \quad \text{for} \quad i = 1, \ldots, N.$$

So V is homoscedastic . It is straightforward to extend this result to the case where W is block-diagonal with B blocks of dimension $(R \times R)$, defined as

$$W_R^{EW} = (w_{R,ij}^{EW}) = \begin{cases} \frac{1}{R-1} & \text{for } i \neq j \\ 0 & \text{for } i = j, \end{cases}$$

where $N = BR$. We therefore conclude that our analysis is applicable in many relevant spatial econometric specifications.

2 The Relative Bias of s^2 in Finite Samples

We have

$$\begin{aligned} E\left(\frac{s^2}{\sigma^2}\right) &= E\left(\frac{1}{\sigma^2(N-K)}u'Mu\right) \\ &= \frac{1}{\sigma^2(N-K)}tr(MV) \\ &= \frac{1}{N-K}tr(M\Sigma). \end{aligned}$$

Watson (1955) and Sathe and Vinod (1974) derive the (attainable) bounds

$$\begin{aligned} &\text{mean of } N-K \text{ smallest eigenvalues of } \Sigma \\ &\leqslant E\left(\frac{s^2}{\sigma^2}\right) \leqslant \\ &\text{mean of } N-K \text{ largest eigenvalues of } \Sigma, \end{aligned} \tag{6}$$

which shows that the bias can be both positive and negative, depending on the regressor matrix X, whatever Σ may be. Finally, Dufour (1986) points out that the inequalities (6) amount to

$$0 \leqslant E\left(\frac{s^2}{\sigma^2}\right) \leqslant \frac{N}{N-K} \tag{7}$$

when no restrictions are placed on X and Σ. Again, these bounds are sharp and show that underestimation of σ^2 is much more of a threat in practise than overestimation.

The problem with Dufour's bounds is that they are unnecessarily wide when extra information on V is available. Here we assume a disturbance covariance matrix V as in (3) and show first that the relative bias of s^2 depends crucially on the interplay between X and W. In particular, irrespective of sample size and of the weighting matrix W, we can always produce a regressor matrix X such that $E(s^2/\sigma^2)$ becomes as close to zero as desired. To see this, let W be symmetric[2] and let

$$W = \sum_{i=1}^{N} \lambda_i \omega_i \omega_i' \tag{8}$$

be the spectral decomposition of W, with the λ_i in increasing order and ω_i the corresponding orthonormal eigenvectors . Now it is easily seen that

$$\lim_{\rho \to 1/\lambda_N} E\left(\frac{s^2}{\sigma^2}\right) = 0 \tag{9}$$

whenever

$$M\omega_N = 0. \tag{10}$$

This follows from

$$V = \sigma_\epsilon^2 \left[\sum_{i=1}^{N} \frac{1}{(1-\rho\lambda_i)^2} \omega_i \omega_i' \right] \tag{11}$$

and

$$\Sigma = \frac{1}{\sigma^2} V = \frac{1}{\sum_{i=1}^{N} \frac{1}{(1-\rho\lambda_i)^2} \omega_{i1}^2} \sum_{i=1}^{N} \frac{1}{(1-\rho\lambda_i)^2} \omega_i \omega_i', \tag{12}$$

where ω_{i1}^2 is the $(1,1)$-element of $\omega_i \omega_i'$ (under homoscedasticity, we could select any diagonal element of $\omega_i \omega_i'$) and

$$\sigma^2 = \sigma_\epsilon^2 \sum_{i=1}^{N} \frac{1}{(1-\rho\lambda_i)^2} \omega_{i1}^2. \tag{13}$$

Multiplying the numerator and denominator of (12) by $(1-\rho\lambda_N)^2$, we obtain

$$\Sigma = \frac{1}{\sigma^2} V = \frac{1}{\sum_{i=1}^{N} \frac{(1-\rho\lambda_N)^2}{(1-\rho\lambda_i)^2} \omega_{i1}^2} \sum_{i=1}^{N} \frac{(1-\rho\lambda_N)^2}{(1-\rho\lambda_i)^2} \omega_i \omega_i', \tag{14}$$

which tends to

---

[2] Notice that for all the homoskedastic examples considered above, row-normalization does not destroy symmetry of W.

$$\frac{1}{\omega_{N1}^2} \omega_N \omega_N' \tag{15}$$

as $\rho \to 1/\lambda_N$. Given W, one can therefore choose X to be $(N \times 1)$ and equal to ω_N. Then, M is by construction orthogonal to ω_N, which implies that $tr(M\Sigma)$ and therefore also $E(s^2/\sigma^2)$ tend to zero as $\rho \to 1/\lambda_N$.

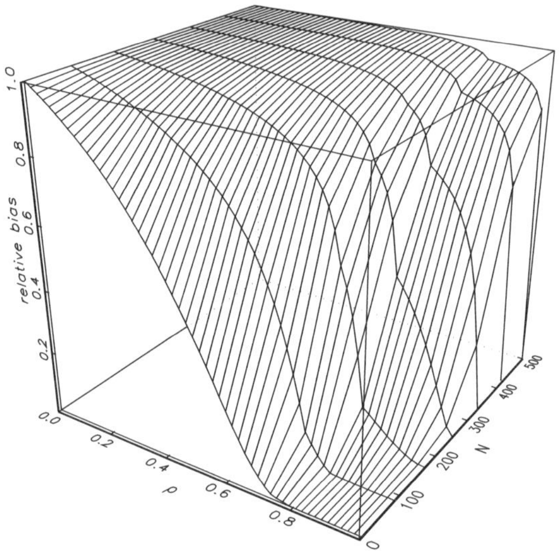

Fig. 1. The relative bias of s^2 as a function of ρ and N

For illustration, consider the following example. The largest eigenvalue λ_N of a row-normalized matrix such as $\tilde{W}/2$ is 1. (This follows immediately from Theorem 8.1.22 of Horn and Johnson (1985).) It is then readily verified that $\omega_N = \iota := (1, \ldots, 1)'$ is (up to the usual multiple) the eigenvector corresponding to λ_N. Now, if $X = \iota$, $M\omega_N = (I - \frac{1}{N}\iota\iota')\iota = 0$. Figure 1 shows the behavior of the relative bias as $\rho \to 1/\lambda_N = 1$. We see that (9) holds for any given N. Also, pointwise in ρ, the relative bias vanishes as $N \to \infty$, as one would expect. We now rigorously establish the latter property.

3 Asymptotic Bias and Consistency

From (7), it is clear that, for any V, the relative upward bias of s^2 must vanish as $N \to \infty$. A sufficient condition for the relative downward bias to disappear as well is that the largest eigenvalue of Σ, μ_N, is

$$\mu_N = o(N). \tag{16}$$

This is so because, using $\sum_{i=1}^{N} \mu_i = \sum_{i=1}^{N-K} \mu_i + \sum_{i=1}^{K} \mu_{i+N-K} = N$, we have
mean of $N - K$ smallest eigenvalues of $\Sigma =$

$$\begin{aligned} &= \frac{N}{N-K} - \frac{1}{N-K} \sum_{i=1}^{K} \mu_{i+N-K} \\ &\geqslant \frac{N}{N-K} - \frac{K}{N-K} \mu_N \end{aligned}$$

and the right-hand side tends to 1 when (16) holds as $N \to \infty$. Condition (16) also guarantees consistency. From (1), we have

$$s^2 = \frac{1}{N} u' M u = \frac{1}{N} u' u - \frac{1}{N} u' H u, \tag{17}$$

where $H = X(X'X)^{-1}X'$. Since $u'u/N \xrightarrow{p} \sigma^2$, it remains to show that

$$\frac{1}{N} u' H u \xrightarrow{p} 0. \tag{18}$$

To this purpose, consider

$$\begin{aligned} E\left(\frac{1}{N} u' H u\right) &= E\left(\frac{1}{N} \varepsilon' \Sigma^{1/2} H \Sigma^{1/2} \varepsilon\right) \qquad (\text{where } \varepsilon = \Sigma^{-1/2} u) \\ &= \frac{\sigma^2}{N} tr(\Sigma^{1/2} H \Sigma^{1/2}) \\ &= \frac{\sigma^2}{N} tr(H\Sigma) \\ &\leqslant \frac{\sigma^2}{N} K \cdot \mu_N, \end{aligned} \tag{19}$$

where the inequality follows from the fact that $H\Sigma$ has rank K (since rank $(H) = K$). Since no eigenvalue of $H\Sigma$ can exceed μ_N, and $H\Sigma$ has exactly K nonzero eigenvalues, the inequality follows from the well known fact that the trace of a matrix equals the sum of its eigenvalues. By assumption, $\mu_N/N \to 0$ as $N \to \infty$, so in view of (19), $E(u'Hu/N) \to 0$. As $u'Hu$ is nonnegative, this in turn implies $u'Hu/N \xrightarrow{p} 0$ and therefore the consistency of s^2.

The crucial condition (16) is a rather mild one; in the present context, it obviously depends on the weighting matrix W. From (4) and (11), we have

$$\mu_N = \frac{\sigma_\epsilon^2}{\sigma^2(1-\rho\lambda_N)^2}, \tag{20}$$

so the condition (16) obtains whenever

$$\sigma^2(1-\rho\lambda_N)^2 N \to \infty. \tag{21}$$

For row-normalized weight matrices , $\lambda_N \equiv 1$ irrespective of N, so (21) holds trivially, provided σ^2 remains bounded away from zero. This in turn follows from the fact that, in view of (13),

$$\sigma^2 \geqslant \frac{\sigma_\epsilon^2}{(1-\rho\lambda_N)^2}\sum_{i=1}^{N}\omega_{i1}^2,$$

where

$$\sum_{i=1}^{N}\omega_{i1}^2 = 1$$

as $\Omega = (\omega_1, \ldots, \omega_N)$ satisfies $\Omega\Omega' = I$.

As another example, consider the "one ahead and one behind" matrix adapted to a "non-circular world" where the $(1, N)$ and $(N, 1)$ entries of $\tilde{W}$ are set to zero, such that after row-normalization ,

$$W' := \begin{pmatrix} 0 & 0.5 & 0 & \cdots & 0 & 0 \\ 0.5 & 0 & \ddots & 0 & \cdots & 0 \\ 0 & \ddots & \ddots & \ddots & \ddots & \vdots \\ \vdots & \ddots & \ddots & 0 & \ddots & 0 \\ 0 & \cdots & 0 & 0.5 & 0 & 0.5 \\ 0 & 0 & \cdots & 0 & 0.5 & 0 \end{pmatrix}.$$

Ord (1975) shows that the eigenvalues of W' are then given by

$$\lambda_i' = \cos\left(\frac{\pi i}{N+1}\right), \; i = 1, \ldots, N,$$

so

$$\lambda_i' \in [-1, 1], \; i = 1, \ldots, N.$$

References

Anselin L (2001) Rao's score test in spatial econometrics. Journal of Statistical Planning and Inference 97(1):113–139

Anselin L, Florax R (1995) Small sample properties of tests for spatial dependence in regression models: Some further results. In: L. Anselin and R. Florax (eds) New Directions in Spatial Econometrics. Springer, Berlin

Case A (1992) Neighborhood Influence and Technological Change. Regional Science and Urban Economics 22(2):491–508

Dufour JM (1986) Bias of s^2 in Linear Regression with Dependent Errors. The American Statistican 40(4):284–285

Horn RA, Johnson CR (1985) Matrix Analysis. Cambridge University Press

Kelejian HH, Prucha IR (1999) A Generalized Moments Estimator for the Autoregressive Parameter in a Spatial Model. International Economic Review 40(2):509–533

Kelejian HH, Prucha IR (2002) SLS and OLS in a spatial autoregressive model with equal spatial weights. Regional Science and Urban Economics 32(6):691–707

Kelejian HH, Prucha IR, Yuzefovich Y (2006) Estimation Problems in Models with Spatial Weighting Matrices Which Have Blocks of Equal Elements. Journal of Regional Science 46(3):507–515

Kiviet J, Krämer W (1992) Bias of s^2 in the Linear Regression Model with Autocorrelated Errors. The Review of Economics and Statistics 74(2):362–365

Krämer W (1991) The Asymptotic Unbiasedness of S^2 in the Linear Regression Model with AR(1)-Disturbances. Statistical Papers 32(1):71–72

Krämer W, Berghoff S (1991) Consistency of s^2 in the Linear Regression Model with Correlated Errors. Empirical Economics 16(3):375–377

Krämer W, Donninger C (1987) Spatial Autocorrelation Among Errors and the Relative Efficiency of OLS in the Linear Regression Model. Journal of the American Statistical Association 82(398):577–579

Lee LF (2004) Asymptotic Distributions of Quasi-Maximum Likelihood Estimators for Spatial Autoregressive Models. Econometrica 72(6):1899–1925

Ord K (1975) Estimation Methods for Models of Spatial Interaction. Journal of the American Statistical Association 70(349):120–126

Sathe S, Vinod H (1974) Bounds on the Variance of Regression Coefficients due to Heteroscedastic or Autoregressive Errors. Econometrica 42(2):333–340

Watson G (1955) Serial Correlation in Regression Analysis I. Biometrika 42(3/4):327–341

Whittle P (1954) On Stationary Processes in the Plane. Biometrika 41(3/4):434–449.

Application of Self-Organizing Maps to Detect Population Stratification

Nina Wawro[1] and Iris Pigeot[2]

[1] Bremen Institute for Prevention Research and Social Medicine, Linzer Strasse 10, 28359 Bremen, Germany wawro@bips.uni-bremen.de
[2] Bremen Institute for Prevention Research and Social Medicine, Linzer Strasse 10, 28359 Bremen, Germany pigeot@bips.uni-bremen.de

1 Introduction

Genetic epidemiology has become a major field of interest at the border of traditional epidemiology and genetics. Genetic data call for specific epidemiological and statistical methods that have to account on the one hand for the family structure within a data set and on the other hand for the paired information of two alleles at each gene locus (genotype). In addition to these obvious dependencies genetic data can be very complex. Often enormous numbers of hypotheses are investigated simultaneously and various levels of data can be thought of, e.g. gene expression data from different pathways as well as data on the protein level. Complexity arises from possible interactions between genes and the environment and from the lack of methods to model biological interactions by statistical interaction terms.

Two main concepts, linkage and association, form the basis of possible analysis. Linkage on the one hand describes the idea that two gene loci are jointly inherited to the next generation more often than one would expect by chance. Association on the other hand can be observed between a phenotypic trait, e.g. a disease, and genetic information at a certain gene locus. This concept is linked to the term statistical dependence and will be in the focus of this paper.

Population-based association studies are increasingly used to investigate genetic risk factors for diseases. In the context of an association study, a statistical association between an allele or genotype at a certain gene locus and the phenotype of the disease under investigation may indicate that the gene is either a causal disease locus itself or that it is in linkage disequilibrium with the disease locus. However, population-based studies might suffer from spurious association due to undetected genetic heterogeneity of the population. This phenomenon is commonly called *population stratification* . Population stratification is statistically determined by trying to identify a number of subpopulations with different allele frequencies in non-coding regions over the whole genome in the population under study. Population stratification has to be detected and accounted for in statistical analysis of association studies, because it is likely to lead to false-positive results (see Cardon and Palmer (2003)).

To deal with the problem of population stratification, different approaches were discussed in the literature (for a review see e.g. Cardon and Palmer (2003) or Pritchard and Donnelly (2001)). Devlin and Roeder (1999) proposed the use of unlinked null-markers, which they called genomic controls (GC). These serve to correct the impact of population stratification on the test statistic via an estimated factor that inflates the observed value of the χ^2-statistic. An alternative approach introduced by Pritchard et al. (2000a,b) is the *structured association* approach, which has a more general aim. While the genomic control approach targets to correct the value of the test statistic, the structured association approach explores the nature of the structure. Roughly speaking, each individual is assumed to possess fractions of ancestry in underlying subpopulations. In the first step, these fractions are estimated by a model-based cluster analysis. In the second step, a test for association called STRAT of a candidate locus and the disease is carried out taking into account the estimated ancestry of the individuals. Power comparisons of the GC approach, the structured association (SA) approach and a family based test, the S-TDT, have been conducted by Wawro et al. (2006).

This paper aims to provide an alternative approach to the model-based cluster analysis that is carried out within the structured association approach. The dependence on distributional assumptions (see Pritchard et al. (2000a)) when identifying the number of underlying subpopulations will be avoided by the application of an exploratory method, namely Self-Organizing Maps (SOMs). The paper organizes as follows. Section 2 will give a brief introduction to the idea of the Self-Organizing Maps. Section 3 investigates the limitations of this proposed method as clustering tool by means of a simulation study. The focus is on the identification of different population structure models and thus on the identification of the number of underlying subpopulations. Some concluding remarks will be given in Section 4 of the paper.

2 Material and Methods

Self-Organizing Maps are a flexible tool to discover hidden structure in a data set. They were developed by Kohonen (2001) in the eighties and today still have a broad application in signal processing or speech recognition, to name only two examples. The following sections give a short introduction to the basic SOM algorithm.

2.1 The Idea of SOMs

Following Kohonen (2001) we present the idea and training algorithm of SOMs. A SOM consists of a number of neurons J, connected to each other by a, usually two-dimensional, grid and thus establishing a relationship between the neurons $N_j, j = 1, \ldots, J$. A simple example of a net is given in Figure 1.

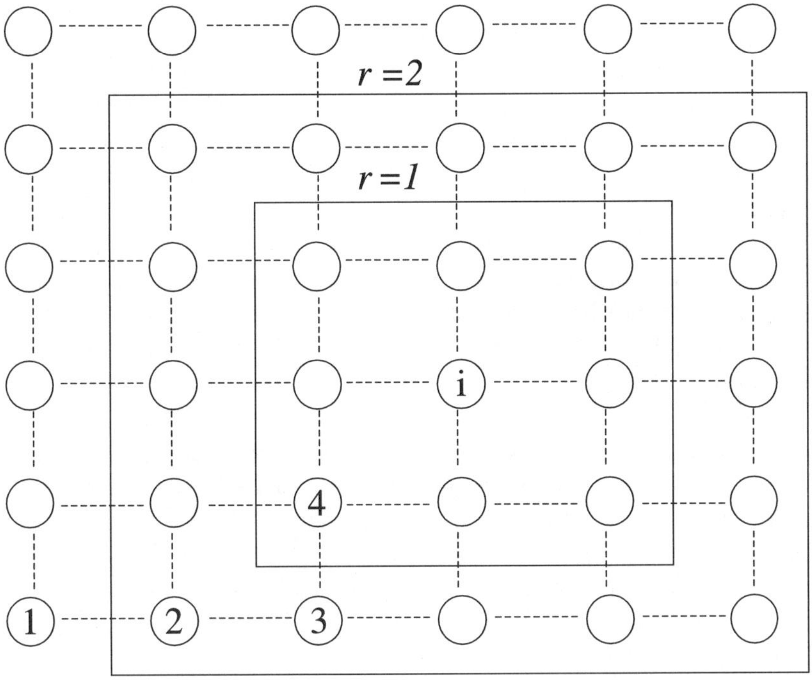

Fig. 1. Example of a Self-Organizing Map with two neighborhoods (radius 1 resp. 2) marked around neuron i

Each neuron has a weight vector w_j. This weight vector will be updated during the training process as follows (see also equation 1). At each training step one data vector $x^{(t)} \in \{x_1, \ldots, x_n\}$ is processed. The weight vector c which has minimum distance to that data vector will be updated. In addition, all weight vectors within a certain neighborhood defined by the neighborhood function h of neuron N_c will be updated as well. This process is formalized by

$$w_j^{(t+1)} = w_j^{(t)} + h_{cj}^{(t)}[x^{(t)} - w_j^{(t)}]. \tag{1}$$

The neighborhood function h_{cj} is fixed, but the shape of the neighborhood has to be chosen in advance, e.g. as bubble or gaussian neighborhood. The learning rate is incorporated in the neighborhood function. Roughly speaking, the weight vectors to be updated are adapted to the data vector by moving them in its direction. The learning rate reflects the magnitude by which the weight vector is moved in this direction at that point in time. Both, the neighborhood function as well as the learning rate are monotone decreasing during the training process. For a more detailed introduction the reader is referred to Kohonen (2001) where issues of convergence of the training process and ordering of the map are discussed as well.

The initialization of the untrained map, that is the untrained weight vectors, can be done in various ways. Most often a random sample of J data vectors as $\{w_j, j \in 1, \ldots, J\}$ or a initialization by random numbers are used.

When the training process has converged, the trained weight vectors can be interpreted as abstraction of the data set which leads to a reduction of the dimensionality of the original data set. In a visual representation of a SOM similar weight vectors should be closer to each other than strongly differing weight vectors.

The input data $x_1, \ldots, x_n$ have to be on a metric scale to allow calculating a meaningful difference in equation 1. This is also necessary when the distances between weight vectors and data vector are calculated by the Euclidean distance, as would be a natural choice. All weight vectors w_j and all data vectors x_i are of the same dimension $1 \times L$, where L is the number of variables of interest. When using SOMs to detect population structure, L represents the number of loci from which genetic information is available.

In this section the original SOM algorithm has been discussed briefly. The extension and adaption of the algorithm to handle genetic data appropriately is presented in a forthcoming paper by Wawro and will not be discussed here.

The following section will focus on the application of SOMs as a clustering tool in the context of population stratification.

2.2 SOMs as a Clustering Tool

Clustering aims at identifying groups that are present in the data set described by a number of attributes. Two main approaches have to be distinguished: (1) fuzzy clustering, where each observation belongs to each possible group with a certain probability, and (2) hard clustering, where this probability is 100% for exactly one of the groups and 0 else. In both cases, a cluster solution consists of all these probabilities for the whole data set. A SOM yields a cluster solution as it divides the data set into a number of groups equal to the number of neurons that build the map. Each data vector has minimum distance to one neuron, which represents the group the data vector is assigned to.

There is a variety of clustering algorithms that range from hierarchical to partitioning, model- or density-based methods. For an overview, the reader is referred to Bock (1974), Everitt (1993) or Jain et al. (1999). In a two step procedure within a SOM framework, first, the weight vectors have to be trained and second the clustering algorithms have to be applied to the trained weight vectors, not on the data vectors of individuals. This two step procedure is more efficient as the weight vectors are representatives of the data set, but with a far smaller dimension. Whenever a weight vector is assigned by a hard clustering algorithm to a certain group, an individual is assigned to the same group if its data vector has minimum distance to this weight vector. This has been proposed by Vesanto and Alhoniemi (2000).

The groups present in the genomic data set are characterized by the genetic background they represent. That is, each group represents a genetically homogenous subgroup of the population characterized by a vector $q = (q_1, \ldots, q_K)$. Here, K represents the number of subpopulations that form the population under investigation and $q_i, i \in \{1, \ldots, K\}$, denotes the fraction of ancestry from subpopulation i. A cluster solution, no matter of fuzzy or hard, can be interpreted as estimate of the genetic background q. This is straightforward, even in a situation where $K > 2$ subpopulations build the basis of the genetic background. Additional information, e.g. known genetic background for a subsample of individuals, is needed to identify the relevant subpopulations.

The crucial task is to identify the correct number of subpopulations K. This has direct impact on the interpretation of a cluster solution. Depending on the method of clustering a variety of indices exists to evaluate the obtained cluster solution, such as the silhouette or the Davies-Bouldin index (see Davies and Bouldin (1979)) for hard solutions and the Xie-Beni index (see Xie and Beni (1991)) or the partition index for fuzzy solutions where in general smaller values of the indices indicate better solutions. A major drawback of all these indices is that they only give reasonable results for $K \geq 2$. Especially in the context of genetic epidemiology it is crucial to identify the situation of 'one group only' as in this case no concerns arise from population stratification. Tibshirani et al. (2001) developed the gap statistic as an index being reasonable for $K \geq 1$.

3 Simulation Study: Application of SOMs in the Context of Genetic Epidemiology

This section outlines the simulation study that we conducted to investigate if SOMs are appropriate to detect population stratification and to estimate the genetic background. This estimate can be used for a stratified analysis or in the framework of the structured association approach.

3.1 Simulation

For setting up the simulation design we had to select a population model, the number of subpopulations, an information model and the sample sizes. We will here present in detail the results of the discrete population model, the corresponding results for the admixed models will be briefly referred to in the discussion. In the following, the simulation of the genetic data is described and a visualization of the data on a SOM is presented. All simulations were carried out using R (R Development Core Team (2004)).

Population Model of Interest

Since the population stratification model of interest is a discrete model the population under investigation can be partitioned into a number K of subpopulations and each individual has its origins in one and only one subpopulation. K was chosen as either two or five. To simulate the population of interest, subsamples of size $n_i, i = 1, \ldots, K$, were drawn from the K subpopulations. Each of the K subpopulations was of size 5000. Subsamples were either of equal size or, more realistically, of unequal size. For two subpopulations sample sizes were chosen either $n_1 = n_2 = 200$ or $n_1 = 100, n_2 = 500$, for five subpopulations either $n_1 = \ldots = n_5 = 100$ or $n_1 = n_2 = 300$ and $n_3 = n_4 = n_5 = 600$. Thus, sample sizes of simulated data sets were increased with an increasing number of subpopulations to cope with the more challenging situation of five subpopulations compared to two.

Genetic Data

Genetic data were simulated at $L = 10$ loci. All loci were micro satellites, i.e., each allele represented the number of repeats (repeat score) of a certain nucleotide sequence. Each genotype was simulated as two independent draws from the repeat score distribution. The repeat scores were assumed to be normally distributed, where the parameters of the normal distribution were varied for each subpopulation. Depending on the choice of parameters, i.e., similar or rather unequal means, the subpopulations can be more or less easily distinguished from each other. Therefore, the different choices of the parameters are referred to as 'information' and are characterized in Table 1. To introduce a random component the means were generated from a uniform distribution on different intervals displayed in the following table.

Table 1. Intervals for the mean and standard deviations to simulate genotypes with different information to distinguish subpopulations

$K = 2$		
	mean	std. deviation
highly inform.	(20;50), (70;100)	20
informative	(50;70), (70;90)	20
uninform.	(40;70), (60;100)	15 resp. 30[1]

$K = 5$		
	mean	std. deviation
highly inform.	(20;30), (50;60), (80;90), (110;120), (140;150)	15
informative	(20;40), (45;65), (70;90), (95;115), (120;145)	15
uninform.	(60;130)	10, 15, 20, 25 resp. 30

The first type of information represents the situation where the mean repeat score in all subpopulations is rather different and the standard deviation of the distribution is the same in all subpopulations. In contrast, means under the second type of information are closer, so these loci are less useful to distinguish the subpopulations. Finally,

[1] Lower standard deviation has been assigned to mean drawn from lower interval.

'uninformative' represents strongly overlapping distributions with minor information to detect population stratification. For illustrative purposes, examples of different distributions for all three information types for $K = 2$ subpopulations are displayed in Figure 2.

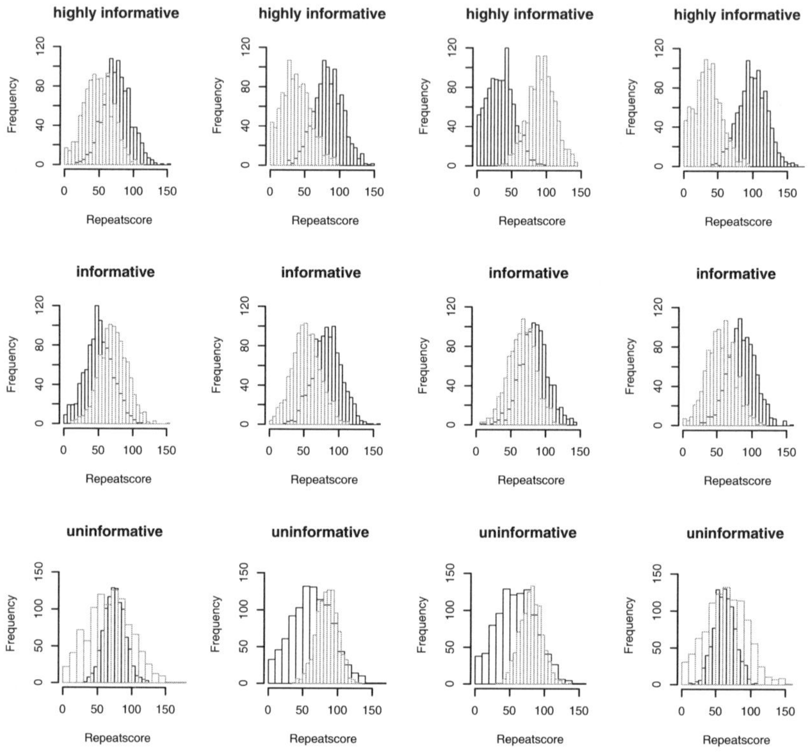

Fig. 2. Repeat score distributions for two subpopulations

In total, three different information models characterized by combinations of the three information types were chosen. Numbers of loci under the respective information model were denoted by L_1, L_2 and L_3 with $L_1 + L_2 + L_3 = L = 10$. That is, in case that 100% of loci are of highly informative type, the information model is called 'no noise'. Information model 'little noise' ('lot of noise') means that 70% (50%) of loci were simulated from the highly informative distribution, 20% (30%) of the loci were drawn from the informative distribution and the remaining 10% (20%) were generated from the uninformative type.

All 12 different settings are summarized in Table 2.

Table 2. Summary of the simulation design

Number of subpopulations	Information model	Sample sizes
2	no, little or lot of noise	$n_1 = n_2 = 200$ $n_1 = 100, n_2 = 500$
5	no, little or lot of noise	$n_1 = \ldots = n_5 = 100$ $n_1 = n_2 = 300$ $n_3 = n_4 = n_5 = 600$

Training of SOMs

Since the population structure to be detected is unknown the appropriate net structure is not known either and a variety of nets has to be trained and compared. For the simulation study described here, the net sizes given in Table 3 have been chosen.

Table 3. Size of SOMs applied

$K = 2$	$K = 5$
8x4	10x5
5x1	10x1
2x1	5x1

The smallest net sizes represent the true situation. Each neuron represents one subpopulation. The medium net size was chosen to investigate whether a certain net size is needed to adequately capture the population structure whereas the largest net size was chosen sufficiently large, assuming that no a priori information on the structure is available. Both, the medium net size and the large net size were used to investigate whether the correct number of subpopulations can be identified based on cluster indices.

All nets were initialized and trained with the function `som` from package som (Version 0.3-4). For this purpose, the following parameters had to be fixed: a Gaussian kernel and rectangular net topography were chosen; the length of the training resp. fine tuning phase was set to 6000 resp. 12000. Data sets generated were standardized in advance of the training. Convergence was ensured by initializing and training the net three times and checking identity of obtained weight vectors, apart from the net being mirrored or rotated.

Visual Inspection and Clustering of Data

After the weight vectors were trained, each data vector was assigned to one weight vector. The assignment was defined by minimum distance. Figure 3 illustrates the visual inspection for the 5×1 net that is represented by a chain of five cells.

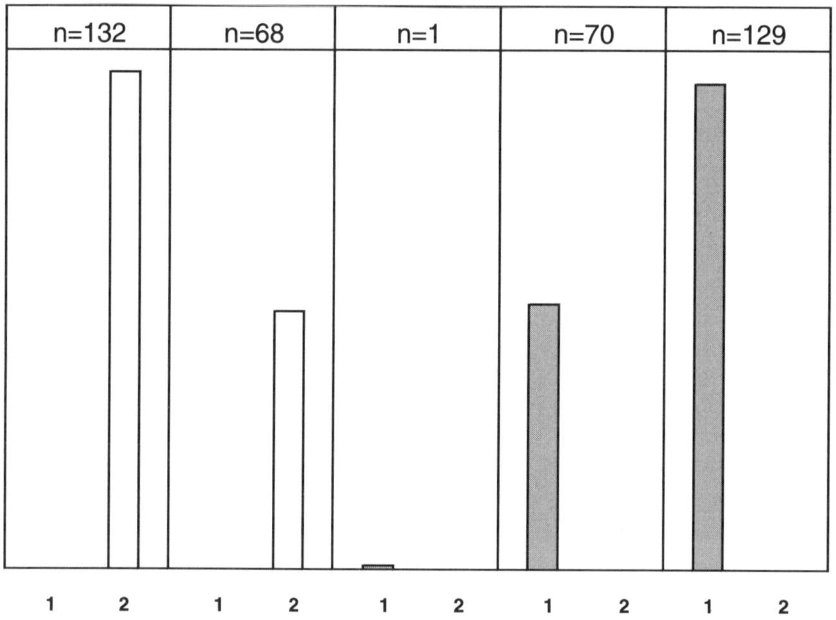

Fig. 3. Assignment of data to weight vectors in 5×1 net by true origin. Equal sample sizes and information model 'little noise'

The two different patterns of bars in each cell represent the two true possible subpopulations of origin. In a realistic setting, this information is of course not available: otherwise application of the method is not needed. Each cell is labeled by the number of data vectors assigned to the respective neuron. According to the general features of SOMs individuals with a similar genetic background should be assigned to weight vectors that are close to each other. Ideally the SOM is partitioned into the different subpopulations.

In Figure 3 all individuals from subpopulation two are assigned to the first or second neuron and all individuals from subpopulation one are assigned to the third or fourth neuron. One individual from subpopulation one is assigned to the medium neuron. Thus, we obtained a nearly perfect partition of the map according to the true population of origin. Further clustering of the weight vectors has led to the desired solution of two subpopulations (see Section 3.2). For five subpopulations and for larger net sizes an analogous visualization showed a comparably promising partition.

In a next step, cluster algorithms were applied to the trained weight vectors. Finally, cluster indices suggesting a number of clusters were calculated as objective criteria to strengthen the above visual impression. Based on hard clustering, we clustered the trained weight vectors by two hierarchical algorithms, namely complete linkage and ward, and the k-means algorithm. Then, each data vector was assigned to the same group as the weight vector to which it had minimum distance. The Davies-Bouldin (DB) index was not calculated for the weight vectors but for the complete data set. As at each loci paired data were simulated we had to work with data matrices instead of data vectors. Thus, the Davies-Bouldin index had to be extended accordingly to be applicable. Let Ω_k denote the set of data matrices $\boldsymbol{x_i}$ and $\mathbf{C_k}$ the cluster center in cluster k. To calculate the mean distance S_k^{DB} of data matrices in cluster k to the cluster center we used the Frobenius norm $d^F(.,.)$ instead of the Euclidean norm:

$$S_k^{DB} = \frac{1}{|\Omega_k|} \sum_{j \in \Omega_k} d^F(\boldsymbol{x_i}, \boldsymbol{C_k}) \, .$$

Analogously, the distance between two clusters i and k is measured by

$$M_{ik}^{DB} = d^F(\boldsymbol{C_i}, \boldsymbol{C_k}) \, .$$

Let

$$DB_{ik} = \frac{S_i^{DB} + S_k^{DB}}{M_{ik}^{DB}} \, , i \neq k \, , \tag{2}$$

and $DB_i = max_{i \neq k} DB_{ik}, i, k, \in \{1, \ldots, K\}$, then the Davies-Bouldin index is defined as

$$\overline{DB} = \frac{1}{K} \sum_{i=1}^{K} DB_i \, . \tag{3}$$

The above hard clustering is appropriate for a discrete population structure where the individuals belong to only one subpopulation. The more realistic situation of an admixed population structure is captured by a fuzzy clustering since here it is allowed that each individual has a certain fraction of ancestry in each subpopulation. The fuzzy clustering was performed in a two step procedure. First, complete linkage clustering was carried out, as the results from the hard clustering approach suggested. Cluster centers obtained from this first step were then used to calculate a so-called membership function of each data matrix $\boldsymbol{x_i}$ to a cluster l by

$$u_l(\boldsymbol{x_i}) = \left(\sum_{k=1}^{K} \left(\frac{d^F(\boldsymbol{x_i}, \boldsymbol{C_l})}{d^F(\boldsymbol{x_i}, \boldsymbol{C_k})} \right)^{1/(m-1)} \right)^{-1} , m > 1 . \quad (4)$$

The Xie-Beni index was calculated on the basis of this membership function to evaluate the cluster solution:

$$XB = \frac{\sum_{i=1}^{N} \sum_{k=1}^{K} u_k(\boldsymbol{x_i})^2 d^F(\boldsymbol{x_i},\boldsymbol{C_k})^2}{N \min_{j \neq l}\{d^F(\boldsymbol{C_j},\boldsymbol{C_l})^2\}} . \quad (5)$$

In the remaining discussion of the results, data are referred to as data vectors since two data vectors belong to each individual, building the data matrix.

3.2 Results

In the following, the results obtained by hard clustering will be described in detail whereas the results obtained by fuzzy clustering will only be briefly discussed.

Results Obtained by Hard Clustering

Table 4 shows the values obtained from the Davies-Bouldin index for $K = 2$ subpopulations and all information models investigated. The index was derived by assigning each data vector to the same group as the weight vector to which it had minimum distance. Please note that possible numbers of clusters depend on the size of the SOM that is to be clustered.

On the largest map, the 8×4 net, the Davies-Bouldin index took the minimum value for two subpopulations, i.e., for the correct number of groups. This held true for all three algorithms applied and all information models. On the medium map of size 5×1 and for little and lot of noise as information models, the Davies-Bouldin index took its minimum value for three groups. Further inspection of the number

Table 4. Davies-Bouldin indices of clustered data for all information models and two subpopulations with equal sample sizes

Net size	Noise	Algorithm	Number of clusters 2	3	4	5	6	7	8
8×4	no	Complete	0.59	1.41	1.85	2.54	2.94	2.94	2.62
		Ward	0.59	1.41	1.85	2.46	3.04	3.08	3.08
		k-means	0.59	1.41	2.39	2.54	4.28	3.09	2.78
8×4	little	Complete	0.88	1.99	2.50	2.15	2.38	2.60	2.38
		Ward	0.88	1.99	2.53	2.71	2.92	2.62	2.39
		k-means	0.88	1.99	1.96	2.55	2.70	2.93	2.92
8×4	lot	Complete	1.02	1.39	1.99	2.26	2.45	2.46	2.46
		Ward	1.02	1.39	1.99	2.28	2.45	2.46	2.46
		k-means	1.02	1.92	2.35	2.67	2.83	2.91	2.71

Net size	Noise	Algorithm	Number of clusters 2	3	4
5×1	no	Complete	0.59	0.59	2.89
		Ward	0.59	0.59	2.89
		k-means	0.59	0.59	3.12
5×1	little	Complete	0.89	0.87	1.83
		Ward	0.89	0.87	1.83
		k-means	0.89	0.87	1.83
5×1	lot	Complete	1.02	0.97	2.04
		Ward	1.02	0.97	2.04
		k-means	1.02	0.97	2.04

Net size	Noise	Algorithm	Number of clusters 2
2×1	no		0.59
	little		0.88
	lot		1.02

of assigned data vectors to the respective clusters on the medium map revealed that one cluster consisted of one individual only. This single observation should therefore be regarded rather as an outlier than as a separate cluster. This is supported by the fact that the minimum value of the Davies-Bouldin index was very close to the value obtained for the true number of two subgroups. For the smallest map representing the true situation that did not need to be clustered further, the

Davies-Bouldin index yielded the same values for the correct number of subpopulations on both the largest and the medium map.

Table 5. Davies-Bouldin indices of clustered data for all information models and two subpopulations with unequal sample sizes

Net size	Noise	Algorithm	Number of clusters 2	3	4	5	6	7	8
8×4	no	Complete	0.60	2.05	2.05	2.05	3.03	2.81	3.02
		Ward	0.60	2.05	3.03	3.03	3.03	2.81	2.79
		k-means	0.60	2.99	2.99	2.83	2.85	2.99	2.94
8×4	little	Complete	0.88	1.98	2.46	2.46	2.64	2.77	2.91
		Ward	0.94	1.98	2.46	2.64	2.77	2.77	2.91
		k-means	0.88	2.15	2.60	2.65	2.81	2.90	2.97
8×4	lot	Complete	0.96	2.40	2.73	2.33	2.24	2.37	2.47
		Ward	0.96	2.40	2.73	2.54	2.24	2.36	2.49
		k-means	0.97	2.63	2.57	2.35	2.56	3.02	2.45
5×1	no	Complete	0.61	0.74	2.42				
		Ward	0.61	0.74	2.42				
		k-means	0.60	0.74	4.34				
5×1	little	Complete	0.94	1.89	2.84				
		Ward	0.94	1.89	2.84				
		k-means	0.94	2.68	2.84				
5×1	lot	Complete	1.04	2.28	3.10				
		Ward	1.04	2.28	3.10				
		k-means	1.04	2.28	2.72				
2×1	no		0.95						
	little		1.37						
	lot		1.57						

Table 5 shows similar results for the situation 'unequal sample sizes' of the two subpopulations. The correct number of subpopulations is detected by the Davies-Bouldin index on the largest and the medium map, irrespectively of the cluster algorithm and the information model applied. On the largest map, the minimum value of the index slightly varied depending on the clustering algorithms. On the true 2×1 map, the index took higher values than on the clustered maps. If no further clustering was carried out about 6-12% of individuals were misclassified.

Summarizing the above results for two subpopulations, the Davies-Bouldin index proved to be an appropriate tool to detect the correct number of clusters. As expected, its value increased when more noise was introduced to the data. The different cluster algorithms only led to slightly varying results.

Let us now investigate a discrete population structure with $K = 5$ subpopulations. Table 6 shows the values obtained from the Davies-Bouldin index for all information models and equal and unequal sample sizes.

As it becomes obvious from Table 6 for many situations the Davies-Bouldin index led to exactly the same value which was caused by 'empty' clusters that contained no data vector. Therefore, collapsing the respective cluster with another cluster did not change the cluster solution. Thus, it is reasonable to choose the minimum number of clusters that gave the same value of the index as the number of clusters.

Irrespectively of the sample size situation, i.e., equal or unequal sample sizes, the Davies-Bouldin index tended to overestimate the number of clusters on the 10×5 map if the k-means algorithm was applied, whereas the correct number of clusters was detected if the hierarchical algorithms were used. All three algorithms led to quite similar values of the index for most numbers of clusters. On the large map, the index increased when noise was introduced to the data.

On the medium map, the Davies-Bouldin index tended to underestimate the number of clusters for the k-means algorithm and unequal sample sizes. Analogously to the large map, the hierarchical algorithms correctly detected the number of clusters. The minimum values of the index did not differ for the different information models and the hierarchical algorithms. This was also the case for unequal sample sizes and the k-means algorithm.

A comparison of the Davies-Bouldin index values on the clustered maps and those obtained from the minimal 5×1 net showed that especially for unequal sample sizes the clustered maps seemed to perform much better. In contrast to the situation of two subpopulations, a higher proportion of misclassifications was observed for the smallest net. However, the assignment derived from e.g. a 10×5 map with 'complete linkage' algorithm and lot of noise was perfect.

Table 6. Davies-Bouldin indices of clustered data for all information models and five subpopulations

			Equal sample sizes Number of clusters							Unequal sample sizes Number of clusters						
Net size	Noise	Algorithm	2	3	4	5	6	7	8	2	3	4	5	6	7	8
10×5	no	Complete	1.21	0.89	0.66	0.42	0.42	0.42	0.42	1.15	0.92	0.62	0.44	0.44	0.52	0.52
		Ward	1.21	0.89	0.68	0.42	0.42	0.42	0.42	1.16	0.73	0.62	0.44	0.86	1.28	1.28
		k-means	1.21	0.75	0.68	0.68	0.42	0.42	1.67	1.16	0.73	0.73	0.44	1.65	1.18	1.52
10×5	little	Complete	1.07	0.99	0.76	0.62	0.62	0.62	0.62	0.99	0.93	0.74	0.61	0.61	0.61	0.61
		Ward	1.07	0.99	0.74	0.62	0.62	0.92	0.92	0.99	0.93	0.71	0.61	0.61	0.61	0.61
		k-means	1.20	0.86	0.74	1.16	0.62	1.05	1.22	0.99	0.99	0.71	1.54	0.61	1.26	0.67
10×5	lot	Complete	0.95	1.16	0.88	0.69	0.76	0.76	1.22	0.95	1.15	0.91	0.69	0.87	0.87	0.87
		Ward	1.56	1.16	0.88	0.69	0.76	0.94	0.94	1.49	1.15	0.91	0.69	0.69	0.69	0.69
		k-means	1.56	1.17	0.91	0.88	0.69	0.92	0.76	1.68	1.15	0.91	1.15	1.13	0.87	1.73

			Number of clusters							Number of clusters						
Net size	Noise	Algorithm	2	3	4	5	6	7	8	2	3	4	5	6	7	8
10×1	no	Complete	1.21	0.89	0.66	0.42	0.42	0.42	0.42	1.15	0.92	0.62	0.44	0.44	0.83	0.85
		Ward	1.21	0.89	0.66	0.42	0.42	0.42	0.42	1.16	0.73	0.62	0.44	0.44	0.83	0.85
		k-means	1.25	0.90	0.82	0.69	1.39	1.11	1.11	1.16	1.02	0.62	1.03	1.78	0.85	1.80
10×1	little	Complete	1.21	0.89	0.66	0.42	0.42	0.42	0.42	1.15	0.92	0.62	0.44	0.44	0.83	0.85
		Ward	1.21	0.89	0.66	0.42	0.42	0.42	0.42	1.16	0.73	0.62	0.44	0.44	0.83	0.85
		k-means	1.25	0.90	0.82	0.69	1.39	1.11	1.11	1.16	1.02	0.62	1.03	1.78	0.85	1.80
10×1	lot	Complete	1.21	0.89	0.66	0.42	0.42	0.42	0.42	1.15	0.92	0.62	0.44	0.44	0.83	0.85
		Ward	1.21	0.89	0.66	0.42	0.42	0.42	0.42	1.16	0.73	0.62	0.44	0.44	0.83	0.85
		k-means	1.25	0.90	0.82	0.69	1.39	1.11	1.11	1.16	1.02	0.62	1.03	1.78	0.85	1.80

continued...

Table 6 (...continued) Davies-Bouldin indices of clustered data for all information models and five subpopulations.

Net size	Noise	Algorithm	Number of clusters 5	Number of clusters 5
5×1	no		0.42	2.08
	little		1.18	0.87
	lot		0.70	1.56

Let us finally investigate whether SOMs are able to identify the underlying population structure. For this purpose, the proportion of correctly classified individuals is presented in Table 7. In general it can be stated that SOMs in combination with the Davies-Bouldin index proved to be a valid tool in genetic epidemiology to detect population structure and to correctly classify individuals. To be somewhat more specific, only small differences in the results for different information models and assuming different sample sizes could be observed. Both hierarchical clustering algorithms performed better than the k-means algorithm and led to nearly equal solutions. Unequal sample sizes led to slightly lower proportions of correctly classified individuals. For two subpopulations, the proportion of correctly classified individuals decreased with increasing noise in the data, whereas this could not be seen for $K = 5$ subpopulations.

Results Obtained by Fuzzy Clustering

For the sake of comparison we applied fuzzy clustering to detect a discrete population structure with $K = 2$ and $K = 5$ subpopulations. Fuzzy clustering needs for its application that the number of subpopulation is correctly identified in advance. In the preceding section, this has been shown to be feasible by applying the hard clustering approach. Thus, we consider the correct number of subpopulations as given by the Davies-Bouldin index.

We then proceeded as follows: the trained weight vectors were clustered by the complete linkage algorithm and cluster centers were computed as they were needed to calculate the membership function for each data vector. The solution was assessed by the Xie-Beni index defined in equation 5 (cf. Table 8).

For equal sample sizes, the Xie-Beni index came up with smaller values than for unequal sample sizes. This held for all information models and showed that balanced groups were easier to separate. For $K = 2$ subpopulations the index increased with increasing noise in the data,

Table 7. Proportion of correctly classified individuals in both discrete population structure models

		$K = 2$					
		equal sample sizes			unequal sample sizes		
Net size	Noise	Complete	Ward	k-means	Complete	Ward	k-means
8×4	no	1	1	1	1	1	1
	little	1	1	1	1	0.99	0.998
	lot	1	1	1	1	1	1
5×1	no	1	1	1	0.998	0.998	1
	little	0.998	0.998	1	0.988	0.988	0.988
	lot	1	1	1	0.986	0.986	0.986

		$K = 5$					
		equal sample sizes			unequal sample sizes		
Net size	Noise	Complete	Ward	k-means	Complete	Ward	k-means
10×5	no	1	1	1	1	1	1
	little	1	1	0.992	1	1	0.875
	lot	1	1	0.8	1	0.875	0.625
10×1	no	1	1	0.8	1	1	0.875
	little	1	1	1	1	1	0.832
	lot	1	1	0.8	0.885	0.885	0.885

Table 8. Xie-Beni index for 'fuzzy' cluster solutions for all information models, net sizes and sample size models for discrete population structure

	$K = 2$				$K = 5$			
	8×4		5×1		10×5		10×1	
Noise	equal sample sizes	unequal sample sizes	equal sample sizes	unequal sample sizes	equal sample sizes	unequal sample sizes	equal sample sizes	unequal sample sizes
no	0.15	0.20	0.20	0.36	0.21	0.22	0.40	0.47
little	0.29	0.32	0.35	0.57	0.27	0.35	0.63	0.45
lot	0.34	0.38	0.42	0.69	0.27	0.37	0.44	0.52

which could not be said to be generally true for $K = 5$ subpopulations. All values of the Xie-Beni index were rather small, indicating a clear separation of the data. This was not surprising as a discrete population structure situation was present.

Let us also investigate in this situation whether the solution based on the fuzzy clustering yielded a reasonable estimate of the genetic background. For this purpose, we inspected the value of the membership function as an estimate of the genetic background for each individual. In contrast to the hard clustering approach, SOMs performed poorly. Even though the values of the Xie-Beni index were small, the estimates strongly deviated from the true values. The true value for each individual is the unit vector, as each individual has its origin in exactly one subpopulation. The histogram in Figure 4 illustrates the distribution of estimates for the fraction of ancestry in subpopulation one for $K = 2$ subpopulations, no noise in the data and equal sample sizes, obtained from a 8×4 net. It can be seen that the fraction of ancestry in subpopulation one was estimated as about 75% for those individuals from subpopulation one and as about 25% for those individuals that have their origins in subpopulation two.

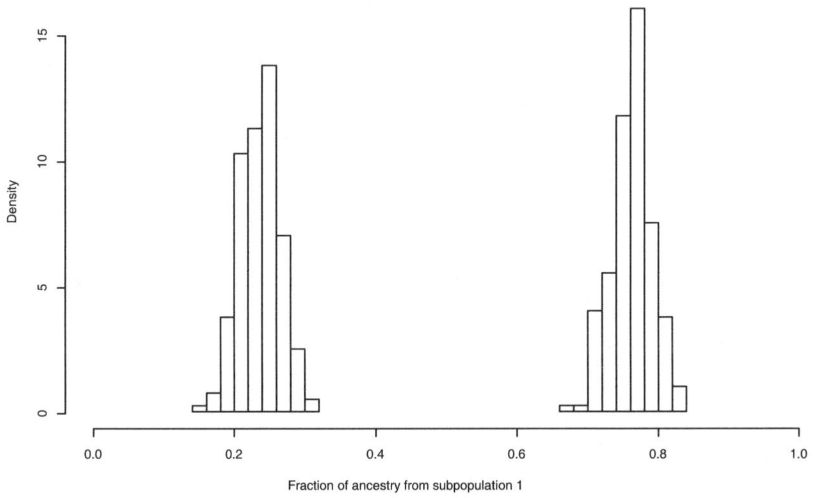

Fig. 4. Histogram representing the fraction of ancestry in subpopulation 1 for $K = 2$ subpopulations, no noise and equal sample sizes

4 Conclusion and Discussion

Summarizing the above results a fuzzy clustering approach cannot be considered as appropriate if the population structure is discrete since it does not lead to a sufficiently precise estimate of the genetic background. This is in contrast to the hard clustering approach which works well in the considered situations. This is expected from the visual inspection of the trained map, incorporating unknown information on the true origins. If such information is not available, visual inspection of a large map may allow to detect a discrete population structure if there are neurons to which no data vectors are assigned. The decision on the number of clusters can be taken based on the Davies-Bouldin index by additionally inspecting the number of assigned data vectors, if necessary. For instance, clusters including not more than one observation should be discarded as these are rather outliers than subpopulations.

An important drawback of the Davies-Bouldin index is that it is not applicable to decide for $K = 1$ cluster. To overcome this problem the gap statistic introduced by Tibshirani et al. (2001) may be used. Tibshirani et al. (2001) note that the statistic works well for non-overlapping clusters. This was confirmed by our simulations, where two resp. five clusters were detected for all noise models investigated. Unfortunately, if no cluster was present in the data, the gap statistic still detected two or three groups. Therefore, the use of the statistic was not pursued any further.

The application of SOMs to detect admixed population structure is beyond the scope of this paper. In this case each individual does not belong to a single subpopulation of origin but possesses fractions of ancestry in some or all subpopulations involved. This situation has been investigated in Wawro (2005). It was possible to detect the correct number of subpopulations involved in the random mating process but the estimates of the genetic background based on the fuzzy algorithm were poor, as has already been observed in the discrete case presented above.

As a conclusion, SOMs can be successfully used to detect a discrete population structure, but are of limited use for admixed population structures. Comparing SOMs with genomic controls or the structured association approach for discrete population structure it has to be pointed out that SOMs require less additional loci to be genotyped to correctly detect the population structure.

References

Bock HH (1974) Automatische Klassifikation. Vandenhoeck and Ruprecht Göttingen

Cardon LR, Palmer LJ (2003) Population stratification and spurious allelic association. The Lancet 361:598–604

Davies D, Bouldin D (1979) A cluster separation measure. IEEE Transactions on Pattern Analysis and Machine Intelligence 1:224–227

Devlin B, Roeder K (1999) Genomic control for association studies. Biometrics 55:997–1004

Everitt BS (1993) Cluster analysis. Edward Arnold, London

Jain AK, Murty MN, Flynn PJ (1999) Data clustering: A review. ACM Computing Surveys 31:264–323

Kohonen T (2001) Self-organizing maps. Springer Series in Information Sciences (30)

Pritchard JK, Donnelly P (2001) Case-control studies of association in structured or admixed populations. Theoretical Population Biology 60:227–237

Pritchard JK, Stephens M, Donnelly P (2000a) Inference of population structure using multilocus genotype data. Genetics 155:945–959

Pritchard JK, Stephens M, Rosenberg NA, Donnelly P (2000b) Association mapping in structured populations. American Journal of Human Genetics 67:170–181

R Development Core Team (2004) R: A language and environment for statistical computing. R Foundation for Statistical Computing, Vienna, Austria, http://www.R-project.org

Tibshirani R, Walther G, Hastie T (2001) Estimating the number of clusters in a data set via the gap statistic. Journal of the Royal Statistical Society. Series B. Statistical Methodology 63:411–423

Vesanto J, Alhoniemi E (2000) Clustering of the self-organizing map. IEEE Transactions on Neural Networks 11: 586–600

Wawro N (2005) Self-Organizing Maps als neue Methode zum Umgang mit Populationsstratifikation in der genetischen Epidemiologie. Dissertation, Universitaet Bremen

Wawro N, Bammann K, Pigeot I (2006) Testing for association in the presence of population stratification: a simulation study comparing the S-TDT, STRAT and the GC. Biometrical Journal 483:420–434

Xie X, Beni G (1991) A validity measure for fuzzy clustering. IEEE Transactions on Pattern Analysis and Machine Intelligence 13:841–847

Optimal Designs for Microarray Experiments with Biological and Technical Replicates

Rashi Gupta[1,2], Panu Somervuo[2], Sangita Kulathinal[1] and Petri Auvinen[2]

[1] Department of Mathematics and Statistics, University of Helsinki, P.O. Box 68 , FIN-00014, Helsinki, Finland rashi.gupta@helsinki.fi sangita.kulathinal@inseed.org

[2] Institute of Biotechnology, University of Helsinki, P.O.Box 56, FIN-00014, Helsinki, Finland pjsomerv@mappi.helsinki.fi petri.auvinen@helsinki.fi

1 Introduction

Microarrays are powerful tools for global monitoring of gene expressions in many areas of biomedical research (Brown and Botstein (1999)). Since the first publication on the statistical analysis of data from microarray experiments (Chen et al. (1997)), considerable amount of research has been carried out regarding such analysis. However, little work has been done on designing microarray experiments despite the fact that designing is the key for optimization of resources and efficient estimation of the parameters of interest.

Microarray experiments consist of large number of steps, as a result various sources of errors and variability crop-in during the experiment which then affect the final outcome. However, the sources of variation in the microarray experiment are yet to be completely understood. The extent to which these sources of variations are known should be considered while designing the experiment so as to obtain quality data and precise results.

Among the various experimental settings, reference designs are the most common (Kerr and Churchill (2001), Vinciotti et al. (2004), Wit and McClure (2004)). They are known to be inefficient as they assign 50% of their resources in measuring a reference condition which is not of any interest but they are mainly used because of their simplicity. Loop designs have also been used for microarray experiments. Usually either reference or loop design is chosen without estimating which would minimize the effect of unwanted variations and maximize the precision of the estimates of the parameters of interest. Though the importance of randomization, blocking and replication has been repeatedly mentioned in literature, unavailability of user-friendly statistical design software makes it difficult to explore the design aspect of the microarray experiments.

The main purpose of this article is to describe approaches for designing microarray experiments considering both technical and biological replicates. Our approach is similar to the ones taken by Churchill (2002); Wit and McClure (2004). The method for searching optimal designs has been implemented in Matlab. In Section 2, we describe the various sources of variations in the microarray experiment. Section 3 describes the model, optimality criteria, and the implementation. In Section 4, we illustrate our approach with examples. The paper concludes with a discussion section.

2 Sources of Variation

The sources of variation in microarrays can be partitioned into biological and technical variations. Biological variation is intrinsic to all organisms and is influenced by genetic or environmental factors. Technical variation is introduced during mRNA extraction from a biological sample, labeling and hybridisation. Additional source of variation is associated with reading the signal from the array.

To understand the biological and technical variations, we consider a microarray experiment where a treatment is applied to mice. The aim of the experiment is to determine how the treatment affects the mice in general. To answer this, we sample individual mice randomly from the population. Sampling more than one mouse is essential in order to draw conclusions that are valid for an entire population and not only for a particular mouse. Selected mice for the experiment are referred to as biological replicates. The variation among the sampled mice refers to the biological variation (between-biological sample variation) of the population they have been sampled from.

If mRNA sample from an individual mouse is hybridised several times on different arrays then the variation introduced is termed as technical variation (within-biological sample variation) and hybridizations with respect to the same mouse are referred as technical replicates. Technical replicates reduce the uncertainty about gene expression in the particular mRNA sample in a study. This is useful in situations where mRNAs are of interest individually.

3 Model, Optimality Criteria, and Implementation

In this section, the method is described for a single gene but the same parameterisation can be employed separately for every gene on the microarray.

3.1 Model

For a gene, the measured intensity can be characterized by the following additive model:

$$\log(Z_{ij(r)}) = \mu^{\phi}_{ij(r)} + \alpha_i + \beta_{j(r)} + \epsilon_{ij(r)}, \tag{1}$$

where $\log(Z_{ij(r)})$ is the log-intensity of the rth replicate of the jth mRNA sample on the ith array ; $\mu^{\phi}_{ij(r)}$ denotes the expected reading of rth replicate of jth sample on i^{th} array; α_i denotes the array effect of ith array; $\beta_{j(r)}$ denotes the mRNA effect of the rth replicate of jth mRNA sample; and $\epsilon_{ij(r)}$ is the model error of the rth replicate of the jth mRNA sample on the ith array with mean zero and variance $Var(\epsilon_{ij(r)}) = \sigma^2$ Note that μ's are the parameters of interest and are described by ϕ depending upon the condition which they represent (see section 4.2 for more details).

An observation $Y_{ij(p)k(q)}$ is the log-ratio of pth replicate of jth sample and q^{th} replicate of kth sample on i^{th} array. It is defined as:

$$\begin{aligned} Y_{ij(p)k(q)} &= \log(Z_{ij(p)}) - \log(Z_{ik(q)}) \\ &= (\mu^{\phi}_{ij(p)} + \alpha_i + \beta_{j(p)} + \epsilon_{ij(p)}) - (\mu^{\phi}_{ik(q)} + \alpha_i + \beta_{k(q)} + \epsilon_{ik(q)}) \\ &= (\mu^{\phi}_{ij(p)} - \mu^{\phi}_{ik(q)}) + (\beta_{j(p)} - \beta_{k(q)}) + (\epsilon_{ij(p)} - \epsilon_{ik(q)}) \\ &= \mu^{\phi}_{ij(p)k(q)} + (\beta_{j(p)} - \beta_{k(q)}) + \epsilon_{ij(p)k(q)}, \end{aligned} \tag{2}$$

where $\epsilon_{ij(p)k(q)}$ is the random error with mean zero and variance $Var(\epsilon_{ij(p)k(q)}) = Var(\epsilon_{ij(p)} - \epsilon_{ik(q)}) = \sigma^2_{\epsilon}$. To simplify the notations, we refer $\epsilon_{ij(p)k(q)} = \epsilon_i$ since the error corresponds to array i.

The log-ratios between different hybridizations could be correlated if they involve technical and biological replicates. Our model in (2) does not involve separate terms for technical and biological variability but both can be interpreted by a single term $\beta_{j(r)}$. Further, we assume that

$$Cov(\beta_{j(p)}, \beta_{j^*(r)}) = \begin{cases} \frac{\sigma_b^2}{2} + \frac{\sigma_t^2}{2}, & \text{if } j = j^*; p = r \\ \frac{\sigma_b^2}{2}, & \text{if } j = j^*; p \neq r \\ 0, & \text{otherwise,} \end{cases} \tag{3}$$

where $\frac{\sigma_b^2}{2}$ and $\frac{\sigma_t^2}{2}$ are interpreted as biological and technical variations.

Similarly, ϵ_i represents the measurement error while measuring $Y_{ij(p)k(q)}$ on i^{th} array . Then

$$Cov(\epsilon_i, \epsilon_{i^*}) = \begin{cases} \sigma_\epsilon^2, & \text{if } i = i^* \\ 0, & \text{otherwise.} \end{cases} \tag{4}$$

Let us consider the covariance structure between two observations coming from slide i and i^*:

$$\begin{aligned} Cov(Y_{ij(p)k(q)}, Y_{i^*j^*(r)k^*(s)}) &= Cov(\mu^{\phi}_{ij(p)k(q)} + (\beta_{j(p)} - \beta_{k(q)}) + \epsilon_i, \\ &\quad \mu^{\phi}_{i^*j^*(r)k^*(s)} + (\beta_{j^*(r)} - \beta_{k^*(s)}) + \epsilon_{i^*}) \\ &= Cov(\beta_{j(p)}, \beta_{j^*(r)}) - Cov(\beta_{j(p)}, \beta_{k^*(s)}) \\ &\quad - Cov(\beta_{k(q)}, \beta_{j^*(r)}) + Cov(\beta_{k(q)}, \beta_{k^*(s)}) \\ &\quad + Cov(\epsilon_i, \epsilon_{i^*}). \end{aligned} \tag{5}$$

To simplify notation, $j(p), j^*(r), k(q), k^*(s)$ can be written without p, r, q, s. When $i = i^*$,

$$Cov(Y_{ijk}, Y_{ijk}) = Var(Y_{ijk}) = \begin{cases} \sigma_t^2 + \sigma_\epsilon^2, & \text{if } j = k \\ \sigma_b^2 + \sigma_t^2 + \sigma_\epsilon^2, & \text{otherwise,} \end{cases} \tag{6}$$

and when $i \neq i^*$,

$$Cov(Y_{ijk}, Y_{i^*j^*k^*}) = \begin{cases} \sigma_b^2, & \text{if } j = j^*; k = k^*; j \neq k; j^* \neq k^* \\ \frac{\sigma_b^2}{2}, & \text{if } j = j^*; k \neq k^*; j \neq k; j^* \neq k^* \\ & \text{or } j \neq j^*; k = k^*; j \neq k; j^* \neq k^* \\ 0, & \text{otherwise.} \end{cases} \tag{7}$$

The above covariance structure can be obtained by assuming a random effects model where $\beta_{j(r)} = \beta_j + u_{j(r)}$ where $u_{j(r)} \sim N(0, \sigma_t^2/2)$ and $\beta_j = w + v_j$ where $v_j \sim N(0, \sigma_b^2/2)$ with the usual assumption of independence between ϵ's, u's, and v's. If there are two populations under study, then w's can be defined differently for them. Note that since we consider log-ratios, the entry of the covariance matrix will be with the sign reversed if the dyes have been swapped.

For each gene, a vector of n observations $y = (y_1, \ldots, y_n)^t$ obtained on n arrays can now be represented equivalently by a linear regression model as

$$y = X\phi + \delta, \tag{8}$$

where X is a $(n \times l)$ design matrix defining the relationship between the values observed in the experiment and a set of l independent parameters ϕ, δ is a $n \times 1$ vector of correlated $\delta_i = (\beta_{j(p)} - \beta_{k(q)}) + \epsilon_i$ with variance-covariance matrix $(Cov(Y_{ij(p)k(q)}, Y_{i^*j^*(r)k^*(s)}))$. We denote this variance-covariance matrix as V. For an experiment with m conditions, one way to choose the parameter vector is $\phi = (\phi_1, \phi_2, \ldots, \phi_l)^t$ where $l = m - 1$, ϕ_1 is a change in the gene expression at the condition 2 and ϕ_j is a change at the condition $j+1$ compared to the condition 1. Any other contrast can be obtained by the relation $\phi_{ij} = \phi_j - \phi_i$. The goal is to obtain an estimate of ϕ separately for each gene. The least square estimator of ϕ is:

$$\hat{\phi} = (X^tV^{-1}X)^{-1}X^tV^{-1}y, \tag{9}$$

and variance-covariance matrix of $\hat{\phi}$ is $(X^tV^{-1}X)^{-1}(= M)$. The standard error $\hat{\phi}_i$ is given by $\sqrt{a_i}$, where a_i is the ith diagonal element of the matrix M (Searle (1971)).

3.2 Optimal Design

In microarray experiments, the parameters of interest are the changes in gene expressions from one condition to another, i.e., ϕ. In order to select an optimal design, we need to choose a design that makes each of the a_i as small as possible so that the standard error of $\hat{\phi}_i$, $i = 1, \ldots, l$ is minimized. When comparing designs over a fixed number of arrays, some of the a_i may be smaller for one design and some may be smaller for other designs. In addition, not all designs provide information about all the parameters. The aim is to select those designs which provide the most accurate estimates for ϕ using some optimality criterion. Here, four definitions are used for comparing and finding optimal designs (each of them may result into different optimal designs):

1. A-optimality: the trace of M $(= A)$ is minimized,
2. G-optimality: the maximum of diagonal elements of M $(= G)$ is minimized,
3. D-optimality: the determinant of M $(= D)$ is minimized (same as maximising the determinant of $X^tV^{-1}X$),
4. E-optimality: the largest eigenvalue of M $(= E)$ is minimized.

To implement these optimality criteria, knowledge of biological and technical variations is required.

3.3 Implementation

We implemented the four optimality criteria defined in the earlier section using a computer program written in Matlab. For a particular parametrization, the inputs required for this program are the number of arrays, possible hybridizations on an array, biological variance, technical variance, and random error. The number of biological replicates available is given along with the list of possible hybridizations. The program then enumerates all possible designs, i.e., assignment of possible hybridizations for all the arrays under consideration. If a particular biological replicate under certain treatment is used more than once, its multiple copies are treated as technical replicates. The program calculates A, D, E and G for each design for which $X^T V^{-1} X$ is non-singular and searches for the A-, D-, E-, and G-optimal designs.

4 Illustrations

4.1 Comparison of Two Groups

When the aim is to compare two groups, the parameter of interest is the difference of the two group means. In case of a single parameter (as in here), the variance of the parameter estimator is directly the A-score. Fig. 1 shows how using five and ten arrays, this parameter can be estimated accurately with different values of biological and technical variance, and with different number of biological replicates. The error variance σ_ϵ^2 was set as 0.1. It is easy to conclude from the figure that increase in the number of biological replicates minimizes the A-score in the presence of biological variation. Also, increase in the number of arrays affects A-score the most when biological and technical variances are large.

4.2 Factorial Design

Consider an experiment with the aim of studying the impact of two drugs. To study gender specific responses, the population was divided into two groups: male (A) and female (B). Each group was given three

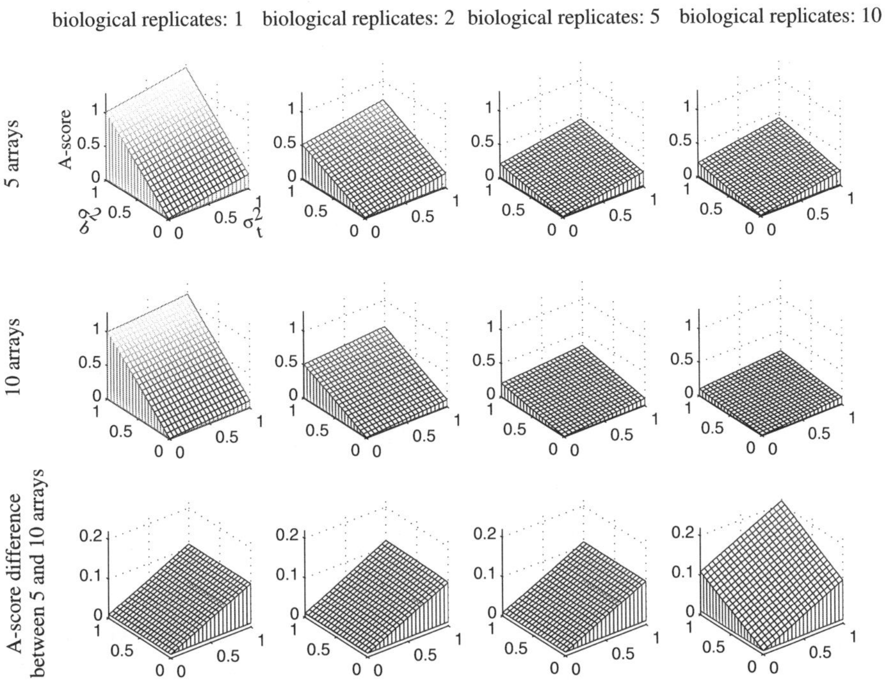

Fig. 1. Comparison of two groups with five and ten arrays for various values of σ_b^2 and σ_t^2. A-score is shown along the vertical axis; σ_b^2 and σ_t^2 are shown along the other two axes.

treatments: no drug (T_0), drug 1 (T_1), and drug 2 (T_2). Two groups and three treatments result in six possible conditions to study. Using two dyes, six conditions result in $\binom{6}{2} = 15$ possible combinations that can be hybridized on each array, see Fig. 2.

Let ϕ_{a0} be the baseline parameter for males and ϕ_{b0} denote the change in the expression levels of females compared to males under treatment T_0. Similarly, ϕ_{a1} and ϕ_{a2} denote changes of expression levels for T_1 and T_2 with respect to T_0, respectively. The interactions between groups and the treatments are represented by ϕ_{b0a1} and ϕ_{b0a2}. Table 1 summarizes the parameterizations for this experiment. Note that the interaction parameters are $\phi_{b0a1} = (B1 - A1) - (B0 - A0)$ and $\phi_{b0a2} = (B2 - A2) - (B0 - A0)$. The parameters of interest are denoted by $\phi = (\phi_{a1}, \phi_{a2}, \phi_{b0}, \phi_{b0a1}, \phi_{b0a2})^t$.

With fifteen combinations and n arrays to hybridize, there are 15^n designs. This includes several equivalent designs, however the program

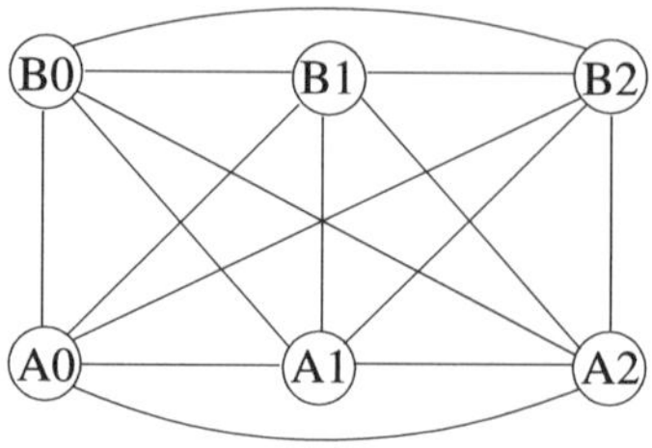

Fig. 2. Fifteen hybridizations using two groups, three treatments, and two dyes.

Table 1. Parametrization for two groups and three treatments experiment.

Group	Intensity for T_0 (log)	Intensity for T_1 (log)	Intensity for T_2 (log)
B	$B0 = A0 + \phi_{b0}$	$B1 = B0 + \phi_{a1} + \phi_{b0a1}$	$B2 = B0 + \phi_{a2} + \phi_{b0a2}$
A	$A0 = \phi_{a0}$	$A1 = A0 + \phi_{a1}$	$A2 = A0 + \phi_{a2}$

enumerate unique designs. Table 2 lists all, unique, and possible designs for different number of arrays along with the time taken to search these designs. In this particular example, five is the smallest number of arrays required in order to estimate all parameters.

Fifteen hybridizations result when there is one biological replicate, two dyes and six conditions. It should be noted that with unlimited number of biological replicates, the complexity of the search is same as with one biological replicate. In general, the number of hybridizations increases quadratically with the number of biological replicates since each individual from one condition may be hybridized together with each individual from another condition. Say, c represent possible hybridizations when one biological replicate and n arrays. Therefore, when there are b biological replicates, the possibilities are cb^{2n}. Thus, possibilities increase rapidly as the number of biological replicates or arrays increase. To present how enormous the possibilities can grow with a small increase in biological replicates, we consider a small example with five arrays. With two biological replicates, the possibilities are 1,000 times more and with three replicates, the possibilities are 60,000 times more than when there is one biological replicate.

The only practical way to search optimal designs in case of such huge possibilities is to divide the search into multiple steps and first find an optimal design for some feasible number of arrays. Then augment this design with additional feasible number of arrays till the available number of arrays are utilized. This approach was used to find an optimal

Table 2. Number of designs for different number of arrays in case of fifteen possible hybridizations per array. Column 'Unique' gives the number of designs ignoring the order of hybridizations, 'Possible' gives the number of designs for which all parameters are estimable, and 'Time' gives the search time in seconds using standard computer.

Arrays	All	Unique	Possible	Time
5	15^5	11,628	1,296	2
6	15^6	38,760	10,140	7
7	15^7	116,280	47,100	23
8	15^8	319,770	168,285	76
9	15^9	817,190	509,545	218
10	15^{10}	1,961,256	1,372,698	549

design with three biological replicates and ten arrays. The optimal design proposed is shown in Fig. 3. Following values were used to search for this optimal design ($\sigma_b^2 = 0.6, \sigma_t^2 = 0.3, \sigma_\epsilon^2 = 0.1$). Proposed optimal design corresponds to the following model:

$$\begin{pmatrix} y_1 \\ y_2 \\ y_3 \\ y_4 \\ y_5 \\ y_6 \\ y_7 \\ y_8 \\ y_9 \\ y_{10} \end{pmatrix} = \begin{pmatrix} 1\,0\,0\,0\,0 \\ 0\,1\,0\,0\,0 \\ 0\,0\,1\,0\,0 \\ 0\,0\,1\,1\,0 \\ 0\,0\,1\,0\,1 \\ 0\,1\,0\,0\,1 \\ 1\,0\,0\,1\,0 \\ 1\,0\,0\,0\,0 \\ 0\,0\,1\,0\,0 \\ 0\,1\,0\,0\,0 \end{pmatrix} \begin{pmatrix} \phi_{a1} \\ \phi_{a2} \\ \phi_{b0} \\ \phi_{b0a1} \\ \phi_{b0a2} \end{pmatrix} + \begin{pmatrix} \delta_1 \\ \delta_2 \\ \delta_3 \\ \delta_4 \\ \delta_5 \\ \delta_6 \\ \delta_7 \\ \delta_8 \\ \delta_9 \\ \delta_{10} \end{pmatrix} \quad (10)$$

Further, to investigate how parameters ($\phi_{a1}, \phi_{a2}, \phi_{b0}, \phi_{b0a1}, \phi_{b0a2}$) get affected by the number of biological replicates, optimal designs with one, three and unlimited number of biological replicates were found (note: unlimited number of biological replicates imply no technical replicates). Variances of the parameter estimators are shown in Fig. 4. It can be seen that for any number of arrays, the difference between the variance of parameter estimators for one and three biological replicates are larger than the differences between three and unrestricted number of biological replicates.

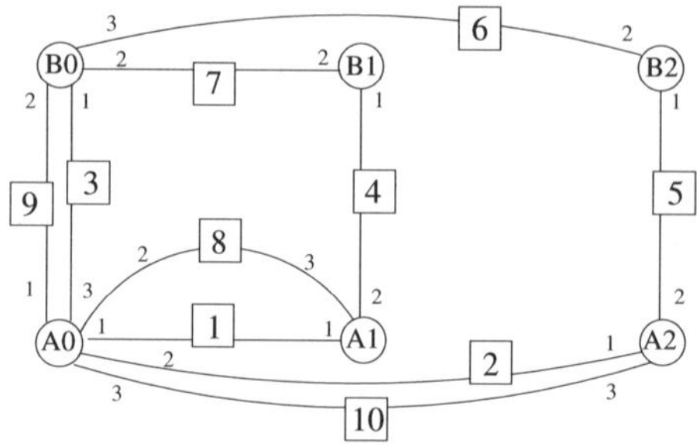

Fig. 3. A-optimal design with ten arrays for two groups, three treatments and two dyes. The maximum number of independent samples available at each condition was limited to be three. Arcs denote hybridisations and numbers beside arcs denote the indices of biological replicates. Numbers inside boxes correspond to the rows of equation (10).

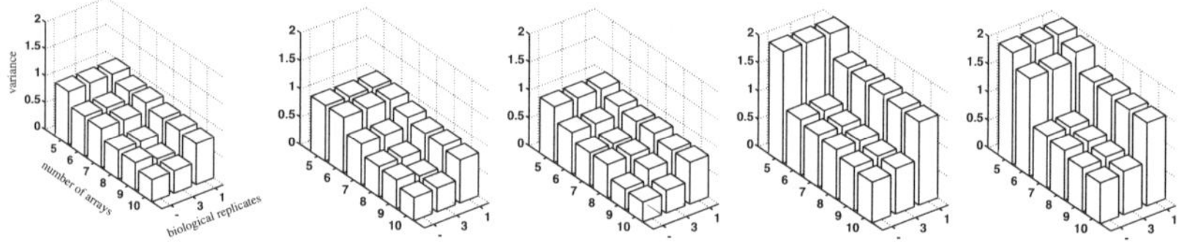

(a) ϕ_{a1} (b) ϕ_{a2} (c) ϕ_{b0} (d) ϕ_{b0a1} (e) ϕ_{b0a2}

Fig. 4. Parameter variances of A-optimal designs in a two-group, three-treatment, two-dye experiment, for different number of arrays and different number of biological replicates available ('-' denotes unrestricted number), $\sigma_b^2 = 0.6$, $\sigma_t^2 = 0.3$, $\sigma_\epsilon^2 = 0.1$. ϕ_{a1} and ϕ_{a2} are the main effects of treatments, ϕ_{b0} is the main effect of group difference, and ϕ_{b0a1} and ϕ_{b0a2} are the interaction terms between the treatments and group.

5 Discussion

In this paper, we propose a linear model to find optimal designs using various optimality criteria and considering both biological and technical replicates. The proposed approach for searching optimal designs has been implemented in Matlab.

We would like to point out that the biological variability needs to be specified in order to search for optimal designs. In practice, this variability is not known and has to be estimated from the data. Some prior knowledge about this variability for the experiment under consideration is obligatory for the program. The proposed design is optimal for the given parameters and it does not guarantee that when the experiment is conducted, data obtained from the experiment have same biological variability. Also, it does not maximize the degrees of freedom required in a statistical test for finding differentially expressed genes.

The role of biological and technical replicates can be understood by an example where the parameter is the difference of two population means. The variance of the parameter is $\frac{2\sigma_b^2}{n} + \frac{4\sigma_t^2}{nr}$ for r technical replicates on n biological replicates where nr represents the number of arrays. (Kerr (2003)). If the number of arrays is fixed, the second term is constant and the first term can be minimized which depends only on the number of biological replicates. So it is obvious that technical replicates can never substitute for biological replication but should be considered to obtain precise information from the assay. Technical replicates also help to detect failed hybridizations. As a result, some arrays should be kept aside to repeat failed hybridizations.

The complexity of matrix inversion in our program grows cubically with the number of parameters. The main source of complexity however is the number of arrays and the restricted number of biological replicates. To design an experiment for a large number of arrays or parameters may become computationally difficult to handle. A sequential approach to design may be more appropriate in such situations. For example, to first search optimal designs using a small number of arrays and extend the existing designs to remaining arrays to reduce the complexity of the problem. Various heuristic approaches could also be introduced.

Although the examples discussed in this paper are restricted to two dyes, our program is generic and can be used for any number of dyes. The idea of using more than two dyes have already been applied and discussed in (Woo et al. (2005)). We also with our program design an experiment with three dyes at Institute of Biotechnology, Helsinki. The aim was to compare RhoG protein G12V (GTP, activated mutant) and T17N (GDP, inactivated mutant) against a control. Restricted number of arrays (10) were used to propose an optimal design. The hybridisations were carried out according to the proposed design. The data were analyzed to test which genes behaved differently in G12V and T17N with respect to the control. The results of this analysis will be published elsewhere.

Acknowledgements

The research of Rashi Gupta was supported by the ComBi graduate school and Panu Somervuo was supported by project MMM Dnro

4655/501/2003. Sangita Kulathinal was supported by the Academy of Finland via its grant number 114786. Petri Auvinen was supported by Institute of Biotechnology.

References

Brown P, Botstein D (1999) Exploring the new world of the genome with DNA microarrays. Nature Genetics (Suppl.) 21:33–37

Chen Y, Dougherty E, Bittner M (1977) Ratio-based decisions and the quantitative analysis of cDNA micro-array images. Journal of Biomedical Optics 2:364–374

Churchill GA (2002) Fundamentals of experimental design for cDNA microarrays. Nature Genetics 32:490–495

Kerr MK, Churchill GA (2001) Experimental design for gene expression microarrays. Biostatistics 2:183–201

Kerr MK (2003) Design considerations for efficient and effective microarray studies. Biometrics 59:822–828

Searle S (1971) Linear Models. New York: Wiley

Vinciotti V, Khanin R, D'Alimonte D, Liu X, Cattini N, Hotchkiss G, Bucca G, de Jesus O, Rasaiyaah J, Smith CP, Kellam P, Wit E (2005) An experimental evaluation of a loop versus a reference design for two-channel microarrays. Bioinformatics 21(4):492–501

Wit EC, McClure JD (2004) Statistics for Microarrays: Design and Analysis and Inference. John Wiley & Sons

Woo Y, Krueger W, Kaur A, Churchill G (2005) Experimental design for three-color and four-color gene expression microarrays. Bioinformatics 21(Suppl 1):i459–i467

Weighted Mixed Regression Estimation Under Biased Stochastic Restrictions

Christian Heumann[1] and Shalabh[2]

[1] Department of Statistics, University of Munich, Akademiestrasse 1, 80799 Munich, Germany chris@stat.uni-muenchen.de

[2] Department of Mathematics and Statistics, Indian Institute of Technology, Kanpur - 208016, India shalab@iitk.ac.in; shalabh1@yahoo.com

1 Introduction

The use of prior information in linear regression analysis is well known to provide more efficient estimators of regression coefficients. Such prior information can be available in different forms from various sources like as past experience of the experimenter, similar kind of experiments conducted in the past, etc. The available prior information sometimes can be expressed in the form of exact, stochastic or inequality restrictions. The methods of restricted regression estimation, mixed estimation (Theil and Goldberger (1961)) and minimax estimation are preferred when prior information is available in the form of exact, stochastic and inequality restrictions, respectively. More details about these estimation procedures can be found in Rao, Toutenburg, Shalabh and Heumann (2008).

When the prior information is available in the form of stochastic restrictions, then in many applications a systematic bias is also present. Such systematic bias can arise from different sources and due to various reasons like personal judgements of the persons involved in the experiment, in testing of general linear hypothesis in linear models when null hypothesis is rejected, in imputation of missing values through regression approach etc. Teräsvirta (1980) and Hill and Ziemer (1983) have given some interesting examples for this type of information. How to incorporate such systematic bias in the estimation procedure is an issue which is addressed in this article. The method of weighted mixed regression estimation is utilized for the purpose. How to choose the weights in this estimation procedure so as to have gain in efficiency under the criterion of mean dispersion error matrix is also addressed.

The plan of the paper is as follows. The model description and the estimation of parameters are discussed in Section 2. The properties of the estimators are derived and analyzed in Section 3. Some conclusions are placed in Section 4.

2 Model Specification and Estimation of Parameters

Consider the linear regression model

$$y = X\beta + \epsilon \tag{1}$$

where y is a $(T \times 1)$ vector of study variable, X is a $(T \times K)$ full column rank matrix of T observations on each of the K explanatory variables, β is a $(K \times 1)$ vector of regression parameters and ϵ is a $(T \times 1)$ vector of random disturbances with $\mathrm{E}(\epsilon) = 0$ and $\mathrm{V}(\epsilon) = \sigma^2 I_T$ where σ^2 is unknown.

The application of Gauss–Markov theory on (1) yields the ordinary least squares estimator (OLSE) of β as

$$\hat{\beta} = S^{-1}X'y \tag{2}$$

where $S = X'X$. The OLSE is the best linear unbiased estimator of β with covariance matrix

$$\mathrm{V}(\hat{\beta}) = \sigma^2 S^{-1} \, .$$

Further we assume that some prior information about the regression coefficients is available which is stochastic in nature and contains systematic bias. We use the framework of linear stochastic restrictions to present the available prior information and systematic bias as

$$r = R\beta + \delta + \phi \tag{3}$$

where r is a $(J \times 1)$ vector and R is a $(J \times K)$ matrix of known elements; δ is a $(J \times 1)$ vector that expresses the unknown systematic but non-stochastic bias in the restrictions; and ϕ is a $(J \times 1)$ vector representing the stochastic nature of prior information. We assume that $\mathrm{E}(\phi) = 0$ and $\mathrm{V}(\phi) = \sigma^2 I_J$. For R, we assume full row rank (if $J < K$) or full column rank (if $J \geq K$).

Note that in many statistical applications, the assumption that the prior information is unbiased, i.e., $\mathrm{E}(r) = R\beta$ is violated. Under those cases, the set up of (3) fits well. Also, the mixed regression estimator (Theil and Goldberger (1961)) which is an unbiased estimator of β

when $\delta = 0$ becomes biased when $\delta \neq 0$. Wijekoon and Trenkler (1995) have used the framework of (3) in pre-test estimation of parameters.

The criterion of mean dispersion error matrix (MDEM) comparison allows a more general view on the properties of estimators in the linear regression model when additional and possibly biased stochastic restrictions are available. A motivation can be given as follows. Using a quadratic loss function

$$L(\hat{\beta}, \beta, A) = (\hat{\beta} - \beta)' A (\hat{\beta} - \beta)$$

where $A \geq 0$ is a symmetric and nonnegative definite matrix, the (quadratic) risk function $R(\hat{\beta}, \beta, A)$ of an estimator $\hat{\beta}$ of β is the expected loss as

$$R(\hat{\beta}, \beta, A) = \mathrm{E}L(\hat{\beta}, \beta, A) = \mathrm{E}(\hat{\beta} - \beta)' A (\hat{\beta} - \beta) \ .$$

A theorem by Theobald (1974) and Trenkler (1985) gives a necessary and sufficient condition that if an estimator is superior over other estimators under the criterion of MDE matrix (MDEM) (often called as MSE–I superiority), then the same estimator remains uniformly superior over other estimators under the criterion of risk function also for all nonnegative definite matrix A. The MDEM superiority means that an estimator $\hat{\beta}_2$ of β is better than an estimator $\hat{\beta}_1$ of β when

$$\Delta(\hat{\beta}_1, \hat{\beta}_2) = \mathrm{M}(\hat{\beta}_1, \beta) - \mathrm{M}(\hat{\beta}_2, \beta) \geq 0 \ , \tag{4}$$

i.e., $\Delta(\hat{\beta}_1, \hat{\beta}_2)$ is nonnegative definite where MDEM of $\hat{\beta}$ is

$$\begin{aligned} \mathrm{M}(\hat{\beta}, \beta) &= \mathrm{E}(\hat{\beta} - \beta)(\hat{\beta} - \beta)' \\ &= \mathrm{V}(\hat{\beta}) + \mathrm{Bias}(\hat{\beta}, \beta)\mathrm{Bias}(\hat{\beta}, \beta)' \ , \end{aligned}$$

covariance matrix of $\hat{\beta}$ is

$$\mathrm{V}(\hat{\beta}) = \mathrm{E}[(\hat{\beta} - \mathrm{E}(\hat{\beta}))(\hat{\beta} - \mathrm{E}(\hat{\beta}))'] \ ,$$

and bias of $\hat{\beta}$ is

$$\mathrm{Bias}(\hat{\beta}, \beta) = \mathrm{E}(\hat{\beta}) - \beta \ .$$

The techniques of MDEM comparisons have been studied and illustrated, e.g. by Trenkler (1981), Teräsvirta (1982), Trenkler and Toutenburg (1990) and Toutenburg and Trenkler (1990). An overview can be found in Rao, Toutenburg, Shalabh and Heumann (2008). In order to incorporate the restrictions (3) in the estimation of parameters, we minimize

$$(y - X\beta)'(y - X\beta) + w\ (r - R\beta)'(r - R\beta)$$

with respect to β where w is the weight lying between 0 and 1 such that $w \neq 0$ ($w = 0$ would lead to OLSE). The resulting estimator of β is given by

$$\hat{\beta}_w = (S + wR'R)^{-1}(X'y + wR'r) = Z_w^{-1}(X'y + wR'r) \qquad (5)$$

where $Z_w = S + wR'R$ and $\hat{\beta}_w$ is termed as weighted mixed regression estimator (WMRE).

3 Properties and Efficiency of WMRE Over OLSE

Now we study the efficiency properties of weighted mixed regression estimator and the dominance conditions for the MDEM superiority of WMRE over OLSE.

The bias of $\hat{\beta}_w$ is

$$\begin{aligned}\text{Bias}(\hat{\beta}_w, \beta) &= \text{E}(\hat{\beta}_w) - \beta \\ &= wZ_w^{-1}R'\delta \end{aligned} \qquad (6)$$

and MDEM of $\hat{\beta}_w$ is

$$\text{M}(\hat{\beta}_w, \beta) = \sigma^2 Z_w^{-1}(S + w^2R'R)Z_w^{-1} + w^2 Z_w^{-1}R'\delta\delta'RZ_w^{-1}\,. \qquad (7)$$

The covariance matrix of $\hat{\beta}_w$ is

$$\text{V}(\hat{\beta}_w) = \sigma^2 Z_w^{-1}(S + w^2R'R)Z_w^{-1}\,. \qquad (8)$$

The difference in the covariance matrices of OLSE and WMRE is

$$\begin{aligned}\text{D}(\hat{\beta}, \hat{\beta}_w) &= \text{V}(\hat{\beta}) - \text{V}(\hat{\beta}_w) \\ &= \sigma^2 S^{-1} - \sigma^2 Z_w^{-1}(S + w^2R'R)Z_w^{-1} \\ &= \sigma^2 Z_w^{-1}\left[Z_w S^{-1} Z_w - S - w^2R'R\right]Z_w^{-1} \\ &= w^2\sigma^2 Z_w^{-1}R'\left[\left(\frac{2}{w} - 1\right)I + RS^{-1}R'\right]RZ_w^{-1}\,. \end{aligned} \qquad (9)$$

The difference in (9) is positive definite when Z_w is positive definite and

$$\left(\frac{2}{w} - 1\right)I + RS^{-1}R' > 0\,, \qquad (10)$$

which is possible as long as $w < 2$.

Now there are two possible cases:

1. When $J < K$, R has full row rank, therefore R' has full column rank and it follows that in this case we can only conclude that
$$\mathrm{D}(\hat{\beta}, \hat{\beta}_w) \geq 0 \; .$$
2. When $J \geq K$, R has full column rank and it is concluded that
$$\mathrm{D}(\hat{\beta}, \hat{\beta}_w) > 0 \; .$$

3.1 Case 1: When $J < K$

Now we study the necessary and sufficient condition for the MDEM superiority of WMRE over OLSE in case when $J < K$. The next theorem presents a necessary and sufficient condition for such superiority.

Theorem 1. *The WMRE $\hat{\beta}_w$ is MDEM superior to OLSE $\hat{\beta}$ in case of $J < K$ if and only if*
$$\rho(w) = \sigma^{-2}\delta' \left[(2w^{-1} - 1)I + Rs^{-1}R'\right]^{-1} \delta \leq 1 \; . \tag{11}$$
Thereby we assume a priori, that $0 < w \leq 1$.

See Toutenburg (1989) for the derivation of (11). Note that for $w = 0$, we get the OLSE and condition (11) is trivial, since $\rho(0) = 0$. Now we show, that $\rho(w)$ is monotone in w.

Theorem 2. *$\rho(w)$ is monotonic increasing in w.*

Proof:

$$\begin{aligned}
&\frac{\partial \rho(w)}{\partial w} = \\
&= \frac{\partial}{\partial w}\sigma^{-2}\delta' \left[(2w^{-1} - 1)I + RS^{-1}R'\right]^{-1} \delta \\
&= \sigma^{-2}\delta' \left\{\frac{\partial}{\partial w} \left[(2w^{-1} - 1)I + Rs^{-1}R'\right]^{-1}\right\} \delta \\
&= -\sigma^{-2}\delta' \left[(2w^{-1} - 1)I + RS^{-1}R'\right]^{-1} \left\{\frac{\partial}{\partial w} \left[(2w^{-1} - 1)I + Rs^{-1}R'\right]\right\} \\
&\quad \times \left[(2w^{-1} - 1)I + RS^{-1}R'\right]^{-1} \delta \\
&= 2w^{-2}\sigma^{-2}\delta' \left[(2w^{-1} - 1)I + RS^{-1}R'\right]^{-2} \delta \; > 0 \; .
\end{aligned} \tag{12}$$

This completes the proof.

Now we derive the sufficient condition for the MDEM superiority of WMRE over OLSE in case when $J < K$. To derive a sufficient dominance condition, we use the following theorems, see e.g. Rao, Toutenburg, Shalabh and Heumann (2008, Theorems A.39 and A.44):

1. If A is a $(n \times n)$ is a symmetric matrix and λ_1 is its maximum eigenvalue (the eigenvalues of a symmetric matrix are all real), then for the quadratic form $h'Ah$,

$$\sup_h \frac{h'Ah}{h'h} = \lambda_1 \, .$$

2. If $A > 0$, then all the eigenvalues of A are positive.

Now, if $\mu_1 \geq \ldots \geq \mu_J > 0$ are the real eigenvalues of positive definite matrix $RS^{-1}R'$, then the eigenvalues of the matrix

$$Q = (2w^{-1} - 1)I + RS^{-1}R' \tag{13}$$

are also all positive as $0 < w \leq 1$.

Applying the spectral decomposition on $RS^{-1}R'$ using Rao, Toutenburg, Shalabh and Heumann (2008, Theorem A.30), the matrix Q in (13) becomes

$$\begin{aligned} Q &= (2w^{-1} - 1)I + P \operatorname{diag}(\mu_j)\ P' \\ &= (2w^{-1} - 1)PP' + P \operatorname{diag}(\mu_j)\ P' \\ &= P\left[\operatorname{diag}(2w^{-1} - 1) + \operatorname{diag}(\mu_j)\right] P' \\ &= P \operatorname{diag}(2w^{-1} - 1 + \mu_j)\ P' \end{aligned} \tag{14}$$

where P is an orthogonal matrix.

So we obtain the condition (11) as

$$\begin{aligned} \rho_w &= \sigma^{-2}\delta' Q^{-1}\delta \\ &= \sigma^{-2}\delta'(P')^{-1} \operatorname{diag}(2w^{-1} - 1 + \mu_j)^{-1}(P')^{-1}\delta \\ &= \sigma^{-2}\tilde{\delta}' \operatorname{diag}\left(\frac{1}{2w^{-1} - 1 + \mu_j}\right)\tilde{\delta} \leq 1 \, , \end{aligned} \tag{15}$$

where

$$\tilde{\delta} = P'\delta.$$

An equivalent transformation of (15) is

$$\frac{\tilde{\delta}' \operatorname{diag}(\frac{1}{2w^{-1}-1+\mu_j})\tilde{\delta}}{\tilde{\delta}'\tilde{\delta}} \leq \sigma^2 \frac{1}{\tilde{\delta}'\tilde{\delta}} \, .$$

Using

$$\tilde{\delta}'\tilde{\delta} = \delta' P P' \delta = \delta'\delta$$

and a result that the eigenvalues of a diagonal matrix are the diagonal elements themselves, we can derive the following condition for the superiority of $\hat{\beta}_w$ over $\hat{\beta}$ with respect to w. The WMRE $\hat{\beta}_w$ is MDEM superior to OLSE $\hat{\beta}$ if

$$\sup_{\tilde{\delta}} \frac{\tilde{\delta}' \operatorname{diag}(\frac{1}{2w^{-1}-1+\mu_j})\tilde{\delta}}{\tilde{\delta}'\tilde{\delta}} \leq \sigma^2 \frac{1}{\tilde{\delta}'\tilde{\delta}}$$

or if

$$\sup_{\tilde{\delta}} \frac{\tilde{\delta}' \operatorname{diag}(\frac{1}{2w^{-1}-1+\mu_j})\tilde{\delta}}{\tilde{\delta}'\tilde{\delta}} \leq \frac{\sigma^2}{\delta'\delta}$$

or if

$$\frac{1}{2w^{-1} - 1 + \mu_J} \leq \frac{\sigma^2}{\delta'\delta}$$

with $2w^{-1} - 1 + \mu_J > 0$ $(0 < w \leq 1)$ and μ_J being the smallest eigenvalue of $RS^{-1}R'$.

This can further be transformed into

$$\frac{2}{w} \geq \sigma^{-2}\ \delta'\delta + 1 - \mu_J\ . \tag{16}$$

Now we have two cases:

1. When $\sigma^{-2}\delta'\delta + 1 - \mu_J \leq 0$, i.e.,

$$\mu_J \geq \sigma^{-2}\delta'\delta + 1\ . \tag{17}$$

   Then we obtain that every $w \in (0, 1]$ can be chosen to obtain the MDEM superiority of $\hat{\beta}_w$ over $\hat{\beta}$.
2. When $\sigma^{-2}\delta'\delta + 1 - \mu_J > 0$, then (16) can be transformed into

$$w \leq \frac{2}{\sigma^{-2}\delta'\delta + 1 - \mu_J}\ . \tag{18}$$

   We again have two subcases:
   - When $2/(\sigma^{-2}\delta'\delta + 1 - \mu_J) \geq 1$, then

$$\mu_J \geq \sigma^{-2}\delta'\delta - 1\ . \tag{19}$$

     Then we obtain that every $w \in (0, 1]$ is selectable to obtain the MDEM superiority of $\hat{\beta}_w$ over $\hat{\beta}$.

- When $2/(\sigma^{-2}\delta'\delta + 1 - \mu_J) < 1$, then

$$w \in \left(0, \frac{2}{\sigma^{-2}\delta'\delta + 1 - \mu_J}\right] \tag{20}$$

can be selected to obtain the MDEM superiority of $\hat{\beta}_w$ over $\hat{\beta}$.

Now we can formulate the following theorem.

Theorem 3. *A sufficient condition for the MDEM superiority of WMRE $\hat{\beta}_w$ over OLSE $\hat{\beta}$ when $J < K$ is given by the following choice of w:*

$$\left.\begin{array}{ll} w \in (0, 1] & \text{if } \sigma^{-2}\delta'\delta \leq 1 + \mu_J \\ w \in \left(0, \frac{2}{\sigma^{-2}\delta'\delta+1-\mu_J}\right] & \text{otherwise .} \end{array}\right\} \tag{21}$$

Now we look for the existence of weight w^* which guarantees the MDEM superiority of WMRE over OLSE in case of $J < K$. This is stated in the following theorem.

Theorem 4.

$$w^* = \frac{1}{1 + \sigma^{-2}\delta'\delta} \tag{22}$$

always fulfills condition (11).

Proof:

It suffices to show that w^* fulfills the condition (21). In case $\sigma^{-2}\delta'\delta \leq 1 + \mu_J$, then this is fulfilled because of $0 < w^* < 1$ and in the other case, we only have to show that

$$w^* = \frac{1}{1 + \sigma^{-2}\delta'\delta} \leq \frac{2}{\sigma^{-2}\delta'\delta + 1 - \mu_J} .$$

This is also fulfilled because $\mu_J > 0$. Therefore, independent of J ($J < K$), there exists a superiority guaranteed by w^*.
This completes the proof.

Next we obtain the distribution of estimated w^* under the assumption of normal distribution when $J < K$.

We assume that the error vectors ϵ and ϕ are independently and normally distributed as follows.

$$\epsilon \sim N\left(0, \sigma^2 I_T\right)$$

and

$$\phi \sim N\left(0, \sigma^2 I_J\right) .$$

In order to find the distribution of estimated w^*, we replace the unknown parameters δ and σ^2 in (22) by their unbiased estimators

$$\widehat{\delta} = r - R\hat{\beta}$$

and

$$\widehat{\sigma}^2 = \frac{1}{T-K}\hat{\epsilon}'\hat{\epsilon} \ ,$$

respectively.

We note that for $\widehat{\delta}$,

$$\begin{aligned} \mathrm{E}(\widehat{\delta}) &= \mathrm{E}(r - R\hat{\beta}) \\ &= \delta \end{aligned} \tag{23}$$

and

$$\begin{aligned} \mathrm{V}(\widehat{\delta}) &= \mathrm{V}(r - R\hat{\beta}) \\ &= \mathrm{V}(\phi) + \mathrm{V}(R(X'X)^{-1}X'\epsilon) \\ &= \sigma^2\left[I_J + RS^{-1}R'\right] \ . \end{aligned} \tag{24}$$

So we obtain

$$\hat{\delta} \sim N\left(\delta, \sigma^2\left[I_J + RS^{-1}R'\right]\right)$$

and it follows that

$$\sigma^{-2}(\hat{\delta} - \delta)'\left[I_J + RS^{-1}R'\right]^{-1}(\hat{\delta} - \delta) \sim \chi^2_J.$$

For the estimation of σ^2, we use its unbiased estimator as $\widehat{\sigma}^2$. We obtain under the normal distribution assumption that

$$\frac{(T-K)\hat{\sigma}^2}{\sigma^2} \sim \chi^2_{T-K} \ .$$

Further we note that $\widehat{\delta}$ and $\widehat{\sigma}^2$ are also independent because $\widehat{\delta}$ depends on $\hat{\beta}$ and; $\hat{\beta}$ and $\widehat{\sigma}^2$ are independent.

Then

$$\begin{aligned} \widehat{w}^* &= \frac{1}{1 + \frac{\widehat{\delta}'\widehat{\delta}}{\widehat{\sigma}^2}} \\ &= \frac{\widehat{\sigma}^2}{\widehat{\sigma}^2 + \widehat{\delta}'\widehat{\delta}} \end{aligned}$$

and it follows that

$$\widehat{w}^* \sim \frac{\frac{\sigma^2}{T-K} Z}{\frac{\sigma^2}{T-K} Z + Y'Y} \tag{25}$$

or

$$\widehat{w}^* \sim \frac{\sigma^2 Z}{\sigma^2 Z + (T-K)Y'Y} \tag{26}$$

where $Z \sim \chi^2_{T-K}$ and Y is distributed as $\hat{\delta} \sim N\left(\delta, \sigma^2\left[I_J + RS^{-1}R'\right]\right)$.

For a better representation of this distribution, we choose another illustration for the quadratic form $Y'Y$ from Mathai and Provost (1992, p. 29, Representation 3.1a.1), with the special case $A = I$.

We use again the spectral decomposition for this as

$$\sigma^2(I_J + RS^{-1}R') = P\text{diag}[\sigma^2(1+\mu_j)]P' \ ,$$

where μ_j, $(j = 1, \ldots, J)$ are the eigenvalues of $RS^{-1}R$.

Then we obtain

$$\left[\sigma^2\left(I_J + RS^{-1}R'\right)\right]^{-1/2} = P\text{diag}\left(\frac{1}{\sigma\sqrt{1+\mu_j}}\right)P'.$$

Further, let

$$\begin{aligned} b' &= P'\left[\sigma^2\left(I_J + RS^{-1}R'\right)\right]^{-1/2}\delta \\ &= P'P\text{diag}\left(\frac{1}{\sigma\sqrt{1+\mu_j}}\right)P'\delta \\ &= \text{diag}\left(\frac{1}{\sigma\sqrt{1+\mu_j}}\right)P'\delta \\ &= \text{diag}\left(\frac{1}{\sigma\sqrt{1+\mu_j}}\right)\tilde{\delta} \end{aligned}$$

so that

$$\begin{aligned} b &= \left(\frac{\tilde{\delta}_1}{\sigma\sqrt{1+\mu_1}}, \ldots, \frac{\tilde{\delta}_J}{\sigma\sqrt{1+\mu_J}}\right)' \\ &= (b_1, \ldots, b_J)' \ . \end{aligned}$$

with $\tilde{\delta} = (\tilde{\delta}_1, \ldots, \tilde{\delta}_J)' = P'\delta$.

Then we obtain

$$Y'Y = \sum_{j=1}^{J} \sigma^2(1+\mu_j)(U_j+b_j)^2$$
$$= \sigma^2 \sum_{j=1}^{J}(1+\mu_j)\left(U_j + \frac{\tilde{\delta}_j}{\sigma\sqrt{1+\mu_j}}\right)^2$$

where

$$U = (U_1, \ldots, U_J)',$$

and we have

$$\mathrm{E}(U) = 0 \quad \text{and} \quad \mathrm{V}(U) = I_J \; .$$

Because Y is assumed to be normally distributed, so U_j's are also independent and standard normally distributed random variables. Therefore $Y'Y$ is a linear combination of independent non-central χ^2–variables.

The distribution of $\hat{w}^*$ from (26) is then

$$\hat{w}^* \sim \frac{\sigma^2\chi^2_{T-K}}{\sigma^2\chi^2_{T-K} + (T-K)\sigma^2\sum_{j=1}^{J}(1+\mu_j)\chi^2_1(b_j^2)} \tag{27}$$

or

$$\hat{w}^* \sim \frac{\chi^2_{T-K}}{\chi^2_{T-K} + (T-K)\sum_{j=1}^{J}(1+\mu_j)\chi^2_1\left(\frac{\tilde{\delta}_j^2}{\sigma^2(1+\mu_j)}\right)}$$

where $\chi^2_1(b_j^2)$ indicates the non-central χ^2–distribution with non-centrality parameter

$$b_j^2 = \frac{\tilde{\delta}_j^2}{\sigma^2(1+\mu_j)} \; , \quad (j = 1, \ldots, J).$$

So the distribution of (27) depends on T, K, J, σ^2, δ and the eigenvalues of $RS^{-1}R'$.

3.2 Case 2: When $J \geq K$

Now we discuss the superiority of WMRE and OLSE over each other when $J \geq K$.

The necessary and sufficient condition for the MDEM superiority of $\hat{\beta}_w$ over $\hat{\beta}$ in case when $J \geq K$ is mentioned in the next theorem.

Theorem 5. *The WMRE $\hat{\beta}_w$ is MDEM superior over OLSE $\hat{\beta}$ in case when $J \geq K$ if and only if*

$$\begin{aligned}
\rho_w &= \text{Bias}(\hat{\beta}_w)'\text{D}(\hat{\beta}, \hat{\beta}_w)^{-1}\text{Bias}(\hat{\beta}_w) \\
&= w^2\delta' R Z_w^{-1} \left\{ w^2\sigma^2 Z_w^{-1} R' \left[\left(\frac{2}{w} - 1\right) I + RS^{-1}R' \right] R Z_w^{-1} \right\}^{-1} \\
&\quad \times Z_w^{-1} R'\delta \\
&= \sigma^{-2}\delta' R \left\{ \left(\frac{2}{w} - 1\right) R'R + R'RS^{-1}R'R \right\}^{-1} R'\delta \\
&\leq 1 . \qquad (28)
\end{aligned}$$

Because R is assumed to have full column rank, so $R'R$ is positive definite and invertible. A transformation of (28) provides

$$\begin{aligned}
\rho_w &= w\sigma^{-2}\delta' R \left[(2 - w) R'R + wR'RS^{-1}R'R \right]^{-1} R'\delta \\
&= w\sigma^{-2}\delta' R \ \left[2R'R - wR'R + wR'RS^{-1}R'R \right]^{-1} R'\delta \qquad (29)
\end{aligned}$$

and we obtain that

$$\rho_w = 0 , \quad \text{if} \quad w = 0 .$$

Now we show the monotonicity of ρ_w with respect to w in case of $J \geq K$. For this, we differentiate ρ_w in (29) with respect to w and use again Rao, Toutenburg, Shalabh and Heumann (2008, Theorems A.94 and A.96).

Let

$$Q_* = \left(\frac{2}{w} - 1\right) R'R + R'RS^{-1}R'R \qquad (30)$$

and so

$$\frac{\partial}{\partial w} Q_* = -\frac{2}{w^2} R'R .$$

Then we obtain for $0 < w \leq 1$:

$$\begin{aligned}
\frac{\partial \rho_w}{\partial w} &= \frac{\partial}{\partial w} \ \sigma^{-2}\delta' R Q_*^{-1} R'\delta \\
&= \sigma^{-2}\delta' R \left\{ \frac{\partial}{\partial w} Q_*^{-1} \right\} R'\delta \\
&= -\sigma^{-2}\delta' R Q_*^{-1} \left\{ \frac{\partial}{\partial w} Q_* \right\} Q_*^{-1} R'\delta \\
&= \frac{2}{w^2} \sigma^{-2}\delta' R Q_*^{-1} R'R Q_*^{-1} R'\delta \ \geq \ 0 . \qquad (31)
\end{aligned}$$

We note from (31) that ρ_w is monotonic increasing in w.

Also $\rho_w = 0$ for $w = 0$. So there exists a w^* which guarantees the MDEM superiority of $\hat{\beta}_w$ over $\hat{\beta}$.

Now we derive the sufficient condition for the MDEM superiority of $\hat{\beta}_w$ over $\hat{\beta}$ in case when $J \geq K$.

As in the case of $J < K$, we tried to obtain a sufficient condition with the help of eigenvalue system. For this case, now we use again the spectral decomposition for positive semi-definite matrix $RS^{-1}R'$. To accomplish this, first we use two times the inversion formula for matrices from Rao, Toutenburg, Shalabh and Heumann (2008, Theorem A.18 (iii)) on ρ_w.

Let

$$q = \frac{2}{w} - 1 \ (\geq 1)$$

and

$$S_R = R'R\ .$$

Then we obtain:

$$\begin{aligned}
\rho_w &= \sigma^{-2}\delta' R\ \left\{qS_R + S_R S^{-1} S_R\right\}^{-1}\ R'\delta \\
&= \sigma^{-2}\delta' R \left\{ \frac{1}{q} S_R^{-1} - \frac{1}{q} S_R^{-1}\ S_R \left[S + S_R\ \frac{1}{q} S_R^{-1}\ S_R \right]^{-1} \right. \\
&\quad \left. \times S_R\ \frac{1}{q} S_R^{-1} \right\} R'\delta \\
&= \frac{1}{q}\sigma^{-2}\delta' R S_R^{-1} R'\delta - \frac{1}{q^2}\sigma^{-2}\delta' R \left[S + \frac{1}{q} S_R \right]^{-1} R'\delta \\
&= \frac{1}{q}\sigma^{-2}\delta' R(R'R)^{-1}R'\delta \\
&\quad - \frac{1}{q^2}\sigma^{-2}\delta' R \left[S^{-1} - S^{-1}R'\ \left(qI + RS^{-1}R'\right)^{-1}\ RS^{-1} \right] R'\delta.
\end{aligned} \tag{32}$$

Now we use the spectral decomposition of $RS^{-1}R' = P\mathrm{diag}(\mu_j)P'$. Thereby $\mu_1 \geq \ldots \geq \mu_J \geq 0$ are again the eigenvalues of $RS^{-1}R'$. In contrast to the case $J < K$, some eigenvalues are zero now, and so in particular, we can assume $\mu_J = 0$.

With this, we obtain

$$\begin{aligned}\rho_w &= \frac{1}{q}\sigma^{-2}\delta' R(R'R)^{-1}R'\delta \\ &\quad -\frac{1}{q^2}\sigma^{-2}\delta' \ P\text{diag}\left(\mu_j - \frac{\mu_j^2}{q+\mu_j}\right) P'\delta \\ &= \frac{1}{q}\sigma^{-2}\delta' R(R'R)^{-1}R'\delta \\ &\quad -\sigma^{-2}\delta' \ P\text{diag}\left(\frac{\mu_j}{q(q+\mu_j)}\right) P'\delta. \end{aligned} \tag{33}$$

With

$$\tilde{\delta} = P'\delta \quad \text{and} \quad \tilde{\delta}'\tilde{\delta} = \delta'\delta \ ,$$

we can derive the following condition. The WMRE $\hat{\beta}_w$ is MDEM superior to the OLSE $\hat{\beta}$, if

$$\frac{\rho_w}{\delta'\delta} = \frac{\frac{1}{q}\delta' R(R'R)^{-1}R'\delta}{\delta'\delta} - \frac{\delta' \ P\text{diag}\left(\frac{\mu_j}{q(q+\mu_j)}\right) P'\delta}{\delta'\delta} \leq \frac{\sigma^2}{\delta'\delta}$$

or

$$\frac{\rho_w}{\delta'\delta} = \frac{\frac{1}{q}\delta' R(R'R)^{-1}R'\delta}{\delta'\delta} - \frac{\tilde{\delta}'\text{diag}\left(\frac{\mu_j}{q(q+\mu_j)}\right)\tilde{\delta}}{\tilde{\delta}'\tilde{\delta}} \leq \frac{\sigma^2}{\delta'\delta} \ . \tag{34}$$

Because the second term in (34) after the minus sign is positive, we consider the worst case, viz., the first and second terms in (34) are maximum and minimum, respectively.

The first term in (34) contains the idempotent matrix $\delta' R(R'R)^{-1}R'\delta$, whose eigenvalues are only zero and one (Rao, Toutenburg, Shalabh and Heumann (2008, Theorem A.61 (i)).

The minimum eigenvalue of the diagonal matrix in the second term of (34) is

$$\frac{\mu_J}{q(q+\mu_J)} \ ,$$

since $\frac{\partial}{\partial x}\frac{x}{(c^2+cx)} > 0$.

So a sufficient condition is

$$\frac{1}{q} - \frac{\mu_J}{q(q+\mu_J)} \leq \frac{\sigma^2}{\delta'\delta}$$

or

$$\frac{1}{2w^{-1}-1+\mu_J} \leq \frac{\sigma^2}{\delta'\delta}$$

is sufficient for $\rho_w \leq 1$.

Because $\mu_J = 0$ surely, this condition simplifies more in contrast to the case when $J < K$. So we obtain the following theorem.

Theorem 6. *A sufficient condition for the MDEM superiority of $\hat{\beta}_w$ against $\hat{\beta}$ in case when $J \geq K$ is*

$$w \leq \min\left\{1\ ;\ \frac{2}{1+\sigma^{-2}\delta'\delta}\right\}\ . \tag{35}$$

4 Conclusions

We have considered the method of weighted mixed regression estimation to incorporate the systematic bias and randomness in the prior information to estimate the regression coefficients in a linear regression model. The weighted mixed regression estimator is derived and its dominance over the ordinary least squares estimator is studied under the criterion of mean dispersion error matrix. The choice of weight is found with which the MDEM dominance of WMRE over OLSE is obtained. We find that the MDEM dominance depends on the range of weight which itself depends on the model settings. The distribution of the estimated weight is obtained. The choice of weight which guarantees the MDEM dominance of WMRE over OLSE is found as well as its distribution is derived which is a function of central and noncentral χ^2–variables.

Acknowledgement

The second author gratefully acknowledges the support from Alexander von Humboldt Foundation (Germany).

References

Hill C, Ziemer RF (1983) Missing regressor values under conditions of multicollinearity. Communications in Statistics (Theory and Methods) A12:2557–2573

Mathai AM, Provost SB (1992) Quadratic Forms in Random Variables: Theory and Application. Marcel Dekker Inc., New York

Rao CR, Toutenburg H, Shalabh, Heumann C (2008) Linear Models and Generalizations – Least Squares and Alternatives (3rd edition). Springer, Berlin Heidelberg New York

Teräsvirta T (1980) Linear restrictions in misspecified linear models and polynomial distributed lag estimation. Research report no. 16, Department of Statistics, University of Helsinki 1–39

Teräsvirta T (1982) Superiority comparisons of homogeneous linear estimators. Communications in Statistics (Theory and Methods) A11:1595–1601

Theil H, Goldberger AS (1961) On pure and mixed estimation in econometrics. International Economic Review 2: 65-78

Theobald CM (1974) Generalizations of mean square error applied to ridge regression. Journal of the Royal Statistical Society B36:103–106

Toutenburg H (1989) Mean-square-error-comparisons between restricted least squares, mixed and weighted mixed estimators. Forschungsbericht 89/12, Universität Dortmund, Germany

Toutenburg H, Trenkler G (1990) Mean square error matrix comparisons of optimal and classical predictors and estimators in linear regression. Computational Statistics and Data Analysis 10:297-305

Trenkler G (1981) Biased Estimators in the Linear Regression Model. Hain, Königstein/Ts

Trenkler G (1985) Mean square error matrix comparisons of estimators in linear regression. Communications in Statistics (Theory and Methods) A14: 2495–2509

Trenkler G, Toutenburg H (1990) Mean-square error matrix comparisons between biased estimators-an overview of recent results. Statistical Papers 31:165–179

Wijekoon P, Trenkler G (1995) Pre-test estimation in the linear regression model under stochastic restrictions. Ceylon Journal of Sciences: Physical sciences 2/1:57–64

Coin Tossing and Spinning – Useful Classroom Experiments for Teaching Statistics

Helmut Küchenhoff

Department of Statistics, University of Munich, Akademiestrasse 1, 80799 Munich, Germany kuechenhoff@stat.uni-muenchen.de

1 Introduction

A dicer's dispute in 1654 led to the creation of the theory of probability by Blaise Pascal and Pierre de Fermat. Later, 350 years after their famous correspondence, throwing a dice is still the standard way of teaching the basic ideas of probability. The second classical example for randomness is tossing of a coin. Famous experiments were run by Buffon (he observed 2048 heads in 4040 coin tosses), Karl Pearson (12012 heads in 24000 coin tosses), and by John Kerrich (5067 heads in 10000 coin tosses) while he was war interned at a camp in Jutland during the second world war. The coin tossing or rolling dice experiments are often performed in the classes to introduce the ideas and concepts of probability theory. In higher classes students sometimes do not find them attractive and get bored. An attempt is made here to illustrate how these experiments can be made interesting by simple extensions. Dunn (2005) proposed some nice variations rolling special types of dice. In this article we focus on repeated coin spinning experiment by several students. We show how planning experiments including the determination of sample size, multiple testing, random effects models, overdispersion, non standard testing and autocorrelation can be illustrated in the context of coin spinning.

2 The Experiments

Shortly after the Euro had been introduced in 12 European countries, it was claimed in an article in The Times (Boyes, Baldwin and Hawkes (2002)) that the new Euro coins are not "fair". There are different coins in all European countries of the Euro zone, but their diameter and weight are identical. There was a report on an experiment in spinning the Belgian Euro coins on a table that resulted in 140 times head (the head of King Albert) out of 250 trials. Therefore the author was asked by a German TV station to conduct a similar experiment with the German Euro coins in his statistics class. In the following chapters we show how different statistical topics can be illustrated with this experiment.

3 Planning and Sample Size

The first step is to clarify the research question and to plan the experiment including the adequate sample size. It was decided to conduct four experiments with the German 1 Euro and 2 Euro coins. Both types of coins are tossed and spun. The main data analysis of the experiment had to be fixed in advance. Here we consider a simple Binomial test with the null hypothesis $H_0 : p = 0.5$ against $H_1 : p \neq 0.5$ where p denotes the probability of getting tails. If we want to conduct a significance test, one has to fix the significance level and the desired power of test. In addition, the concept of the minimal relevant effect should be addressed. Substantive considerations and data from former studies have to be taken into account. In our example, the data from the Belgian Euro experiment gave a proportion of 0.56 for "head". We started the calculation by a minimal relevant effect of 0.05. Note that the minimal relevant effect is not the expected effect. A good explanation for a minimal relevant effect of 0.05 is that we are are not interested in deviations from $p = 0.5$ which are less than 0.05. Using the normal approximation of the Binomial distribution, this yields a sample size of $n = 783$ for a type one error of $\alpha = 0.05$ and a power of $1 - \beta = 0.8$ in a two–sided test. Since four different variations of the experiment are conducted, the issue of multiple testing emerges in a natural way. Using the Bonferroni inequality, the significance level can be chosen to be $0.05/4$. This translates to a sample size of $n = 1112$. In general, the sample size is a function of three variables– significance level, power and minimal relevant effect. A nice interactive tool for calculating the sample size and illustrating this relationship has been provided by Ruth Lenth (`www.stat.uiowa.edu/~rlenth/Power/`). Determining a sensible sample size is always a balancing of these three input variables and effort (costs) for the experiment. This can be illustrated in our case by Table 1. One can argue similarly for a higher power of 0.95 or for a higher or lower minimal relevant effect. The sample size of the experiments was fixed as $n = 800$ per experiment, which corresponds to a minimal effect of 0.06 using a type one error of $\alpha = 0.054$.

Table 1. Sample size determination using a Binomial test of the null hypothesis $H_0 : p = 0.5$.

Significance level (α)	Power ($1-\beta$)	Minimal effect	Minimal sample size (n)
0.05	0.8	0.05	783
0.0125	0.8	0.05	1112
0.0125	0.8	0.06	771
0.05	0.9	0.05	1048
0.05	0.8	0.1	194

Further aspects to be possibly addressed are – the model assumptions, the adequacy of the approximation (the calculations are based on the normal approximation of the Binomial, which may not be suitable in cases of small sample size), and the possibility to conduct a sample size analysis by Monte Carlo simulations for complex study designs.

4 Conducting the Experiments and Simple Analysis

Before conducting the experiments, the exact conditions should be fixed in a protocol (comparable to study protocols in medical and epidemiological studies). In our case, 16 students conducted the experiment with 50 times tossing and then spinning the coins. They received a standard form to fill in their results. There were clear rules regarding the validity of a trial, e.g. when spinning the coins on a table and the coin fell on the ground, this trial was not valid and had to be repeated. Furthermore, a stable spinning of the coin had to be achieved to produce a valid trial.

The results of the experiments are detailed in Table 2.

Table 2. Analysis of the experiments with 800 trials and p–values correspond to testing $H_0 : p = 0.5$ by a simple Binomial test.

	1 Euro (tossing)	1 Euro (spinning)	2 Euro (tossing)	2 Euro (spinning)
Number of tails	406	411	395	495
proportion of tails	0.51	0.51	0.49	0.62
p–value	0.70	0.45	0.75	< 0.001

It can be seen that there was only a significant result for spinning the 2 Euro coins. The estimated probability for tails was 0.62. This type of unexpected result makes the experiment exciting for the students. One immediate step is to check the correctness of data, experiment and plausibility of the result from physical point of view. There was one student with 48 tails out of 50. Now a discussion about outliers in a data analysis was conducted. Since there was no indication of cheating, the observation was not excluded from the main analysis. For a sensitivity analysis, these 50 trials were left out, still leading to a rejection of the null hypothesis. Concerning the plausibility of the result, some literature had to be consulted. There is wide agreement that there are no effects or only small effects when tossing the coin. Basically the result is deterministic when the initial conditions are known. Typically the distribution of the initial conditions in such an experiment lead to a fair coin. For details, we refer to the interesting paper by Gelman and Nolan (2002). Those authors conclude "the biased coin has long been a part of statistical folklore, but it does not exist in the form in which it is imagined." In a recent paper, Diaconis, Holmes and Montgomery (2007) show that there is a small effect (the probability for heads is estimated to be 0.51) for a coin which is flipped starting from heads. So finding no effect in the tossing experiments is not a surprise. The physics for coin spinning for the ideal coin is still under discussion (called as Euler's disk, see Moffatt (2000)). For spinning of different coins there are different reports on deviations from the probability of 0.5, see Gelman and Nolan (2002) and Paulos (1995). So our result does not contradict current physical knowledge and seems to be plausible.

5 Checking Assumptions

The next step in a statistical analysis is to check the assumptions. We focus on the spinning experiment of the 2 Euro coins. Two basic assumptions of a Binomial experiment are the independence of trials and constant probability. Since in our case, every experiment was conducted by one person with tossing of one coin 50 times, so the question arises that whether there is a dependence between the probabilities and the person. Denoting by I the number of persons, and by J_i the number of trials for each person, a statistical model can be given by

$$P(Y_{ij} = 1) = p_i, \quad i = 1, \ldots, I, \quad j = 1, \ldots, J_i. \tag{1}$$

with p_i being the individual probability for one person. The question is whether the p_i's are equal, i.e., we want to test the null hypothesis $H_0 : p_1 = p_2 = \ldots = p_I$. This can be done by different methods.

5.1 Contingency Table

Table 3. Results for the classroom experiment of 2 Euro coin spinning

Person No.	1	2	3	4	5	6	7	8	9	10	11	12	13	14	15	16	Sum
Number of head	23	29	24	18	12	20	22	27	12	14	17	20	13	18	34	2	305
Number of tails	27	21	26	32	38	30	28	23	38	36	33	30	37	32	16	48	495

The most simple way is to arrange the data in a $(I \times 2)$ contingency table, see Table 3, and to use a standard χ^2–test for independence, see e.g. Toutenburg (2005). Then the question of independence of outcome and person is addressed. Here the test yields a rejection of the null hypothesis (p–value < 0.001).

5.2 Logistic Regression

Independence is a symmetric concept, but in the case of our experiment there is a clear indication of the possible effect. The person possibly affects the result of the experiment and therefore a regression type analysis can be performed. The standard for binary data is the logistic regression model, which can be used with a categorical covariate:

$$P(Y_{ij} = 1) = G(\tau_i), \quad i = 1, \ldots, I, \quad j = 1, \ldots, J_i \tag{2}$$

Here, G is the function $t \rightarrow 1/(1 + \exp(-t))$. Then the null hypotheses can be tested by likelihood ratio test or by a score test. In our case, this also gives a clear indication for a person effect ($p < 0.001$). Note that the score test of logistic regression model is identical to the χ^2–test in the contingency table mentioned earlier.

5.3 Random Effects

In both approaches discussed earlier, the p_i and τ_i are fixed unknown parameters (fixed effects) which may not be suitable in this context, see Toutenburg (2002, p. 111). So the concept of a random effect can be introduced. Then in model (2), τ_i is not treated as a parameter but as a random variable with the assumption $\tau_i \sim N(\tau, \sigma_\tau^2)$. The existence of this random effect can be assessed by testing the null hypothesis $H_0 : \sigma_\tau^2 = 0$. Note that this is a nonstandard test problem, since the null hypothesis is on the border of the parameter space. This causes some difficulties and there are still problems using these types of tests in more complex situations; see Greven, Crainiceanu, Küchenhoff and Peters (2008), Scheipl, Greven and Küchenhoff (2008) for this problem in the context of linear mixed models. Here the model can be fitted using maximum likelihood method and a likelihood ratio test can be performed using a mixture of a χ^2-distribution with one degree of freedom and a point mass of 0.5 in 0, provided the number of experimenters is large enough to allow for a good asymptotic approximation. So even current research topics can be addressed in this rather simple experiment.

5.4 Overdispersion and Goodness of Fit

Another topic is overdispersion, which is a common problem in logistic regression, see e.g. Rao, Toutenburg, Shalabh and Heumann (2008). The idea can be illustrated by looking at the variance of the number of "tails" for each person, which is denoted by X_i, $i = 1, \ldots, I$. Assuming no person effect, but a possible bias in the coin, the X_i are Binomial with parameters $n = 50$ and probability p. The expected variance of the X_i is estimated as $n \cdot p \cdot (1-p) = 50 \cdot 0.6 \cdot 0.4 = 11.8$, while the observed variance is 59.6 which is considerably high. This can be explained by a variation between the persons which adds on to the Binomial variance. Here the overdispersion factor is estimated to be 5.6. The effect can also be confirmed by a significance test, see e.g. Tutz (2000).

An alternative is to test the hypothesis $X_i \sim \text{Binomial}(50, p)$ by a goodness of fit test. This can be performed by χ^2-test or Kolmogorov-Smirnov test, see e.g. Toutenburg (2005).

5.5 Autocorrelation

Another possible deviation from the Binomial assumption is the autocorrelation, i.e., that there is some dependence between two subsequent trials. A simple way of checking this is to look at the conditional distribution of tail given the result of the trial before. In the case of spinning, we obtained 315 times 'tail' out of 488 trials (64.6%) when the preceding spin was tails. When there was a head in the preceding trial then the proportion of tails was 57.4% (126/296). The exact test of Fisher for comparing two proportions shows that the difference is significant

(p-value $= 0.049$). This looks plausible, but there is problem associated with it – the observed correlation could be partly induced by the person effect mentioned earlier. This can be illustrated in our experiment by the extreme case of a person with 48 tails out of 50. Here the correlation between two consecutive trials is very obvious and is simply due to the person effect. So one has to find a method to check for autocorrelation taking the person effect into account. One possible way is to extend the logistic regression model (2) is by a possible effect of the preceding trial.

$$P(Y_{ij} = 1) = G(\beta_i + \gamma Y_{i,j-1}), \quad i = 1, \ldots, I, \quad j = 1, \ldots, J_i. \tag{3}$$

Indeed, the effect is no longer significant when the person effect is taken into account. The discussion can be continued by using a random effects model and testing for autocorrelation, e.g. modeled by a AR(1), see Molenberghs and Verbeke (2005). Furthermore, a Bayesian approach for the analysis of this experiment can also be used.

6 Correction of the Main Analysis

Since one main assumption for the simple Binomial model is not fulfilled, the analysis should be redone including the person effect. Again there are simple and more complex solutions. First we have to reformulate the concept of a fair coin. A random person i has a certain probability P_i for tails. Then the null hypothesis is given by $H_0 : E(P_i) = 0.5$.

6.1 Normal Approximation

First the concept of a mixture distribution has to be introduced. The number of tails X_i given P_i is Binomial $X_i \sim B(50, P_i)$. Assuming that P_i is normal, the marginal distribution of X is Binomial–normal mixture. The Binomial part can be approximated by a normal distribution, so X_i is approximately normal with identical variances under the null hypothesis. Therefore the t–test for null hypothesis $H_0 : E(X_i) = 50{\cdot}0.5$ can be conducted. As may be expected, the result is still significant, but the p-value is much higher ($p = 0.008$) than in the simple Binomial test. One could also discuss nonparametric alternatives like sign test and signed rank test. The beta Binomial can be mentioned as a further alternative. Here the individual probabilities are not assumed to be normal, but have a beta distribution. The beta distribution has the support $[0, 1]$ and therefore – unlike the normal distribution – excludes negative probabilities or probabilities greater than 1.

6.2 Logistic with Overdispersion and GEE

As mentioned above, correction for overdispersion is a common technique used in logistic regression. For our experiment we can use a logistic model without covariates but with possible overdispersion due to the person effect. Another similar possibility is a GEE model (see Toutenburg (2002)), which also takes the correlation between the trials of one person into account. Here the correlation between two trials of the same person is assumed to be constant (exchangeable variance structure). Both methods still give a significant effect ($p = 0.005$ for logistic regression with overdispersion and $p = 0.0213$ for GEE).

7 Discussion and Other Experiments

There have been other proposals for further classroom experiments. Schuster (2006) describes the following game: A two-player game consists of repeated tosses (spins) of a coin, until one of the pre-specified sequence of three outcomes occurs (for example, Tails, Heads, Tails (THT)). Player 2 chooses his or her sequence after knowing the choice of Player 1. This is a nice possibility of calculating elementary probabilities and finding some surprising results.

Another well–known strategy for comparing randomness of tossing a coin with human intuition of randomness is to ask one group of students to do a coin tossing experiment (say 10 trials) and the other group should just write down a "random" sequence of zeros and ones. Than the number of changes in the sequence is Binomial $B(n, 0.5)$ assuming independence and a fair coin. Students in the second group usually have too much changes in their "random sequences". One could organize the experiment such that the supervisor only receives the results of the two groups and has to decide which is the group who really tossed the coins.

We have shown that many different statistical methods can be illustrated by planning and evaluating a classroom experiment of coin spinning. The main point is the possible existence of a person effect, i.e., the possibility that every student could have a different probability for tails in spinning experiments. Note that checking and interpreting the model assumptions is straightforward, which makes the experiment very attractive for teaching. For checking and correcting the person effect, we have given an elementary solution (contingency table and t-test) as well as a more complex solution (testing for a random effect, GEE model). Of course this can be enhanced by further aspects like sample size determination using the information that there is a person effect ("how many persons and how many experiments per person?") or finding differences between male and female students. When one person uses different coins one could try to fit a model with a random person and a random coin effect etc.

The described experiment was replicated in a live experiment in a TV show and the result that the German 2 Euro coin is not fair in a spinning experiment was confirmed (501 tails out of 800 spinning experiments). Again, a clear person effect was found. Summarizing, the 2 Euro spinning experiment possibly leads to surprising results and can serve as a good motivating example for discussing different statistical topics. This makes the experiment very attractive for teaching statistics.

Acknowledgement

I would like to thank the students who took part in the experiment.

References

Boyes R, Baldwin T, Hawkes N (2002) Heads I win, tails you lose. The Times (London)

Diaconis P, Holmes S, Montgomery R (2007) Dynamical bias in the coin toss. Society for Industrial and Applied Mathematics 49:211–235

Dunn PK (2005) We can still learn about probability by rolling dice and tossing coins. Teaching Statistics 27:37–41

Gelman A, Nolan D (2002) You can load a die, but you can't bias a coin. The American Statistician 56:308–311

Greven S, Crainiceanu C, Küchenhoff H, Peters A (2008) Restricted likelihood ratio testing for zero variance components in linear mixed models. Journal of Graphical and Computational Statistics (Appearing)

Shuster JJ (2006) Using a two-player coin game paradox in the classroom. The American Statistician 60:68–70

Moffatt HK (2000) Euler's disk and its finite-time singularity. Nature 404:833–834

Molenberghs G, Verbeke G (2005) Models for Discrete Longitudinal Data. Springer, New York

Paulos JA (1995) A Mathematician Reads the Newspaper. Basic Books

Rao CR, Toutenburg H, Shalabh, Heumann C (2008) Linear Models and Generalizations – Least Squares and Alternatives (3rd edition). Springer Verlag, Berlin, Heidelberg, New York

Scheipl F, Greven S, Küchenhoff H (2008) Size and power of tests for a zero random effect variance or polynomial regression in additive and linear mixed models. Computational Statistics and Data Analysis (Appearing)

Toutenburg H (2002) Statistical Analysis of Designed Experiments. Springer, New York

Toutenburg H (2005) Induktive Statistik. Eine Einführung mit SPSS für Windows. Springer, Heidelberg

Tutz G (2000) Die Analyse kategorialer Daten. Oldenbourg

Linear Models in Credit Risk Modeling

Thomas Nittner

UBS AG[†], Credit & Country Risk Controlling, Pelikanstrasse 6/8, P.O. Box, 8098 Zurich, Switzerland thomas.nittner@ubs.com

1 Introduction

Linear models have been used in various applications. Credit risk analysis is an important area which relies on linear regression models. The objective of this article is to illustrate briefly the role of linear models in credit risk analysis.

Although media told us about the subprime–crisis (caused by US mortgage business) almost every day especially in early autumn and its impact on European market, credit risk modeling still (or even more) is one major part of daily banking business. The Basel–II Capital Accord created a lot of new jobs with a main focus on implementing statistically driven models, especially in assessing the creditworthiness of banking clients (single obligor name basis) as well as in quantifying credit portfolio risk. This covers many business segments, e.g. consumer credits, mortgages (for self–used properties as well as for income producing real estate), lombard loans or loans for corporate clients of different sizes. For each of these business segments, one could think about developing own statistical models on single obligor level which may provide the probability of default (PD) for each client. This PD might be translated into a credit rating as being done by the three big rating agencies Moody's KMV, Standard & Poor's and Fitch.[1] However, the PD does

[†] The opinions expressed in this presentation are my own and do not represent the ones of UBS AG or any affiliates. The risk control principles presented are not necessarily used by UBS AG or any of its affiliates. This presentation does not provide a comprehensive description of concepts and methods used in risk control at UBS.

[1] In general, one can differ between two approaches: models explicitly estimating a PD for each client which then is mapped into a rating class or models yielding a score which is translated to a rating; then, each rating class is assigned a PD

not only reflect the risk that the client does not meet his obligations within a given time frame (usually one year), e.g. he does not pay interest rates for more than 90 days but together with the so–called loss given default (LGD) and the exposure at default (EAD), it determines the expected loss (EL) within this time horizon. The expected loss is a statistical measure of the loss distribution. Consider a portfolio of n clients, then the expected loss of the corresponding portfolio is given by

$$\mathrm{EL} = \frac{1}{n} \sum_{i=1}^{n} \mathrm{PD}_i \cdot \mathrm{LGD}_i \cdot \mathrm{EAD}_i \,. \tag{1}$$

The expected loss determined by (1) is an essential measure which characterizes the (unknown) loss distribution of a given portfolio.[2] We focus on the modeling of loss distributions and on one of its major inputs on obligor level, the probability of default in this paper. The reason why we focus on the estimation of the PD instead of LGD or EAD is mainly driven by the fact that it is much easier to understand PD than LGD or EAD models.

The linearity in the model can enter into the models because of the assumed functional form of regression model, e.g. a linear model or a logit link in logistic regression model. But here we are more interested in studying the linearity implied by linear correlation.

However, correlation is a crucial aspect within this context in general. On the one hand, we want to avoid correlation between the independent covariates within a regression–like context when developing a rating model on obligor level. On the other hand, we need to capture the correlation when estimating a loss distribution. This does not affect the two models themselves as they are separate models but not modules within a common framework. It is interesting to know how important correlation is and how it enters into the model development from a practical point of view.

After a short note on measuring linear correlation, we give some ideas about the development of rating models from a practical point of view. The section about estimating the loss distributions will be more theoretical and is followed by some notes on the simulation of losses.

(e.g. average PD of the rating class). Here, it was implicitly assumed to follow the first approach.

[2] Actually, we do not know the loss distribution but only the expected loss on obligor level, $\mathrm{EL} = \mathrm{PD} \cdot \mathrm{LGD} \cdot \mathrm{EAD}$. Later, it will be shown that the loss distribution is simulated by simulating defaults.

2 Measuring Linear Correlation

Given two vectors of continuous variables, say X and Y, the (linear) Pearson's correlation coefficient is measured by

$$\rho(X,Y) = \frac{\mathrm{Cov}(X,Y)}{\sqrt{\mathrm{Var}(X)}\sqrt{\mathrm{Var}(Y)}} \tag{2}$$

see e.g. Toutenburg (2005). In general, $-1 \leq \rho(X,Y) \leq 1$, and when $\rho(X,Y) = 0$ then X and Y are said to be uncorrelated.

Assuming a response vector y to be dependent on a single factor X in the context of a univariate linear regression model

$$y = a + bX + \epsilon$$

and further defining the coefficient of determination R^2 by

$$R^2 = \frac{\mathrm{SSQ}_{\mathrm{REG}}}{\mathrm{SSQ}_{\mathrm{TOT}}}, \tag{3}$$

where $\mathrm{SSQ}_{\mathrm{REG}} = \sum_{i=1}^{n}(\hat{y}_i - \bar{y})^2$ and $\mathrm{SSQ}_{\mathrm{TOT}} = \sum_{i=1}^{n}(y_i - \bar{y})^2$ are the sum of squares due to regression and total, respectively, allows us to translate the coefficient of determination of the classical linear regression model into the linear correlation coefficient, cf. for example Toutenburg (2003, pp. 56)

$$R^2 = \rho^2 .$$

3 Avoiding Correlation: Credit Rating Models

As already mentioned, credit rating models assess the creditworthiness of a client and provide estimates for the PD based on historical data. Here, we want to illustrate and give an idea how a rating model can be developed and how correlation among the independent factors often can be avoided in a simple case.

3.1 Data Preparation

Now assume that we are interested in developing a rating model for a bank with its special focus on mortgage business for self–used properties. Further assume that the bank provides us the data on client level which contain annually updated information[3] about

[3] This is a quite unrealistic assumption as information for private clients usually is not updated on a regular base; however, within this context it does not affect our further explanations.

- the mortgage contract itself, e.g. mortgage value, interest rate, amortization,
- the property funded, e.g. market value, ancillary costs,
- information about the financial situation of the client, e.g. tax statement,
- a rating based on the above information.[4]

Assume that this information is contained in the data—each row containing the information mentioned above for a given time stamp. We may call it as client analysis. Additionally, we assume that the data covers a time period of some years such that for each client analysis, a default flag can be mapped. For example, one can measure default for a given client analysis at time t by the rating of consecutive analysis until time $t+1$. In case a client defaults within the period $]t; t+1]$ induced by a corresponding rating, then the analysis at time t is flagged as default and all future analysis are excluded from the data (absorbing state of default).

The banks are generally interested in a tailor–made PD based on the data we are provided with. Before explaining that how and where the correlation analysis enters into model building, we summarize some necessary steps, namely,

- data cleaning, e.g. assure absorbing default,[5] data consistency checks, e.g. check whether the value of total assets is not smaller than the value of the property,
- factor definition (formula, hypothesis about direction with respect to default) and factor calculation,
- factor transformation, e.g. remove non–linear relationships with the default indicator to increase e.g. discriminatory power.

These steps are necessary from a practical point of view.[6]

3.2 Factor Calculation and Univariate Analysis

Usually the list of factors (we call it factor longlist) is built from the input data which itself and within the context of retail business may be classified into

---

[4] This rating might be a calculated rating based on the current rating model in place or it might be a rating of the credit officer in charge, due to an override.

[5] That means that whenever a counterparty defaults, it is set to default to all future time periods; this does not necessarily have to be the case in practice as clients may recover and 'come back' as non–defaults

[6] At this point in time, a cleaned development sample and a set of factors which can be used to explain the risk of default exist.

- *income* positions like salary, bonus or other regular income,
- *expense* positions like interest payments, amortizations or alimonies,
- *asset* positions like deposits, life insurance policies or real estate, and, finally,
- *debt* positions like mortgages, loans or other credits.

Combining single positions to ratios provides a factor longlist. To give some examples within this context, one may think about

- loan–to–value, defined by mortgage value/market value of a property
- affordability, defined by expenses/income
- break–even, defined by (salary - expenses)/expenses.

Though today's computer performance is quite advanced, but estimating multivariate models from a huge factor longlist can not be recommended. Instead, the factor longlist can be reduced by running univariate analyses controlling for discriminatory power, missing data portion or other distribution relevant measures. Pre–defined cut–off values often reduces the longlist to at least half of its length. Some first multivariate analyses further shorten this list such that some dozens of factors remain.[7] This 'shortlist' then is used to estimate the models which can be close to the final model. As our main focus here is on the aspect of correlation, we refer to Henking, Bluhm und Fahrmeir (2006) who give a very good and extensive overview about the development of rating models. At this point, we would like to remark why (highly) correlated factors should not enter the final model.

1. From a statistical point of view, correlations between the independent variables of a regression model can affect the parameter estimates or makes even them not reliable.[8] Note that correlation is a necessary but not sufficient condition in the context of multicollinearity and that other measures, e.g. the condition index should be used to detect multicollinearity, (cf. Belsley, Kuh and Welsch (2004)).
2. In terms of the Basel–II validation process for A–IRB correlations and the problem of multicollinearity is an essential topic to be covered.
3. The quality of rating models is not only determined by the statistical measures like goodness–of–fit or discriminatory power but

[7] One could for example run models for all possible combinations and check which of the factors enter the different models most often. In addition, one might check this for forward, backward and stepwise selection procedures.

[8] see e.g. Toutenburg (2003) pp. 112, within the context of linear regression models.

also how they are accepted by people using it daily, i.e., the credit officers. As already it is very difficult to think in a bivariate setup, it is even more difficult to incorporate interactions between factors in a multivariate setup.

The latter point should not be underestimated as rating models considered as black boxes often can be asked to be revised shortly after their implementation. Thus, correlations between factors should be avoided at least up to a certain amount. But let's go back to the way how to avoid correlation.

3.3 Multivariate Analysis—Dealing with Correlations

Remember that we have a shortlist of factors entering candidate models. Calculating bivariate correlations among the factors of the list often shows that not only factors of the same group as defined earlier are highly correlated. In order to avoid correlated factors being selected, we propose to

1. define new factor groups driven by correlations, say up to 6 or 8 groups;
2. take into account business–relevant knowledge as well, as also different drivers from a business point of view should be contained in the model;
3. determine a range for the number of factors of each group which could be selected;
4. generate all possible combinations of factors based on the previously defined ranges and a given total of factors allowed to enter the model, e.g. all possible models for candidate models containing 6, 7, 8 or 9 factors;
5. estimate all these candidate models of the given size;
6. rank these models based on reliable measures in terms of
   a) goodness–of–fit and
   b) discriminatory power.

For each model size, we may now have a few dozens of best models. A first business challenging as well as checking factor weights (none should seriously dominate, none should show signs deviating from its hypothesis) may further reduce the candidates. As we just tried to avoid correlations on a bivariate level, we additionally should check collinearity diagnostics like variance inflation factor or condition indices. Often a handful of models remain and a more extensive round of business challenging based on a representative sample of daily business cases but also untypical cases might help to come up with a final proposal.

4 Asking for Correlation: Credit Portfolio Models

4.1 Introduction

The PD as one main input for the calculation of the expected loss on client as well as portfolio level contains information only about the individual default probability. However, in the economic cycle, downturns and upturns have different implications on groups of clients in the sense of common defaults within a given period. Within this context, we talk about correlated defaults. Schönbucher (2000) gives a good overview about modeling the dependent defaults with a special focus on factor models; he further illustrates the impact of default correlation on the joint probability of default and the conditional probability of default. He shows that default correlation dominates both the probabilities. Correlation substantially matters, especially in the tails of the loss distribution. Assume for example a portfolio of 100 obligors all having a PD of 5% and the same exposure of say EUR 1. Schönbucher (2000) shows that without any asset correlation, the 99.9%–percentile (VaR, Value–at–Risk) of the loss distribution equals EUR 13 but already slightly increasing the asset correlation to 10% or 20% yields a VaR of EUR 27 or EUR 41, respectively. Loss distributions show a right–skewed picture which is driven by correlated defaults in the tail of the distribution meaning large portfolio losses with small probabilities. Obviously, correlation has an influence on the shape of the loss distribution. But how to measure default correlation? Though there exists a variety of methods to estimate the default correlation, e.g. Frey and McNeil (2003), in most of the cases, time series of sufficient length (covering at least one full business cycle) do not exist and the resulting estimates are quite poor. Time series of PDs might not be a problem but remember that here we need time series of defaulted loans. Thus, default correlation must be tackled by a different concept.

4.2 Concept

Best practice in modeling default correlation is an underlying factor model basically starting at the generalization of the correlation coefficient introduced in (2) to the Bernoulli case, i.e., $X = 1$ in case of default and $X = 0$ in case of non–default. Thus,

$$\rho_{ij}^{\mathrm{D}} = \frac{p_{ij} - p_i p_j}{\sqrt{p_i(1-p_i)}\sqrt{p_j(1-p_j)}}, \tag{4}$$

with p_{ij} denoting the joint default probability; p_i and p_j denote the marginal PD of two obligors X_i and X_j, respectively. The crucial point in determining or calculating the default correlation ρ_{ij}^{D} is the joint probability of default p_{ij}. At this point, the famous asset value process introduced by Merton (1974) enters into the concept. Here, default is assumed to be triggered by the change in (correlated) asset values. It is assumed that the default for an obligor i occurs if and only if his assets R_i lie below his liabilities c_i, i.e., as soon as

$$R_i < c_i \,. \tag{5}$$

This implies that the obligor specific probability of default is given by

$$p_i = P(R_i < c_i) \,. \tag{6}$$

In the most simple case, R_i is assumed to follow a one–factor model, namely,

$$R_i = \sqrt{\nu_i}\; Z + \sqrt{1-\nu_i}\; \epsilon_i \,, \tag{7}$$

assuming Z to be i.i.d. and $\epsilon_1, \ldots, \epsilon_n \sim \mathrm{N}(0,1)$ and Z is called as a systematic factor. Thus, $\mathrm{E}(R_i) = 0$, $\mathrm{Var}(R_i) = 1$ and R_i follow standard normal distributions, too. Further, the correlation between two asset values is given by

$$\rho^{\mathrm{A}}_{R_i R_j} = \sqrt{\nu_i}\sqrt{\nu_j} \,. \tag{8}$$

The cut–off value for the asset values with respect to liabilities is

$$c_i = \phi^{-1}(p_i) \,, \tag{9}$$

where ϕ^{-1} denotes the inverse of the normal density function. Usually, the marginal default probabilities are known for each obligor and it remains to calculate the joint probability of default for two obligors,

$$p_{ij} = P(X_i = 1, X_j = 1) \,,\; i \neq j \,. \tag{10}$$

Applying equation (6) and the fact that (R_i, R_j) are following a bivariate normal distribution yields

$$\begin{aligned} P(X_i = 1, X_j = 1) &= P(R_i < c_i, R_j < c_j) \\ &= \phi_2(c_i, c_j; \sqrt{\rho^{\mathrm{A}}_{ij}}) \,, \end{aligned} \tag{11}$$

where ϕ_2 denotes the density function of a bivariate normal distribution. Remember that c_i and c_j are known because we assumed the

PD for both the obligors to be known as well. Hence, to determine the joint probability of default, we use the asset correlation based on the underlying process or the assumed factor model, respectively. To obtain the default correlations—which typically is much smaller than asset correlation—one has to

(I) estimate the parameters of model (7),
(II) calculate the joint probability of default based on equation (11),
(III) use p_{ij} to calculate the default correlation by using (4).

Of course, (II) and (III) is an exercise even feasible after a glass of wine but step (I) needs some further explanation.

Basically, we want to focus on the two possibilities to estimate a model of type (7). Though here we assume the most simple approach of a one–factor model, it will give an understanding in depth of how it can work for even more complex models. We refer to Bluhm, Overbeck and Wagner (2003) who give an extensive overview over the whole subject of credit risk modeling and to Pitts (2004) who proposes an extended version of the Merton–style model.

Generally speaking, the question is not only how to estimate the parameters of the regression(–like) model introduced in (7) but how to gather the information about the underlying asset value process and the systematic factor Z. In the following two subsections, we introduce two approaches to make a portfolio model work – the simple classical regression model briefly and a more complicated approach in more detail. The next section will describe on how the predicted asset values are used to simulate losses and a final loss distribution.

4.3 Linear Regression: Fixed Effects

The (classical) linear regression model assumes all inputs to be observed and to minimize the sum of squares due to the necessary stochastic error term ϵ. Transferring these assumptions to the one–factor model implies, that the response vector (asset values) as well as the systematic factor Z are to be observed in order to be able to apply estimation techniques for linear regression, i.e., least squares or maximum–likelihood methods of estimation. Though asset values, or, better log returns of asset values are difficult to observe, we assume them to be given, e.g. by

- using time series of equity values (share prices) and their volatilities (see e.g. Bluhm, Overbeck and Wagner (2003)), or

- using data offered by rating agencies, e.g. Moody's KMV which have own models for asset values.

Given these asset values or their logarithmized returns, one would have to search for data potentially describing the response for the underlying portfolio; these explanatory factors for example could be industry indices or country indices. In the most simple case, the parameters of a linear model are estimated and used to predict asset values based on given indices for all firms contained in the portfolio. Bluhm, Overbeck and Wagner (2003), e.g., describe the three–level factor model of Moody's KMV and the whole background from a conceptual as well as from a theoretical point of view.

4.4 Linear Regression: Random Effects

An extension of the classical (linear) regression model is the random effect model. They are often used in the context of clustered data or (time) dependent observations in general (see e.g. Fahrmeir and Tutz (2001), Jiang (2007)). Here, the parameters are stochastic and assumed to be draws of an underlying distribution. Its main idea within this context of course is to estimate correlations of asset values based on a set of covariates. Assume log asset returns y_{ijt} (say, log–returns) to follow

$$y_{ijt} = \alpha_{ij} + \beta_{jt} + \epsilon_{ijt} \, , \quad \epsilon_{ijt} \sim N(0;\tau^2) \, . \tag{12}$$

Here, $i = 1,\ldots,I$ denote firms, $j = 1,\ldots,J$ denote, e.g., branches and $t = 1,\ldots,T$ denote time periods of the data being observed.

Thus, log–returns of the firm i doing business in branch j ($\beta_{jt} \sim N(0;\sigma_j^2)$) depends on a firm–branch–specific intercept α_{ij} and a random effect β_{jt} for branch j in period t. The latter one describes the *systematic* effect, the error term is the remaining *idiosyncratic* effect. As we here assume β_{jt} to be a random effect, it could be specified further by a single factor model

$$\beta_{jt} = \gamma_t \delta_j + \zeta_{jt} \, , \tag{13}$$

where we assume $\gamma_t \sim N(0,1), \zeta_{jt} \sim N(0,\kappa^2)$ and $\mathrm{E}(\gamma_t\zeta_{jt}) = \mathrm{E}(\zeta_{jt}\zeta_{j't}) = 0 \ \forall \ j \neq j'$. Thus, it holds that $\mathrm{E}(\beta_{jt}) = 0$ and $\mathrm{Var}(\beta_{jt}) = \sigma_j^2 = \delta_j^2 + \kappa^2$. Further, correlation between two branch effects β_{jt} and $\beta_{j't'}$ is given by

$$\text{corr}(\beta_{jt}, \beta_{j't'}) = \frac{\delta_j \delta_{j'}}{\sqrt{\delta_j^2 + \kappa^2}\sqrt{\delta_{j'}^2 + \kappa^2}} \,. \tag{14}$$

This functionality can be translated into the covariance structure for y_{ijt} which then can be estimated by statistical software packages, e.g. by using SAS PROC MIXED. The resulting estimate for the covariance matrix of the branch effects is then used for the simulation of the loss distribution. Further, the variance of the idiosyncratic effect, τ^2—which e.g. can further carry firm–specific cluster effects as for example legal entity—has to be estimated. All the estimated parameters will be used in the simulation of losses. Though it is a short paper, Pitts (2004) gives a good overview over the source of this model, its statistical estimability and how this can be achieved in practice.

4.5 Simulating Losses

After the parameters have been estimated, we can start to simulate losses on counterparty level. The simulation can be split into the following steps performed for the counterparties $i = 1, \ldots, n$:

1. Simulate correlated log–returns following equation (12) for each counterparty given
   - its branch
   - its domicile (country), and, if necessary,
   - its legal form.
2. Check whether the predicted log asset return is below the cut–off determined by the firms' PD, namely, $\phi^{-1}(\text{PD})$.
   a) If the log asset return is below the cut–off, the firm is assumed to default and its loss is calculated by $\text{L} = \text{EAD} \cdot \text{LGD}$.
   b) If the log–return is above the cut–off, the firm did not default and its loss is zero.
3. Update the portfolio loss at step (i) according to $L(i) = L(i-1) + L, i = 1, \ldots, n, L(0) = 0$.
4. Repeat steps (1)–(3) R times and store the losses $L_r, r = 1, \ldots, R$.

Having calculated the losses for all counterparties yields the portfolio loss within one simulation experiment. Repeating steps (1)–(3) R times, e.g. 10^6 times, gives 10^6 portfolio losses. These losses can be used to draw a loss distribution and to calculate its entire measures such as the expected loss, the unexpected loss (standard deviation of the loss distribution) its VaR, and, another measure which often is reported, the expected tail loss also known as expected shortfall; the latter is the conditional expectation given that the loss exceeded a given percentile. There exists a variety of portfolio models, especially of the companies Credit Suisse (Credit Risk$^+$), McKinsey (Credit Portfolio View), Riskmetrics (CreditMetrics) or Moody's KMV (Portfolio Manager). Bluhm, Overbeck and Wagner (2003) give a short overview over the main differences between the various models with a special focus on Credit Risk$^+$. Henking, Bluhm und Fahrmeir (2006) give some simple examples of portfolios and describe how to simulate losses as well.

5 Summary

In this paper we connected linear models and credit risk. A linear measure for correlation was used to show how correlation plays an important role in today's credit risk models illustrated by choosing rating models and portfolio models. Of course, measuring LGD and EAD was left aside and correlation is present here, too. We know from practical data, for example, that default rates and loss rates are correlated as well, see e.g., Schürmann (2004). But, we better should celebrate the guy we are writing for instead of worrying about what we could have missed. Cheers Helge!

References

Basel Committee on Banking Supervision (2004) International Convergence of Capital Measurement and Capital Standards. June 2004

Belsley D A, Kuh E, Welsch R E (2004) Regression Diagnostics. Wiley

Bluhm C, Overbeck L, Wagner C (2003) Credit Risk Modelling. Chapman & Hall/CRC

Henking A, Bluhm C, Fahrmeir L (2006) Kreditrisikomessung. Springer–Verlag

Fahrmeir L, Tutz G (2001) Multivariate Statistical Modelling Based on Generalized Linear Models. Springer

Frey R, McNeil A J (2006) Dependent Defaults in Models of Portfolio Credit Risk. Preprint ETHZ

Jiang J (2007) Linear and Generalized Linear Mixed Models and Their Applications. Springer

Merton A (1974) On the rpicing of corporate debt: the risk structure of interest rates. Journal of Finance 29: 449–470

Pitts A (2004) Correlated defaults: let's go back to the data. Risk, June 2004: 75–79

Rao CR, Toutenburg H, Shalabh, Heumann C (2008) Linear Models: Least Squares and Generalizations (3rd edition). Springer, Berlin Heidelberg New York

Schönbucher P J (2000) Factor Models for Portfolio Credit Risk. Working Paper

Schürmann T (2004) What Do We Know About Loss Given Default, in: Credit Risk Models and Management (2nd edition). London UK, Risk Books

Toutenburg H (2003) Lineare Modelle. Physica–Verlag, Heidelberg

Toutenburg H (2005) Induktive Statistik. Springer–Verlag, Berlin

Index

Printing: Krips bv, Meppel, The Netherlands
Binding: Stürtz, Würzburg, Germany